Arc Routing

MOS-SIAM Series on Optimization

This series is published jointly by the Mathematical Optimization Society and the Society for Industrial and Applied Mathematics. It includes research monographs, books on applications, textbooks at all levels, and tutorials. Besides being of high scientific quality, books in the series must advance the understanding and practice of optimization. They must also be written clearly and at an appropriate level for the intended audience.

Series Volumes

Corberán, Ángel and Laporte, Gilbert, *Arc Routing: Problems, Methods, and Applications*

Toth, Paolo and Vigo, Daniele, *Vehicle Routing: Problems, Methods, and Applications, Second Edition*

Beck, Amir, *Introduction to Nonlinear Optimization: Theory, Algorithms, and Applications with MATLAB*

Attouch, Hedy, Buttazzo, Giuseppe, and Michaille, Gérard, *Variational Analysis in Sobolev and BV Spaces: Applications to PDEs and Optimization, Second Edition*

Shapiro, Alexander, Dentcheva, Darinka, and Ruszczyński, Andrzej, *Lectures on Stochastic Programming: Modeling and Theory, Second Edition*

Locatelli, Marco and Schoen, Fabio, *Global Optimization: Theory, Algorithms, and Applications*

De Loera, Jesús A., Hemmecke, Raymond, and Köppe, Matthias, *Algebraic and Geometric Ideas in the Theory of Discrete Optimization*

Blekherman, Grigoriy, Parrilo, Pablo A., and Thomas, Rekha R., editors, *Semidefinite Optimization and Convex Algebraic Geometry*

Delfour, M. C., *Introduction to Optimization and Semidifferential Calculus*

Ulbrich, Michael, *Semismooth Newton Methods for Variational Inequalities and Constrained Optimization Problems in Function Spaces*

Biegler, Lorenz T., *Nonlinear Programming: Concepts, Algorithms, and Applications to Chemical Processes*

Shapiro, Alexander, Dentcheva, Darinka, and Ruszczyński, Andrzej, *Lectures on Stochastic Programming: Modeling and Theory*

Conn, Andrew R., Scheinberg, Katya, and Vicente, Luis N., *Introduction to Derivative-Free Optimization*

Ferris, Michael C., Mangasarian, Olvi L., and Wright, Stephen J., *Linear Programming with MATLAB*

Attouch, Hedy, Buttazzo, Giuseppe, and Michaille, Gérard, *Variational Analysis in Sobolev and BV Spaces: Applications to PDEs and Optimization*

Wallace, Stein W. and Ziemba, William T., editors, *Applications of Stochastic Programming*

Grötschel, Martin, editor, *The Sharpest Cut: The Impact of Manfred Padberg and His Work*

Renegar, James, *A Mathematical View of Interior-Point Methods in Convex Optimization*

Ben-Tal, Aharon and Nemirovski, Arkadi, *Lectures on Modern Convex Optimization: Analysis, Algorithms, and Engineering Applications*

Conn, Andrew R., Gould, Nicholas I. M., and Toint, Phillippe L., *Trust-Region Methods*

Arc Routing

Problems, Methods, and Applications

Edited by

Ángel Corberán
University of Valencia
Burjassot, Valencia
Spain

Gilbert Laporte
HEC Montréal
Montréal, Quebec
Canada

Society for Industrial and Applied Mathematics
Philadelphia

Mathematical Optimization Society
Philadelphia

10 9 8 7 6 5 4 3 2 1

(continued)

Figure 1.7b used courtesy of Filep László.
Figure 1.9a used with permission from the Institute for Operations Research and the Management Sciences and the family of Lester Randolph Ford Jr.
Figure 1.9b used with permission from the University of Warsaw.
Figure 1.11 used courtesy of Ya-Xiang Yuan.
Figure 1.12a used with permission from Jay VanRensselaer/Johns Hopkins University.
Figures 1.12b and 1.13a used with permission from Jack Edmonds.
Figures 1.13b and 1.15b used with permission from Bruce L. Golden.
Figure 1.14a used with permission from Christos H. Papadimitriou.
Figure 1.14b used with permission from Greg Frederickson.
Figure 1.15a used with permission from Lawrence Bodin.
Figure 1.16 used with permission from H. Psaraftis.
Figure 3.1 used with permission from Nikolaj Van Omme.
Figures 3.2-3 used with permission from John Wiley and Sons.
Figures 3.4-5 and 6.1 used with permission from Elsevier.
Figure 6.2 used with permission from Carlos Balaguer.
Figure 6.3a used courtesy of Redsail Tech Co., Ltd.
Figure 6.3b used with permission from Serge Defever.
Figures 6.7 and 11.1 used with permission from Palgrave Macmillan.
Figures 8.1-6 and Tables 8.1-8 used with permission from Dino Ahr and VDM Verlag.
Figures 13.1-3 used with permission from RouteSmart.
Figure 14.1 used with permission from the Institute for Operations Research and the Management Sciences.
Figures 15.1-2 used with permission from the Municipality of Sexial.
Figures 16.1-3 used with permission from Aftenposten.
Figure 16.4 used with permission from MIT Operations Research Center.
Figures 16.6-10 used with permission from Distribution Innovation.

Publisher	David Marshall
Acquisitions Editor	Sara Murphy
Developmental Editor	Gina Rinelli
Managing Editor	Kelly Thomas
Production Editor	Ann Manning Allen
Copy Editor	Teresa Wilson
Production Manager	Donna Witzleben
Production Coordinator	Cally Shrader
Compositor	Scott Collins
Graphic Designer	Lois Sellers

Library of Congress Cataloging-in-Publication Data

Arc routing (Corberán)

Arc routing : problems, methods, and applications / edited by Ángel Corberán, University of Valencia, Burjassot, Valencia, Spain, Gilbert Laporte, HEC Montréal, Quebec, Montréal, Canada.

pages cm. – (MOS-SIAM series on optimization ; 20)

Includes bibliographical references and index.

ISBN 978-1-611973-66-2

1. Operations research–Mathematics. 2. Physical distribution of goods–Mathematics. 3. Traffic flow–Mathematical models. 4. Theory of distributions (Functional analysis) I. Corberán, Ángel, editor. II. Laporte, Gilbert, editor. III. Title.

T57.6.A727 2014

511'.5–dc23

2014035318

List of Contributors

Dino Ahr
Institut für Informatik
Ruprecht-Karls-Universität Heidelberg
Germany
dino.ahr@gmail.com

Claudia Archetti
Dipartimento Metodi Quantitativi
Universitá di Brescia
Italy
archetti@eco.unibs.it

José M. Belenguer
Departamento de Estadística e
Investigación Operativa
Universitat de València
Spain
jose.belenguer@uv.es

Enrique Benavent
Departamento de Estadística e
Investigación Operativa
Universitat de València
Spain
enrique.benavent@uv.es

René van Bevern
Institut für Softwaretechnik und
Theoretische Informatik
Technische Universität Berlin
Germany
rene.vanbevern@tu-berlin.de

James F. Campbell
College of Business Administration
University of Missouri-St. Louis
USA
campbell@umsl.edu

Ángel Corberán
Departamento de Estadística e
Investigación Operativa
Universitat de València
Spain
angel.corberan@uv.es

Richard Eglese
Department of Management Science
Lancaster University Management School
United Kingdom
R.Eglese@lancaster.ac.uk

Gianpaolo Ghiani
University of Salento
Lecce, Italy
gianpaolo.ghiani@unisalento.it

Bruce Golden
Department of Decision, Operations and
Information Technologies
Robert H. Smith School of Business
University of Maryland
USA
BGolden@rhsmith.umd.edu

Geir Hasle
SINTEF ICT
Oslo, Norway
geir.hasle@sintef.no

Stefan Irnich
Chair of Logistics Management, Gutenberg School of Management and Economics
Johannes Gutenberg University Mainz
Germany
irnich@uni-mainz.de

André Langevin
Polytechnique Montréal
Montréal, Canada
andre.langevin@polymtl.ca

Gilbert Laporte
HEC Montréal
Montréal, Canada
gilbert.laporte@cirrelt.ca

M. Cândida Mourão
Instituto Superior de Economia e Gestão
Centro de Investigação Operacional
Universidade de Lisboa
Portugal
cmourao@iseg.utl.pt

Luc Muyldermans
Nottingham University Business School
Nottingham University
United Kingdom
luc.muyldermans@nottingham.ac.uk

Rolf Niedermeier
Institut für Softwaretechnik und Theoretische Informatik
Technische Universität Berlin
Germany
rolf.niedermeier@tu-berlin.de

Gu Pang
Newcastle University Business School
Newcastle University
United Kingdom
gu.pang@ncl.ac.uk

Nathalie Perrier
Polytechnique Montréal
Montréal, Canada
n.perrier@polymtl.ca

Leonor S. Pinto
Instituto Superior de Economia e Gestão
CEMAPRE
Universidade de Lisboa
Portugal
lspinto@iseg.utl.pt

Isaac Plana
Departamento de Matemáticas para la Economía y la Empresa
Universitat de València
Spain
isaac.plana@uv.es

Christian Prins
Institut Charles Delaunay
Université de Technologie de Troyes
France
christian.prins@utt.fr

Gerhard Reinelt
Institut für Informatik
Ruprecht-Karls-Universität Heidelberg
Germany
gerhard.reinelt@informatik.uni-heidelberg.de

José M. Sanchis
Departamento de Matemática Aplicada
Universidad Politécnica de Valencia
Spain
jmsanchis@mat.upv.es

Manuel Sorge
Institut für Softwaretechnik und Theoretische Informatik
Technische Universität Berlin
Germany
manuel.sorge@tu-berlin.de

M. Grazia Speranza
Dipartimento Metodi Quantitativi
Universitá di Brescia
Italy
speranza@eco.unibs.it

Daniele Vigo
Department of Electrical, Electronic and Information Engineering "Guglielmo Marconi"
Universitá di Bologna
Italy
daniele.vigo@unibo.it

Edward Wasil
Department of Information Technology
Kogod School of Business
American University
USA
ewasil@american.edu

Mathias Weller
Laboratory of Informatics, Robotics, and Microelectronics of Montpellier
University Montpellier II
France
mathias.weller@lirmm.fr

Contents

List of Figures

List of Tables

Preface

The projects of reediting the Toth and Vigo book on vehicle routing and of editing a book on arc routing germinated during the ROUTE Conference in Sitges, Spain, in June 2011. The first edition of the vehicle routing book had been highly successful, and it was then felt that the evolution of the field over the past 10 years justified a significantly revamped reedition. This led Corberán and Laporte (while exploring the cellars of the Codorníu Winery during the conference excursion) to think up a proposal for a similar arc routing book that would be produced in parallel with the second edition of the vehicle routing book, with a similar structure and the same format. Again, the last major edited book on arc routing had been published more than 10 years before and the field had evolved considerably since then. Both proposals were presented to SIAM in the summer and were accepted. Today we are proud to offer to the research community two up-to-date collections of scientific contributions written by specialists in various areas of vehicle routing and arc routing. The two books are entitled *Vehicle Routing: Problems, Methods and Applications*, Paolo Toth and Daniele Vigo, editors, and *Arc Routing: Problems, Methods and Applications*, Ángel Corberán and Gilbert Laporte, editors, both published by SIAM.

The vehicle routing book contains 15 chapters. A few of these are amalgamations or significantly revised versions of chapters published in the first edition, while most of the others are entirely new. The first chapter offers an overview of the field of the Vehicle Routing Problem (VRP) and its main variants. The remainder of the book is made up of three parts: the Capacitated VRP, important variants of the VRP, and applications. The first part contains two chapters on classical and new exact algorithms, as well as a chapter on heuristics. The second part surveys several variants: the VRP with Time Windows, pickup and delivery problems for goods or people transportation, stochastic VRPs, and miscellaneous variants. The third part is devoted to applications and covers the VRP with profits, real-time and dynamic VRPs, software and emerging technologies, ship routing, VRP applications in disaster relief, and green vehicle routing.

The arc routing book is new and contains 16 chapters. It opens with a chapter on historical perspectives, followed by three main parts: arc routing problems with a single vehicle, arc routing problems with several vehicles, and applications. The first part starts with a chapter on complexity, which is followed by four chapters on the Chinese Postman Problem and on the Rural Postman Problem. The second part contains four chapters on the Capacitated Arc Routing Problem and two on arc routing problems with min-max and profit maximization objectives. The last part covers some of the most important arc routing applications, including meter reading, salt spreading, snow removal, garbage collection, and newspaper delivery.

We thank all authors for the quality of their contributions, as well as all referees who carefully reviewed the chapters and Claudio Gambella for his help in editing the final manuscript of the VRP book. Thanks are also due to Dr. Thomas Liebling, Ms. Elizabeth Greenspan, Ms. Ann Manning Allen, and Ms. Sarah J. Murphy from SIAM for their support and encouragement.

Ángel Corberán, Universitat de València
Gilbert Laporte, HEC Montréal
Paolo Toth, Università di Bologna
Daniele Vigo, Università di Bologna

May 2014

Chapter 1

A Historical Perspective on Arc Routing

Ángel Corberán
Gilbert Laporte

1.1 ▪ Introduction

The study of arc routing is rooted in the seminal work of the Swiss mathematician Leonhard Euler who was called to study the famous Königsberg bridges problem in the 18th century while he was chair of mathematics at the St. Petersburg Academy of Sciences. The study of this problem led Euler to lay the foundation for modern graph theory. Recent accounts of the scientific career of Euler and of the history of the bridges of Königsberg are provided in Assad [2] and in Gribkovskaia, Halskau, and Laporte [20], respectively. A number of 19th century mathematicians also showed some interest in arc routing problems, namely, in relation to feasibility problems arising in graph traversals. However, the study of modern arc routing truly started in 1960 with the first publication on the Chinese postman problem. Over the following years and to this day, arc routing has evolved into a rich research area, firmly embedded within the broader domain of combinatorial optimization. This introductory chapter identifies the pioneer researchers in the field of arc routing and some of its main modern contributors.

1.2 ▪ Origins

We will first summarize some of the early works related to graph traversals. In particular, the Königsberg bridges problem is revisited and some comments on figure drawing and maze traversal are also included. The reader is referred to the excellent book by Biggs, Lloyd, and Wilson [6] for more details on these fascinating topics.

1.2.1 ▪ The Königsberg bridges problem

"In Königsberg in Prussia, there is an island A, called *the Kneiphof*; the river which surrounds it is divided into two branches, as can be seen in the figure (see Figure 1.2), and these branches are crossed by seven bridges a, b, c, d, e, f, and g. Concerning these bridges, it was asked whether anyone could arrange a route in such a way that he would

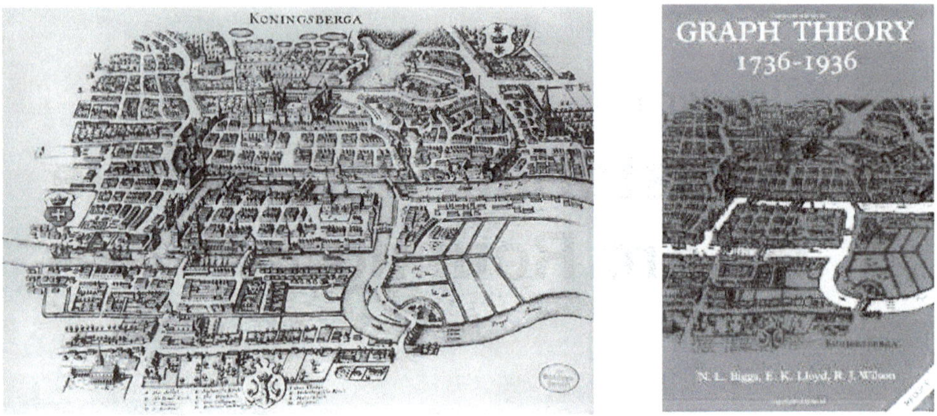

Figure 1.1. *Königsberg in the 18th Century. Front cover of the Biggs, Lloyd, and Wilson book (Oxford University Press).*

cross each bridge once and only once. I was told that some people asserted that this was impossible, while others were in doubt; but nobody would assert that it could be done."

This is how Euler [16] described the well-known Königsberg bridges problem (see Biggs, Lloyd, and Wilson [6] for a full English translation of the original paper).

And Euler continued: "From this, I have formulated the general problem: whatever be the arrangement and division of the river into branches, and however many bridges there be, can one find out whether or not it is possible to cross each bridge exactly once?"

Note that Euler wrote about finding a route traversing all the bridges exactly once, but it is not yet clear whether he was asking about the existence of a "closed" or of an "open" route. While some authors formulate the Königsberg bridges problem as that of finding an open route crossing each bridge exactly once (see, for instance, Assad and Golden [3], Biggs, Lloyd, and Wilson [6], Hopkins and Wilson [25], and Grötschel and Yuan [21]), others also require that the route should return to the starting point (see Wilson and Watkins [36], Eiselt, Gendreau, and Laporte [14], Zaragoza [38], Ahr [1], Gribkovskaia, Halskau, and Laporte [20], and Wøhlk [37]).

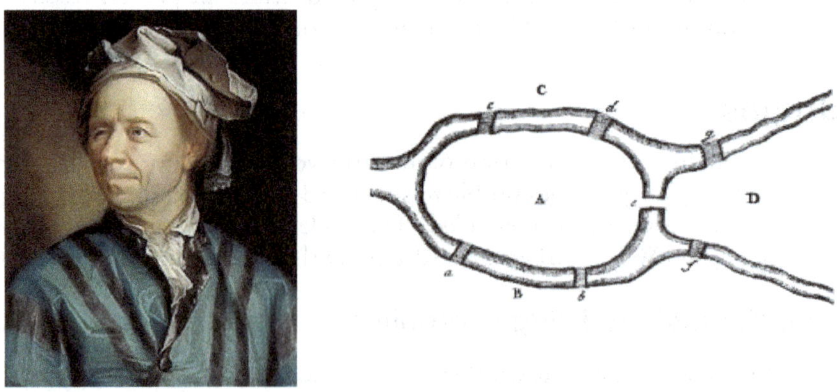

Figure 1.2. *Leonard Euler (1707–1783) and his drawing of the river Pregel and the seven bridges.*

At the end of [16], Euler pointed out: "So whatever arrangement may be proposed, one can easily determine whether or not a journey can be made, crossing each bridge once, by the following rules. If there are more than two areas to which an odd number of bridges lead, then such journey is impossible. If, however, the number of bridges is odd for exactly two areas, then the journey is possible if it starts in either of these areas. If, finally, there are no areas to which an odd number of bridges leads, then the required journey can be accomplished starting from any area."

It is now clear that Euler used the term "route" in a broad sense and interpreted the problem as that of finding a walk, closed or not, traversing each bridge exactly once. However, the "closed" version of the problem has an important advantage because of its direct relationship with Eulerian graphs. Recall that a connected graph G is said to be Eulerian (or unicursal) if there exists a closed walk including all the edges of G exactly once. Therefore, the closed version of the problem can be stated as follows: Is the graph representing the bridges of Königsberg Eulerian? Since the four vertices of the graph representing the Königsberg areas have odd degrees, the answer is no, and there is no open or closed route crossing exactly once each of the seven bridges of Königsberg.

1.2.2 ▪ Drawing figures

A closely related problem is that of drawing a given figure with the minimum number of strokes. In 1809, Poinsot [33] used arguments similar to those of Euler to prove that figures consisting of n points, where each point is joined to every other point, can be drawn in only one stroke if n is odd, but not if n is even. Note that these figures (see Figure 1.4) correspond to what we call complete graphs, and complete graphs are Eulerian for odd values of n.

Figure 1.3. *Louis Poinsot (1777–1859) and Johann Benedict Listing (1808–1882).*

In 1847, Listing, first a student and later a collaborator of Gauss, wrote a book on topology [28] which included some parts about figure drawing. Among other interesting comments, he pointed out that Figure 1.5a has eight odd-degree vertices and cannot therefore be drawn with less than four lines. In a sense, this sentence contains the key idea for solving what will later become known as the Chinese postman problem: to link the odd-degree vertices of the graph.

Finally, let us mention that not so many years ago, schoolchildren from many countries used to play a game consisting of drawing some figures with just one stroke. For instance, it is mentioned at http://www.mathematische-basteleien.de/house.html that the

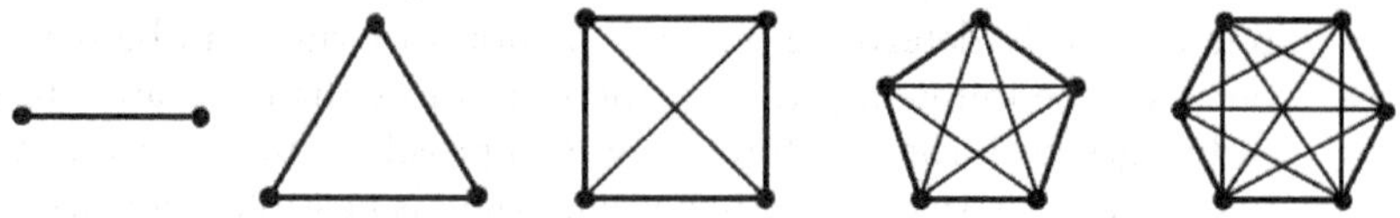

Figure 1.4. *Complete graphs with two to six vertices.*

House of Santa Claus is a very old German drawing game for small children: "You have to draw a house (see Figure 1.5b) in one line. You must not lift your pencil while drawing. You must not repeat a line. During drawing you have to say "Das ist das Haus des Nikolaus (This is Santa Claus' house)." You must speak a syllable of this sentence to every straight line. Girls must know "Wer dies nicht kann, kriegt keinen Mann (Who can't do this drawing, will get no man)."

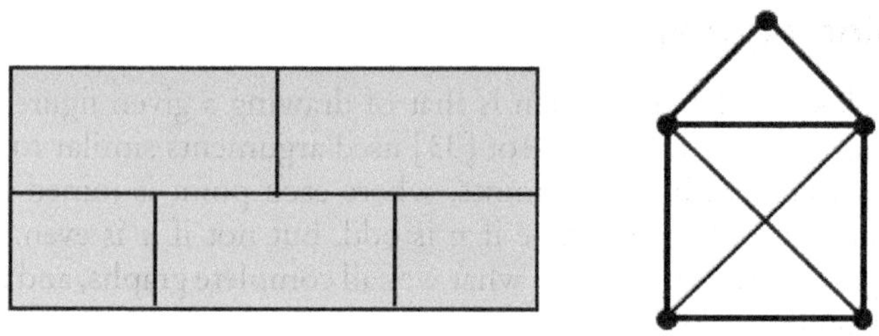

(a) (b)

Figure 1.5. *Listing's drawing and the house of Santa Claus.*

1.2.3 ▪ Mazes and labyrinths

At the end of 19th century, several researchers became interested in the problem of escaping from a maze. In his book *Theorie der endlichen und unendlichen Graphen*, König [26] devotes a chapter to this problem (*Das Labyrinthenproblem*), in which he discusses in detail the algorithms of Wiener, Trémaux, and Tarry. König states the problem as follows: "A labyrinth can be identified with a graph, more precisely a finite and connected graph; the vertices of this graph correspond in the labyrinth to the branch points and the endpoints of the blind alleys."

"It is now required to give a method for reaching a specified point, which is usually treated as the center of the labyrinth. Since it is entirely arbitrary which point is treated as the center, the problem is only to be solved by specifying a way to reach every point of the labyrinth. If the plan of the whole labyrinth is known, there is no difficulty."

And König continues: "We will first assume that if during the journey one arrives at a vertex P, it is known for every edge ending at P whether it has already been traversed; besides this we assume that (like Theseus with the aid of Ariadne's thread) one can at any moment retrace in the opposite direction the whole of the path already covered (with the same repetitions). Under these assumptions the problem was solved by Wiener, who was the first ever to treat the age-old problem of the labyrinth as a mathematical one."

Figure 1.6. *Left: Hever castle. Right: Hampton Court maze.*

Figure 1.7. *Gaston Tarry (1843–1913) and Dénes König (1884– 1944).*

A description of the method proposed by Wiener follows, as well as some comments on it: "The method of Wiener cannot be called economical, since in general the edges have to be walked through a great many times. This is the more noticeable in that there is a journey through all the edges of the graph in which no edge is traversed more than twice."

Note that if every edge in the graph representing the labyrinth is duplicated, one obtains another graph with even-degree vertices. Since this graph is Eulerian, there exists a closed walk traversing every edge of the original graph twice, although, as pointed out by Biggs, Lloyd, and Wilson [6], this is only an argument proving the existence of a solution, and not a practical algorithm.

As noted by König [26], Trémaux proposed an algorithm which assumed that "on arrival during the journey at any vertex P, it is known for every edge ending at P whether it has already been walked, and if so how many times (the directions in which the edges have been walked are not assumed to be known here)." However, it was Tarry [34] who proposed the simplest procedure to escape from a labyrinth, just assuming that "every time we arrive at any vertex P it is known for every edge PK ending at P whether it has been traversed in the direction of K; and we can always determine the edge by which P was reached for the first time (the "entry edge" of P)." The method proposed by Tarry guarantees that every edge is traversed at most once in each direction. It works as follows: "If you come via an edge PQ to a vertex Q, continue the journey along any edge QR that has not yet been traversed in the direction of R; but do not choose QR to be the entry edge of Q unless all other edges QK ending at Q have already been traversed in the direction of K."

1.3 ▪ Characterizations of Eulerian graphs

We now describe the necessary and sufficient conditions an undirected, directed, or mixed graph G has to satisfy in order to be Eulerian. Obviously, undirected graphs need to be connected, whereas directed and mixed graphs need to be strongly connected. In what follows we assume that these conditions hold.

1.3.1 ▪ Undirected Eulerian graphs

Euler stated necessary conditions for an undirected graph G to be Eulerian. He also pointed out they are sufficient without actually providing a proof of it. Sufficiency was proved in 1873 in a posthumous paper by Hierholzer [24]. Hence,

An undirected connected graph G is Eulerian if and only if every vertex of G has even degree.

At the end of his celebrated paper, Euler mentions the problem of constructing a Eulerian tour if one exists: "When it has been determined that such a journey can be made, one still has to find how it should be arranged. ... I do not think it worthwhile to give any further details concerning the finding of the routes." One can therefore assume that Euler thought this was a trivial issue. It was also Hierholzer [24] who proposed a linear-time algorithm to find a Eulerian walk in an undirected Eulerian graph G. This algorithm is described in Chapter 2.

A second, and very interesting, characterization of undirected Eulerian graphs was given in 1912 by Veblen [35]:

An undirected connected graph G is Eulerian if and only if it is the disjoint union of some cycles.

1.3.2 ▪ Directed Eulerian graphs

Clearly, the evenness of the vertices of a directed graph $G = (V, A)$ is again a necessary condition for the graph to be Eulerian. However, as can be seen in Figure 1.8a, which shows an even but non-Eulerian graph, it is not sufficient. A sufficient condition for a directed graph G to be Eulerian was given by König [26]: Every vertex of G must be symmetric; i.e., the number of arcs entering and leaving each vertex must be equal. Hence,

A directed and strongly connected graph G is Eulerian if and only if every vertex of G is symmetric.

Hierholzer's algorithm for undirected graphs can be easily adapted to determine a Eulerian walk of a directed Eulerian graph.

1.3.3 ▪ Mixed Eulerian graphs

A vertex of a mixed graph $G=(V,E,A)$ is said to have even degree if it is incident to an even number of edges and arcs. Again, the evenness of the vertices of G is a necessary but not a sufficient condition for a mixed graph to be Eulerian. However, it is easy to see that evenness together with symmetry are sufficient conditions for unicursality. But are evenness and symmetry also necessary conditions for a mixed graph to be Eulerian? The answer is obviously no. For instance, the mixed graph shown in Figure 1.8b is not symmetric although is clearly an Eulerian graph.

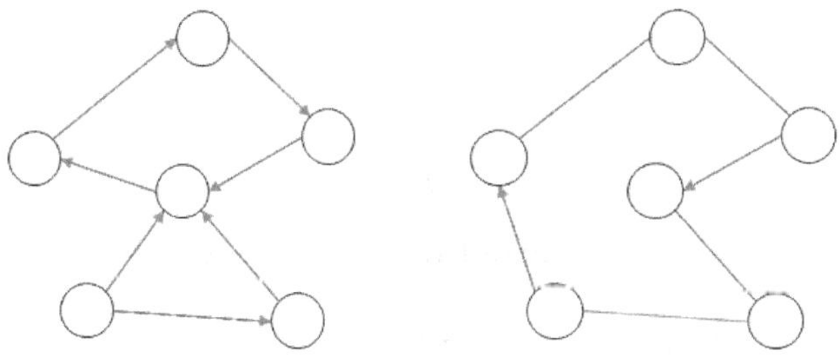

(a) (b)

Figure 1.8. *Non-Eulerian directed graph and Eulerian mixed graph.*

Figure 1.9. *Lester Randolph Ford Jr. and Delmer Ray Fulkerson (1924– 1976).*

The next question is then, "Are there necessary and sufficient conditions for a mixed graph G to be Eulerian?" Ford and Fulkerson answered this question in 1962 in their celebrated book *Flows in Networks* [17]: Every vertex of G must be even and graph G must be balanced, i.e., for every subset of vertices $S \subseteq V$, the difference between the number of arcs leaving S and the number of arcs entering S must be less than or equal to the number of edges incident with S. This last condition is generally known as the "balanced-set condition." Hence,

A mixed and strongly connected graph G is Eulerian if and only if G is even and balanced.

Figure 1.10 shows two even graphs. The first one does not satisfy the balanced-set condition and is not Eulerian, whereas the second one is a balanced and therefore Eulerian graph.

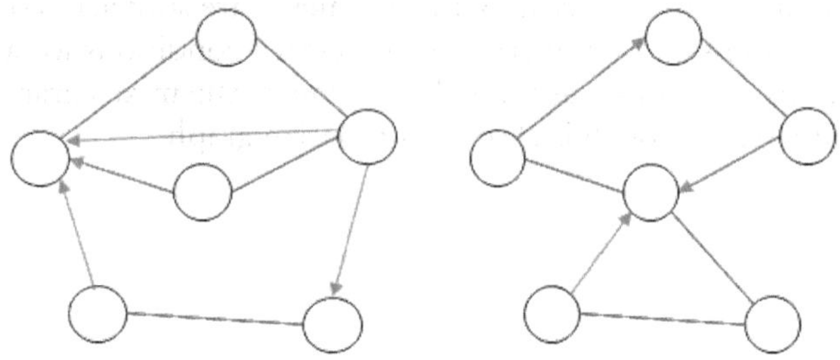

(a) (b)

Figure 1.10. *Unbalanced and balanced mixed graphs.*

Note that checking whether a mixed graph G satisfies the balanced-set condition is a more complex task than checking whether it is even or symmetric since a condition must be verified for every subset of vertices. Fortunately, Ford and Fulkerson [17] also proposed an efficient algorithm which, applied to an even graph, either transforms the mixed graph into a completely directed symmetric (and therefore Eulerian) graph or fails, in which case the graph is not balanced (nor Eulerian). Let us also mention here that Nobert and Picard [30] presented a polynomial-time procedure that identifies a violated balanced-set inequality in a mixed graph if one exists (see also Benavent, Corberán, and Sanchis [5] for an explicit description and a proof of the procedure).

It is interesting to note that the balanced-set condition applied to a mixed graph without any edge (a directed graph) reduces to the symmetry condition of each subset of vertices, a condition that is satisfied if all the vertices are symmetric. Therefore, the necessary and sufficient conditions for a directed graph to be Eulerian can be obtained from the corresponding conditions for mixed graphs.

Finally, as was observed by Zaragoza [38], the characterization of Veblen for undirected Eulerian graphs in terms of cycle decompositions also applies to directed and mixed graphs.

1.4 ▪ The emergence of optimization

The Königsberg bridges problem posed the question of whether there exists a closed walk traversing each of the bridges exactly once. In general terms, the question is "Given an undirected and connected graph G, is there a closed walk traversing all its edges exactly once?" The answer to this decision problem is "Yes" or "No." There is no optimization involved in it.

It was Meigu Guan (Mei-Ko Kwan) who in 1960 introduced optimization to the problem. Guan is a Chinese mathematician who was working at that time at the Shandong Normal University. It is commonly assumed that Guan's work originated when he was working at a post office during the Chinese Cultural Revolution. However, according to Grötschel and Yuan [21], it seems that like many other scientists in China, Guan was encouraged to solve real-life problems during the Great Leap Forward movement (1958–1960), which attempted to transform the country from an agrarian to a modern economy.

The problem stated by Guan in a paper published in Chinese in 1960 and in English in 1962 [22] is as follows:

When the author was plotting a diagram for a mailman's route, he discovered the following problem: "A mailman has to cover his assigned segment before returning to the post office. The problem is to find the shortest walking distance for the mailman."

Figure 1.11. *Meigu Guan. Courtesy of Ya-Xiang Yuan [21].*

This problem is widely known as the Chinese postman problem (CPP). It seems that Alan Goldman mentioned it to Jack Edmonds when Edmonds was a member of Goldman's operations research group created in 1961 at the US National Bureau of Standards. It is not known whether Goldman suggested the name "Chinese postman problem" to Edmonds or whether it was Edmonds who coined the name. However, it seems to have first appeared in the title of an abstract by Edmonds [11] for the 27th National Meeting of the Operations Research Society of America (May 1965): "The Chinese Postman's Problem."

The CPP can also be interpreted as the problem of finding a subset of edges with minimum length which, added to the original graph, produces a Eulerian graph. Guan noted that a graph has an even number of odd-degree vertices and that a Eulerian graph can be obtained by replicating some edges in order to connect the odd-degree vertices. Based on this observation, Guan [22] proposed an algorithm for the CPP. Unfortunately, Guan's procedure is nonpolynomial.

A problem closely related to the CPP is the generalized Königsberg bridges problem, studied by Bellman and Cooke [4]: "In connection with a given graph, we ask for the paths which traverse every edge at least once and for which the number of repetitions of edges is a minimum." Also in this case, the procedure proposed by the authors was nonpolynomial: "The (dynamic programming) procedure outlined can require a large number of comparisons. The number is difficult to estimate."

In the above-mentioned abstract (see [11]), Edmonds writes "We present an algorithm ... (that) is *good* in the sense that the amount of work in applying it is at worst moderately algebraic, relative to the size of the graph, rather than exponential. It combines two earlier known algorithms: (1) the well-known *shortest path* algorithm, (2) a recent algorithm for *maximum matching*." Of course, Edmonds was referring to his celebrated polynomial algorithm for the minimum cost matching problem [12], and the procedure for the CPP

Figure 1.12. *Alan J. Goldman (1933–2010) and Jack Edmonds.*

Figure 1.13. *Ellis L. Johnson. Nicos Christofides and Bruce L. Golden.*

mentioned consists of constructing first a complete graph G' whose vertices are the odd-degree vertices of the original graph G and whose edge costs are those of shortest paths linking them in G, and computing a perfect matching of minimum cost in G'.

The full details and many other outstanding results, including the complete description of the CPP polyhedron and several algorithms for the directed and mixed versions of the CPP, were published in 1973 in a seminal paper coauthored with Ellis Johnson [13]. Also in 1973, Christofides [8] proposed the same two-step procedure to solve the CPP.

While the directed version of the CPP can also be solvable in polynomial time (Edmonds and Johnson [13]), the CPP defined on a mixed graph (MCPP) is NP-hard. It was Papadimitriou [32] who proved this important result in 1976. However, Edmonds and Johnson proposed a polynomial-time algorithm for the MCPP defined on even graphs. Based on this algorithm, Frederickson [18] proposed two heuristics, each having a worst-case ratio of 2 for this problem, although applying both heuristics and selecting the best solution leads to a worst-case ratio of 5/3. Since the CPP defined on a windy graph (Minieka [29]) contains the MCPP as a special case, it is also an NP-hard problem (Brucker [7], Guan [23]).

In closing this section, we mention two important extensions of the CPP. In the first, called the Rural Postman Problem (RPP), some arcs or edges of the graph require service

Figure 1.14. *Christos H. Papadimitriou and Greg N. Frederickson.*

Figure 1.15. *Larry D. Bodin and Bruce L. Golden.*

while the others do not but may be traversed in the solution. The aim is to determine a minimum cost traversal of the required arcs and edges of the graph. This problem was introduced by Orloff [31] and was later proved to be NP-hard by Lenstra and Rinnooy Kan [27]. Golden and Wong [19] introduced an extension of the RPP, called the Capacitated Arc Routing Problem, in which demands are associated with the arcs and edges, and these must be optimally collected by a set of capacitated vehicles based at a depot.

Many different formulations and exact and heuristics algorithms have been proposed for the above-mentioned problems and their extensions, but they will be the subject of other chapters in this book. As a complement we refer the interested readers to the following surveys and books: Eiselt, Gendreau, and Laporte [14] and [15], Assad and Golden [3], Dror [10], Wøhlk [37], and Corberán and Prins [9].

1.5 ▪ Arc routing today

Over the past 35 years or so, arc routing has evolved into a mature research field. The initial impetus took place in the late 1970s and early 1980s under the scientific leadership of Christofides at Imperial College and of Bodin and Golden at the University of Maryland. Today, several research groups in various parts of the world are active in arc routing. Over the past years, major developments have occurred in the design of exact

Figure 1.16. *Attendants at WARP1: Juan P. Riquelme-Rodrígues, José M. Sanchis, Richard Eglese, Enrico Bartolini, Lukas Bach, André Langevin, Kevin Gaze, Gilbert Laporte, Ángel Corberán, Stefan Irnich, Bruce Golden, José M. Belenguer, Sanne Wøhlk, Manuel Sorge, Elisabeth Gussmagg-Pfliegl, Geir Hasle, Ana C. Nunes, Thais Ávila, Enrique Benavent, Leonor Pinto, Luc Muyldermans, Elena Fernández, Marcus Poggi, Stefan Ropke, Harilaos Psaraftis, and Jörg Kalcsics (Claudia Archetti, Alexander Butsch, Demetrio Laganá, and Vittorio Maniezzo are not in the photo).*

integer linear programming algorithms based on branch-and-cut, column generation, and their hybridization. In parallel, we have witnessed the development of powerful metaheuristics for arc routing problems, as well as the emergence of new complexity results. Several of the newer findings were presented at the 1st Workshop on Arc Routing Problems (WARP1) held in Copenhagen in May 2013 (see Figure 1.16).

1.6 ▪ Arc routing tomorrow

The development of exact and heuristic algorithms for arc routing problems has now reached a very high level of sophistication. This means that most classical variants described in this book can now be solved quite accurately for ever increasing sizes and within reasonable computation time limits. At the same time, the state-of-the-art methodologies are becoming increasingly difficult to reproduce and to teach. We may even have reached a point where it may be desirable to devise simpler algorithms, even if this is likely to result in a small loss of accuracy. In terms of applications, it is highly likely that new sophisticated variants of arc routing problems will emerge in the coming years, namely, in the areas of arc routing problems with profits, arc routing problems with synchronization components, and combinations of node and arc routing problems, as well as stochastic and dynamic problems. The latter class of problems is likely to become increasingly relevant in a context where vehicles are now often equipped with communication technologies which allow real-time interactions and decisions to be made.

Acknowledgments

Ángel Corberán wishes to thank the Ministerio of Economía y Competitividad (project MTM2012-36163-C06-02) of Spain and the Generalitat Valenciana (project PROMETEO/2013/049) for their support.

Bibliography

[1] D. AHR, *Contributions to Multiple Postmen Problems*, Ph.D. thesis, University of Heidelberg, Heidelberg, Germany, 2004.

[2] A.A. ASSAD, *Leonhard Euler: A brief appreciation*, Networks, 49 (2007), pp. 190–198.

[3] A.A. ASSAD AND B.L. GOLDEN, *Arc routing methods and applications*, in Network Routing, M.O. Ball, T.L. Magnanti, C.L. Monma, and G.L. Nemhausser, eds., Handbooks of Operations Research and Management Science 8, North-Holland, Amsterdam, 2000, pp. 375–483.

[4] R. BELLMAN AND K.L. COOKE, *The Königsberg bridges problem generalized*, Journal of Mathematical Analysis and Applications, 25 (1969), pp. 1–7.

[5] E. BENAVENT, Á. CORBERÁN, AND J.M. SANCHIS, *Linear programming based methods for solving arc routing problems*, in Arc Routing: Theory, Solutions and Applications, M. Dror, ed., Kluwer, Boston, 2000, pp. 231–275.

[6] N.L. BIGGS, E.K. LLOYD, AND R.J. WILSON, *Graph Theory 1736–1936*, Clarendon Press, Oxford, UK, 1998.

[7] P. BRUCKER, *The Chinese postman problem for mixed graphs*, in Proceedings of the International Workshop, Bad Honnef, 1980, Lecture Notes in Computer Science 100, Springer, Berlin, 1981, pp. 354–366.

[8] N. CHRISTOFIDES, *The optimum traversal of a graph*, Omega, 1 (1973), pp. 719–732.

[9] Á. CORBERÁN AND C. PRINS, *Recent results on arc routing problems: An annotated bibliography*, Networks, 56 (2010), pp. 50–69.

[10] M. DROR, *Arc Routing : Theory, Solutions and Applications*, Kluwer, Boston, 2000.

[11] J. EDMONDS, *The Chinese postman's problem*, Operations Research Bulletin, 13 (1965), pp. B73–B77.

[12] ——, *Paths, trees and flowers*, Canadian Journal of Mathematics, 17 (1965), pp. 449–467.

[13] J. EDMONDS AND E.L. JOHNSON, *Matching, Euler tours and the Chinese postman problem*, Mathematical Programming, 5 (1973), pp. 88–124.

[14] H.A. EISELT, M. GENDREAU, AND G. LAPORTE, *Arc routing problems, part 1: The Chinese postman problem*, Operations Research, 43 (1995), pp. 231–242.

[15] ——, *Arc routing problems, part 2: The rural postman problem*, Operations Research, 43 (1995), pp. 399–414.

[16] L. EULER, *Solutio problematis ad geometriam situs pertinentis*, Commentarii Academiae Scientiarum Imperialis Petropolitanae, 8 (1736), pp. 128–140.

[17] L.R. FORD AND D.R. FULKERSON, *Flows in Networks*, Princeton University Press, Princeton, NJ, 1962.

[18] G.N. FREDERICKSON, *Approximation algorithms for some postman problems*, Journal of the Association for Computing Machinery, 26 (1979), pp. 538–554.

[19] B.L. GOLDEN AND R.T. WONG, *Capacitated arc routing problems*, Networks, 11 (1981), pp. 305–315.

[20] I. GRIBKOVSKAIA, Ø. HALSKAU, AND G. LAPORTE, *The bridges of Königsberg, a historical perspective*, Networks, 49 (2007), pp. 199–203.

[21] M. GRÖTSCHEL AND Y.-X. YUAN, *Euler, Mei-Ko Kwan, Königsberg, and a Chinese postman*, Documenta Mathematica, Extra Volume ISMP (2012), pp. 43–50.

[22] M.G. GUAN, *Graphic programming using odd or even points*, Chinese Mathematics, 1 (1962), pp. 237–277.

[23] M.G. GUAN, *On the windy postman problem*, Discrete Applied Mathematics, 9 (1984), pp. 41–46.

[24] C. HIERHOLZER, *Über die Möglichkeit, einen Linienzug ohne Wiederholung und ohne Unterbrechung zu umfahren*, Mathematische Annalen, 6 (1873), pp. 30–32.

[25] B. HOPKINS AND R.J. WILSON, *The truth about Königsberg*, The College Mathematics Journal, 35 (2004), pp. 198–207.

[26] D. KÖNIG, *Theorie der endlichen und unendlichen Graphen*, Akademische Verlagsgessellschaft, Leipzig, 1936.

[27] J.K. LENSTRA AND A.H.G. RINNOOY KAN, *On general routing problems*, Networks, 6 (1976), pp. 273–280.

[28] J.B. LISTING, *Vorstudien zur Topologie*, Vandenhoeck und Ruprecht, Göttingen, 1847.

[29] E. MINIEKA, *The Chinese postman problem for mixed networks*, Management Science, 25 (1979), pp. 643–648.

[30] Y. NOBERT AND J.-C. PICARD, *An optimal algorithm for the mixed Chinese postman problem*, Networks, 27 (1996), pp. 95–108.

[31] C.S. ORLOFF, *A fundamental problem in vehicle routing*, Networks, 4 (1974), pp. 35–64.

[32] C.H. PAPADIMITRIOU, *On the complexity of edge traversing*, Journal of the Association for Computing Machinery, 23 (1976), pp. 544–554.

[33] L. POINSOT, *Mémoire sur les polygones et les polyèdres*, Journal de l'École Polytechnique, 4 (1810), pp. 16–49.

[34] G. TARRY, *Le problème des labyrinthes*, Nouvelles Annales de Mathématiques, 14 (1895), pp. 187–190.

[35] O. VEBLEN, *An application of modular equations in analysis situs*, Annals of Mathematics, 2 (1913), pp. 86–94.

[36] R.J. WILSON AND J.J. WATKINS, *Graphs: An Introductory Approach*, Wiley, New York, 1990.

[37] S. WØHLK, *A decade of capacitated arc routing*, in The Vehicle Routing Problem: Latest Advances and New Challenges, B.L. Golden, S. Raghavan, and E.A. Wasil, eds., Springer, New York, 2008, pp. 29–48.

[38] F.J. ZARAGOZA, *Postman Problems on Mixed Graphs*, Ph.D. thesis, University of Waterloo, Waterloo, Ontario, Canada, 2003.

Part I

Arc Routing Problems with a Single Vehicle

Chapter 2

The Complexity of Arc Routing Problems

René van Bevern
Rolf Niedermeier
Manuel Sorge
Mathias Weller

2.1 ▪ Introduction

This chapter is devoted to surveying aspects of computational complexity for three central arc routing problems (and their corresponding variants):

- CHINESE POSTMAN, where one asks for a minimum-cost tour traversing all edges of a graph at least once;
- RURAL POSTMAN, which generalizes CHINESE POSTMAN in the sense that *only a subset* of the edges has to be visited; and
- CAPACITATED ARC ROUTING, representing the most general arc routing problem in this chapter, allows more than one vehicle to be used to traverse the edges.

With the exception of the basic version of CHINESE POSTMAN, almost all of these problems are NP-hard; that is, they are computationally intractable problems with respect to the worst-case running times of existing solving algorithms (see Garey and Johnson [41]). Due to their practical importance, however, it is of interest to search for ways to cope with their computational intractability. In theoretical computer science—where strong emphasis is laid on *provable* performance bounds—there are basically two lines of attack:

- polynomial-time approximation algorithms [6, 92, 95], where one trades solution optimality for efficient, polynomial running time instead of exponential, as to be expected for finding exact solutions; and
- parameterized algorithmics and complexity analysis [24, 34, 75], where one seeks to identify small problem-specific parameters that influence the seemingly unavoidable exponential-time behavior. Then, algorithms for finding exact solutions in a running time that is exponential only in the parameter value can be designed.

Besides classical complexity classification along the lines of NP-hardness, we will discuss for all problems aspects of polynomial-time approximability as well as parameterized complexity. Notably, parameterized complexity classification was not subject of the previous surveys [25, 31, 30, 63, 83], and it is still a very young field in arc routing with numerous challenging research questions. Indeed, throughout the chapter we will feature several open questions targeted at stimulating future research.

The outline of our presentation is as follows. Different from the older survey by Dror [25], we follow a problem-wise organization. That is, for each of the described arc routing problems we discuss it and some of its interesting variants in terms of classical computational complexity, polynomial-time approximability, and parameterized complexity. As to the problems, we basically follow a strategy from easier to more demanding problems; that is, typically the later a problem comes, the harder (in computational terms) or more general it is. Before doing so, however, in the remainder of this introductory section we briefly describe the central concepts used throughout this chapter.

2.1.1 ▪ Eulerian graphs and notational conventions

We use standard notation for directed and undirected graphs. In particular, *graphs* are pairs (V,E) of a set V of *vertices* and a set E of *edges* (unordered pairs of vertices). Instead of a set of edges, directed graphs have a set A of ordered pairs of vertices called *arcs*. Mixed graphs contain both edges and arcs (a mixed graph is a triplet (V,E,A)). *Multigraphs* may contain multiple copies of each edge/arc, that is, the set of edges/arcs is a multiset. Let $G=(V,E)$ and let $u \in V$. A vertex $v \in V$ is said to be *adjacent* to u if $\{u,v\} \in E$. The number of vertices that are adjacent to a vertex u is denoted by $\deg_G(u)$ and called the *degree* of u in G. The vertex u is called *even* or *balanced* in G if $\deg_G(u)$ is even. A *path* from u to w in G is a sequence $(u = v_0, v_1, v_2, \ldots, v_\ell = w)$ of vertices in V such that $\{v_i, v_{i+1}\} \in E$ for all $0 \le i < \ell$. Its *length* is ℓ, and it *traverses* or *visits* the edges $\{v_i, v_{i+1}\}$ for all $0 \le i < \ell$. If such a path exists in G, then u is said to be *connected to* w. A *cycle* or *tour* is a path whose first and last vertices are identical. A subgraph G' of a graph G is *connected* if each two vertices of G' are connected in G'. If G' is maximal with respect to the number of edges, then it is called a *connected component* of G.

In directed graphs, slight variations of these concepts exist. For a vertex u its *incoming arcs* are all arcs of $A \cap (V \times \{u\})$ and its *indegree* is the number of such arcs in G. The *outgoing arcs* of u and the *outdegree* of u are defined analogously. We call u *balanced* if its indegree equals its outdegree.

The concept of paths and cycles trivially extends to directed graphs. A subgraph G' of a directed graph G is called *strongly connected* if each two vertices are connected by a directed path. It is *weakly connected* or simply *connected* if the undirected (multi)graph resulting from ignoring the arc directions is connected.

A *postman tour* in a graph is a tour visiting each edge of a certain set of edges at least once. An *Eulerian tour* in a graph is a tour visiting each edge exactly once. A graph that admits an Eulerian tour is called *Eulerian*. It is common knowledge that a graph is Eulerian if and only if it is connected and each vertex is balanced. This statement and the definitions of tours are analogous for directed graphs. A mixed graph $G=(V,E,A)$, however, is Eulerian if and only if the graph G' resulting from ignoring the directions of the arcs is connected, the degree of each vertex in G' is even, and for each subset $S \subseteq V$ of vertices, the number of arcs entering S minus the number of arcs leaving S is at most the number of edges between S and $V \setminus S$ (see Ford and Fulkerson [35]).

2.1.2 ▪ Classical complexity

In order to examine the computational complexity of a problem, we first have to define what a computational problem is. We will consider two types of problems:

A *decision problem* Q is a set of instances that fulfill some property. For example, the EVEN problem is the set of all even nonnegative integers. Equivalently, one can formulate the EVEN problem as "Given a number x, is x an even nonnegative integer?" In the context of this chapter, instances of decision problems are often pairs of a graph G and some number k. For example, the VERTEX COVER problem is the set of all pairs (G,k), such that the undirected graph G can be covered by at most k vertices. Herein, a graph is covered by a vertex set Z if all edges are incident to a vertex in Z. Equivalently we may define VERTEX COVER by the question "Given an undirected graph G and a number k, can G be covered by at most k vertices?"

An *optimization problem* Q is a function that, given some instance, outputs a feasible solution that minimizes (or maximizes) a measurement m_Q. For example, the *minimization variant* of the VERTEX COVER problem is "Given an undirected graph G, find a vertex cover Z of G such that $m_{\text{VERTEX COVER}}(Z) := |Z|$ is minimum." In the context of this chapter, the measurement m_Q is usually $m_Q(Z) = |Z|$ for vertex subsets Z of a graph or $m_Q(Z) = \sum_{e \in Z} c(e)$ for edge subsets Z of a graph with cost-function $c: E \to \mathbb{Q}$.

We say that a decision problem Q can be *solved* in some time $T(n)$ if there is an algorithm that, given an instance x of length $|x| = n$ bits, determines whether $x \in Q$ using at most $T(n)$ computation steps. Likewise, a minimization or maximization problem Q can be *solved* in time $T(n)$ if there is an algorithm that, given an instance x, computes a feasible solution Y for x in $T(n)$ computation steps such that $m_Q(Y)$ is minimum or maximum, respectively. Classical complexity theory investigates a zoo of interesting classes of decision problems (see Aaronson [2]), but we will consider only two of them here. The first class, called P, is the class of all decision problems that can be solved by a deterministic Turing machine in n^c time for some constant $c \in \mathbb{N}$. The second class, called NP, is the class of all decision problems that can be solved by a *non*deterministic Turing machine in n^c time for some constant $c \in \mathbb{N}$ [5, 36, 15, 79].

P vs. NP. The question of whether P = NP is both amazing and popular [36, 42, 45, 70, 79] since many problems faced by people in many sciences and businesses are easily seen to be in NP but no algorithm solving them "quickly," that is, in (deterministic) polynomial-time, is known.

Deciding whether or not P = NP is one of the seven famous Millennium Prize Problems stated by the Clay Mathematics Institute [1]: A solution to each of these problems is promised to be rewarded with a US$1,000,000 prize. The persistent inability to decide whether P = NP gave rise to the concept of polynomial-time reduction (see Arora and Barak [5]). We say that a decision problem A can be (polynomial-time) reduced to a decision problem B if there is some algorithm running in polynomial time and, given an instance x of A, produces an instance y of B such that $x \in A$ if and only if $y \in B$. This allowed the concept of so-called "hard problems" for the class NP: A problem Q is said to be NP-*hard* if all problems in NP can be reduced to it. If Q is also contained in NP, then it is called NP-*complete*. In the last decades, thousands of problems have been shown to be NP-hard (mostly by using the transitivity of reducibility that allows showing NP-hardness of a problem Q by reducing any NP-hard problem to Q). A large subset of these problems has been listed by Garey and Johnson [41]. Since it is widely believed, we will assume P $\neq$ NP in the remainder of this chapter.

Coping with NP-hardness. Many real-world problems have been shown to be NP-hard and, thus, do not admit polynomial-time algorithms. Since we can use algorithms that solve optimization problems to solve decision problems, this also implies that there are no polynomial-time algorithms to minimization or maximization problems corresponding to NP-hard decision problems. Since we cannot simply refuse to solve these problems, ways of coping with this computational hardness have been developed. One possibility is to accept a feasible, yet not necessarily optimal, solution that can be shown to be reasonably close to the optimum. Many problems have been shown to admit such *approximation algorithms* [6, 92, 95]. Another approach to coping with NP-hard problems exploits the fact that the instances produced by reductions are pathological for most applications; that is, aspects that make the considered problem hard are often unlikely to be seen in practice. Thus, it might be possible to design an algorithm that is fast for instances that are practically relevant but takes exponential time for other instances. A theoretical approach incorporating this thought is *Parameterized Complexity* [24, 34, 75]. Another theoretical approach in this direction is the (relatively new) technique of *Smoothed Analysis* introduced by Spielman and Teng [89, 90] that diverges from the concept of worst-case analysis, which "is often the source of discrepancy between the theoretical evaluation of an algorithm and its practical performance" [90]. Here, the performance of algorithms is measured as expected running time when input instances are exposed to a slight random perturbation, thus eliminating the influence of pathological worst-case instances. Unfortunately, there seem to be no works considering routing problems under smoothed analysis yet.

2.1.3 ▪ Complexity of approximation

As mentioned, one approach of coping with NP-hardness is to give up the search for an optimal solution and accept a slightly less accurate answer [6, 92, 95]. For instance, VERTEX COVER cannot be solved in polynomial time, but we can approximate the answer to the minimization variant in polynomial time. More precisely, with OPT_G denoting the optimum solution of the minimization variant of VERTEX COVER on input G, a feasible solution (that is, a vertex cover) Z for G such that $1 \leq |Z|/|\mathrm{OPT}_G| \leq 2$ can be computed in polynomial time: Find an edge $\{u, v\}$ in G such that neither u nor v are selected, select both u and v, and repeat until no such edge can be found.

In general, for a minimization problem Q, we define $\mathrm{OPT}(x)$ as the optimal solution for an instance x. Then, a polynomial-time algorithm that, given an instance x of Q, computes a feasible solution Z such that $1 \leq m_Q(Z)/m_Q(\mathrm{OPT}(x)) \leq d(|x|)$ is called a *factor $d(|x|)$ approximation* for Q. Likewise, for a maximization problem Q, a factor $d(|x|)$ approximation algorithm computes a feasible solution Z with $1 \leq m_Q(\mathrm{OPT}(x))/m_Q(Z) \leq d(|x|)$. Most commonly, the function d is constant and, thus, factor d approximations are called *constant-factor* approximations. However, some problems are known not to admit constant-factor approximations [6, 92, 95].

2.1.4 ▪ Parameterized complexity

Parameterized algorithmics [24, 34, 75] is an approach to find optimal solutions for NP-hard problems in reasonable time. The idea is to accept the seemingly inevitable combinatorial explosion, but to confine it to one aspect of the problem, the *parameter*.

Consider a decision problem Q. If there is an algorithm A such that, for all $p \in \mathbb{N}$ and all instances x of Q whose parameter is p, A decides whether $x \in Q$ in polynomial time, then we say Q is polynomial-time solvable for constant parameter values. The class of all such problems is called XP. Algorithms for problems in XP run in time $n^{f(p)}$ where p is the parameter and f is some computable function. For practical applications, however, such algorithms often still take too much time. Thus, the following refinement rose: A decision problem Q is called *fixed-parameter tractable* (FPT) with respect to a parameter p if there is an algorithm solving any instance x of Q in $f(k)\cdot|x|^c$ time for some computable function f and a constant c. For example, choosing the parameter "size k of the desired vertex cover" for VERTEX COVER, Chen, Kanj, and Xia [18] presented an algorithm running in $O(1.2738^k + k|V|)$ time, which can solve instances with small desired solution sizes fast. Note that the crucial difference between XP and FPT is that, while problems in both classes are solvable in polynomial time for constant parameter values, in the case of XP the degree of the corresponding polynomial may depend on the parameter, while in the case of FPT it may not.

Like in classical complexity, there is a concept of hardness in parameterized complexity. A class of problems that are conjectured to not be FPT is $W[1]$. A decision problem Q is called $W[1]$-hard with respect to the parameter p if Q being FPT with respect to p implies that all problems in $W[1]$ are also FPT. Similarly to the concept of NP-hardness, this can be achieved by a reduction concept [24, 34, 75].

Often, finding a good parameter that captures aspects that make a problem hard is not an easy task [32, 60, 76]. Multiple techniques exist to support the parameter search. For example, "deconstructing intractability" is the technique of analyzing the characteristics of proofs of computational worst-case hardness—proofs that often exploit parameter values unlikely to occur in practice. Other techniques include "distance from triviality" and "above guarantee" parameterizations [17, 48, 61, 68, 76].

Problem kernelization. An important property of parameterized design and analysis of algorithms is that it provides means of measuring the effectiveness of preprocessing. Consider, for example, some algorithm that, given an instance (G,k) of VERTEX COVER, produces a simpler instance of VERTEX COVER in polynomial time. In classical complexity, simpler means smaller. Then, however, if such an algorithm existed, we could solve VERTEX COVER in polynomial time by iterating the described algorithm. By introducing the parameter k, however, an algorithm is possible that produces smaller instances until some threshold measured in k is reached. For example, the books of Downey and Fellows [24], Flum and Grohe [34], and Niedermeier [75] present polynomial-time algorithms that, given an instance (x,k) of VERTEX COVER, produce an equivalent instance (x',k') of VERTEX COVER with at most $2k'$ vertices. In general, a *kernelization* (or *problem kernel*) of a decision problem Q is a polynomial-time algorithm that, given an instance x with parameter p of Q, computes an instance x' with parameter p' of Q such that, for some functions $f, g : \mathbb{N} \to \mathbb{N}$,

- $x \in Q$ if and only if $x' \in Q$,
- $|x'| \le g(p)$, and
- $p' \le f(p)$.

The function g is then called the *size* of the problem kernel and constitutes a measure of effectiveness of the kernelization algorithm. A common method to develop such preprocessing algorithms is to provide polynomial-time executable *data reduction rules* [14, 49, 66]. Kernelization is a core tool of parameterized algorithmics and perhaps the concept in parameterized algorithmics with the widest practical relevance (see Guo and Niedermeier [49] and Hüffner, Niedermeier, and Wernicke [56]). In the recent past, Bodlaen-

Table 2.1. *Complexity of the* CHINESE POSTMAN (CP) *problem.*

CP variant	Classical complexity
Undirected	$O(\|V\|^3)$-time algorithm [29]
Directed	$O(\|V\|^3)$-time algorithm [29]
	$O((\|E\|-\|V\|)\|V\|^2)$-time algorithm [65]
Mixed	NP-complete [78]
	$O(\|V\|^3)$-time solvable if each vertex has even degree [29]
Windy	NP-complete [47]
	in P in some special cases [47, 97, 101]
k-Hierarchical	NP-complete [27]
	$O(k\|V\|^4)$-time solvable if precedence relation linear [62]
Min-sum k postmen	$O(\|V\|^3)$-time algorithm with post office vertex [80, 102], otherwise NP-complete [91, 52]
Min-max k postmen	NP-complete [40]
CP variant	**Approximation**
Mixed	$O(\max\{\|V\|^3, \|A\|(\max\{\|A\|,\|E\|\})^2\})$-time factor 3/2 [85]
Windy	factor 3/2 [84]
Min-max k postmen	$O(\|V\|^3)$-time factor $(2-1/k)$ [40]
CP variant	**Parameterized complexity**
Mixed	$O(2^{\|E\|} \cdot \|V\|^3)$-time algorithm [50]
	in FPT with respect to $\|A\|$ [50]
	in XP with respect to treewidth [33]
Min-sum k postmen	in FPT with respect to k without post office vertex [51, 52]

der et al. [15] and Fortnow and Santhanam [37] developed methods to show that certain problems do not admit polynomial-size kernels. Most of these techniques build on the hypothesis that the polynomial hierarchy does not collapse to the second level (which implies that P $\neq$ NP).

2.2 ▪ The Chinese Postman Problem

The CHINESE POSTMAN problem is to find a minimum cost tour in a graph with edge costs such that each edge is traversed at least once. As CHINESE POSTMAN is one of the most central arc routing problems, there are many interesting studies from a complexity-theoretic perspective. In its original form, this problem is solvable in polynomial time. As usual with problems motivated by real-world applications, there are many variants and possible side constraints. However, even side constraints that seem innocuous make CHINESE POSTMAN NP-hard. Here, we review the computational complexity of CHINESE POSTMAN and some of its variants. As mentioned above, NP-hardness is not a reason for despair but rather stirs further investigation. Thus, we try to explain the combinatorial structure in the problems that is used to show hardness and reflect on the impact on practice. For an overview of the results that are presented in this section, see Table 2.1.

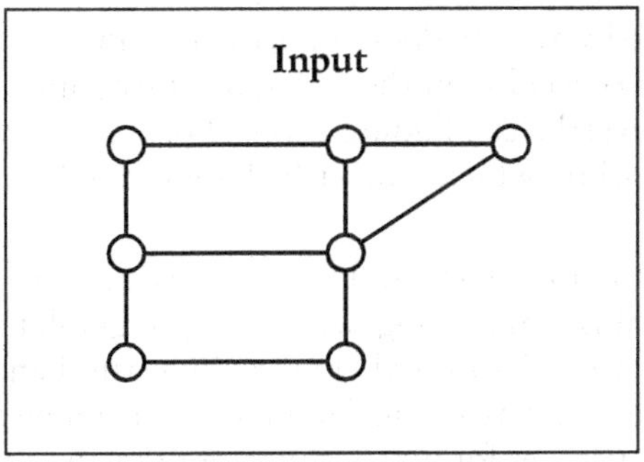

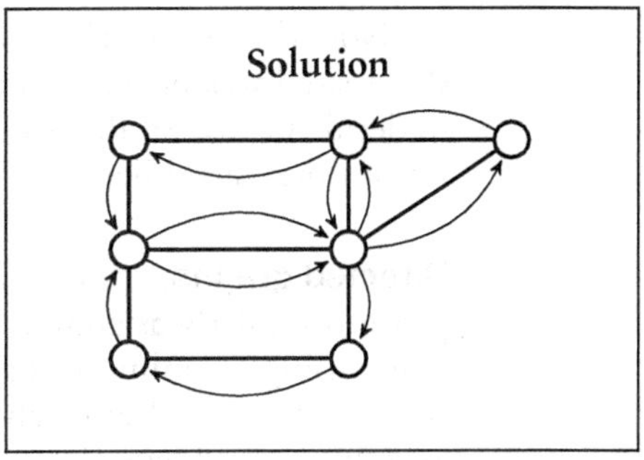

Figure 2.1. *Instance and solution for the* CHINESE POSTMAN *problem.*

2.2.1 ▪ The classical Chinese Postman Problem

Edmonds [28] and Guan [46] originally introduced CHINESE POSTMAN as the following problem:

CHINESE POSTMAN
Input: A connected, undirected graph $G = (V, E)$ with edge cost $c(e) \geq 0$ for every $e \in E$ and a maximum cost $c_{\max}$.
Question: Is there a tour that traverses every edge in E at least once and has cost at most $c_{\max}$?

Figure 2.1 depicts on the left an input graph with edge costs proportional to the length in the drawing and a minimum cost solution on the right.

Tractability. Edmonds and Johnson [29] showed that CHINESE POSTMAN is solvable in polynomial time. A strategy to efficiently decide CHINESE POSTMAN is based on the observation that any desired tour induces an Eulerian supergraph of G by multiplying every edge of G according to the number of times it is traversed by the tour. Obviously, the obtained supergraph is a multigraph. In the reverse direction, every such Eulerian supergraph also induces a postman tour of the same cost. CHINESE POSTMAN thus reduces to the problem of finding an Eulerian supergraph G' such that the cost of the edges in G' does not exceed $c_{\max}$.

Finding such an Eulerian supergraph can be done as follows. Recall that a graph is Eulerian if and only if it is connected and each vertex has even degree. By prerequisite, the input graph G is already connected, and it remains only to add edges in order to give every vertex even degree. It can be shown that this can be achieved by solving an equivalent minimum-cost perfect matching problem in an auxiliary graph G^* with edge costs: The graph G^* is complete and contains only the odd-degree vertices of G. Every edge is priced according to a shortest path between its endpoints in the graph G. The basic ideas for proving the equivalence are as follows. First, show that the excess edges in a minimum-cost Eulerian supergraph of G form paths between odd-degree vertices. Second, show that shortest paths according to two matching edges in G^* do not share any edge. Then one can easily transfer a matching in G^* to an Eulerian supergraph of G and vice versa. To construct the graph G^*, one can employ an all-pairs shortest path algorithm using $O(|V|^3)$ time and then solve the corresponding matching problem on G^* in $O(\sqrt{|V|} \cdot |E|)$ time (see Micali and Vazirani [73] for the matching running time). Thus, CHINESE POSTMAN is solvable in $O(|V|^3)$ time, and, therefore, it is in P.

Note that, in the above described way, one does not actually find a postman tour of the desired weight but ensures that one exists in the Eulerian supergraph. This suffices to decide CHINESE POSTMAN. Nevertheless, finding a tour if one exists can be done in linear time using Hierholzer's algorithm, which is described, for example, by Berge [12].

Directed graphs. If the input graph for CHINESE POSTMAN is directed, then one can proceed similarly as above. The goal is again to augment the input graph to an Eulerian graph using a minimum-cost set of arcs. A directed graph is Eulerian if and only if it is strongly connected and each of its vertices has an equal number of incoming and outgoing arcs. It can be shown that mending vertices for which this is not the case can be reduced to a minimum-cost network flow problem (see Edmonds and Johnson [29]) or solved using linear programs that have integral optimal solutions (see Beltrami and Bodin [8] and Christofides [19]). This yields running times of $O(|V|^3)$ and $O(|E| \cdot |V|^2)$, respectively. Lin and Zhao [65] later gave an alternative algorithm based on a network flow formulation that has $O((|E|-|V|) \cdot |V|^2)$ running time.

In practice, the plain CHINESE POSTMAN problem is often augmented with additional constraints or amended to better reflect its application. Below we consider the complexity of some of these variants. As we will see, it is easy to step from polynomial-time solvable into NP-hard terrain. Nevertheless, for practical scenarios there often are ways that allow one to preserve *efficient* solvability to some extent.

2.2.2 ▪ Mixed graphs

In practical instances it can happen that both (undirected) edges and (directed) arcs occur. For example, in snow plowing, it can happen that some streets are narrow enough such that they can be cleaned by traversing them only once in either direction, but other streets have to be plowed in both directions. In this case, we have to solve the MIXED CHINESE POSTMAN problem, as defined by Edmonds and Johnson [29]:

MIXED CHINESE POSTMAN
Input: A strongly connected, mixed graph $G=(V,E,A)$ with cost $c(e) \geq 0$ for every $e \in E \cup A$ and a maximum cost $c_{\max}$.
Question: Is there a (directed) tour that traverses every edge in E and every arc in A at least once and has cost at most $c_{\max}$?

If we assume P $\neq$ NP, then MIXED CHINESE POSTMAN is not polynomial-time solvable as we will see below. Nevertheless, there are some tractable special cases.

Note that the question of whether a given mixed graph admits a postman tour of any cost is easy to decide: One simply has to check whether the graph is strongly connected. This changes drastically if we restrict the postman tour to traverse each arc exactly once or if we restrict it to traverse each edge exactly once. As proved by Zaragoza Martínez [98, 99, 100], it is NP-hard to decide whether such tours exist. This result has some applications to hardness of approximability, which we will touch on below.

2.2.2.1 ▪ Hardness

MIXED CHINESE POSTMAN has been shown NP-complete by Papadimitriou [78]. The main difficulty in solving MIXED CHINESE POSTMAN lies in choosing orientations for the (undirected) edges when we are given a tight budget for our tour and can only afford to traverse each edge once. We then have to orient the edges and add some further arcs in

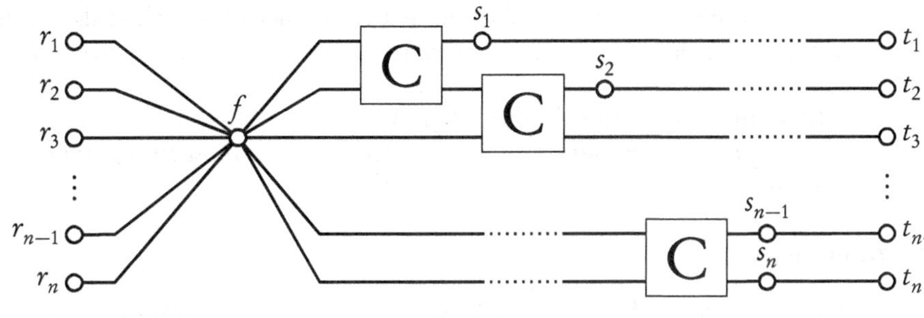

Figure 2.2. *Variable gadget for reducing* 3SAT *to* MIXED CHINESE POSTMAN.

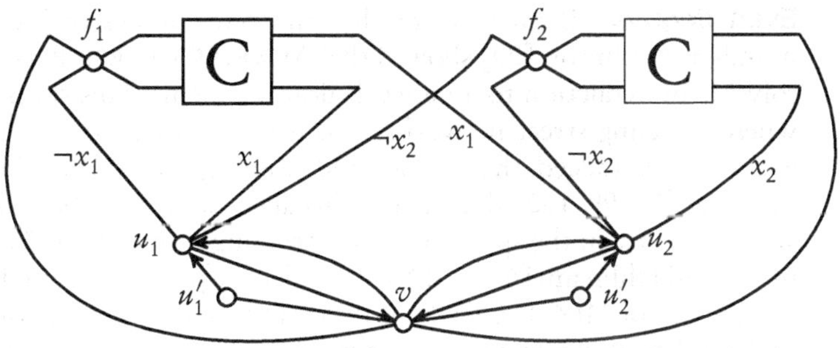

Figure 2.3. *Example for NP-hardness reduction of* MIXED CHINESE POSTMAN.

order to obtain a directed Eulerian graph, that is, to make every vertex balanced. If there are multiple edges incident to one vertex, it is not an easy task to determine the correct orientation of each edge.

In fact, one can exploit this difficulty to reduce 3SAT to MIXED CHINESE POSTMAN; we sketch some ideas of the reduction here. Suppose we can construct a mixed graph with edge costs as shown in Figure 2.2. This graph will represent a variable of the instance of 3SAT. The boxes labeled C represent graphs with the following property. They can be made Eulerian—up to the vertices that connect to the rest of the graph—by adding arcs of cost at most two only in two ways: by orienting every outgoing edge to the right or to the left (see Papadimitriou [78] for details). Hence, the variable gadget of Figure 2.2 has the property that, if we are given a budget of at most $2(n-1)$, we can orient all of its edges either to the right or to the left. We can use this property to represent the two possible truth values via the orientations.

The main idea for representing clauses is shown via an example in Figure 2.3 (reproduced from Papadimitriou [78]). It shows an instance of MIXED CHINESE POSTMAN constructed from the trivial 3SAT formula $(\neg x_1 \vee x_1 \vee \neg x_2) \wedge (x_1 \vee \neg x_2 \vee x_2)$. At the top, there are two variable gadgets corresponding to x_1 and x_2. The two clauses are represented by the pairs of vertices u_1, u_1' and u_2, u_2' and their connections to v and the variable gadgets. The vertex v is a dummy vertex that takes surplus connections from the variable gadgets which are not needed to represent them in clauses.

Consider the vertex u_1 and orientations of its incident edges. If the budget for our tour and the costs of the incoming arcs of u_1 are chosen such that we can traverse each of them only once, then, at least one of the edges incident to u_1 has to be oriented toward it. Otherwise, we could not balance u_1. By the way the edges are connected to u_1 and to the variable gadgets, the orientation corresponds to satisfying the clause that u_1 represents.

Papadimitriou [78] also showed that the reduction sketched above can be modified to yield NP-hardness even for very restricted instances as stated in the following theorem.

Theorem 2.1 (Papadimitriou [78]). MIXED CHINESE POSTMAN *is* NP-*complete, even if the input graph is planar, each vertex has degree at most three, and each edge and arc has cost one.*

2.2.2.2 ▪ Tractability

Despite the above hardness result even for very restricted instances, there are some tractable special cases of MIXED CHINESE POSTMAN.

Even degree. If every vertex has an even number of incident edges and arcs, Edmonds and Johnson [29] showed that MIXED CHINESE POSTMAN is polynomial-time solvable by reduction to a network flow problem. This special case is interesting since when modeling street networks, it is expected that many vertices obtain exactly four edges or arcs. Indeed, the hardness reductions given by Papadimitriou [78] and Zaragoza Martínez [98, 99, 100] all construct instances with an unbounded number of odd-degree vertices. Thus, in the spirit of "deconstructing intractability" (see Komusiewicz, Niedermeier, and Uhlmann [61] and Niedermeier [76]), it would also be interesting to know the complexity of MIXED CHINESE POSTMAN when there are only a few vertices with an odd number of incident edges or arcs. To the best of our knowledge, this is currently an open question.

Challenge 1. *Devise an algorithm for* MIXED CHINESE POSTMAN *that witnesses fixed-parameter tractability with respect to the parameter "number of odd-degree vertices," or show that such an algorithm is unlikely to exist.*

Few edges or arcs. When the number of (undirected) edges in the input graph is small, MIXED CHINESE POSTMAN can be solved efficiently. This is because, in an optimum postman tour, every edge is traversed either forward, backward, or in both directions. Observe that, once we know which one is true for every edge, we can replace the edges by (directed) arcs and solve the resulting instance of directed CHINESE POSTMAN in polynomial time. Thus, guessing one of the three options for every edge, we obtain an algorithm running in $O(3^{|E|} \cdot |V|^3)$ time. Gutin, Jones, and Sheng [50] showed how this running time can be improved to $O(2^{|E|} \cdot |V|^3)$ using network flows. It is reasonable to assume that $|E|$ is small in some practical applications, for example, in snow plowing, where it is rare that a street can be cleaned by traversing it only once. However, the reverse is true for some other applications, for example, in police patrols, where the direction of traversal does not matter except for one-way streets. If the number of one-way streets is small, then an FPT algorithm with respect to the parameter number $|A|$ of arcs might be handy. Indeed, Gutin, Jones, and Sheng [50] described such an algorithm. The procedure includes an application of the "treewidth reduction" technique [71] and dynamic programming, iteratively computing partial solutions until a complete one is obtained. As a consequence of the generality of treewidth reduction, the number of partial solutions one has to keep track of in the dynamic program can be rather large, of order $\Omega(2^{2^{|A|}})$. FPT algorithms are widely regarded as promising for practice if their running time is "single exponential," that is, $O(c^k \cdot n^d)$ for some constants c, d, parameter k, and input size n. Hence, an interesting open question is whether Gutin, Jones, and Sheng's algorithm can be improved to such a running time, at least on the practically relevant planar graphs.

Challenge 2. *Give an algorithm for planar* MIXED CHINESE POSTMAN *with* $O(c^{|A|} \cdot |V|^d)$ *running time for constants* c, d*; or show that such an algorithm is unlikely to exist.*[1]

Small treewidth. Furthermore, Fernandes, Lee, and Wakabayashi [33] proved that MIXED CHINESE POSTMAN is in XP with respect to the parameter "treewidth of the input graph." That is, it can be solved in polynomial time if the treewidth is constant. Intuitively, treewidth measures the tree-likeness of graphs. For example, trees have treewidth one, and cycles have treewidth two (see Bodlaender and Koster [16] for a recent survey on treewidth). However, the exponent in the polynomial of the running time depends exponentially on the treewidth. Thus, this result is of mainly theoretical interest. It also raises the question whether MIXED CHINESE POSTMAN is FPT with respect to the treewidth.

Further tractable cases. There are also two polynomial-time solvable special cases for MIXED CHINESE POSTMAN that can be obtained by transforming it into WINDY CHINESE POSTMAN, which we will consider in Subsection 2.2.3.

2.2.2.3 ▪ Approximability

Motivated by the NP-hardness of MIXED CHINESE POSTMAN, the question was pursued of how close one can get to an optimal solution using only polynomial time. This was successful in that Raghavachari and Veerasamy [85] developed an algorithm that approximates the cost of the optimal postman tour within a factor of 3/2, improving the previously known factor 3/2 approximation algorithm, given by Frederickson [39], that was limited to planar graphs. However, Zaragoza Martínez [98, 99, 100] proved that it is not possible to polynomial-time-approximate the cost of *deadheading* within any constant factor, where deadheading means traversing arcs or edges more than once.

A factor 2 approximation. The origin of the currently best known approximation algorithm for MIXED CHINESE POSTMAN is a heuristic proposed by Edmonds and Johnson [29]. It is based on the following sufficient condition for the existence of an Eulerian tour: If in a mixed graph each vertex has even degree—a vertex in a mixed graph has even degree if it has an even number of incident arcs and edges—and each vertex has an equal number of incoming and outgoing arcs, then the graph has an Eulerian tour, as noted by Ford and Fulkerson [35]. Edmonds and Johnson's heuristic first produces even degree by ignoring the direction of the arcs and using a matching method analogous to that used for solving the undirected CHINESE POSTMAN. Using a network flow formulation, each vertex in the resulting graph is then given an equal number of incoming and outgoing arcs in a cost-optimal way. Frederickson [39] showed that the second step can be carried out such that it preserves even degrees. In this way, he obtained a factor 2 approximation algorithm for the tour cost. Roughly, the argument to show this is as follows: Let us duplicate each edge and arc in the input graph G to obtain a multigraph G_2. In G_2, each vertex has even degree, and the above algorithm will solve the resulting instance optimally. Duplicating an optimum solution for G yields a feasible solution for G_2, and, hence, we have that the algorithm gives a solution for G_2 of cost at most twice the optimum of G. Then, one can show that the edges added to G due to the matching are a subset of the duplicated edges in G_2 and, furthermore, that producing equal in- and outdegrees for each vertex after the matching is at most as costly in G as in G_2. Thus, we obtain that the algorithm

[1] In an earlier version of this chapter, the authors asked whether MIXED CHINESE POSTMAN is FPT with respect to $|A|$. This question was also featured in a survey talk by Sorge [86]. Subsequently, Gutin, Jones, and Sheng [50] resolved the challenge, showing fixed-parameter tractability.

produces a solution of cost at most the one in G_2, which is at most twice the optimum for G.

A mixed strategy for factor 5/3. Frederickson [39] showed that a "reverse" algorithm also gives a factor 2 approximation, that is, an algorithm that first gives every vertex equal number of incoming and outgoing arcs and then produces even degree for every vertex while preserving the first property. He observed that the two algorithms perform optimally on each others' worst-case examples, and, in fact, he analyzed their solution cost to be at most $C^* + 2C_{\leftrightarrows}$ and $2C^* - C_{\leftrightarrows}$ for the reverse algorithm. Here, C^* is the cost of the optimal postman tour and $C_{\leftrightarrows}$ is the cost of the step that produces equal in- and out-degree. Thus, the solution cost guarantees coincide when $C_{\leftrightarrows} = C^*/3$, and, hence, one of the guarantees is always at most $5C^*/3$, giving a factor $5/3$ approximation algorithm when choosing the best solution.

Improving the mixed strategy to factor 3/2. The approximation factor of the above algorithm has later been improved by Raghavachari and Veerasamy [85]. They adopted the algorithms as presented by Frederickson [39] and made a modification to get a solution cost guarantee of $C^* + C_{\leftrightarrows}$ instead of $C^* + 2C_{\leftrightarrows}$ for the first of the two algorithms. Now the guarantees for the solution cost coincide when $C_{\leftrightarrows} = C^*/2$, and, hence, one gets a factor $3/2$ approximation algorithm. The crucial observation for the better bound is that a lower bound on the optimal solution used by Frederickson can be tightened to also include $C_{\leftrightarrows}$, which then "swallows" one of the $C_{\leftrightarrows}$ summands in the solution cost bound above.

Theorem 2.2 (Raghavachari and Veerasamy [85]). MIXED CHINESE POSTMAN *can be factor* $3/2$ *approximated in* $O(\max\{|V|^3, |A|(\max\{|A|, |E|\})^2\})$ *time.*

Inapproximability of the deadheading cost. Since a postman tour always includes the edges and arcs of the input graph, one might wonder whether one can approximate the cost of additional edge and arc traversals, also called *deadheading* cost, rather than the whole tour cost. However, Zaragoza Martínez [98, 100] proved that this is not possible within any constant factor and in polynomial time. He obtained this result by proving NP-hardness of deciding whether a given mixed graph has a postman tour that traverses each edge exactly once. The proof is by reduction from the NP-hard NOT ALL EQUAL SATISFIABILITY problem. Given this hardness result, consider a mixed graph $G = (V, E, A)$ and set the cost $c_a = 0$ for all $a \in A$ and the cost $c_e = 1$ for all $e \in E$. Then, G admits a postman tour that traverses each edge at most once if and only if it has a postman tour with zero deadheading cost. Thus, any constant-factor approximation algorithm could decide whether such a tour exists and, thus, cannot run in polynomial time.

2.2.3 ▪ Windy costs

Another natural variant of the CHINESE POSTMAN problem arises when considering asymmetric costs; that is, it may be harder to traverse a street in one direction than it is in the reverse one, modeling roads on a slope, or blowing wind. Minieka [74] formulated this as the following problem:

WINDY CHINESE POSTMAN
Input: A connected, undirected graph $G = (V,E)$ with two cost values $c(u,v)$, $c(v,u) \geq 0$ for every edge $\{u,v\} \in E$ and a maximum cost $c_{\max}$.
Question: Is there a (directed) tour of cost at most $c_{\max}$ that traverses every edge in E at least once?

Hardness. Unfortunately, this problem is NP-hard, as Guan [47] showed: MIXED CHINESE POSTMAN can be transformed into WINDY CHINESE POSTMAN in polynomial time. The idea for this transformation is to keep each (undirected) edge and to replace each pair of arcs $(u,v),(v,u)$ by an edge $\{u,v\}$ with the appropriate costs. Furthermore, replace each arc (u,v), for which (v,u) is not present, by an edge $\{u,v\}$ with cost $c(u,v)$ according to the original cost of (u,v) and very high cost $c(v,u) > c_{\max}$. Hence, the arc (v,u) cannot be traversed by any tour of cost at most $c_{\max}$. This gives a one-to-one correspondence between optimal solutions, and Theorem 2.1 implies that WINDY CHINESE POSTMAN is NP-hard even on planar graphs with maximum degree three.

Tractability. On the positive side, WINDY CHINESE POSTMAN is solvable in polynomial time if the cost of traversing each cycle in the input graph is the same regardless of the direction of traversal (see Guan [47]). WINDY CHINESE POSTMAN is also polynomial-time solvable if the input graph is Eulerian (see Win [97]).

Furthermore, Win [96, 97] started the line of research of classifying so-called *windy postman perfect* graphs. These are graphs for which a certain linear programming formulation yields integer solutions and, consequently, for which the corresponding integer linear program can be solved in polynomial time. In this regard, Zaragoza Martínez [101] more recently proved that, in particular, series-parallel graphs are windy postman perfect.

Approximability. Exploiting the polynomial-time solvability of WINDY CHINESE POSTMAN on Eulerian graphs, Win [97] obtained a factor 2 approximation algorithm. A simplified version of his idea works as follows: First, double all edges in the input graph, which is now Eulerian. Then, solve the resulting instance optimally. Since following the optimal tour for the input graph twice gives a feasible solution for the graph with the doubled edges, this gives a factor 2 approximation. Raghavachari and Veerasamy [84] later gave a factor 3/2 approximation algorithm by combining a mixed strategy similar to the one used for the approximation algorithm for MIXED CHINESE POSTMAN and a modification of Win's linear programming formulation for WINDY CHINESE POSTMAN.

2.2.4 ▪ Edge hierarchies and precedence relations

In practical applications, one sometimes has to solve a variant of CHINESE POSTMAN where some edges have to be traversed before other edges. For example, in snow plowing, main streets should be cleared before secondary streets and, in turn, secondary streets should be cleared before residential streets. Dror, Stern, and Trudeau [27] formalized this in the following problem. Let $G = (V,E)$ be an undirected graph, and let $P = \{E_1,\ldots,E_k\}$ be an edge partition, that is, $\bigcup_{i=1}^{k} E_i = E$ and $E_i \cap E_j = \emptyset$ for all $1 \leq i,j \leq k$. Call the sets E_i *priority classes*. Furthermore, let $R \subseteq P \times P$ be a partial ordering. We say that a tour in G *respects* R if for every $(E_i,E_j) \in R$ each edge in E_i is traversed before any edge in E_j is traversed.

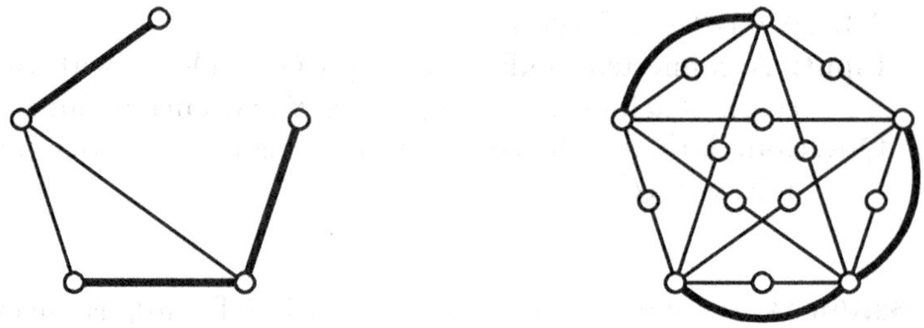

Figure 2.4. *Example for NP-hardness of* HIERARCHICAL CHINESE POSTMAN *even with two priority classes.*

HIERARCHICAL CHINESE POSTMAN

Input: A connected (undirected) graph $G = (V, E)$ with edge cost $c(e) \geq 0$ for every $e \in E$ and a maximum cost $c_{\max}$. Additionally, an edge-partition $P = \{E_1, \ldots, E_k\}$ and a partial ordering relation $R \subseteq P \times P$.

Question: Is there a tour that traverses every edge in E at least once, has cost at most $c_{\max}$, and respects R?

As we will see, this problem is NP-hard in general. However, there are three factors that seem to influence its complexity. Namely, whether the priority classes are connected; if so, their number; and whether the precedence relation is a linear ordering.

We note that a different HIERARCHICAL CHINESE POSTMAN formulation has been proposed by Alfa and Liu [4] for which, to our knowledge, there are no complexity-theoretic results yet.

Hardness with two priority classes. HIERARCHICAL CHINESE POSTMAN is NP-hard even when there are just two priority classes. This follows from observations by Dror, Stern, and Trudeau [27] via a reduction from the NP-hard RURAL POSTMAN problem, which we will consider more closely in Section 2.3. In UNDIRECTED RURAL POSTMAN one is given an undirected Graph $G = (V, E)$ with costs c_e for each $e \in E$ and a set of required edges $R \subseteq E$. The task is to find a minimum-cost tour that traverses each edge in R at least once.

To reduce RURAL POSTMAN to HIERARCHICAL CHINESE POSTMAN, first create a complete graph G' on the vertices V. Price each edge $\{u, v\}$ according to a shortest path between the vertices u and v in G. Then, replace each edge by a path with two edges and split the original cost equally between both new edges. Place all edges constructed so far in the priority class E_1'. This is the first step. In the second and final step, simply add all edges in R to the constructed graph G' and price them according to their original cost. The edge partition is $\{E_1', R\}$, and the precedence relation is defined by $E_1' < R$. An example of the reduction is shown in Figure 2.4 with a RURAL POSTMAN instance to the left, where bold lines represent required edges, and the corresponding HIERARCHICAL CHINESE POSTMAN instance on the right. There, thin lines represent edges in one priority class, and bold lines represent edges in a second priority class.

We claim that every rural postman tour in G with cost C implies a hierarchical Chinese postman tour in G' with cost $C + C_{E_1'}$ and vice versa. Here, $C_{E_1'}$ is the cost of the edges in E_1'. Without loss of generality, assume that $|V|$ is odd. Then, the graph constructed in the first step is Eulerian. Thus, given a rural postman tour T in G of cost C, we obtain a hierarchical Chinese postman tour in G' of cost $C + C_{E_1'}$ by first traversing all edges in E_1' and then tracing T in G', following edges in R whenever possible and pairs of edges corresponding to shortest paths, otherwise. Conversely, every hierarchical Chinese postman tour has to first traverse each edge in E_1'. It is not hard to see that the tour can be partitioned into a tour for E_1' and a tour corresponding to a rural postman tour for G: Consider a hierarchical Chinese postman tour T for G' and the Eulerian multigraph M_T it induces by multiplying each edge according to how often it is traversed. Remove the edge set E_1' from M_T to obtain M_T'. Up to isolated vertices, the graph M_T' is still Eulerian: Since $|V|$ is odd, removing E_1' removes an even number of edges for each vertex, resulting in even degrees. Connectedness of M_T' follows, because for two vertices that are incident to required edges, there is a path between them in T *after* the initial traversal of E_1'. Thus, T can be partitioned into a tour for E_1' and the tour implied by M_T'.

Hardness with connected priority classes. The above considerations express the intuition that one has to solve RURAL POSTMAN as a subproblem if one of the priority classes is not connected. However, as Dror, Stern, and Trudeau [27] proved, HIERARCHICAL CHINESE POSTMAN remains NP-hard even if we demand that the priority classes be connected. Using their ideas, we modify the above sketch of a hardness proof to get hardness for connected priority classes: Instead of putting all required edges in R into one priority class, we have a priority class R_i for each connected component induced by the required edges. The new precedence relation is defined by $E_1' < R_i$ for all R_i. It is still the case that each edge in E_1' has to be traversed before any required edge. But the new precedence relation does not restrict in which order a tour traverses the required edges. Thus, the correctness follows as above.

Influence of the number of priority classes on complexity. Note that if the priority classes are connected, then it is necessary to have many priority classes in the above reduction: The standard hardness reduction for RURAL POSTMAN is from HAMILTONIAN CYCLE and creates one connected component in the required edges for each vertex in the original instance (see Section 2.3). This raises the question of whether HIERARCHICAL CHINESE POSTMAN is still a hard problem when each priority class is connected and their number is constant or small, that is, whether HIERARCHICAL CHINESE POSTMAN is in XP or FPT with respect to the parameter "number of priority classes." We conjecture that HIERARCHICAL CHINESE POSTMAN is indeed in XP. Note that, by the above reduction from RURAL POSTMAN, solving the FPT question in the affirmative would also imply that RURAL POSTMAN is FPT with respect to the number of connected components in the required edges. This is an intriguing and currently open question (see Challenge 4).

Tractability with linear precedence relations. If one is given a *linear* instead of a partial ordering of the edge partition, then HIERARCHICAL CHINESE POSTMAN with connected priority classes becomes polynomial-time solvable [27, 43, 62]. The central idea to get a polynomial-time algorithm here is to consider for each of the partition classes the subproblem of traversing every one of its edges. Dror, Stern, and Trudeau [27] showed that the optimal route that enters such a partition class at a given vertex u, traverses every one of its edges, and then leaves the class at another given vertex v can be computed in polynomial time. Call u the *entry point* and v the *exit point*. One now has to find a pair of a "good" entry point and a "good" exit point for each of the edge partition classes. Dror, Stern, and Trudeau [27] solved this problem by modeling it as a shortest-path problem on a graph whose vertices correspond to pairs of entry and exit points. Their algorithm gives a running time upper bound of $O(k|V|^5)$. More recently, alternative algorithms with running times $O(k^3|V|^3)$ and $O(k|V|^4)$ have been given by Ghiani and Improta [43] and Korteweg and Volgenant [62], respectively.

Korteweg and Volgenant [62] also showed that the complexity upper bound of $O(k|V|^4)$ holds for two modifications of HIERARCHICAL CHINESE POSTMAN with a linear precedence relation and connected priority classes: One variant distinguishes between servicing and traversing costs such that only servicing has to respect the precedence relation, and the other variant seeks to minimize the completion cost of the first priority class, then the completion cost of the second priority class, and so on, instead of minimizing the overall tour cost.

2.2.5 ▪ Multiple postmen

A natural variant of CHINESE POSTMAN is to search for k tours in an edge-weighted graph G, such that each edge in G is traversed by at least one of the postmen. More formally, this can be stated as the following problem with objective function ω:

MIN-ω-k-CHINESE POSTMAN
Input: A connected, undirected graph $G=(V,E)$ with cost $c(e)\geq 0$ for every $e\in E$ and a maximum cost $c_{\max}$. Furthermore, a distinguished vertex $v\in V$ (the post office).
Question: Are there k tours T^* each starting and ending in v, such that none of the tours are empty, each edge in E is traversed by at least one of the tours, and $\omega(T^*)\leq c_{\max}$?

When focusing on the undirected case, the complexity of k-CHINESE POSTMAN highly depends on which objective function ω we choose: Minimizing the sum of the tour costs yields the polynomial-time solvable MIN-SUM-k-CHINESE POSTMAN [80, 102], whereas minimizing the maximum tour cost yields the NP-hard MIN-MAX-k-CHINESE POSTMAN [40]. This also holds if we substitute a directed graph for G. If both arcs and edges are present, then, clearly, both variants have MIXED CHINESE POSTMAN as special case and are NP-hard.

Minimizing the sum of the tour costs. Polynomial-time solvability of MIN-SUM-k-CHINESE POSTMAN has been proved by Zhang [102] and, independently, by Pearn [80], who observed that a lower bounding procedure for CAPACITATED ARC ROUTING proposed by Benavent et al. [10] solves MIN-SUM-k-CHINESE POSTMAN optimally. The main observation is that, if there is a postman tour for one postman that visits the post office v at least k times, then it can be split into k tours of the same cost. The converse is also true. Using this observation, one can solve MIN-SUM-k-CHINESE POSTMAN by introducing copies of the post office v as needed and then applying a matching technique similar to the one for the classical CHINESE POSTMAN problem. DIRECTED MIN-SUM-k-CHINESE POSTMAN can be solved using similar ideas [80, 102]. Pearn [80] also observed that some polynomial-time solvable special cases of MIXED CHINESE POSTMAN and WINDY CHINESE POSTMAN transfer to their k postmen variants.

It is interesting to note that the post office v is crucial to the polynomial-time solvability of undirected MIN-SUM-k-CHINESE POSTMAN. In fact, the problem becomes NP-hard if there is an unbounded number of postmen, and if there is no post office, that is, each tour can start and end at an arbitrary vertex of the graph. Thomassen [91] obtained this hardness result for the undirected case by reduction from a graph coloring problem. Simpler NP-hardness reductions for both directed and undirected input graphs were given more recently by Gutin, Muciaccia, and Yeo [52].

Of course, having an unbounded number of postmen is not realistic in many practical scenarios. Fortunately, MIN-SUM-k-CHINESE POSTMAN without a post office becomes tractable if the number of postmen is small: Gutin, Muciaccia, and Yeo [52] described how to compute a problem kernel with $O(k^2 \log k)$ vertices in the undirected case. Their procedure either obtains k tours of the input graph or, if this is not possible, guarantees that there are only a few vertices of degree at least three or degree one. Vertices of degree two can then be removed by a simple reduction rule that shortens long paths, preserving k tours if they are present. This then yields a bound on all vertices in the problem kernel. Using an exhaustive search procedure on the problem kernel then gives an FPT algorithm for finding the k tours. The directed case seems more complicated, but also this case is FPT with respect to k: Using a theorem from Ramsey theory, Gutin et al. [51] observed that either one can extract k tours or the input graph has some specific structure that lends itself to dynamic programming; in this way they obtained an FPT algorithm.[2] The dependence of the running time on the parameter k is large, however, and it remains to engineer algorithms with faster running times.

Minimizing the maximum tour cost. Frederickson, Hecht, and Kim [40] showed that MIN-MAX-k-CHINESE POSTMAN is NP-hard via a simple reduction from k-PARTITION. Intuitively, the reduction exploits that a difficulty in solving MIN-MAX-k-CHINESE POSTMAN lies in deciding which arc is to be traversed by which vehicle. An analogous argument also holds if the input graph is directed. On the positive side, Ahr [3] observed that MIN-MAX-k-CHINESE POSTMAN is polynomial-time solvable on paths, cycles, and cliques. It would be interesting to know whether his results can be generalized to broader graph classes. Motivated by the NP-hardness, Frederickson, Hecht, and Kim [40] gave a factor $(2-1/k)$ approximation algorithm that runs in $O(|V|^3)$ time. The basic idea is to first find a postman tour for a single postman and then split it into k tours in a clever way.

2.2.6 ▪ Further variants

We now briefly state some results for further interesting variants of CHINESE POSTMAN.

Turn constraints. In some practical applications, it is necessary to forbid certain ways to traverse crossings. For example, this is due to traffic rules or, in the case of snow plowing, to avoid depositing snow on the crossing. Unfortunately, such turn constraints make CHINESE POSTMAN NP-hard, as proved by Benavent and Soler [11].

Traversing at least one edge out of a subset. A slight generalization of the constraint that each edge in the graph has to be traversed makes CHINESE POSTMAN NP-hard: Given a number of subsets of the edges, traverse at least one edge of each subset. This was observed by Dror and Haouari [26].

Maximizing the benefit from servicing streets. It might be beneficial to service a street multiple times, for example, when routing street sweepers. This can be modeled as another variant of CHINESE POSTMAN, but as Pearn and Wang [82] proved, this also results in an NP-hard problem.

[2] In an earlier version of this chapter, the authors asked whether MIN-SUM-k-CHINESE POSTMAN without a post office is FPT with respect to k. This question was also featured in a survey talk by Sorge [86]. Subsequently, Gutin et al. [51] resolved the challenge, showing fixed-parameter tractability.

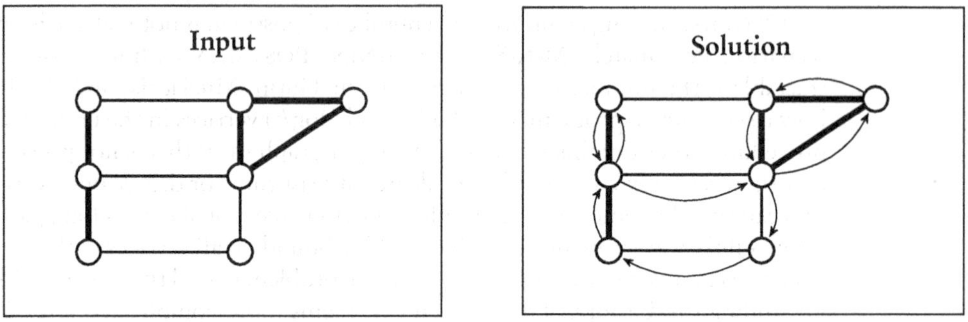

Figure 2.5. *Instance and solution for the* RURAL POSTMAN *problem.*

2.3 ▪ The Rural Postman Problem

The CHINESE POSTMAN problem we considered in the previous section can be used to model the task of delivering mail by some postman to houses in each street. In rural areas, however, not all streets may have houses to deliver mail to. If we take into account that a postman can use those streets to get his deliveries done more quickly, we end up with the following problem:

RURAL POSTMAN
Input: A connected (undirected) graph $G = (V,E)$ with edge costs $c(e) \geq 0$ for each $e \in E$, a set $R \subseteq E$ of *required edges*, and a maximum cost c_{max}.
Question: Is there a tour in G that traverses every edge in R at least once and costs at most c_{max}?

An input graph for RURAL POSTMAN with required edges drawn boldly and edge costs proportional to edge lengths in the drawing is depicted on the left in Figure 2.5, and a corresponding minimum-cost solution is pictured on the right.

Table 2.2 summarizes known results in approximation and parameterized complexity, omitting classical complexity, since all considered variants of RURAL POSTMAN are NP-complete if not *all* edges are required, which would make RURAL POSTMAN equivalent to CHINESE POSTMAN.

2.3.1 ▪ Classical complexity

By reducing the NP-hard HAMILTONIAN CYCLE problem to RURAL POSTMAN, Lenstra and Rinnooy Kan [64] showed RURAL POSTMAN to be NP-hard. HAMILTONIAN CYCLE asks whether a given graph admits a closed walk visiting each vertex exactly once. The reduction consists of attaching a loop to each vertex in the given graph $G = (V,E)$ and requiring a visit to all loops. All edges are given cost one, and the total cost of the desired tour is $2|V|$. If loops are not allowed in the graph, then we can simply subdivide them, that is, replace each loop by two edges and join them to a new vertex, and set the total cost of the desired tour to $3|V|$. It is not hard to see that removing the required edges from the postman tour leaves a tour of cost $|V|$ visiting all vertices of G at least once. Since all edges have cost one, no such tour visits a vertex twice. Similarly, HAMILTONIAN CYCLE on directed graphs can be reduced to RURAL POSTMAN on directed graphs, and, thus, NP-hardness for directed, mixed, and windy versions of RURAL POSTMAN follows from this proof.

Postman problems are very similar to graph problems related to the Eulerian property. In fact, a closed walk (such as a postman tour) in a graph can be seen as an Eulerian

Table 2.2. *Complexity of* RURAL POSTMAN (RP) *with respect to the parameters: $|R|$ is the number of required arcs, γ is the number of connected components induced by the required arcs, b is the number of imbalanced vertices, c_{max} is the maximum allowed tour cost, and k is the number of nonrequired arcs in a solution postman tour.*

RP variant	Approximation
Undirected	$O(\|V\|^3)$-time factor 3/2 [9, 39, 13] $O(\|V\|^3)$-time factor 3/2 for more general problem [57, 13]
Mixed	Require the triangle inequality to hold: factor 15/8 for symmetric [21] factor 9/5 if exactly the directed arcs are required [40]
RP variant	**Parameterized complexity**
Undirected	$O(\|V\|^{2\gamma}/\gamma! \cdot n)$-time algorithm [38] $2\|R\|$-vertex problem kernel $O(2^{3\|R\|} \cdot \|R\|^2 + \|V\|^3)$-time algorithm $O(2^\gamma \cdot (c_{max} + \|V\|)^d)$-time randomized algorithm, $d = \text{const.}$ [53] results for directed graphs conjectured to hold [23, 87]
Directed	$O(4^k \cdot \|V\|^3)$-time algorithm [23] $O(4^{\gamma \log(b\gamma^2)} \cdot \|V\|^4)$-time algorithm [87] $O(2^\gamma \cdot (c_{max} + \|V\|)^d)$-time randomized algorithm, $d = \text{const.}$ [53] no polynomial-size problem kernel with respect to γ and k [87]

multigraph. Thus, RURAL POSTMAN can be transformed into the following problem and vice versa by considering the required edges as edges of a multigraph that is to be extended to form an Eulerian cycle (see also Höhn, Jacobs, and Megow [55] and Dorn et al. [23]):

EULERIAN EXTENSION
Input: A multigraph $G = (V, E)$ with costs $c(\{u, v\}) > 0$ for all $u, v \in V$ and a maximum cost c_{max}.
Question: Is there a multiset E' of pairs $\{u, v\}$ with $u, v \in V$ such that adding the edges in E' to G makes G Eulerian and $c(E') \le c_{max}$?

Höhn, Jacobs, and Megow [55] considered RURAL POSTMAN as EULERIAN EXTENSION on a special class of directed graphs where vertices are pairs of numbers and inserting an arc from (x_1, y_1) to (x_2, y_2) is only possible if $x_1 \le x_2$ and $y_1 \le y_2$. They called this variant TWO-DIMENSIONAL EULERIAN EXTENSION and proved it to be NP-hard.

A special case of RURAL POSTMAN on mixed graphs is the STACKER CRANE problem, which requires visiting all directed arcs of the input mixed graph. Frederickson, Hecht, and Kim [40] showed that also this problem remains NP-hard.

2.3.2 ▪ Approximability

Despite the fact that the closely related TRAVELING SALESMAN problem presumably cannot be approximated to within any constant factor of the optimum, constant-factor approximations for special cases of TRAVELING SALESMAN serve as the basis for constant-factor approximation algorithms for RURAL POSTMAN. For example, the famous algorithm by Christofides [20] approximates TRAVELING SALESMAN to within a factor of 3/2 on graphs with triangle inequality (the input is required to be a complete graph such that, for each of three vertices u, v, and w, the inequality $c(\{u,v\})+c(\{v,w\}) \geq c(\{u,w\})$ holds). A similar approach is also used by Frederickson [39] and Jansen [57] to obtain constant-factor approximations for RURAL POSTMAN on instances respecting the triangle inequality. As pointed out by van Bevern et al. [13], this is sufficient for obtaining constant-factor approximations also for instances without triangle inequality.

In particular, polynomial-time algorithms of Frederickson [39] and Jansen [57] compute postman tours that are at most 50% longer than optimal tours. Both algorithms are based on the above-mentioned $O(|V|^3)$-time factor 3/2 approximation algorithm of Christofides [20]. While Frederickson's algorithm [39] is designed for RURAL POSTMAN, Jansen's algorithm [57] works also for the GENERAL ROUTING problem, where, aside from required edges that have to be visited at least once, we are also given a set of required vertices, each of which has to be visited *exactly* once.

In the following, we briefly sketch an adaptation of Jansen's algorithm [57] which is simplified and fitted to approximate RURAL POSTMAN to within a factor of 3/2. Given a complete graph $G = (V,E)$ with edge costs $c(\{u,v\}) \geq 0$ for each $u,v \in V$, and a set $R \subseteq E$ of required edges, it computes an Eulerian multigraph containing all edges of R. To this end, it connects the connected components induced by R in a treelike manner and then matches odd-degree vertices, such that inserting paths between them requires minimum additional cost. Next, we describe this four-step algorithm in more detail:

Step 1: Compute the graph $G_R := (\bigcup_{e\in R} e, R)$, and determine its connected components $\mathscr{C} := \{C_1, C_2, \ldots\}$.

Step 2: Compute the complete graph K on the vertex set $\mathscr{C}$ such that for each $i \neq j$ the cost of the edge $\{C_i, C_j\}$ equals the cost of a cheapest edge in G between any vertex of C_i and any vertex of C_j.

Step 3: Compute a minimum-cost spanning tree of K, and let G'_R denote the multigraph that results from adding to G_R the edges of G corresponding to the edges of the computed spanning tree.

Step 4: With V_{odd} denoting the vertices that have odd degree in G'_R, compute a minimum-cost matching M in $G[V_{\text{odd}}]$, and add the edges of M to G'_R.

Theorem 2.3 (Frederickson [39], Jansen [57]). RURAL POSTMAN *on undirected graphs can be approximated to within a factor of* 3/2 *in* $O(|V|^3)$ *time.*

In the following, we briefly sketch the proof of Theorem 2.3. For ease of presentation, we assume the input graph satisfies the triangle inequality, but, as pointed out by van Bevern et al. [13], the theorem also holds without triangle inequality.

Let c_{opt} denote the cost of an optimal solution for the given instance, and let $c(G'_R)$ and $c(M)$ denote the sum of costs of all edges in G'_R and in the matching M, respectively. Then, $c(G'_R) \leq c_{\text{opt}}$ since each optimal solution spans all connected components of G_R and visits all required edges, which cannot be done more cheaply than at cost $c(G'_R)$. Furthermore, one can show that the cost of the minimum-cost matching M in $G[V_{\text{odd}}]$ is at most $c_{\text{opt}}/2$: Since a vertex with odd degree in G'_R is also present in G_R, an optimal solution also visits all vertices in V_{odd}. Modifying an optimal tour by skipping over all vertices that are not in V_{odd}, we get a traveling salesperson tour in $G[V_{\text{odd}}]$, where *skipping over* v means replacing a subpath (u,v,w) of the tour by the subpath (u,w). Since the triangle inequality holds, the cost of this tour is at most c_{opt}. Furthermore, this tour can be partitioned into two matchings; the smaller one costs at most $c_{\text{opt}}/2$. Since M is a minimum-cost matching, the claimed factor of 3/2 follows. The running time of the algorithm is dominated by the computation of the edge costs of K in Step 2. This can be done in $O(|V|^3)$ time using the Floyd–Warshall algorithm. It is unknown whether RURAL POSTMAN on directed graphs can be approximated to within a constant factor, even if the input respects the directed triangle inequality.

For solving RURAL POSTMAN on *symmetric* mixed graphs, where for each arc (u,v) in the input, there is also an arc (v,u) with equal cost, Chyu [21] developed an algorithm achieving an approximation factor of 15/8 if the triangle inequality holds. STACKER CRANE (the input is a mixed graph, and the required arcs are exactly the directed arcs of the input) can be approximated within a factor of 9/5 if the triangle inequality holds, as shown by Frederickson, Hecht, and Kim [40]. Although the latter is also based on Christophides's algorithm, it loses a factor of 6/5 compared to the factor 3/2 that is achieved by Frederickson's [39] and Jansen's [57] algorithms. It is interesting to investigate whether this deficiency can be overcome.

Challenge 3. *Can the gap between the approximation factors of Christophides's algorithm and known approximation algorithms for the mentioned special cases of* MIXED RURAL POSTMAN *be overcome? Is* MIXED RURAL POSTMAN, *in general, constant-factor approximable?*

2.3.3 ▪ Parameterized complexity

As exhibited in the previous section, where an approximation algorithm for TRAVELING SALESMAN was adapted to work for RURAL POSTMAN, node routing algorithms can help to solve arc routing problems: Pearn, Assad, and Golden [81] showed that the former can easily be transformed into the latter. Since this transformation of RURAL POSTMAN to TRAVELING SALESMAN creates instances with at most $3|R|$ vertices, it can be used for parameterized algorithms by combining it with the well-known dynamic programming algorithm by Held and Karp [54], yielding an algorithm that solves RURAL POSTMAN in $O(2^{3|R|} \cdot |R|^2 + |V|^3)$ time, where R is the set of required edges. RURAL POSTMAN also admits a simple problem kernel containing $2|R|$ vertices which can be constructed as follows: First, for all vertex pairs $\{u,v\}$, set $c'(\{u,v\})$ to the cost of a shortest path between u and v in the input graph. Then, for all $\{u,v\} \in R$, decrease $c_{\max}$ by $c(\{u,v\}) - c'(\{u,v\})$, remove all vertices that are not incident with required edges, and insert all edges $\{u,v\}$ for which $c'(\{u,v\}) \neq \infty$. Since only vertices that are incident with required edges remain, the constructed graph together with the cost function c' and the modified $c_{\max}$ is a problem kernel containing at most $2|R|$ vertices.

Normally, solution tours for RURAL POSTMAN traverse additional, nonrequired arcs to connect the required arcs. The number of additional arcs needed in an optimal solution can be considered dual to the number of required arcs. The number of additional arcs also fits the profile of an "above guarantee" parameter (such parameters are discussed by Mahajan and Raman [68] and Mahajan, Raman, and Sikdar [69]), since each solution is guaranteed to contain the arcs of R. Considering this parameterization, Dorn et al. [23] developed a dynamic programming algorithm that, given a directed instance of RURAL POSTMAN and an integer k, computes the minimum cost of a tour visiting all required arcs plus at most k additional arcs. The algorithm runs in $O(4^k \cdot |V|^3)$ time.

Theorem 2.4 (Dorn et al. [23]). RURAL POSTMAN *is FPT with respect to the number k of arcs that an optimal solution visits in addition to the required arcs.*

The algorithm of Dorn et al. is tailored to solve the EULERIAN EXTENSION problem for directed multigraphs with arc-insertion costs, which is equal to the RURAL POSTMAN problem. In the EULERIAN EXTENSION instance, the required arcs form connected components. The idea of the algorithm for EULERIAN EXTENSION is to make copies of these connected components such that an optimal solution can be assumed to visit each copy exactly once. Then, an optimal solution can be computed using dynamic programming, much like the algorithm of Held and Karp [54] for TRAVELING SALESMAN.

Pondering the NP-hardness proof of RURAL POSTMAN by Lenstra and Rinnooy Kan [64], one quickly notices that it produces very specific instances. Notably, the number γ of connected components induced by the required arcs is large, making the number of connected components a critical parameter for complexity analysis, as already implied by Orloff [77] in 1976. While Frederickson [38] gave an algorithm that runs in $O(|V|^{2\gamma}/\gamma! \cdot |V|)$ time, it is yet unknown whether RURAL POSTMAN is FPT with respect to γ.

Challenge 4. *Is* RURAL POSTMAN *FPT with respect to the number γ of connected components induced by the required arcs?*

Sorge et al. [87, 88] attacked this problem, showing that RURAL POSTMAN with respect to the parameter γ is "FPT-equivalent" to an unstudied, yet promising, NP-hard matching variant. As a side-result, Sorge et al. obtained an algorithm that solves RURAL POSTMAN in $O(4^{\gamma \log(b\gamma^2)} \cdot |V|^4)$ time, where b denotes the number of imbalanced vertices, that is, vertices whose indegree does not equal their outdegree. More recently, Gutin, Wahlström, and Yeo [53] and Marx and Pilipczuk [72] independently gave randomized algorithms for the above-mentioned matching variant. Both represent the instances as a potentially large polynomial in which certain coefficients indicate whether a solution exists or not. Testing for these coefficients can then be done in FPT time by using dynamic programming and exploiting the Schwartz–Zippel lemma.[3] The resulting algorithms may fail to recognize a solution, but the probability for this can be made arbitrarily small with reasonable running time overhead. Gutin, Wahlström, and Yeo [53] also applied the approach sketched above directly to RURAL POSTMAN; from their proof, one obtains the following.

Theorem 2.5 (Gutin, Wahlström, and Yeo [53]). RURAL POSTMAN *can be solved in $O(2^\gamma \cdot (c_{\max} + |V|)^{\ell+d})$ time for a given real $\ell > 0$ and some constant $d > 0$, reporting a yes-instance as a no-instance with probability at most $1/(c_{\max} + |V|)^\ell$.*

Their approach works for both directed and undirected variants of RURAL POSTMAN. As derandomizations for the above type of algorithm seem currently elusive, Challenge 4 remains an interesting open problem.

Regarding preprocessing of RURAL POSTMAN instances, Sorge et al. [87] showed that, unless the polynomial hierarchy collapses to the second level, there is no polynomial-size problem kernel with respect to the parameter k as defined above. Since $k \geq \gamma$, this result also holds for the parameter γ. However, considering the practical relevance of polynomial-time preprocessing, it is important to find a parameter for which, even in combination with γ, a problem kernel can be constructed.

[3]The Schwartz–Zippel lemma asserts that, when assigning random numbers to the variables in a nonzero polynomial, it is unlikely to evaluate to 0.

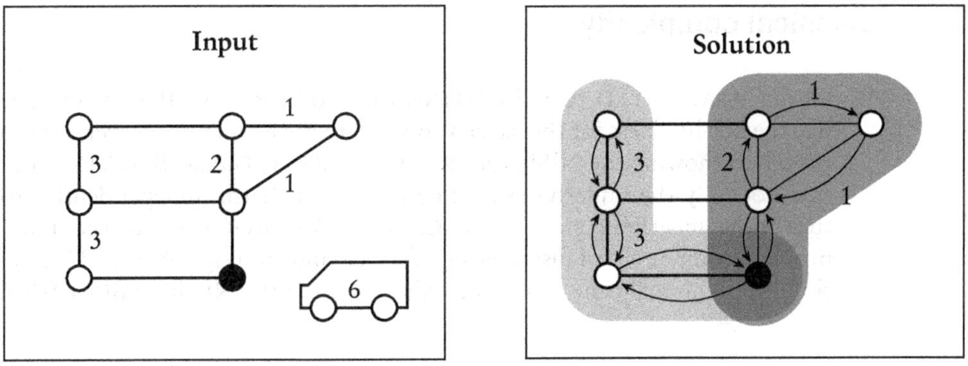

Figure 2.6. *Instance and solution for the* CAPACITATED ARC ROUTING *problem.*

2.4 ▪ The Capacitated Arc Routing Problem

The most general arc routing problem considered in this chapter is CAPACITATED ARC ROUTING, introduced by Golden and Wong [44]. It generalizes both of the previously considered problems, CHINESE POSTMAN and RURAL POSTMAN, and, speaking in terms of the mail delivery example, adds the possibility of serving mail to a set of required streets by multiple vehicles, each of which can only carry a limited number of parcels. In CAPACITATED ARC ROUTING, we are given an undirected graph, each edge of which is associated with a cost and some edges with a demand and a special vertex, called the *depot*. The depot houses an unlimited number of vehicles, each of a given capacity, that can *serve* edges or merely *traverse* them. An example is shown in the left part of Figure 2.6: The depot vertex v_0 is filled in black, edge demands are shown as labels next to the edges, and edge costs are proportional to the edge lengths in the drawing. All vehicles have capacity six. The task is then to serve each edge according to its demand by exactly one vehicle, where the sum of the demands of edges served by a single vehicle may not exceed its capacity. Moreover, the overall cost of the traversed edges is to be minimized. On the right in Figure 2.6 an optimal solution is shown. The route of one vehicle is shown on light gray background; the route of the other vehicle on dark gray background. Formally, the problem can be posed as follows:

CAPACITATED ARC ROUTING

Input: An undirected graph $G = (V, E)$ with a vehicle depot vertex $v_0 \in V$, with edge costs $c(e) \geq 0$ and edge demands $d(e) \geq 0$ for every $e \in E$, a vehicle capacity W, and a maximum cost $c_{\max}$.

Question: Is there a set $\mathscr{C}$ of cycles in G, each corresponding to the route of one vehicle and each passing through the depot vertex v_0, such that

1. $\sum_{C \in \mathscr{C}} \sum_{e \in C} c(e) \leq c_{\max}$,
2. each edge e with $d(e) > 0$ is traversed by at least one cycle and *served* by exactly one vehicle, and
3. the sum of demands of edges served by a single vehicle is at most W?

By giving each edge a demand of 1 and letting vehicles have a capacity equal to the number of edges in the graph, we obtain CHINESE POSTMAN. Giving only a subset of edges a demand of 1, we obtain RURAL POSTMAN.

2.4.1 ▪ Classical complexity

Since CAPACITATED ARC ROUTING can model RURAL POSTMAN, hardness results like the NP-hardness for the latter also apply to the former. However, using a different approach to showing the NP-hardness of CAPACITATED ARC ROUTING, one can observe that not only the underlying routing problem is hard. Independently from the structure of the underlying graph, CAPACITATED ARC ROUTING also contains the NP-hard problem component of distributing edge demands among vehicles. This can be seen by a simple polynomial-time many-one reduction from the NP- hard BIN PACKING problem:

BIN PACKING
Input: A bin size W, a number of bins B, and a list $w_1,\dots,w_n$ of integers.
Question: Is there a partition $S_1 \cup \dots \cup S_B$ of $\{w_1,\dots,w_n\}$ such that $\sum_{w\in S_k} w \leq W$ for all $1 \leq k \leq B$?

To reduce BIN PACKING to CAPACITATED ARC ROUTING, an approach similar to the hardness proof of Golden and Wong [44] can be employed, where we here use a more restricted output graph structure to infer more hardness results: Simply create a path $(v_1,\dots,v_{n+2})$, where v_{n+2} is the depot vertex. Let $d(\{v_{n+1},v_{n+2}\}) = W$ and $c(\{v_{n+1},v_{n+2}\}) = 1$. For all $k \leq n$, let $d(\{v_k,v_{k+1}\}) = w_k$ and $c(\{v_k,v_{k+1}\}) = 0$. It follows that the resulting graph can be served by $B+1$ vehicles of capacity W if and only if the given items $\{w_1,\dots,w_n\}$ can be packed into B bins (one vehicle is needed to serve edge $\{v_{n+1},v_{n+2}\}$). That is, the resulting graph can be served with cost $2(B+1)$ if and only if the input BIN PACKING instance is a yes-instance, since each vehicle has to traverse the edge $\{v_{n+1},v_{n+2}\}$ twice. Combining this reduction with known hardness results for BIN PACKING by Jansen et al. [59], we obtain the following theorem.

Theorem 2.6. CAPACITATED ARC ROUTING *is* NP-*hard even if all edges are required to be visited, the requested maximum cost and number of required vehicles is a constant, and the maximum vertex degree in the graph is at most two.*

Theorem 2.6 contrasts the polynomial-time solvability of RURAL POSTMAN in the case that all edges are required. It is easy to execute the same reduction of BIN PACKING to CAPACITATED ARC ROUTING with directed graphs. Thus, the NP-hardness results hold in the same way for generalizations of CAPACITATED ARC ROUTING that allow for directed edges, multiple depots, or a requested time interval for each edge, or that allow forbidding certain turns of vehicles (for example, to properly remove snow from crossroads).

Relations to node routing problems. One approach to obtaining exact (rather than heuristic) solutions for CAPACITATED ARC ROUTING is to transform CAPACITATED ARC ROUTING into the CAPACITATED VEHICLE ROUTING problem, as defined by Dantzig and Ramser [22], and then apply state-of-the-art algorithms to the latter.

CAPACITATED VEHICLE ROUTING

Input: A set V of vertices with a vehicle depot vertex $v_0 \in V$, with costs $c(\{u,v\}) \geq 0$ for all $u,v \in V$ and vertex demands $d(v) \geq 0$ for every $v \in V$, a vehicle capacity W, and a maximum cost $c_{\max}$.

Question: Is there a partition $C_1 \uplus \cdots \uplus C_\ell = \{v \in V \setminus \{v_0\} \mid d(v) > 0\}$ and for each C_i a permutation σ_i such that, for $1 \leq i \leq \ell$, it holds that $\sum_{v \in C_\ell} d(v) \leq W$ and such that

$$\sum_{i=1}^{\ell}\Big(c(\{v_0,\sigma_{i,1}\})+c(\{v_0,\sigma_{i,|C_i|}\})+\sum_{j=1}^{|C_i|-1} c(\{\sigma_{i,j},\sigma_{i+1,j}\})\Big) \leq c_{\max}?$$

In order to solve CAPACITATED ARC ROUTING using algorithms for CAPACITATED VEHICLE ROUTING, the first transformation from CAPACITATED ARC ROUTING to CAPACITATED VEHICLE ROUTING by Pearn, Assad, and Golden [81] has been subsequently improved to yield smaller instances by Baldacci and Maniezzo [7] and by Longo, de Aragão, and Uchoa [67].

Golden and Wong [44] give a straightforward transformation from CAPACITATED VEHICLE ROUTING to CAPACITATED ARC ROUTING, which, however, requires the input VEHICLE ROUTING instance to respect the triangle inequality: Given a VEHICLE ROUTING instance, split each vertex with positive demand into two, and join them by a zero-cost edge whose demand corresponds to the demand of the original vertex. If the input VEHICLE ROUTING instance respects the triangle inequality, this transformation is correct since a solution for the output CAPACITATED ARC ROUTING instance can be transformed into an equal-cost solution of the original VEHICLE ROUTING instance. Since a vertex in VEHICLE ROUTING may be visited at most once, whereas the found CAPACITATED ARC ROUTING solution may visit the corresponding edges several times, this transformation is not correct *in general*. However, if the triangle inequality of the VEHICLE ROUTING instance holds, then tours that visit a vertex twice can be shortened.

2.4.2 ▪ Approximability

The NP hardness of CAPACITATED ARC ROUTING even in very special cases tempts us to search for approximate solutions. However, using a reduction from PARTITION similar to the proof of Theorem 2.6, Golden and Wong [44] showed that computing a factor 3/2 approximation for CAPACITATED ARC ROUTING is NP-hard, even if the cost function respects the triangle inequality. On the positive side, there are factor $(7/2-3/W)$ approximations for CAPACITATED ARC ROUTING by Jansen [58] and Wøhlk [94].

Theorem 2.7 (Golden and Wong [44], Jansen [58], Wøhlk [94], van Bevern et al. [13]). *Computing a factor* 3/2 *approximation for* CAPACITATED ARC ROUTING *is* NP-*hard. A factor* $(7/2-3/W)$ *approximation can be computed in polynomial time.*

In fact, the algorithms by Jansen [58] and Wøhlk [94] assume the triangle inequality to hold. Moreover, in the literature, the following claim can also be found [44, 93, 94]: If the cost function in a CAPACITATED ARC ROUTING instance is not required to respect the triangle inequality, then computing a factor α approximation for CAPACITATED ARC ROUTING is NP-hard for any $\alpha > 0$. However, van Bevern et al. [13] showed that any factor α approximation for CAPACITATED ARC ROUTING with triangle inequality also yields a factor α approximation for CAPACITATED ARC ROUTING without triangle inequality. Thus, the approximation factor given in Theorem 2.7 can also be achieved without the triangle inequality.

It is unknown whether CAPACITATED ARC ROUTING on directed graphs is constant-factor approximable. Herein, it would be sufficient to find a polynomial-time constant-factor approximation for the case where the directed triangle inequality holds [13].

Challenge 5. *Find a constant-factor approximation for* CAPACITATED ARC ROUTING *on directed graphs, or prove that any constant-factor approximation would imply* P = NP.[4]

2.4.3 ▪ Parameterized complexity

The parameterized complexity of CAPACITATED ARC ROUTING is still an untouched field. If we do not require that the input respect the triangle inequality, then Theorem 2.6 rules out the parameters "maximum cost $c_{\max}$," pathwidth, maximum number of vehicles, and maximum degree of the input graph with regard to fixed-parameter tractability, since CAPACITATED ARC ROUTING remains NP-hard even if all these parameters are constant. From parameterized considerations about BIN PACKING by Jansen et al. [59] and Theorem 2.6, we obtain that, even if we were to consider the demands of edges to be encoded in unary, CAPACITATED ARC ROUTING is $W[1]$-hard with respect to the parameter $c_{\max}$ or, analogously, with respect to the number of vehicles. It is yet unclear whether, with respect to these parameters, CAPACITATED ARC ROUTING is in XP.

Challenge 6. BIN PACKING *is in* XP *but W[1]-hard with respect to the parameter "number of required bins" if the input integers are encoded in unary [59]. This implies W[1]-hardness for* CAPACITATED ARC ROUTING *with respect to the maximum cost $c_{\max}$ if the edge demands are encoded in unary. Is* CAPACITATED ARC ROUTING *in* XP *under these conditions?*

If we require that the input satisfy the triangle inequality, then the NP-hardness proof initially given by Golden and Wong [44] shows that CAPACITATED ARC ROUTING even remains NP-hard on graphs of diameter one (the diameter of a graph is the maximum length of a shortest path between two vertices). Hence, the single parameter "diameter" is also ineffective. In their reduction, the parameters "maximum cost $c_{\max}$" and "number of required vehicles" also remain constant. In contrast, the parameters "maximum degree" and "treewidth" in the instances created by this reduction may be large, but, with respect to these parameters, we can find a trivial problem kernel: If the input CAPACITATED ARC ROUTING instance satisfies the triangle inequality, then it is a no-instance or every vertex in the input graph is adjacent to the depot vertex. This is because every vertex v has to be reachable from the depot vertex by some path, and, therefore, v has an edge to the depot vertex that has at most the length of this path. Since all vertices are adjacent to the depot vertex, they are also adjacent to each other, forming a clique. The number of vertices in the clique, and, therefore, the number of vertices in the input graph, is then bounded linearly by the maximum degree and the treewidth, yielding a problem kernel with a linear number of vertices.

It turns out that, with respect to most parameters, CAPACITATED ARC ROUTING remains hard or, if we require the triangle inequality, becomes trivial. Hence, one of the main challenges considering the parameterized complexity of CAPACITATED ARC

[4]In an earlier version of this chapter, the authors asked whether CAPACITATED ARC ROUTING is constant-factor approximable without triangle inequality. Subsequently, van Bevern et al. [13] answered this question affirmatively.

ROUTING seems to be the task of identifying meaningful and small parameters that make the problem tractable. Wøhlk [93] raises the point that restrictions on the demands may help in coping with the hardness of the problem: For example, on paths or cycles with uniform demands the problem can be solved in polynomial time. Also, the BIN PACKING problem used in the NP-hardness reduction for CAPACITATED ARC ROUTING above becomes polynomial-time solvable if the number of bins is a constant and the integers are at most polynomial in the input size.

2.5 ▪ Conclusion and outlook

Classifying the computational complexity of a problem lies at the heart of developing and also justifying lines of attack for solving it. Indeed, many methods for solving arc routing problems are based on heuristics and mathematical programming; this is justified by the fact that most of these problems are NP-hard. A proof of NP-hardness, however, does not mean the end of useful theoretical (and practically useful) work on the considered problem. That is, it may be possible to identify meaningful polynomial-time solvable special cases, to derive efficient approximation algorithms, or to identify problem-specific parameters to be exploited in the spirit of fixed-parameter tractability. Thus, throughout this chapter we featured several challenging open problems whose answering would help to obtain a further refined view on the computational complexity of important arc routing problems.

Clearly, the selected open problems could be extended by numerous ways of systematically identifying research challenges. Let us mainly focus on parameterized complexity analysis here, the "freshest approach" discussed in this chapter. A fruitful way of identifying parameters that have significant influence on the computational complexity of a problem is to revisit known hardness reductions proving NP-hardness or the like. In these proofs, one then checks whether they rely on certain parameter values that are necessarily unbounded to make the proof work. For instance, the NP-hardness proof for RURAL POSTMAN by Lenstra and Rinnooy Kan [64] heavily relies on the fact that the number of connected components is large. Indeed, Frederickson [38] showed the problem to be polynomial-time solvable for a constant number of connected components, rendering this a very interesting parameter and leading to the challenging open question whether RURAL POSTMAN is fixed parameter tractable with respect to this parameter. A partial answer has been given by Gutin, Wahlström, and Yeo [53], who showed that a *randomized* FPT algorithm exists. Obviously, the same approach for identifying parameters can be pursued for other problems as well and also has been termed "deconstructing intractability" [32, 61, 76].

A further way to spot interesting parameterizations for arc routing problems is to inspect real-world data and to measure various network parameters (such as vertex degree, diameter, number of connected components, etc.) in order to see whether some of them adopt small values in applications. If so, then this strongly motivates parameterized complexity investigations with respect to the parameters identified in this way. This is also known as data-driven parameterization.

We believe that, although classical complexity classification of most arc routing problems is in a sense settled, the advent of parameterized complexity analysis paves the way for a prospective theory-based exploration of the computational complexity landscape of arc routing problems. With the research challenges featured in this chapter, we hope to help kick off such a development.[5]

[5]Indeed, after a survey talk by Sorge [86] based on an earlier version of this chapter, several new results on the parameterized complexity of arc routing problems appeared and were included [50, 51, 52, 53, 72].

Acknowledgments

We are grateful to an anonymous referee whose comments greatly helped to improve the presentation of this chapter, and we thank Greg N. Frederickson and Sanne Wøhlk for making their Ph.D. works accessible.

René van Bevern and Manuel Sorge gratefully acknowledge support provided by the DFG grant DAPA (NI 369/12), and Mathias Weller gratefully acknowledges support provided by the DFG grant DARE (NI 369/11). This work was performed while all authors were with Institut für Softwaretechnik und Theoretische Informatik, TU Berlin, Germany.

Bibliography

[1] *Clay Mathematics Institute*, 2013. http://www.claymath.org/millennium-problems.

[2] S. AARONSON, *The complexity zoo*, 2013. https://complexityzoo.uwaterloo.ca/Complexity_Zoo.

[3] D. AHR, *Contributions to Multiple Postmen Problems*, Ph.D. thesis, Ruprecht-Karls-Universität Heidelberg, Heidelberg, Germany, 2004.

[4] A.S. ALFA AND D.Q. LIU, *Postman routing problem in a hierarchical network*, Engineering Optimization, 14 (1988), pp. 127–138.

[5] S. ARORA AND B. BARAK, *Computational Complexity: A Modern Approach*, Cambridge University Press, Cambridge, UK, 2009.

[6] G. AUSIELLO, P. CRESCENZI, G. GAMBOSI, V. KANN, A. MARCHETTI-SPACCAMELA, AND M. PROTASI, *Complexity and Approximation: Combinatorial Optimization Problems and Their Approximability Properties*, Springer, New York, 1999.

[7] R. BALDACCI AND V. MANIEZZO, *Exact methods based on node-routing formulations for undirected arc-routing problems*, Networks, 47 (2006), pp. 52–60.

[8] E.J. BELTRAMI AND L.D. BODIN, *Networks and vehicle routing for municipal waste collection*, Networks, 4 (1974), pp. 65–94.

[9] E. BENAVENT, V. CAMPOS, A. CORBERÁN, AND E. MOTA, *Análisis de heurísticos para el problema del cartero rural*, Trabajos de Estadística y de Investigación Operativa, 36 (1985), pp. 27–38.

[10] ——, *The capacitated arc routing problem: Lower bounds*, Networks, 22 (1992), pp. 669–690.

[11] E. BENAVENT AND D. SOLER, *The directed rural postman problem with turn penalties*, Transportation Science, 33 (1999), pp. 408–418.

[12] C. BERGE, *Graphs*, North-Holland, Amsterdam, 1985.

[13] R. VAN BEVERN, S. HARTUNG, A. NICHTERLEIN, AND M. SORGE, *Constant-factor approximations for capacitated arc routing without triangle inequality*, Operations Research Letters, 42 (2014), pp. 290–292.

[14] H.L. BODLAENDER, *Kernelization: New upper and lower bound techniques*, in Proceedings of the 4th International Workshop on Parameterized and Exact Computation (IWPEC '09), Lecture Notes in Computer Science 5917, Springer, New York, 2009, pp. 17–37.

[15] H.L. BODLAENDER, R.G. DOWNEY, M.R. FELLOWS, AND D. HERMELIN, *On problems without polynomial kernels*, Journal of Computer and System Sciences, 75 (2009), pp. 423–434.

[16] H.L. BODLAENDER AND A.M.C.A. KOSTER, *Combinatorial optimization on graphs of bounded treewidth*, The Computer Journal, 51 (2008), pp. 255–269.

[17] L. CAI, *Parameterized complexity of vertex colouring*, Discrete Applied Mathematics, 127 (2003), pp. 415–429.

[18] J. CHEN, I.A. KANJ, AND G. XIA, *Improved upper bounds for vertex cover*, Theoretical Computer Science, 411 (2010), pp. 3736–3756.

[19] N. CHRISTOFIDES, *The optimum traversal of a graph*, Omega, 1 (1973), pp. 719–732.

[20] N. CHRISTOFIDES, *Worst case analysis of a new heuristic for the traveling salesman problem*, Management Science Research Rept. 388, Carnegie-Mellon University, Pittsburgh, PA, 1976.

[21] C.-C. CHYU, *A mixed-strategy heuristic for the mixed arc routing problem*, Journal of the Chinese Institute of Industrial Engineers, 18 (2001), pp. 68–76.

[22] G.B. DANTZIG AND J.H. RAMSER, *The truck dispatching problem*, Management Science, 6 (1959), pp. 80–91.

[23] F. DORN, H. MOSER, R. NIEDERMEIER, AND M. WELLER, *Efficient algorithms for Eulerian extension and rural postman*, SIAM Journal on Discrete Mathematics, 27 (2013), pp. 75–94.

[24] R.G. DOWNEY AND M.R. FELLOWS, *Fundamentals of Parameterized Complexity*, Springer, New York, 2013.

[25] M. DROR, *Arc Routing: Theory, Solutions, and Applications*, Kluwer Academic Publishers, Dordrecht, The Netherlands, 2000.

[26] M. DROR AND M. HAOUARI, *Generalized Steiner problems and other variants*, Journal of Combinatorial Optimization, 4 (2000), pp. 415–436.

[27] M. DROR, H. STERN, AND P. TRUDEAU, *Postman tour on a graph with precedence relation on arcs*, Networks, 17 (1987), pp. 283–294.

[28] J. EDMONDS, *The Chinese postman problem*, Operations Research, Supplement 1 (1975), pp. B73–B77.

[29] J. EDMONDS AND E.L. JOHNSON, *Matching, Euler tours and the Chinese postman*, Mathematical Programming, 5 (1973), pp. 88–124.

[30] H.A. EISELT, M. GENDREAU, AND G. LAPORTE, *Arc routing problems, part II: The rural postman problem*, Operations Research, 43 (1995), pp. 399–414.

[31] H.A. EISELT, M. GENDREAU, AND G. LAPORTE, *Arc routing problems, part I: The Chinese postman problem*, Operations Research, 43 (1995), pp. 231–242.

[32] M.R. FELLOWS, B.M.P. JANSEN, AND F.A. ROSAMOND, *Towards fully multivariate algorithmics: Parameter ecology and the deconstruction of computational complexity*, European Journal of Combinatorics, 34 (2013), pp. 541–566.

[33] C.G. FERNANDES, O. LEE, AND Y. WAKABAYASHI, *Minimum cycle cover and Chinese postman problems on mixed graphs with bounded tree-width*, Discrete Applied Mathematics, 157 (2009), pp. 272–279.

[34] J. FLUM AND M. GROHE, *Parameterized Complexity Theory*, Springer, New York, 2006.

[35] L.R. FORD AND D.R. FULKERSON, *Flows in Networks*, Princeton University Press, Princeton, NJ, 2010.

[36] L. FORTNOW, *The Golden Ticket: P, NP, and the Search for the Impossible*, Princeton University Press, Princeton, NJ, 2013.

[37] L. FORTNOW AND R. SANTHANAM, *Infeasibility of instance compression and succinct PCPs for NP*, Journal of Computer and System Sciences, 77 (2011), pp. 91–106.

[38] G.N. FREDERICKSON, *Approximation Algorithms for NP-hard Routing Problems*, Ph.D. thesis, Faculty of the Graduate School of the University of Maryland, College Park, MD, 1977.

[39] ——, *Approximation algorithms for some postman problems*, Journal of the ACM, 26 (1979), pp. 538–554.

[40] G.N. FREDERICKSON, M.S. HECHT, AND C.E. KIM, *Approximation algorithms for some routing problems*, SIAM Journal on Computing, 7 (1978), pp. 178–193.

[41] M.R. GAREY AND D.S. JOHNSON, *Computers and Intractability: A Guide to the Theory of NP-Completeness*, W.H. Freeman, San Francisco, CA, 1979.

[42] W.I. GASARCH, *Guest column: The second P =?NP poll*, ACM SIGACT News, 43 (2012), pp. 53–77.

[43] G. GHIANI AND G. IMPROTA, *An algorithm for the hierarchical Chinese postman problem*, Operations Research Letters, 26 (2000), pp. 27–32.

[44] B.L. GOLDEN AND R.T. WONG, *Capacitated arc routing problems*, Networks, 11 (1981), pp. 305–315.

[45] O. GOLDREICH, *P, NP, and NP-Completeness*, Cambridge University Press, Cambridge, UK, 2010.

[46] M. GUAN, *Graphic programming using odd or even points*, Chinese Mathematics, 1 (1962), pp. 273–277.

[47] ——, *On the windy postman problem*, Discrete Applied Mathematics, 9 (1984), pp. 41–46.

[48] J. GUO, F. HÜFFNER, AND R. NIEDERMEIER, *A structural view on parameterizing problems: Distance from triviality*, in Proceedings of the 1st International Workshop on Parameterized and Exact Computation (IWPEC '04), Lecture Notes in Computer Science 3162, Springer, New York, 2004, pp. 162–173.

[49] J. GUO AND R. NIEDERMEIER, *Invitation to data reduction and problem kernelization*, ACM SIGACT News, 38 (2007), pp. 31–45.

[50] G. GUTIN, M. JONES, AND B. SHENG, *Parameterized complexity of the k-arc Chinese postman problem*, Computing Research Repository, arXiv:1403.1512 (2014).

[51] G. GUTIN, M. JONES, B. SHENG, AND M. WAHLSTRÖM, *Parameterized directed k-Chinese postman problem and k arc-disjoint cycles problem on Euler digraphs*, in Proceedings of the 40th International Workshop on Graph-Theoretic Concepts in Computer Science (WG '14), Lecture Notes in Computer Science, Springer, 2014. Accepted for publication.

[52] G. GUTIN, G. MUCIACCIA, AND A. YEO, *Parameterized complexity of k-Chinese postman problem*, Theoretical Computer Science, 513 (2013), pp. 124–128.

[53] G. GUTIN, M. WAHLSTRÖM, AND A. YEO, *Parameterized rural postman and conjoining bipartite matching problems*, Computing Research Repository, arXiv:1308.2599 (2013).

[54] M. HELD AND R. M. KARP, *A dynamic programming approach to sequencing problems*, Journal of the Society for Industrial and Applied Mathematics, 10 (1962), pp. 196–210.

[55] W. HÖHN, T. JACOBS, AND N. MEGOW, *On Eulerian extensions and their application to no-wait flowshop scheduling*, Journal of Scheduling, 15 (2012), pp. 295–309.

[56] F. HÜFFNER, R. NIEDERMEIER, AND S. WERNICKE, *Techniques for practical fixed-parameter algorithms*, The Computer Journal, 51 (2008), pp. 7–25.

[57] K. JANSEN, *An approximation algorithm for the general routing problem*, Information Processing Letters, 41 (1992), pp. 333–339.

[58] K. JANSEN, *Bounds for the general capacitated routing problem*, Networks, 23 (1993), pp. 165–173.

[59] K. JANSEN, S. KRATSCH, D. MARX, AND I. SCHLOTTER, *Bin packing with fixed number of bins revisited*, Journal of Computer and System Sciences, 79 (2013), pp. 39–49.

[60] C. KOMUSIEWICZ AND R. NIEDERMEIER, *New races in parameterized algorithmics*, in Proceedings of the 37th International Symposium on Mathematical Foundations of Computer Science (MFCS '12), Lecture Notes in Computer Science 7464, Springer, New York, 2012, pp. 19–30.

[61] C. KOMUSIEWICZ, R. NIEDERMEIER, AND J. UHLMANN, *Deconstructing intractability—a multivariate complexity analysis of interval constrained coloring*, Journal of Discrete Algorithms, 9 (2011), p. 137–151.

[62] P. KORTEWEG AND T. VOLGENANT, *On the hierarchical Chinese postman problem with linear ordered classes*, European Journal of Operational Research, 169 (2006), pp. 41–52.

[63] J. van Leeuwen and R.B. Tan, *Compact routing methods: A survey*, in Proceedings of the 1st International Colloquium on Structural Information and Communication Complexity (SIROCCO '94), Carleton University Press, Ottawa, Ontario, Canada, 1994, pp. 99–110.

[64] J.K. Lenstra and A.H.G. Rinnooy Kan, *On general routing problems*, Networks, 6 (1976), pp. 273–280.

[65] Y. Lin and Y. Zhao, *A new algorithm for the directed Chinese postman problem*, Computers & Operations Research, 15 (1988), pp. 577–584.

[66] D. Lokshtanov, N. Misra, and S. Saurabh, *Kernelization—preprocessing with a guarantee*, in The Multivariate Algorithmic Revolution and Beyond—Essays Dedicated to Michael R. Fellows on the Occasion of His 60th Birthday, Lecture Notes in Computer Science 7370, Springer, New York, 2012, pp. 129–161.

[67] H. Longo, M.P. de Aragão, and E. Uchoa, *Solving capacitated arc routing problems using a transformation to the CVRP*, Computers & Operations Research, 33 (2006), pp. 1823–1837.

[68] M. Mahajan and V. Raman, *Parameterizing above guaranteed values: MaxSat and MaxCut*, Journal of Algorithms, 31 (1999), pp. 335–354.

[69] M. Mahajan, V. Raman, and S. Sikdar, *Parameterizing above or below guaranteed values*, Journal of Computer and System Sciences, 75 (2009), pp. 137–153.

[70] J. Markoff, *Prizes aside, the P-NP puzzler has consequences*, The New York Times, October 7th, 2009. Available online at http://www.nytimes.com/2009/10/08/science/Wpolynom.html.

[71] D. Marx, B. O'Sullivan, and I. Razgon, *Finding small separators in linear time via treewidth reduction*, ACM Transactions on Algorithms, 9 (2013), p. 30.

[72] D. Marx and M. Pilipczuk, *Everything you always wanted to know about the parameterized complexity of subgraph isomorphism (but were afraid to ask)*, Computing Research Repository, arXiv:1307.2187 (2013).

[73] S. Micali and V. V. Vazirani, *An $O(\sqrt{|v|}|E|)$ algorithm for finding maximum matching in general graphs*, in Proceedings of the 27th Annual IEEE Symposium on Foundations of Computer Science (FOCS '86), IEEE, Washington, DC, 1980, pp. 17–27.

[74] E. Minieka, *The Chinese postman problem for mixed networks*, Management Science, 25 (1979), pp. 643–648.

[75] R. Niedermeier, *Invitation to Fixed-Parameter Algorithms*, Oxford University Press, Oxford, UK, 2006.

[76] ——, *Reflections on multivariate algorithmics and problem parameterization*, in Proceedings of the 27th International Symposium on Theoretical Aspects of Computer Science (STACS '10), vol. 5 of LIPIcs, Schloss Dagstuhl—Leibniz-Zentrum fuer Informatik, 2010, pp. 17–32.

[77] C.S. Orloff, *On general routing problems: Comments*, Networks, 6 (1976), pp. 281–284.

[78] C.H. PAPADIMITRIOU, *On the complexity of edge traversing*, Journal of the ACM, 23 (1976), pp. 544–554.

[79] ——, *Computational Complexity*, Addison–Wesley, Reading, MA, 1994.

[80] W.L. PEARN, *Solvable cases of the k-person Chinese postman problem*, Operations Research Letters, 16 (1994), pp. 241–244.

[81] W.L. PEARN, A. ASSAD, AND B.L. GOLDEN, *Transforming arc routing into node routing problems*, Computers & Operations Research, 14 (1987), pp. 285–288.

[82] W.L. PEARN AND K.-H. WANG, *On the maximum benefit Chinese postman problem*, Omega, 31 (2003), pp. 269–273.

[83] N. PERRIER, A. LANGEVIN, AND J.F. CAMPBELL, *A survey of models and algorithms for winter road maintenance. Part IV: Vehicle routing and fleet sizing for plowing and snow disposal*, Computers & Operations Research, 34 (2007), pp. 258–294.

[84] B. RAGHAVACHARI AND J. VEERASAMY, *Approximation algorithms for the asymmetric postman problem*, in Proceedings of the 10th Annual ACM–SIAM Symposium on Discrete Algorithms (SODA '99), SIAM, Philadelphia, 1999, pp. 734–741.

[85] B. RAGHAVACHARI AND J. VEERASAMY, *A 3/2-approximation algorithm for the mixed postman problem*, SIAM Journal on Discrete Mathematics, 12 (1999), pp. 425–433.

[86] M. SORGE, *Some algorithmic challenges in arc routing*, Talk at NII Shonan Seminar 18, Shonan, Japan, May 2013.

[87] M. SORGE, R. VAN BEVERN, R. NIEDERMEIER, AND M. WELLER, *From few components to an Eulerian graph by adding arcs*, in Proceedings of the 37th International Workshop on Graph-Theoretic Concepts in Computer Science (WG '11), Lecture Notes in Computer Science 6986, Springer, New York, 2011, pp. 307–319.

[88] ——, *A new view on rural postman based on Eulerian extension and matching*, Journal of Discrete Algorithms, 16 (2012), pp. 12–33.

[89] D.A. SPIELMAN AND S.-H. TENG, *Smoothed analysis of algorithms: Why the simplex algorithm usually takes polynomial time*, Journal of the ACM, 51 (2004), pp. 385–463.

[90] ——, *Smoothed analysis: An attempt to explain the behavior of algorithms in practice*, Communications of the ACM, 52 (2009), pp. 76–84.

[91] C. THOMASSEN, *On the complexity of finding a minimum cycle cover of a graph*, SIAM Journal on Computing, 26 (1997), pp. 675–677.

[92] V.V. VAZIRANI, *Approximation Algorithms*, Springer, New York, 2001.

[93] S. WØHLK, *Capacitated Arc Routing Problem, Theory and Solution*, Ph.D. thesis, University of Southern Denmark, Odense, Denmark, 2002.

[94] ——, *An approximation algorithm for the capacitated arc routing problem*, The Open Operational Research Journal, 2 (2008), pp. 8–12.

[95] D.P. WILLIAMSON AND D.B. SHMOYS, *The Design of Approximation Algorithms*, Cambridge University Press, Cambridge, UK, 2011.

[96] Z. WIN, *Contributions to Routing Problems*, Ph.D. thesis, Universität Augsburg, Augsburg, Germany, 1987.

[97] ——, *On the windy postman problem on Eulerian graphs*, Mathematical Programming, 44 (1989), pp. 97–112.

[98] F. J. ZARAGOZA MARTÍNEZ, *Postman Problems on Mixed Graphs*, Ph.D. thesis, University of Waterloo, Waterloo, Ontario, Canada, 2003.

[99] ——, *Complexity of the mixed postman problem with restrictions on the arcs*, in Proceedings of the 3rd International Conference on Electrical and Electronics Engineering (ICEEE'06), IEEE, Washington, DC, 2006, pp. 1–4.

[100] ——, *Complexity of the mixed postman problem with restrictions on the edges*, in Proceedings of the Seventh Mexican International Conference on Computer Science (ENC'06), IEEE, Washington, DC, 2006, pp. 3–10.

[101] ——, *Series-parallel graphs are windy postman perfect*, Discrete Mathematics, 308 (2008), pp. 1366–1374.

[102] L. ZHANG, *Polynomial algorithms for the k-Chinese postman problem*, in Proceedings of the IFIP 12th World Computer Congress on Algorithms, Software, Architecture—Information Processing '92, Volume 1, IFIP Transactions A-12, North-Holland, Amsterdam, 1992, pp. 430–435.

Chapter 3

The Undirected Chinese Postman Problem

Gilbert Laporte

3.1 ▪ Introduction

The *Chinese Postman Problem* (CPP) is arguably the most central problem in arc routing. In this chapter we review the undirected version of the CPP, as well as four of its variants: the generalized CPP, the cumulative CPP, the hierarchical CPP, and the CPP with time windows. Another variant, the CPP with profits, will be treated in Chapter 12.

3.2 ▪ The undirected Chinese Postman Problem

The undirected CPP takes its name from the Chinese mathematician Meigu Guan who introduced it more than 50 years ago (Guan [21]). The problem is defined on an undirected graph $G = (V,E)$, where $V = \{1,\dots,n\}$ is the vertex set and $E = \{(i,j) : i,j \in V, i < j\}$ is the edge set. With each edge (i,j) is associated a travel cost c_{ij}. The problem is to determine a least-cost closed traversal of all edges of the graph. When G is connected and all vertices have an even degree, the graph is Eulerian (or unicursal), which means that there exists a CPP solution traversing each edge only once. As we saw in Chapter 1, Euler [11] observed that connectedness and evenness are two necessary conditions for unicursality, and Hierholzer [23] showed that these conditions were also sufficient.

In order for a feasible CPP solution to exist, the graph must of course be connected. The difficulty consists of determining an optimal solution when not all vertices have an even degree. The problem amounts to determining a least-cost *augmentation* of the graph that will render all degrees even, i.e., identifying a subset of edges that will be traversed more than once. In fact, it is never optimal to traverse the same edge more than twice since, if this were the case, pairs of traversals could be removed without disconnecting the graph and without changing degree parities. A least-cost augmentation is obtained by determining a minimum cost matching of the odd-degree vertices of G, where the matching cost of i and j is the cost of a shortest chain $(i,\dots,j)$ in G.

The algorithm proposed by Guan [21] in his seminal paper effectively solves a matching problem while ensuring that the added chains do not intersect. Guan proved the following two necessary and sufficient conditions for optimality: (1) The solution has no redundancy, which means that at most one edge linking two given vertices in the original graph is duplicated; (2) the cost of the added edges does not exceed half the solution cost. Note that this property lies at the heart of the Christofides [8] heuristic for the *Traveling Salesman Problem* (TSP), which has a worst-case performance ratio of 3/2 whenever the TSP cost matrix is symmetric and satisfies the triangle inequality.

A natural way of formulating the augmentation problem as a linear integer program is to define binary x_{ij} variables equal to 1 if and only if a copy of edge (i,j) appears in the augmentation. Let $T \subseteq V$ be the set of odd-degree vertices of V, and let $\delta(i)$ be the set of edges incident to vertex i. The formulation is then

$$\text{(CPP1)} \qquad \text{minimize} \sum_{(i,j)\in E} c_{ij}x_{ij} \tag{3.1}$$

$$\text{s.t.} \qquad \sum_{(i,j)\in\delta(i)} x_{ij} = \begin{cases} 1 \pmod 2 & \text{if } i \in T, \\ 0 \pmod 2 & \text{if } i \in V\backslash T, \end{cases} \tag{3.2}$$

$$x_{ij} = 0 \text{ or } 1, \qquad (i,j)\in E. \tag{3.3}$$

This model can be solved as a perfect matching problem on the graph (T, E_T), where $E_T = \{(i,j) : i,j \in T, i < j\}$ and the matching cost d_{ij} associated with $(i,j) \in E_T$ is the length of a shortest chain between i and j. The matching problem is solvable in $O(|T|^3)$ time (Lawler [26]). However, lower complexity algorithms that exploit data structures are also available (see, e.g., Galil, Micali, and Gabow [18] and Derigs and Metz [9]).

The paper by Edmonds and Johnson [13] provides the first mathematical programming treatment of the CPP. It proposes a compact formulation of the problem which does away with the modulo conditions. In what follows, $S \subset V$ is said to be odd if it is a set consisting of an odd number of odd-degree vertices. Let $E(S) = \{(i,j) : i \in S, j \in V\backslash S$, or $i \in V\backslash S, j \in S\}$. Using the same x_{ij} variables as in CPP1, the problem can be formulated as

$$\text{(CPP2)} \qquad \text{minimize} \sum_{(i,j)\in E} c_{ij}x_{ij} \tag{3.4}$$

$$\text{s.t.} \qquad \sum_{(i,j)\in E(S)} x_{ij} \geq 1, \quad S \subset V, S \text{ is odd}, \tag{3.5}$$

$$x_{ij} \geq 0, \quad (i,j)\in E, \tag{3.6}$$

$$x_{ij} \text{ integer}, \quad (i,j)\in E. \tag{3.7}$$

In this formulation, constraints (3.5) are referred to as *blossom inequalities* (or *odd-cut* inequalities). They state that in order to achieve evenness, at least one copy of an edge incident to an odd set S must be part of the augmented graph. Edmonds and Johnson prove that the polyhedron of solutions to (3.5) and (3.6) is equal to the convex hull of solutions of the problem and that blossom inequalities can be separated in polynomial time. They solve the problem through an adaptation of the blossom algorithm for matching problems (Edmonds [12]). However, given the efficiency of today's integer linear programming solvers, the use of a specialized algorithm for the matching problem may be less justified.

Once a Eulerian graph has been obtained, determining a Eulerian cycle on it can be achieved by applying Hierholzer's algorithm [23], for example (Edmonds and Johnson [13] described it under the name "end-pairing algorithm"):

Step 1. Starting from an arbitrary vertex v_1, construct a cycle $(v_1, \dots, v, v_2, \dots, v_1)$ by iteratively traversing untraversed edges until v_1 is reached.

Step 2. If all edges have been traversed, stop.

Step 3. Construct a second cycle starting from an edge (v, u_1) incident to the first cycle. Merge the two cycles $(v_1, \dots, v, v_2, \dots, v_1)$ and $(v, u_1, \dots, u_2, v)$ into the single cycle $(v_1, \dots, v, u_1, \dots, u_2, v, v_2, \dots, v_1)$, and go to Step 2.

The complexity of the Hierholzer algorithm is $O(|E|)$. An alternative algorithm was later proposed by Fleury [17]:

Step 1. Consider a vertex v_i. Starting with v_i, traverse an edge (v_i, v_j) that is not a bridge (an edge whose removal disconnects the graph), unless it is an end-edge. Remove (v_i, v_j) from the graph.

Step 2. If all edges have been removed, stop.

Step 3. Set $v_i := v_j$, and go to Step 1.

The main difficulty in Fleury's algorithm is to determine whether an edge is a bridge. If the Tarjan [31] $O(|E|)$ algorithm is used to this end, then this algorithm can run in $O(|E|^2)$ time. Fleury was apparently unaware of Hierholzer's more efficient algorithm. In his paper, he wrote, "Nobody has yet shown how to determine, without trial and error, a continuous traversal without repetitions" (Fleury [17, pp. 257–258], author's translation). Other graph traversal algorithms are provided in Fleischner [16].

3.3 ▪ Variants

The CPP variants are mostly inspired from their counterparts for the TSP and are typically NP-hard.

3.3.1 ▪ The generalized CPP

In the Generalized CPP, introduced by Dror and Haouari [10], the edge set is partitioned into $E = \{E_1, \dots, E_p\}$. The aim is to determine a least-cost cycle containing at least one edge from each element of the partition. The authors do not actually solve this problem, but it can readily be transformed into a generalized TSP, and then into an asymmetric TSP using the technique proposed by Blais and Laporte [4]. This is achieved by first replacing each edge by two opposite arcs, one for each traversal direction, and then replacing each such arc by a vertex. All vertices of the transformed graph are then connected by arcs whose costs are those of shortest paths on the original graph. This gives rise to an asymmetric Generalized TSP which is then transformed into an asymmetric TSP over the same number of vertices, as suggested by Noon and Bean [29].

3.3.2 ▪ The cumulative CPP

The *Cumulative CPP* (CCPP) is rooted in the Cumulative TSP which is also known as the Traveling Repairman Problem (Afrati et al. [1]), the Deliveryman Problem (Lucena [27], Fischetti, Laporte, and Martello [15]), and the Minimum Latency Problem (Blum et al. [5]). In the Cumulative TSP, a salesman based at a depot must complete a Hamiltonian tour over a set of customers so as to minimize the sum of arrival times at the customer locations. The CPP counterpart of this problem for arc routing was introduced by van Omme [32]. The problem is defined on an undirected graph, and one vertex is designated as the depot. The aim is to determine an Eulerian cycle starting and ending at the depot and minimizing the sum of service completions of all edges, which implies that if an edge is traversed several times, it is advantageous to service it during the first traversal. Note that in this definition, the return time to the depot is not included in the objective function whenever some deadheading takes place after servicing the last edge. This problem is strongly NP-hard and is reducible to a version of the Cumulative TSP. However, when the problem is defined on a linear or on a circular graph, it can be solved in $O(n^2)$ time. The CCPP arises naturally in several applications such as snow removal.

The author has developed eight models which were theoretically and empirically compared. The best of these, which we now describe, was retained to conduct the final computational experiments. As for the CPP, the cumulative problem is defined on an undirected graph $G=(V,E)$, where $|V|=n$ and $|E|=m$. This graph is expanded into an auxiliary mixed graph $\bar{G}=(\bar{V},\bar{E}\cup\bar{A})$. In $\bar{G}$, the edges of E are "exploded," i.e., represented as if they were mutually disjoint. Let $\bar{E}$ be this set of exploded edges, and note that the number of vertices incident to the edges of $\bar{E}$ is equal to $2n$. Let $\tilde{V}$ be the set of these vertices, and let $\bar{V}=\tilde{V}\cup\{s,t\}$, where s is a starting depot and t is an ending depot. To define $\bar{A}$, create arcs between the vertices of $\tilde{V}$ incident to different edges, from s to the two vertices of $\tilde{V}$ corresponding to the depot in G, and from the vertices of $\tilde{V}$ to t. This construction is illustrated in the small graph of Figure 3.1.

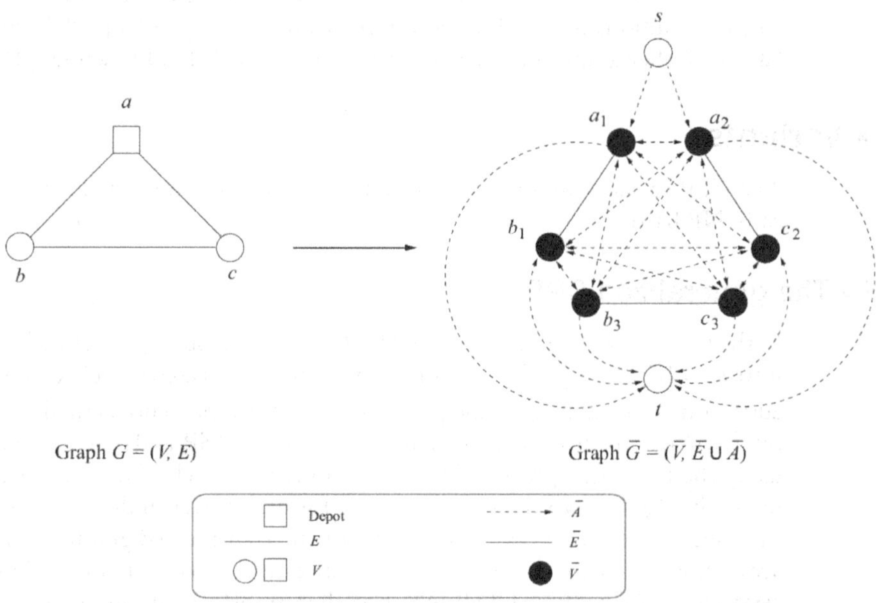

Figure 3.1. *Construction of the graph $\bar{G}$, from van Omme [32].*

To define the cost structure, denote by $*v$ the vertex of V corresponding to vertex $v \in \bar{V}$. Thus, in Figure 3.1 $*a_1 = a$. Given $v \in \tilde{V}$, the vertex $\bar{v} \in \tilde{V}$ is the other vertex of edge $(v,\bar{v}) \in \bar{E}$. For example, in the graph of Figure 3.1, $\bar{a}_1 = b_1$. Let c_{uv} be the cost of edge $(u,v) \in E$, and let d_{uv} be the cost of a shortest chain between u and v in G. For any edge $(i,j) \in \bar{E}$, let $\bar{c}_{ij} = d_{*i,*j} + c_{*j,*\bar{j}}$.

The variables of the model are defined as follows:

(1) If $i,j \in \tilde{V}$, let

$$y_{ij}^k = \begin{cases} 1 & \text{if arc } (i,j) \text{ or arc } (j,i) \text{ is traversed in position } k \text{ (which} \\ & \text{means that edge } (*j,*\bar{j}) \text{ is visited in position } k \text{ in } G), \\ 0 & \text{otherwise.} \end{cases}$$

(2) Otherwise, let

$$y_{sv}^1 = \begin{cases} 1 & \text{if } v \in \tilde{V} \text{ is the first vertex visited after } s \text{ in } \bar{G}, \\ 0 & \text{otherwise,} \end{cases}$$

$$y_{vt}^{m+1} = \begin{cases} 1 & \text{if } v \in \tilde{V} \text{ is the last vertex visited before } t \text{ in } \bar{G}, \\ 0 & \text{otherwise.} \end{cases}$$

The model is then

$$\text{(CCPP)} \qquad \text{minimize} \sum_{k=1}^{m} \sum_{(i,j)\in\bar{A}} (m-k+1)\bar{c}_{ij} y_{ij}^k$$

$$\text{(3.8)} \qquad \text{s.t.} \quad y_{sv}^1 + \sum_{k=2}^{m} \Big(\sum_{(v,i)\in\bar{A}} y_{vi}^k + \sum_{(j,v)\in\bar{A}} y_{jv}^k \Big) + j_{vt}^{m+1} = 1, \qquad v \in \bar{V},$$

$$\text{(3.9)} \qquad \sum_{(i,j)\in\bar{A}} y_{ij}^k = 1, \qquad k \in \{1,\dots,m+1\},$$

$$\text{(3.10)} \qquad \sum_{(p,i)\in\bar{A}} y_{pi}^k = \sum_{(\bar{i},q)\in\bar{A}} y_{\bar{i}q}^{k+1}, \qquad i \in \bar{V}, k \in \{1,\dots,m\},$$

$$\text{(3.11)} \qquad y_{ij}^k = 0 \text{ or } 1, \qquad i,j \in \bar{V}, k \in \{1,\dots,m\}.$$

In this formulation, constraints (3.8) force every edge of E to be serviced. Constraints (3.9) state that for each position k a shortest path must be visited. Constraints (3.10) are flow conservations constraints which guarantee the creation of a tour.

This model contains $O(m^3)$ variables and $O(m^2)$ constraints. Even on relatively small graphs, these numbers can be quite large. However, van Omme notes that most of the variables are equal to zero in an optimal solution. He therefore proposes several preprocessing operations to eliminate many of the variables.

The problem was solved by means of an implicit enumeration heuristic. Several versions of a branch-and-cut algorithm were also developed and compared. Among six families of cuts developed, three were retained: odd set inequalities, even set inequalities, and generalized co-circuit inequalities. Two branch-and-price-and-cut algorithms were also developed and applied to a streamlined version of the CCPP model (with a reduced number of variables), including this time five families of cuts.

The algorithms were tested on five graphs with $10 \le n \le 20$ and $30 \le m \le 45$. Results indicate that the branch-and-price-and-cut algorithm is two to four times faster than the branch-and-cut algorithm, and two to 133 times faster than CPLEX applied to the streamlined CCPP model.

3.3.3 • The hierarchical CPP

In some applications such as snow plowing (Alfa and Liu [2], Haslam and Wright [22]), garbage collection (Bodin and Kursh [6]) and flame cutting (Manber and Israni [28]), constraints may be imposed on the order in which some *clusters* of edges (or arcs) must be visited. Here we restrict our attention to undirected graphs. This problem, called the *Hierarchical CPP* (HCPP), was introduced by Dror, Stern, and Trudeau [11]. It is defined on a graph $G=(V,E)$, where V includes a depot d. The edges are partitioned into $\{E_1,\dots,E_p\}$, where p is the number of clusters. An order relation $\prec$ is imposed on the elements of the partition. This means that if $E_h \prec E_k$, then all edges of E_h must be serviced before any edge of E_k. As shown by Dror, Stern, and Trudeau, the HCPP is NP-hard. However, when the subgraphs $G_k=(V_k,E_k)$ induced by the edges of E_k are connected for each k and a total order relation $E_1 \prec \dots \prec E_p$ is prespecified, then the problem can be solved in polynomial time through the construction of an auxiliary graph $\bar{G}$. Using such a transformation, Dror, Stern, and Trudeau developed an $O(p|V|^5)$ algorithm. This complexity was later reduced to $O(p^3|V|^3)$ by Ghiani and Improta [19] and to $O(p|V|^4)$ by Korteweg and Volgenant [24].

3.3.3.1 ▪ The Dror, Stern, and Trudeau algorithm

Dror, Stern, and Trudeau [11] proved a necessary and sufficient condition for a feasible HCPP solution to exist. Let G_k be the subgraph induced by the edges of E_k, and let F_k be the subgraph induced by the edges of $E_1 \cup \dots \cup E_k$. Then there exists a feasible HCPP solution if and only if F_k is connected for $k=1,\dots,p$.

The graph $\bar{G}$ is constructed as follows. Let $\bar{V}_1=\bar{V}_{p+1}=\{d\}$. For $k=2,\dots,p$, let $\bar{V}_k$ be the set of *entry vertices* of G_k, i.e., the vertices of V_k incident to an edge of $E_1 \cup \dots \cup E_{k-1}$.

Step 1. Construct a $(p+1)$-bipartite graph $\bar{G}$ with layers of vertices $\bar{V}_1,\dots,\bar{V}_{p+1}$. Define an arc (u,v) from every $u \in \bar{V}_k$ and $v \in \bar{V}_{k+1}$ for $k=1,\dots,p$. The cost $\bar{c}_{uv}$ of arc (u,v) is equal to that of a shortest Eulerian path from u to v in F_k.

Step 2. Construct a shortest path from the copy of the depot in $\bar{V}_1$ to that of the depot in $\bar{V}_{p+1}$.

Step 3. Connect the paths that correspond to the edges of G belonging to the shortest path in $\bar{G}$, which yields an optimal HCPP solution in G.

To illustrate, consider the graph G depicted in Figure 3.2, with $d=1$, $E_1=\{(1,2),(2,3),(2,4)\}$, $E_2=\{(3,5),(4,6),(4,7),(5,6),(6,7)\}$, and $E_3=\{(1,8),(2,8)\}$, with $E_1 \prec E_2 \prec E_3$. The graph $\bar{G}$ corresponding to G (Figure 3.2) is depicted in Figure 3.3. The optimal solution is given by the shortest path $(1,3,2,1)$ of cost 30 in $\bar{G}$, corresponding to the optimal solution $(1,2,4,2,3,5,6,7,4,6,4,2,8,1)$ in G.

The $O(p|V|^5)$ complexity of the Dror, Stern, and Trudeau algorithm is determined by Step 1, which requires the solution of $O(p|V|^2)$ matching problems of complexity $O(|V|^3)$. Korteweg and Volgenant [24] reduce this complexity to $O(p|V|^4)$ by exploiting the relationship between the solutions of successive Eulerian path computations.

3.3.3.2 ▪ The Ghiani and Improta algorithm

The Ghiani and Improta [19] algorithm solves a single matching problem on an auxiliary graph G^* containing $O(p|V|)$ vertices, and the authors show that this operation is the most complex step of their procedure. Hence, the overall time complexity of their algorithm is $O(p^3|V|^3)$, which can yield significant savings with respect to the previous two algorithms since in practice p is usually small compared with $|V|$.

We now outline this algorithm. In a solution, after the edges of $E_{k-1} (k=2,\dots,p)$ have been serviced, one must traverse a path $\mathscr{P}_k$ in F_{k-1} linking V_{k-1} to V_k in order to

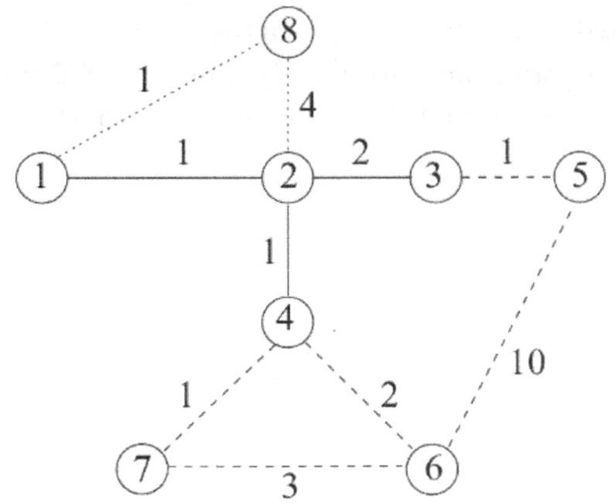

Figure 3.2. *Graph G, from Dror, Stern, and Trudeau [11].*

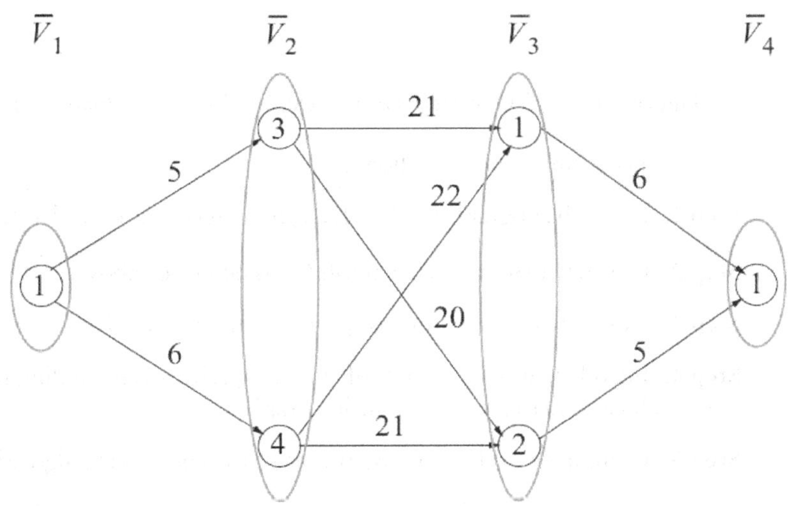

Figure 3.3. *Graph $\bar{G}$ corresponding to graph G (Figure 3.2), from Dror, Stern, and Trudeau [11].*

serve the first edge of E_k. The path $\mathscr{P}_k$ contains a subpath $(v''_{k-1}, \ldots, v'_k)$, where $v''_{k-1} \in V''_{k-1}$, the set of vertices of V_{k-1} incident to at least one edge of $E_1 \cup \cdots \cup E_{k-2} \cup E_k$, and $v'_{k-1} \in V'_{k-1}$, the set of vertices of V_k incident to at least one edge of $E_1 \cup \cdots \cup E_{k-1}$. Let $P = P^L \cup P_1 \cup \cdots \cup P_p$, where P^L is a set of paths $\{(v''_1, \ldots, v''_2), \ldots, (v''_p, \ldots, d)\}$ linking V_k to V_{k+1} in $F_k(k = 1, \ldots, p-1)$ and V_p to d in F_p; P_k is a set of paths in F_k connecting two vertices in $V_k(k = 1, \ldots, p)$.

The graph G^* contains $p+2$ layers $V_0^* = \{d\}$, $V_1^*, \ldots, V_p^*$, $V_{p+1}^* = \{d\}$, where V_k^* contains as many copies of each vertex $v \in V_i$ as the maximum number of paths in $P^L \cup P_k$ incident to v. Each copy v_k of a given vertex $v \in V_k$ is connected to its copies by a zero cost edge and to the copies in V_k^* of the other vertices $v' \in V_i$ by an edge of cost equal to that of a shortest path from v to v' in F_k. In addition, if $v \in V'_k$, v_k is connected to the copies in V_{k-1}^* of vertices v' in V_{k-1}^* by an edge of cost equal to that of a shortest path from v to v' in F_{k-1}, plus (if $k \neq 1$ and $k \neq p+1$) a large constant M. Finally, if $v' \in V''_k$,

v_k is linked to the copies in V^*_{k+1} of the vertices $v' \in V'_{k+1}$ by an edge of cost equal to that of shortest path from v to v' in F_k, plus (if $k \neq 0$ and $k \neq p$) a large constant M. The graph G^* corresponding to G (Figure 3.1) is depicted in Figure 3.4.

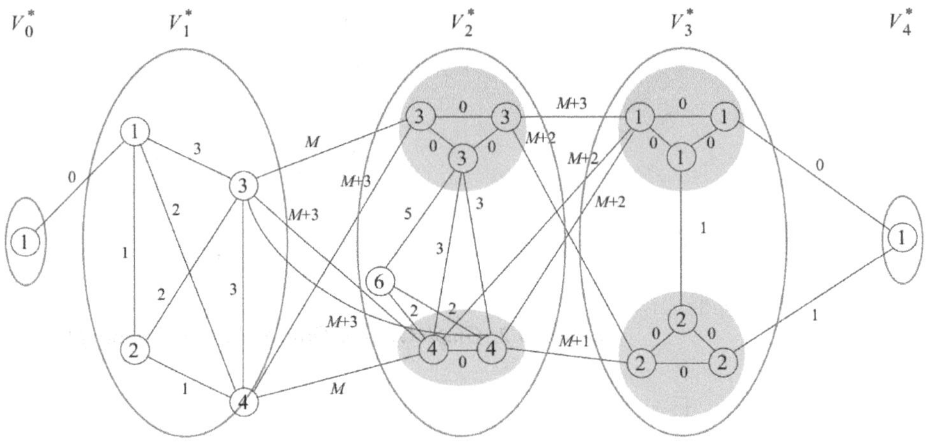

Figure 3.4. *Graph G^* corresponding to graph G (Figure 3.1), from Ghiani and Improta [19].*

The algorithm is then as follows:

Step 1. For each subgraph $G_k (k = 1, \ldots, p)$, determine V'_k and V''_k.

Step 2. Construct the auxiliary graph G^* as described above.

Step 3. Compute a minimum cost perfect matching on G^*.

Step 4. Introduce in G the set P of shortest paths corresponding to the solution found in Step 3. Let G' be the resulting graph.

Step 5. Compute a cycle on G' by means of the end-pairing algorithm.

The complexity of this algorithm is determined as follows. Step 1 requires $O(p|V|^2)$ operations. Step 2 requires $O(p|V|^3)$ operations since it involves the computation of p shortest paths on graphs having at most $|V|$ vertices. The solution of the matching problem in Step 3 requires $O(p^3|V|^3)$ operations since each of the p_2 layers of G^* has at most $3|V|$ vertices. Finally, Step 5 can be executed in $O(|E|)$ time. The overall complexity is therefore that of Step 3.

3.3.3.3 ▪ Transformation into a Rural Postman Problem

Cabral et al. [7] consider a version of the HCPP in which the vertex set contains a depot d, the edge set is partitioned into $E = \{E_1, \ldots, E_p\}$, and a total relation $\prec$ is imposed on the elements of the partition, but these may not be connected. Each edge e has three associated lengths (traversal times): t_e^3 is the time of servicing e, t_e^2 is the time of traversing e if it has not yet been serviced, and t_e^1 is the time of traversing e if it has already been serviced. Typically $t_e^3 \geq t_e^2 \geq t_e^1$, but there exist applications, such as snow removal, where $t_e^2 > t_e^3$ (often $t_e^2 = \infty$), in which case feasibility is not guaranteed. The authors consider two different objectives. The first is a *makespan* objective which consists of minimizing the return time at the depot once all edges have been traversed. The second is a *hierarchical* objective in which the aim is to minimize $\sum_{k=1}^{p} M_k T_k$, where T_k is the time at which all edges of E_k have been serviced and $M_1 \gg \cdots \gg M_p = 1$, which means that the edges of E_k must be serviced before those of $E_{k+1} (k = 1, \ldots, p-1)$. In practice, the notation "$\gg$" means that in any feasible solution, the inequality $M_k T_k > \sum_{h=k+1}^{p} M_h T_h$ must be satisfied for $k = 1, \ldots, p-1$. This inequality is satisfied if it holds when T_k is replaced by a lower bound and each T_h is replaced by an upper bound.

The authors show that for either objective, the HCPP can be cast as a Rural Postman Problem (RPP) (see Chapter 5) through a graph transformation, the advantage being that good algorithms are readily available for the RPP (see, e.g., Ghiani and Laporte [20]). The transformation works as follows. Let $\alpha(e)$ denote the cluster of edge $e \in E$. Define the modified edge lengths

$$\bar{t}_e^k = \begin{cases} t_e^1 & \text{if } \alpha(e) \leq k, \\ t_e^2 & \text{if } \alpha(e) > k, \end{cases}$$

and let s_{ij}^k be the length of a shortest path in G between vertices i to j with respect to the modified edge lengths. The problem is solved on a $(p+2)$-layered multigraph $\tilde{G} = (\tilde{V}, \tilde{E})$. The vertex set $\tilde{V}$ is partitioned into $\tilde{V} = \{\tilde{V}_0, \ldots, \tilde{V}_{p+1}\}$, where $\tilde{V}_0 = \tilde{V}_{p+1} = \{d\}$, and $\tilde{V}_h = V_h$. For any two vertices i and j belonging to the same layer k $(k = 1, \ldots, p)$, define a nonrequired edge of length $\sum_{h=1}^{k} M_h s_{ij}^k$; if $(i,j) \in E_k$, also define a required edge of length $\sum_{h=1}^{k} M_h t_e^3$ between i and j. For any two vertices i and j with $i \in \tilde{V}_{k-1}$ and $j \in \tilde{V}_k$ $(k = 1, \ldots, p)$, define a nonrequired edge of length $\sum_{h=1}^{k-1} M_h s_{ij}^k + M$, where $M \gg M_1$. Finally, define a nonrequired edge of length M between the vertices of $\tilde{V}_p$ and the vertex of $\tilde{V}_{p+1}$, as well as a required edge of cost 0 linking the two copies of the depot. Figure 3.5 depicts this graph construction on a small example.

Solving the HCPP can become rather time consuming as the instance size grows. Therefore, in order to obtain good solutions quickly on larger instances, the authors propose two heuristics. The best one is a decomposition heuristic which works as follows. For $k = 1, \ldots, p$, using the RPP transformation, solve the problem on the subgraph induced by the edges of $E_1 \cup \cdots \cup E_k$ and declare all traversed edges already serviced for the following iterations. The authors performed computational experiments on randomly generated instances with $p = 3$ and $30 \leq |V| \leq 150$. They could easily solve most instances within 15 minutes with the makespan objective, but they could not solve any instance with $|V| \geq 60$ within the same time under the hierarchical objective. However, they showed that using either objective does not make much difference as far as the objective function value is concerned. They also showed that the decomposition heuristic is quick and yields very small optimality gaps.

3.3.4 ▪ The CPP with time windows

The undirected CPP with time windows is defined on a graph $G = (V, E)$, where a time window $[\ell_e, u_e]$ is imposed on the start and completion of service of every edge $e \in E$. This NP-hard problem was introduced by Aminu and Eglese [3]. It arises naturally in several arc routing applications related to street sweeping and snow removal, for example. In several cities, parking interdictions are enforced on some streets during certain periods in order to allow such operations to take place. These no-parking periods define the time windows.

The authors have developed two exact *constraint programming* (CP) algorithms for the CPP with time windows. The first one operates directly on G and assigns a position

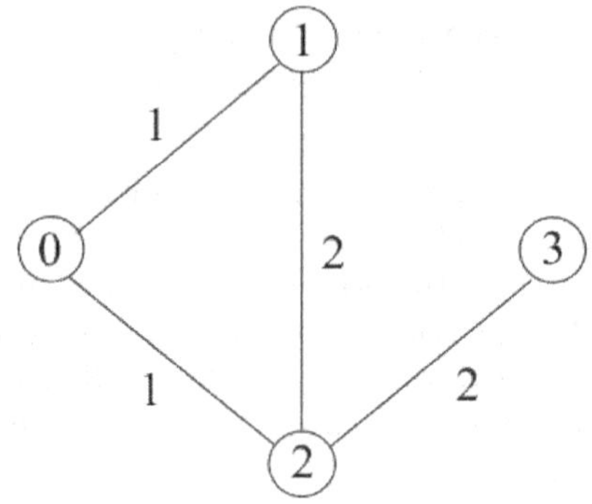

(a) Graph G showing edge priorities

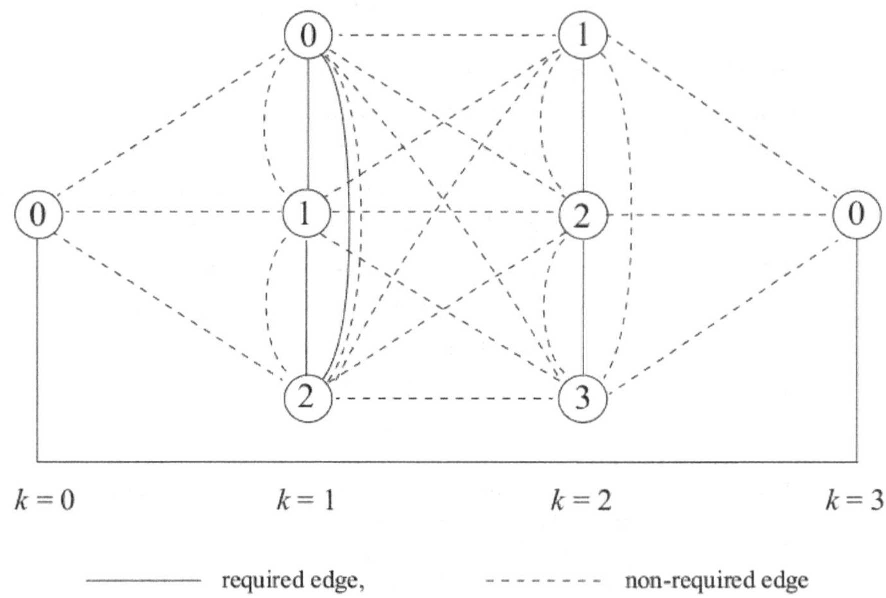

(b) Graph $\tilde{G}$

Figure 3.5. *Transformation of the HCPP into an RPP, from Cabral et al. [7].*

to each edge in the solution. The CP mechanism operates in a branch-and-bound fashion and gradually restricts the domains of the position variables. The second CP implementation is carried out on a transformed graph $\bar{G}$ whose vertex set $\bar{V}$ is the union of $|E|$ clusters of two vertices, one for each of the two end vertices of every $e \in E$. This graph transformation is a simplification of the procedure proposed by Pearn, Assad, and Golden [30], but it replaces each edge with two vertices instead of three. Arcs are then created between the vertices of $\bar{G}$ to represent shortest paths in G, which correspond to costs. Solving a clustered TSP on $\bar{G}$ (see, e.g., Laporte and Palekar [25]) solves the CPP with time windows. The authors have compared two CP implementations on graphs containing between seven and 69 vertices and have shown that the second CP implementation is superior to the first one.

Bibliography

[1] F. AFRATI, S. COSMADAKIS, S. PAPADIMITRIOU, S. PAPAGEORGIOU, AND N. PAPAKOSTANTINOU, *The complexity of the traveling repairmen problem*, RAIRO Informatique Théorique et Applications, 20 (1986), pp. 79–87.

[2] A. S. ALFA AND D. Q. LIU, *Postman routing problem in a hierarchical network*, Engineering Optimization, 14 (1988), pp. 127–138.

[3] U. F. AMINU AND R. W. EGLESE, *A constraint programming approach to the Chinese postman problem with time windows*, Computers & Operations Research, 33 (2006), pp. 3423–3431.

[4] M. BLAIS AND G. LAPORTE, *Exact solution of the generalized routing problem through graph transformations*, The Journal of the Operational Research Society, 54 (2003), pp. 906–910.

[5] A. BLUM, P. CHALSANI, D. COPPERSMITH, B. PULLEYBLANK, P. RAGHAVAN, AND M. SUDAN, *The minimum latency problem*, in STOC'94, Proceedings of the 26th Annual ACM Symposium on the Theory of Computing, ACM, New York, 1994, pp. 163–171.

[6] L. D. BODIN AND S. J. KURSH, *A computer-assisted system for the routing and scheduling of street sweepers*, Operations Research, 26 (1978), pp. 525–537.

[7] E. A. CABRAL, M. GENDREAU, G. GHIANI, AND G. LAPORTE, *Solving the hierarchical Chinese postman problem as a rural postman problem*, European Journal of Operational Research, 155 (2004), pp. 44–50.

[8] N. CHRISTOFIDES, *Worst-case analysis of a new heuristic for the travelling salesman problem*, Tech. Report 388, Graduate School of Industrial Administration, Carnegie-Mellon University, Pittsburgh, PA, 1976.

[9] U. DERIGS AND A. METZ, *Solving (large scale) matching problems*, Mathematical Programming, 50 (1991), pp. 113–121.

[10] M. DROR AND M. HAOUARI, *Generalized Steiner problems and other variants*, Journal of Combinatorial Optimization, 4 (2000), pp. 415–436.

[11] M. DROR, H. STERN, AND P. TRUDEAU, *Postman tour on a graph with precedence relation on arcs*, Networks, 17 (1987), pp. 283–294.

[12] J. EDMONDS, *Paths, trees and flowers*, Canadian Journal of Mathematics, 17 (1965), pp. 449–467.

[13] J. EDMONDS AND E. L. JOHNSON, *Matching, Euler tours and the Chinese postman*, Mathematical Programming, 5 (1973), pp. 88–124.

[14] L. EULER, *Solutio problematis ad geometriam situs pertinentis*, Commentarii Academiae Scientarum Petropolitanae, 8 (1736), pp. 128–140.

[15] M. FISCHETTI, G. LAPORTE, AND S. MARTELLO, *The delivery man problem and cumulative matroids*, Operations Research, 41 (1993), pp. 1055–1064.

[16] M. FLEISCHNER, *Eulerian Graphs and Related Topics*, Annals of Discrete Mathematics 2, North-Holland, Amsterdam, 1991.

[17] M. FLEURY, *Deux problèmes de géométrie de situation*, Journal de Mathématiques Élémentaires, 2 (1883), pp. 257–261.

[18] Z. GALIL, S. MICALI, AND H. GABOW, *An O(EV* log *V) algorithm for finding a maximal weighted matching in general graphs*, SIAM Journal on Computing, 15 (1986), pp. 120–130.

[19] G. GHIANI AND G. IMPROTA, *An algorithm for the hierarchical Chinese postman problem*, Operations Research Letters, 26 (2000), pp. 27–32.

[20] G. GHIANI AND G. LAPORTE, *A branch-and-cut algorithm for the undirected rural postman problem*, Mathematical Programming, 87 (2000), pp. 467–481.

[21] M. GUAN, *Graphic programming using odd and even points*, Chinese Mathematics, 1 (1962), pp. 273–277.

[22] E. HASLAM AND J. R. WRIGHT, *Application of routing technologies to rural snow and ice control*, Transportation Research Records, 1304 (1991), pp. 202–211.

[23] C. HIERHOLZER, *Uber die Möglichkeit, einen Linienzug ohne Wiederholung und ohne Unterbrechnung zu umfahren*, Mathematische Annalen, 6 (1873), pp. 30–32.

[24] P. KORTEWEG AND T. VOLGENANT, *On the hierarchical Chinese postman problem with linear ordered classes*, European Journal of Operational Research, 169 (2006), pp. 41–52.

[25] G. LAPORTE AND U. PALEKAR, *Some applications of the clustered traveling salesman problem*, The Journal of the Operational Research Society, 53 (2002), pp. 972–976.

[26] E. L. LAWLER, *Combinatorial Optimization: Networks and Matroids*, Holt, Reinhart & Winston, New York, 1976.

[27] A. LUCENA, *Time-dependent traveling salesman problem—the deliveryman case*, Networks, 20 (1990), pp. 753–763.

[28] U. MANBER AND S. ISRANI, *Pierce point minimization and optimal torch path determination in flame cutting*, Journal of Manufacturing Systems, 3 (1984), pp. 81–89.

[29] C. E. NOON AND J. C. BEAN, *An efficient transformation of the generalized traveling salesman problem*, INFOR, 31 (1993), pp. 39–44.

[30] W.-L. PEARN, A. A. ASSAD, AND B. L. GOLDEN, *Transforming arc routing into node routing problems*, Computers & Operations Research, 14 (1987), pp. 285–288.

[31] R. E. TARJAN, *A note on finding the bridges of a graph*, Information Processing Letters, 1 (1974), pp. 160–161.

[32] N. VAN OMME, *Le problème du postier chinois cumulatif*, Ph.D. thesis, Département de mathématiques et de génie industriel, Polytechnique Montréal, Montréal, Canada, 2011.

Chapter 4

The Chinese Postman Problem on Directed, Mixed, and Windy Graphs

Ángel Corberán
Isaac Plana
José María Sanchis

4.1 ▪ Introduction

Let $G=(V,E,A)$ be a mixed graph, consisting of a set of vertices V, a set of (undirected) edges E, and a set of (directed) arcs A. For simplicity, the term *link* will be used to refer to both edges and arcs indistinctly. Note that if $E=\emptyset$, G is a directed graph (all the links are arcs that must be traversed in a specified direction), while if $A=\emptyset$, it is an undirected graph (all the links are edges that can be traversed in both directions at the same cost). Usually, the links in undirected, directed, and mixed graphs have a nonnegative cost associated with them. If G is an undirected graph and there are two costs associated with each edge, representing the cost of traversing it in each possible direction, G is called a windy graph. Note that, since any arc (i,j) with cost c_{ij} can be modeled as an edge joining i and j with the same cost from i to j and infinite cost in the opposite direction, a mixed graph can be considered as a special case of a windy graph.

The *Chinese Postman Problem* (CPP) consists of finding a minimum cost tour (a closed walk) traversing every link of G at least once. This well-known problem (Guan [20]) was shown to be solvable in polynomial time for undirected and directed graphs. If the CPP is defined over a mixed graph $G=(V,E,A)$, the problem of traversing all the edges and arcs in G at total minimum cost is called the *Mixed Chinese Postman Problem* (MCPP). Papadimitriou [32] showed the NP-hardiness of the MCPP. If $G=(V,E)$ is a windy graph, the problem of traversing all the edges in G at total minimum cost is called the *Windy Postman Problem* (WPP). Since the MCPP is a particular case of the WPP, it is easy to see that the WPP is also an NP-hard problem (Brucker [2], Guan [21]).

If an (undirected, directed, mixed, or windy) graph G is *Eulerian*, there is a tour passing through each link of G exactly once. Obviously, this tour is optimal. Furthermore, this (Eulerian) tour can be easily computed (see, for instance, Eiselt, Gendreau, and Laporte [14]), and hence graph G can itself be considered as the CPP solution. If an (undirected, directed, mixed, or windy) graph G is not Eulerian, the CPP can be formulated as the problem of finding a set of copies of links at minimum cost, such that when added to G, an Eulerian graph is obtained. As before, the "augmented" graph formed by G, plus the added copies of the links, can be considered as the solution of the problem.

For the undirected CPP, an Eulerian graph can be obtained from G by adding edges to match odd-degree vertices (Christofides [4], Edmonds and Johnson [13]). This can be done by finding a minimum cost perfect matching (see Chapter 3).

In this chapter we study the directed CPP and the two NP-hard versions of the CPP, the MCPP and the WPP, as well as some variants.

4.2 ▪ The directed Chinese Postman Problem

Given a strongly connected directed graph $G = (V,A)$, with nonnegative costs c_{ij} for each $(i,j) \in A$, the *Directed Chinese Postman Problem* (DCPP) is the problem of finding a minimum cost tour passing through each arc in A at least once.

In order to formulate the problem we introduce the following notation. For each vertex i, let $\delta^+(i)$ denote the set of arcs leaving i, $\delta^-(i)$ the set of arcs entering i, and $s_i = |\delta^-(i)| - |\delta^+(i)|$. For each arc $(i,j) \in A$, let x_{ij} be the number of copies of each arc that must be added to G in order to obtain an Eulerian graph. Edmonds and Johnson [13] showed that the DCPP can be formulated as the following linear program (LP):

$$\text{(4.1)} \quad \text{(DCPP)} \quad \min \sum_{(i,j)\in A} c_{ij} x_{ij}$$

$$\text{(4.2)} \quad \sum_{(i,j)\in\delta^+(i)} x_{ij} - \sum_{(j,i)\in\delta^-(i)} x_{ji} = s_i \qquad \forall i \in V,$$

$$\text{(4.3)} \quad x_{ij} \geq 0 \qquad \forall (i,j) \in A.$$

This corresponds to a minimum cost flow problem on G with supplies s_i for vertices i with $s_i > 0$, and demands $-s_i$ for vertices i with $s_i < 0$ (Liebling [26], Edmonds and Johnson [13]).

4.3 ▪ The mixed Chinese Postman Problem

In this section we give a formal definition of the MCPP, present some notation and two different formulations, and discuss its heuristic and exact solution.

Given a strongly connected mixed graph $G = (V,E,A)$ with a vertex set V, an edge set E, an arc set A, and a nonnegative cost c_e for each $e \in E \cup A$, the MCPP consists of finding a minimum-cost tour passing through each link $e \in E \cup A$ at least once.

Given $S \subset V$, $\delta^+(S) = \{(i,j) \in A : i \in S, j \in V \setminus S\}$, $\delta^-(S) = \{(i,j) \in A : i \in V \setminus S, j \in S\}$, $\delta(S)$ denotes the set of edges with exactly one endpoint in S, and $\delta^*(S) = \delta(S) \cup \delta^+(S) \cup \delta^-(S)$. Given a vertex i, d_i^- (indegree) denotes the number of arcs entering i, d_i^+ (outdegree) is the number of arcs leaving i, and d_i (degree) is the number of links incident with i. Note that $d_i = |\delta^*(\{i\})|$. A mixed graph $G = (V,E,A)$ is called *even* if all its vertices have even degree, it is called *symmetric* if $d_i^- = d_i^+$ for each vertex i, and it is said to be *balanced* if, given any subset S of vertices, the difference between the number of arcs directed from S to $V \setminus S$, $|\delta^+(S)|$, and the number of arcs directed from $V \setminus S$ to S, $|\delta^-(S)|$, is no greater than the number of (undirected) edges joining S and $V \setminus S$, $|\delta(S)|$.

It is well known that a mixed graph G is Eulerian if and only if G is even and balanced (Ford and Fulkerson [16]). Notice that if G is even and symmetric, then G is also balanced (and Eulerian). Moreover, if G is even, the MCPP can be solved exactly in polynomial time (Edmonds and Johnson [13]). The procedure, as described by Minieka [27], works as follows:

Even MCPP Algorithm

(1) Given an even and strongly connected mixed graph, $G = (V,E,A)$, let A_1 be the set of arcs obtained by arbitrarily assigning a direction to the edges in E and with the same costs. Compute $s_i = d_i^- - d_i^+$ for each vertex i in the directed graph $(V, A \cup A_1)$. A vertex i with $s_i > 0$ ($s_i < 0$) will be considered as a source (sink) with supply (demand) s_i ($-s_i$). Note that as G is an even graph, all supplies and demands are even numbers (zero is considered an even number).

(2) Let A_2 be the set of arcs in the opposite direction to those in A_1 and with the costs of the corresponding edges, and let A_3 be a set of arcs parallel to those of A_2 at zero cost.

(3) To satisfy the demands s_i of all the vertices, solve a minimum cost flow problem in the graph $(V, A \cup A_1 \cup A_2 \cup A_3)$, in which each arc in $A \cup A_1 \cup A_2$ has infinite capacity and each arc in A_3 has capacity 2. Let x_{ij} be the optimal flow.

(4) For each arc (i,j) in A_3 do: If $x_{ij} = 2$, then orient the corresponding edge in G from i to j (the direction, from j to i, assigned to the associated edge in step 1 was "wrong"); if $x_{ij} = 0$, then orient the corresponding edge in G from j to i (in this case, the orientation given in step 1 was "right"). Note that the case $x_{ij} = 1$ is impossible, as all flow values through arcs in A_3 are even numbers.

(5) Augment G by adding x_{ij} copies of each arc in $A \cup A_1 \cup A_2$. The resulting graph is even and symmetric.

4.3.1 • Heuristic algorithms

When the mixed graph $G = (V,E,A)$ is not even, the above algorithm is the basis for two heuristics suggested by Edmonds and Johnson [13] and developed and improved by Frederickson [17]. Algorithm MIXED1 would be equivalent to first transforming G into an even graph and then applying the previous *Even MCPP Algorithm*. Specifically,

Algorithm MIXED1

(1 Evendegree) Let $G = (V,E,A)$ be a strongly connected mixed graph. Ignoring arc directions, solve a minimum-cost matching of odd-degree vertices and augment G by adding all links used for the matching solution. The resulting graph $G' = (V,E',A')$ is an even graph.

(2 Inoutdegree) With supplies and demands computed in graph (V,A'), solve a minimum cost flow problem in G' to obtain a symmetric graph $G'' = (V,E'',A'')$. (Frederickson points out that after this step some vertices in G'' may have odd degree.)

(3 Evenparity) Let V_O be the set of odd-degree nodes in G''. Identify cycles consisting of alternating paths in $A'' \setminus A$ and E'', with each path starting and ending at vertices in V_O. The paths in $A''\setminus A$ will be formed without considering the arc directions. As the cycles are covered, their arcs are either duplicated or deleted and their edges are directed, so the resulting graph becomes even, while maintaining the symmetry for each vertex.

Steps 1 and 2 above were first proposed by Edmonds and Johnson [13], who show that there is a minimum cost solution to the flow problem in step 2 which preserves the property that each vertex is of even degree. But, as noted by Frederickson [17], this argument merely proves the existence of such a solution. Step 3 above is a simple linear-time algorithm proposed by Frederickson for performimg this task.

Algorithm MIXED2 can be considered as the reverse version of MIXED1. It first solves a minimum-cost flow problem in G, with supplies and demands computed in graph (V,A), to obtain a symmetric graph (V,E',A'). In a second step, it solves the (undirected) CPP on each connected component of subgraph (V,E') to finally obtain an even and symmetric graph.

Frederickson [17] showed that both algorithms, MIXED1 and MIXED2, have a worst-case ratio of 2, but the *Mixed Algorithm*, which consists of applying both algorithms and selecting the best tour obtained, has a worst-case ratio of 5/3. Hertz and Mittaz [23] provide some illustrations and other details on the above heuristic procedures.

Twenty years later, Raghavachari and Veerasamy [34] proposed a modification to Frederickson's Mixed Algorithm with a better worst-case ratio of 3/2:

Modified Mixed Algorithm

- Given $G = (V,E,A)$, solve a minimum-cost flow problem in G with supplies and demands computed in the graph (V,A). The resulting graph is (V,E',A'), where $E' \subseteq E$ are the edges of G that were not oriented (used) by the flow solution and $A' \supseteq A$ are the arcs that satisfy indegree equal to outdegree at each vertex.
- Before running Evendegree of MIXED1 algorithm, reset the costs of all arcs and edges in A' to 0, forcing Evendegree of MIXED1 to duplicate edges and arcs in A' whenever possible.
- Use the original costs for the rest of the MIXED1 algorithm.
- There are no changes in the MIXED2 algorithm.

In addition to its good theoretical behavior, the Modified Mixed Algorithm also shows good performance in practice, especially in graphs with a medium-high percentage of arcs, as was shown by Corberán, Martí, and Sanchis [6]. The following table, taken from [6], reports the percentage deviation from a lower bound (obtained with the cutting plane-algorithm described by Corberán, Romero, and Sanchis [11]) of the three constructive methods in three sets of randomly generated instances. Each set contains 75 instances of 50, 100, and 200 vertices, respectively, and a number of links randomly chosen between $2|V|$ and $3|V|$. Each set is divided into three groups of 25 instances according to the percentage of generated arcs: 30%, 50%, and 70%. The average number of links for each set is shown in the column *Links*.

The computational results, obtained in negligible computing times, show that Modified MIXED1 is significantly better than the original MIXED1 algorithm, except for those instances with a lower percentage of arcs. Moreover, its performance improves slightly when the percentage of arcs increases, while in the original MIXED1 the quality of the solutions clearly deteriorates as this percentage increases. On the other hand, MIXED2 presents the worst results, except for those instances with a large number of arcs.

Table 4.1. *Computational comparison of constructive heuristics for the MCPP.*

$\lvert V \rvert$	Links	Arcs	Mixed 1 Gap (%)	Mixed 2 Gap (%)	M. Mixed 1 Gap (%)
50	132	30%	1.93	16.63	3.18
		50%	7.10	20.41	3.15
		70%	11.25	11.18	1.84
100	257	30%	2.71	17.99	2.75
		50%	6.93	19.83	2.51
		70%	11.10	10.16	1.83
200	510	30%	3.14	18.92	3.39
		50%	7.62	19.67	2.62
		70%	11.15	9.43	2.07

A greedy randomized adaptive search procedure heuristic for the MCPP is also presented by Corberán, Martí, and Sanchis [6]. Other heuristics and improvement procedures for the MCPP are described in Pearn and Chou [33] and Yaoyuenyong, Charnsethikul, and Chankong [41].

4.3.2 ▪ Problem formulations

The MCPP, as well as all other arc routing problems defined on mixed graphs, can be formulated in two different ways, depending on whether only one or two variables associated with each edge of the graph are used. In the first case, the only variable x_e associated with edge e represents the number of times this edge is traversed (in any direction) by the solution. In the second case, two variables per edge are used, representing the number of times it is traversed in the corresponding direction. Both formulations have been widely used in the literature and have been compared from the theoretical and computational point of view in Corberán, Mota, and Sanchis [8].

4.3.2.1 ▪ Formulation F1

Formulation F1 is based on the necessary and sufficient condition given by Ford and Fulkerson [16] for a mixed graph to be Eulerian. Let x_e represent the number of times a link $e \in E \cup A$ appears in the tour. For a subset of links F, $x(F) = \sum_{e \in F} x_e$. Thus, the MCPP can be formulated as

$$\text{(4.4)} \quad \text{(MCPP-F1)} \quad \min \sum_{e \in E} c_e x_e + \sum_{(i,j) \in A} c_{ij} x_{ij}$$

$$\text{(4.5)} \quad \text{s.t.} \quad x(\delta^*(i)) \equiv 0 \bmod 2 \qquad \forall i \in V,$$

$$\text{(4.6)} \quad x(\delta^+(S)) - x(\delta^-(S)) \leq x(\delta(S)) \qquad \forall S \subset V,$$

$$\text{(4.7)} \quad x_e \geq 1 \text{ and integer} \qquad \forall e \in E \cup A,$$

where (4.7) implies that all the links are in the solution, (4.5) ensures that the resulting graph will be even, and (4.6) ensures that the balanced-set conditions will be satisfied. This formulation was proposed by Nobert and Picard [30] and was also used in Corberán, Romero, and Sanchis [11] and Corberán, Mejía, and Sanchis [7].

Note that constraints (4.5) are not linear, and there is no linear expression of them without using integer variables. The effect of constraints (4.5) can be partially achieved by the following odd-cut inequalities (see Corberán, Romero, and Sanchis [11]):

$$x(\delta^*(S)) \geq |\delta^*(S)|+1 \quad \forall S \subset V \text{ with } |\delta^*(S)| \text{ odd.} \tag{4.8}$$

However, the addition of these inequalities does not guarantee the evenness of the solution as constraints (4.5) do.

Inequalities (4.8) and (4.6), although exponential in number, can be separated in polynomial time. Padberg and Rao [31] proved that violated odd-cut inequalities can be identified in polynomial time. The equivalent result for balanced-set inequalities was proved by Nobert and Picard [30], and a direct proof and a description of the method was presented in Benavent, Corberán, and Sanchis [1].

Now consider a set $S \subset V$ such that $\delta(S) = \emptyset$. Then the balanced-set condition corresponding to S and $V \setminus S$ implies the so-called *symmetry equation* $x(\delta^+(S)) = x(\delta^-(S))$. Hence, the above formulation includes an equation associated with each set $S \subset V$ with $\delta(S) = \emptyset$. Although most of these equations will be linearly dependent, if we consider the q subgraphs of G induced by the edges, it can be shown that any q–1 of the corresponding symmetry equations are linearly independent. In Corberán, Romero, and Sanchis [11] these equations,

$$x(\delta^+(K_i)) = x(\delta^-(K_i)), \quad i = 1, \ldots, q, \tag{4.9}$$

are referred to as *the system equations*, where $K_1, \ldots, K_q$ denote the sets of vertices of the connected components of the graph (V, E).

Let $\text{MCPP}_{F1}(G)$ be the convex hull of all the vectors $x \in \mathbb{R}^{E \cup A}$ satisfying (4.5) to (4.7). Although $\text{MCPP}_{F1}(G)$ has not been studied directly, a generalization of it, the Mixed General Routing Problem (MGRP) polyhedron, has been studied in Corberán, Romero, and Sanchis [11] and Corberán, Mejía, and Sanchis [7] (see also Chapter 6 of this book). Therefore, it is known that $\text{MCPP}_{F1}(G)$ is an unbounded polyhedron and $\dim(\text{MCPP}_{F1}(G)) = |E \cup A| - q + 1$, and that the following inequalities are facet-inducing for $\text{MCPP}_{F1}(G)$:

- trivial inequalities: $x_e \geq 1 \ \forall e \in E \cup A$,
- balanced-set inequalities (4.6),
- odd-cut inequalities (4.8), and
- Odd zigzag inequalities described in Corberán, Plana, and Sanchis [10] (see Section 4.4.2).

4.3.2.2 • Formulation F2

Formulation F2 uses two variables x_{ij} and x_{ji} associated with each edge $e = (i, j)$, representing the number of times edge e is traversed in the corresponding direction. As has been said, this formulation is based on the sufficient conditions for a mixed graph to be Eulerian, and it reads as follows:

$$\text{(4.10) (MCPP-F2)} \quad \min \sum_{e=(i,j)\in E} c_e(x_{ij}+x_{ji}) + \sum_{(i,j)\in A} c_{ij}x_{ij}$$

$$\text{(4.11)} \quad \text{s.t.} \quad x_{ij}+x_{ji} \geq 1 \qquad \forall e=(i,j)\in E,$$

$$\text{(4.12)} \quad x(\delta^+(i)) - x(\delta^-(i)) + \sum_{(i,j)\in\delta(i)} (x_{ij}-x_{ji}) = 0 \qquad \forall i \in V,$$

$$\text{(4.13)} \quad x_{ij} \geq 1 \text{ and integer} \qquad \forall (i,j)\in A,$$

$$\text{(4.14)} \quad x_{ij}, x_{ji} \geq 0 \text{ and integer} \qquad \forall e=(i,j)\in E,$$

where (4.11) and (4.13) imply that all the edges and arcs are in the solution, and (4.12) implies that the symmetry conditions will be satisfied. Note that integrality and symmetry conditions imply that the resulting graph will be even. This formulation was proposed by Kappauf and Koehler [24] and is also used in Christofides et al. [5] and Ralphs [36].

Moreover, it is shown in Kappauf and Koehler [24], Ralphs [36], and Win [38] that the components of the extreme points of the polyhedron associated with the linear relaxation of F2 have values of either 0.5 or a nonnegative integer, and that the fractional values occur only on some edges $e=(i,j)$, where $x_{ij}=x_{ji}=0.5$. As with the WPP (Win [38]), this result could be the basis for a simple heuristic algorithm consisting of rounding the variables of value 0.5 to 1. This algorithm could provide good feasible solutions even for large-size instances (mainly on those with a small number of edges) in very short computing times.

Let $\text{MCPP}_{F2}(G)$ be the convex hull of all the vectors $x \in \mathbb{R}^{E\cup A}$ satisfying (4.11) to (4.14). As happens with $\text{MCPP}_{F1}(G)$, $\text{MCPP}_{F2}(G)$ has not been studied directly, but again a generalization of it, the WPP polyhedron in this case, has been (partially) described in Win [38], Grötschel and Win [18], and Corberán et al. [9]. From the results obtained for the WPP, it can be deduced that $\text{MCPP}_{F2}(G)$ is an unbounded polyhedron, with dimension $2|E|+|A|-|V|+1$, and that the following inequalities are facet-inducing:

- trivial inequalities: $x_{ij} \geq 1 \; \forall (i,j)\in A$ and $x_{ij}, x_{ji} \geq 0 \; \forall e=(i,j)\in E$,
- traversing inequalities (4.11),
- odd-cut inequalities (4.8),
- k-wheel inequalities proposed in Win [38] (see Section 4.4.2), and
- Odd zigzag inequalities presented in Corberán, Plana, and Sanchis [10] (see also Section 4.4.2).

4.3.2.3 ▪ Formulations comparison

It is easy to see that (integer) formulations F1 and F2 are equivalent, but what can be said about their linear programming relaxations? In Corberán, Mota, and Sanchis [8] the lower bounds for the MCPP obtained using LP-based methods are compared. Remember that constraints (4.5) in F1 are not linear, and there is no linear expression of them without using integer variables. Hence, these constraints are removed, and odd-cut inequalities (4.8) are added to F1.

On the other hand, note that once the integrality conditions have been removed from F2, symmetry conditions alone no longer guarantee the evenness of the vertices. Hence, constraints (4.8) are also added to F2.

Therefore, let us call the linear relaxation of F1 LP_{F1},

$$\begin{aligned}
& x_{ij} \geq 1 && \forall (i,j) \in A, \\
& x_e \geq 1 && \forall e \in E, \\
(4.6) \quad & x(\delta^+(S)) - x(\delta^-(S)) \leq x(\delta(S)) && \forall S \subset V, \\
(4.8) \quad & x(\delta^*(S)) \geq |\delta^*(S)| + 1 && \forall S \subset V \text{ with } |\delta^*(S)| \text{ odd},
\end{aligned}$$

and the linear relaxation of F2 LP_{F2},

$$\begin{aligned}
& x_{ij} \geq 1 && \forall (i,j) \in A, \\
& x_{ij}, x_{ji} \geq 0 && \forall (i,j) \in E, \\
(4.11) \quad & x_{ij} + x_{ji} \geq 1 && \forall (i,j) \in E, \\
(4.8) \quad & x(\delta^*(S)) \geq |\delta^*(S)| + 1 && \forall S \subset V \text{ with } |\delta^*(S)| \text{ odd}, \\
(4.12) \quad & x(\delta^+(i)) - x(\delta^-(i)) + \sum_{(i,j) \in \delta(i)} (x_{ij} - x_{ji}) = 0 && \forall i \in V.
\end{aligned}$$

Since constraints (4.6) and (4.8) in LP_{F1} and (4.8) in LP_{F2} are exponential in number, only a subset of them can be explicitly added to the LPs to be solved. Consider the following initial LP relaxation, $LP0_{F1}$, corresponding to LP_{F1} containing

- the system equations (4.9),
- constraints $x_{ij} \geq 1 \ \forall (i,j) \in A$ and $x_e \geq 1 \ \forall e \in E$,
- one balanced-set inequality (4.6) for each "unbalanced" vertex, i.e., a vertex i for which $|\delta(i)| < \big| |\delta^+(i)| - |\delta^-(i)| \big|$, and
- one odd-cut inequality (4.8) for each odd "balanced" vertex,

and $LP0_{F2}$ (corresponding to LP_{F2}) as the LP containing

- constraints $x_{ij} \geq 1 \ \forall (i,j) \in A$ and $x_{ij}, x_{ji} \geq 0 \ \forall (i,j) \in E$,
- constraints (4.11),
- one odd-cut inequality (4.8) for each odd-degree vertex, and
- one symmetry equation (4.12) for each vertex.

In Corberán, Mota, and Sanchis [8] it is shown that $LP0_{F2}$ is stronger than $LP0_{F1}$. Regarding the relaxations LP_{F1} and LP_{F2}, note that although they include an exponential number of inequalities, both types of inequalities can be separated in polynomial time. Therefore, the computational effort to solve LP_{F1} and LP_{F2} is, at least in theory, comparable. Moreover, it is shown in [8] that both relaxations produce the same bound.

4.3.3 ▪ Exact algorithms

Several exact procedures have been proposed for the optimal solution of the MCPP. Christofides et al. [5] proposed a branch-and-bound algorithm, using two different lower bounds obtained by relaxing two types of constraints in a Lagrangian manner, able to solve instances with up to 50 vertices, 85 arcs, and 39 edges. Grötschel and Win [19] presented a cutting-plane algorithm for the WPP that was used to solve nine MCPP instances with up to 172 vertices, 116 arcs, and 154 edges. Nobert and Picard [30] were able to solve 148 instances out of 180 instances with a cutting-plane algorithm. The sizes of the instances were in the range $10 \leq |V| \leq 169$, $2 \leq |A| \leq 2896$, and $15 \leq |E| \leq 1849$. Finally, in Corberán et al. [9] a branch-and-cut algorithm for the exact solution of the

WPP is presented. This algorithm, which will be described in Section 4.4.3, was applied to solve MCPP instances and was able to solve 17 out of 24 instances with $|V| = 3000$, $1097 \leq |A| \leq 6742$, and $1992 \leq |E| \leq 6799$ optimally in less than 15 minutes. The average gap obtained on the seven unsolved instances was less than 1%. As far as we know, this is the best exact method for solving the MCPP.

4.4 ▪ The windy postman problem

In this section we define the WPP, present some notation and its formulation, and discuss its heuristic and exact solution.

The WPP can be defined as follows. Given an undirected and connected graph $G = (V, E)$ with two nonnegative costs c_{ij} and c_{ji} associated with each edge $(i, j) \in E$ corresponding to the cost of traversing it from i to j and from j to i, respectively, the WPP is to find a minimum cost tour on G traversing each edge at least once. This problem was introduced by Minieka [28], who wrote *"This* [costs symmetry] *is hardly a good assumption when one direction might be uphill and the other downhill, when one direction might be with the wind and the other against the wind or when fares are different depending on direction."* The WPP contains the MCPP as a special case arising when, for each edge $(i, j) \in E$, either $c_{ij} = c_{ji}$ or $\max\{c_{ij}, c_{ji}\} = \infty$. Therefore, it is easy to see that the WPP is NP-hard (Brucker [2], Guan [21]), although it can be solved in polynomial time if G is even (Win [39]), if the cost of the two opposite orientations of every cycle in G is the same (Guan [21]), or if G is a series-parallel graph (Zaragoza [43]).

If G satisfies that the cost of the two possible orientations of every cycle is the same, the WPP is solved as an undirected CPP where each edge (i, j) has cost $(c_{ij} + c_{ji})/2$. In what follows, we describe the polynomial-time algorithm proposed by Win [39] to find a WPP tour on an even (Eulerian) graph $G = (V, E)$. It is based on the Edmonds and Johnson algorithm for the MCPP on even graphs [13].

Win's algorithm for Eulerian graphs

1 For each edge $(i, j) \in E$, if $c_{ij} \leq c_{ji}$, then orient (i, j) from i to j; otherwise, orient (i, j) from j to i. In this way we obtain a directed graph, $D_G = (V, A)$.

2 Construct a directed graph $D' = (V, A')$ with arc set A', arc costs, arc capacities, and node demands defined as follows:

- For each arc $(i, j) \in A$ there are three arcs in A', namely, one copy of (i, j) at cost c_{ij} and infinite capacity, one copy of (j, i) at cost c_{ji} and infinite capacity, and one additional copy of (j, i), denoted by $(j, i)'$, at cost $(c_{ij} - c_{ji})/2$ and capacity two (such arcs are called "artificial" arcs).
- For each node $i \in V$, the demand is $d_i = d^+(i) - d^-(i)$ in D_G (negative demands are interpreted as supplies).

3 Find a minimum-cost flow in D' satisfying the demands and supplies of the vertices.

4 Denote by y^*_{ij}, y^*_{ji}, and $y^*_{(ji)'}$ the flow values along the arcs (i, j), (j, i), and $(j, i)'$, respectively. For every three arcs $(i, j), (j, i), (j, i)' \in A'$ proceed as follows: If $y^*_{(ji)'} = 0$, then put $y^*_{ij} + 1$ copies of (i, j) (and no copy of (j, i)) in A''; otherwise (i.e., if $y^*_{(ji)'} = 2$), put $y^*_{ji} + 1$ copies of (j, i) (and

no copy of (i,j)) in A''. $D'' = (V,A'')$ is Eulerian, and any Eulerian directed walk in D'' is an optimal WPP tour of G.

4.4.1 ▪ Heuristic algorithms

As in the MCPP, the above algorithm for even graphs is the basis for a heuristic procedure for the general case (Win [39]). It consists of transforming first the original graph into an even one, and then applying the previous algorithm for Eulerian graphs.

Win's algorithm for general graphs

1 Transform G into an Eulerian graph.

- For each edge $e=(i,j)\in E$, define $\bar{c}_e=(c_{ij}+c_{ji})/2$.
- Identify odd nodes of G, and compute the shortest path distances between every pair of them using distances $\bar{c}_e$.
- Find a minimum-cost perfect matching of the odd nodes.
- Add to E one additional copy of each edge that lies on the shortest path joining two matched odd nodes. The resulting graph, say, $G'=(V,E')$, is Eulerian.

2 Apply Win's algorithm for Eulerian graphs to G' with the original cost function c_e.

Win [39] proved that this algorithm has a worst-case bound with a value of twice the optimum, which is approachable. Win [38] proposed another heuristic entailing rounding up the optimal solutions to the linear relaxation of the WPP formulation presented below. This algorithm was based on the following result: The values of the variables in any optimal LP solution are integer or 0.5, and $x_{ij}=0.5$ if and only if $x_{ji}=0.5$ (Kappauf and Koehler [24], Win [38]). Then, it can be proved that forcing $x_{ij}=x_{ji}=1$ for these variables leads to a feasible WPP solution. Win [38] also proved that this heuristic again has a worst-case ratio of 2 and the bound is approachable. Based on their algorithm for the MCPP, Raghavachari and Veerasamy [35] modify Win's LP algorithm to get a new one with a better worst-case ratio of 3/2.

4.4.2 ▪ Problem formulation

Let $G=(V,E)$ be the graph for which a minimum-cost WPP tour has to be determined. For $S_1,S_2\subseteq V$, $(S_1:S_2)$ denotes the set of edges with one endpoint in S_1 and the other in S_2. Given $S\subseteq V$, $\delta(S)=(S:V\setminus S)$ and $E(S)=(S:S)$.

Let x_{ij} be the number of times edge (i,j) is traversed from i to j in a WPP tour. For $F\subseteq E$, we define $x(F)=\sum_{(i,j)\in F}(x_{ij}+x_{ji})$, and for $S_1,S_2\subset V$, we define

$$x(S_1,S_2) = \sum_{i\in S_1, j\in S_2} x_{ij}.$$

Note that $x(S_1:S_2)=x(S_1,S_2)+x(S_2,S_1)$.

The ILP formulation of the WPP in Win [38] and Grötschel and Win [19] is

$$\text{(4.15)} \qquad \text{(WPP)} \quad \min \sum_{(i,j)\in E} (c_{ij}x_{ij} + c_{ji}x_{ji})$$

$$\text{(4.16)} \qquad \text{s.t.} \quad x_{ij} + x_{ji} \geq 1 \qquad \forall (i,j)\in E,$$

$$\text{(4.17)} \qquad \sum_{(i,j)\in\delta(i)} (x_{ij} - x_{ji}) = 0 \qquad \forall i \in V,$$

$$\text{(4.18)} \qquad x_{ij}, x_{ji} \geq 0 \text{ and integer} \qquad \forall (i,j)\in E,$$

where conditions (4.16) imply that each edge will be traversed at least once and conditions (4.17) force the (directed) graph represented by the tour to be symmetric. The above system includes an equation associated with each vertex. The $|V|$ equations (4.17) will be referred to as the *symmetry equations*, and any $|V|-1$ of them are linearly independent.

Let $\text{WPP}(G) \subseteq \mathbb{Z}^{2|E|}$ be the convex hull of vectors satisfying (4.16) to (4.18). In Win [38] and Grötschel and Win [18], it is shown that WPP(G) is an unbounded polyhedron, with dimension $2|E|-|V|+1$, and that trivial and traversing inequalities are facet-inducing.

In addition to these inequalities, other families of inequalities that are not present in the formulation are known.

Odd-cut inequalities

Odd-cut inequalities were presented in Win [38] and shown to be facet-inducing for WPP(G). As has been said, they are based on the fact that any edge cutset must be traversed an even number of times:

$$\text{(4.19)} \qquad x(\delta(S)) \geq |\delta(S)|+1 \quad \forall\, S \subset V, \ |\delta(S)| \text{ odd.}$$

k-wheel inequalities

A wheel consists of a cycle (called the rim), every node of which is incident with a common node (called the center). An edge joining the center and a node on the rim is called a spoke. A k-wheel, denoted by W_k, is a wheel whose rim contains k nodes. Figure 4.1(a) shows a 3-wheel W_3 in which each edge has been oriented in such a way that the edges of the rim do not form a directed cycle and all the spokes are oriented to the center. Then, the 3-wheel inequality

$$\text{(4.20)} \qquad x_{12} + x_{13} + x_{14} + x_{32} + x_{42} + x_{43} \geq 3$$

is valid and facet-inducing for the corresponding WPP polyhedron (Win [38]).

This inequality can be generalized by replacing each vertex by a connected subgraph and each arc by a subset of parallel arcs with the same direction in the following way (see Figure 4.1b): Consider a partition of the set of vertices V into four subsets, M^1, M^2, M^3, and M^4, where each M^i contains an odd number of odd vertices and each subgraph $G(M^i)$ is connected, and let $x(A')$ be the set of variables corresponding to the traversal of the edges in the direction depicted in Figure 4.1(b). The 3-wheel inequality is then

$$\text{(4.21)} \qquad x(A') \geq \frac{1}{2}\Big(|(M^1:M^3)| + |(M^2:M^4)| + |(M^1:M^4)| + |(M^2:M^3)|\Big) + 1.$$

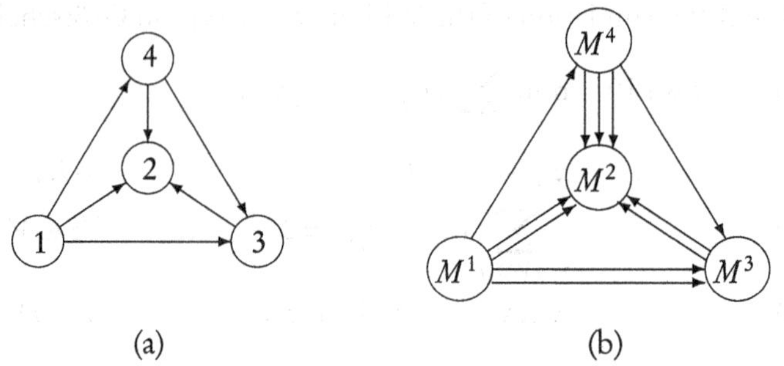

Figure 4.1. *3-wheel inequalities.*

The 3-wheel inequalities can be extended to the more general k-wheel inequalities, which contain an odd number k of nodes in the rim (see details in Win [38]).

Odd zigzag inequalities

When solving the WPP, it is not unusual to find fractional solutions containing closed walks, resembling a zigzag, with $x_{ij} + x_{ji} = 1.5$, such as the vector represented in Figure 4.2(a). This kind of solution violates another family of facet-inducing inequalities for the WPP, the Odd zigzag inequalities, presented in Corberán, Plana, and Sanchis [10]. These inequalities generalize the 3-wheel inequalities.

As with the k-wheel inequalities, consider a partition of the set of vertices V into four subsets, M^1, M^2, M^3, and M^4, where each M^i contains an odd number of odd vertices. Let us define $\mathcal{H} = (M^1 : M^2) \cup (M^3 : M^4)$ (horizontal edges) and $\mathcal{D} = (M^2 : M^3) \cup (M^1 : M^4)$ (diagonal edges). Note that $\mathcal{H} \cup \mathcal{D} = \delta(M^1 \cup M^3)$. Let us assume we have a subset of edges $\mathcal{F} \subset (\mathcal{H} \cup \mathcal{D})$ satisfying $|\mathcal{H} \setminus \mathcal{F}| + |\mathcal{D} \cap \mathcal{F}| = |\mathcal{D} \setminus \mathcal{F}| + |\mathcal{H} \cap \mathcal{F}|$, or, equivalently,

$$|\mathcal{H}| + |\mathcal{D}| = 2|\mathcal{H} \cap \mathcal{F}| + 2|\mathcal{D} \setminus \mathcal{F}|$$

(see Figure 4.2(b), where edges in $\mathcal{F}$ are represented in bold lines).

In Corberán, Plana, and Sanchis [10], it is shown that the following Odd zigzag inequality, whose coefficients are shown in Figure 4.2(b), is valid and facet-inducing:

$$\begin{aligned} x(\delta(M^1 \cup M^2)) + 2x(M^2, M^1) + 2x(M^4, M^3) + 2x(F_{zz}) \\ \geq |(M^1 : M^3)| + |(M^2 : M^4)| + |(M^1 : M^4)| + |(M^2 : M^3)| + 2|\mathcal{H} \cap \mathcal{F}| + 2, \end{aligned} \tag{4.22}$$

where $x(F_{zz})$ denotes the variables associated with the edges in $\mathcal{F}$ in the direction given by the zigzag, i.e., in the direction $M^1 \to M^2 \to M^3 \to M^4 \to M^1$.

Set $\mathcal{F}$ can be understood in the following way. A fractional solution, such as the one in Figure 4.2(a), defines an orientation of the edges in $\delta(M^1 \cup M^3)$. Given any such orientation, set $\mathcal{F}$ is defined by all the edges that have been oriented in the opposite direction to the zigzag, i.e., in the direction $M^4 \to M^3 \to M^2 \to M^1 \to M^4$. In particular, set $\mathcal{F}$ in Figure 4.2(b) is defined from the orientation associated with the fractional solution shown in Figure 4.2(a). Other sets $\mathcal{F}$ can be defined to obtain valid inequalities, but only the one shown in Figure 4.2(b) has an associated inequality violated by the fractional solution.

Note that in the special case when $\mathcal{F} = \emptyset$, the Odd zigzag inequalities are equivalent to the 3-wheel inequalities, and hence they induce the same facets. However, note that

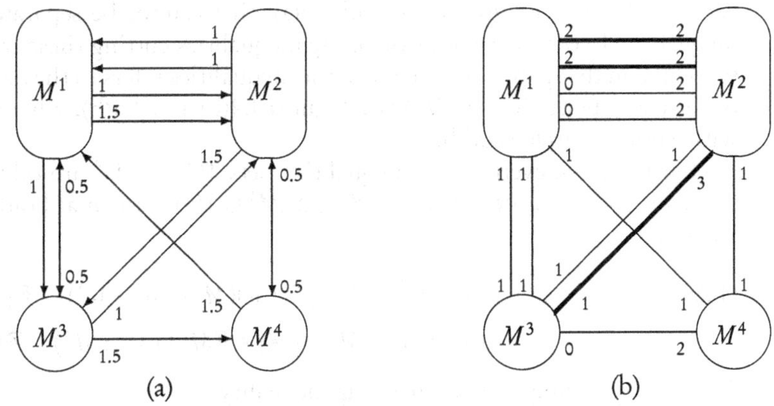

Figure 4.2. *Fractional solution and Odd zigzag inequality.*

when $\mathscr{F} \neq \emptyset$, Odd zigzag inequalities are a different class of facet-inducing inequalities for the WPP. In fact, as noted in Corberán, Plana, and Sanchis [10], it can be seen that the fractional solution shown in Win's Thesis (see Figure 4.3), which satisfies all the 3-wheel inequalities, violates the Odd zigzag inequality defined by node sets $M^1 = \{6\}$, $M^2 = \{3,4,5\}$, $M_3 = \{2\}$, $M_4 = \{1\}$, and edge set $F = \{e_{46}\}$, which is depicted in Figure 4.3. It can be seen that the Odd zigzag inequality is $F(x) \geq 8$, while the fractional solution x^* satisfies $F(x^*) = 7$.

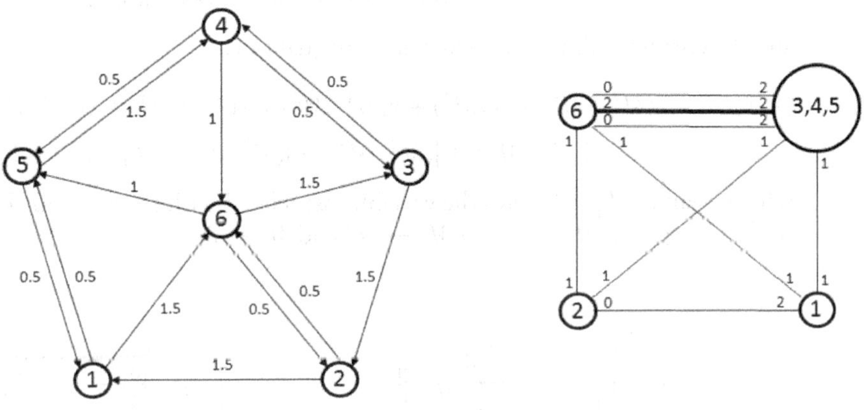

Figure 4.3. *Fractional solution from Win's Thesis and the Odd zigzag inequality.*

It is worth noting that, unlike other Arc Routing Polyhedra, WPP(G) has facet-inducing inequalities, such as the Odd zigzag inequalities presented above, that are not strictly configuration inequalities (Naddef and Rinaldi [29]) since they can have variables x_{uv}, x_{st}, with $u, s \in B_i$ and $v, t \in B_j$, with different coefficients.

Even-even and Odd-odd zigzag inequalities

In Corberán et al. [9] the authors describe some fractional solutions having four edges with value 1.5 forming a nondirected path joining two odd nodes with two even nodes (such as those in Figures 4.4(a) and 4.5(a), that cannot be separated by an Odd zigzag inequality. Two new versions of zigzag inequalities cutting these fractional solutions are presented in that paper. The name of the inequalities refers to the degree of the two shores of the edge cutset $\delta(M^1 \cup M^3)$ (see Figures 4.4(b) and 4.5(b), where a set M^i represented with a double circle is odd).

For the Even-even zigzag inequalities, sets M^2 and M^4 are odd, sets M^1 and M^3 are even, and $\mathscr{F}$ is a subset of edges, $\mathscr{F} \subset \delta(M^1 \cup M^3)$ (shown as bold lines in Figure 4.4a), satisfying

$$|(M^1 : M^2 \cup M^4) \setminus \mathscr{F}| + |(M^2 \cup M^4 : M^3) \cap \mathscr{F}| = |(M^1 : M^2 \cup M^4) \cap \mathscr{F}| + |(M^2 \cup M^4 : M^3) \setminus \mathscr{F}|. \tag{4.23}$$

The corresponding Even-even zigzag inequality is

$$x(\delta(M^3)) + 2x(M^2 : M^4) + 2x(M^2, M^1) + 2x(M^4, M^1) + 2x(F_{zz}) \geq |(M^1 : M^3)| + 2|(M^2 : M^4)| + |(M^2 : M^3)| + |(M^3 : M^4)| + 2|\delta(M^1) \cap \mathscr{F}| + 2, \tag{4.24}$$

where $x(F_{zz})$ denotes the variables associated with the edges in $\mathscr{F}$ in the direction of the zigzag, i.e., $M^1 \to M^2 \to M^3$ and $M^1 \to M^4 \to M^3$. The coefficients of the variables are also shown in Figure 4.4(b).

For the Odd-odd zigzag inequalities (see Figure 4.5(b)), sets M^2 and M^3 are odd, sets M^1 and M^4 are even, set $\mathscr{F} \subset \delta(M^1 \cup M^3)$ satisfies the condition

$$|(M^1 : M^2 \cup M^4) \setminus \mathscr{F}| + |(M^3 : M^4) \setminus \mathscr{F}| + |(M^2 : M^3) \cap \mathscr{F}| + 1 = |(M^1 : M^2 \cup M^4) \cap \mathscr{F}| + |(M^3 : M^4) \cap \mathscr{F}| + |(M^2 : M^3) \setminus \mathscr{F}|, \tag{4.25}$$

and the corresponding Odd-odd zigzag inequality is

$$x(M^2 \cup M^4, M^1 \cup M^3) + x(M^1 : M^3) + x(M^2 : M^4) + x(M^3, M^2) + x(F_{zz}) \geq |(M^1 : M^3)| + |(M^2 : M^4)| + |(M^2 : M^3) \setminus \mathscr{F}| + |\mathscr{F}| + 1, \tag{4.26}$$

where, again, $x(F_{zz})$ denotes the variables associated with the edges in $\mathscr{F}$ in the direction of the zigzag, i.e., $M^1 \to M^2 \to M^3 \to M^4$ and $M^1 \to M^4$.

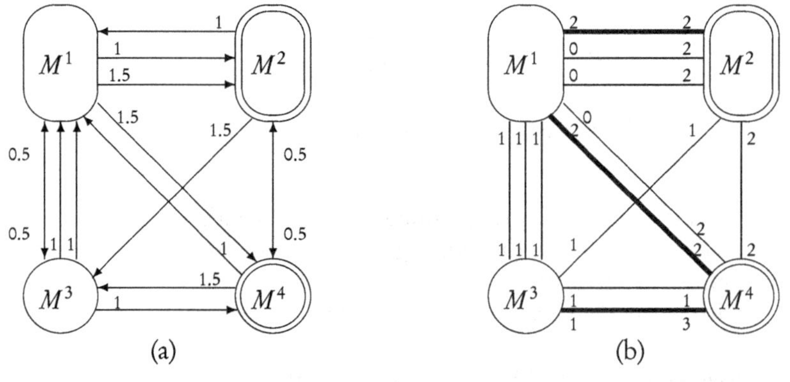

Figure 4.4. *A fractional solution and an Even-even zigzag inequality.*

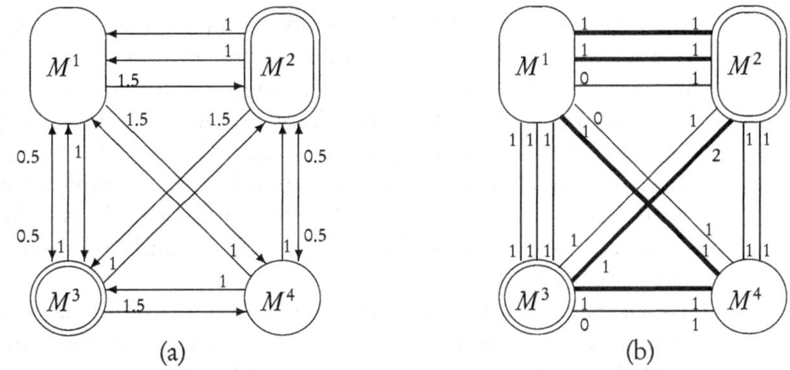

Figure 4.5. *A fractional solution and an Odd-odd zigzag inequality.*

4.4.3 ▪ Exact algorithms

In Corberán et al. [9] a branch-and-cut algorithm for the WPP is presented. This procedure uses heuristic and exact procedures for separating violated odd-cut inequalities and incorporates heuristic separation algorithms for all the families of zigzag inequalities. Moreover, since in the same paper it is shown that all zigzag inequalities are Chvátal–Gomory inequalities of rank at most 2, a polynomial time algorithm for identifying maximally violated mod-k inequalities is added. Note that a mod-k cut (see Caprara, Fischetti, and Letchford [3]) is a rank-1 Chvátal–Gomory inequality in which the multipliers are restricted to the set $\{0, \frac{1}{k}, \ldots, \frac{k-1}{k}\}$. Basically, the method consists of solving a system of congruences where the system is composed of all the binding inequalities of the last LP plus an inequality for each symmetry equation. Moreover, in order to get good upper bounds that decrease the size of the search tree, a heuristic algorithm based on the information given by the fractional solution x^* obtained after processing each node has been implemented.

At the time of writing this chapter, this is the best exact algorithm that has been published for the WPP, and it is capable of solving very large WPP instances with up to 3000 nodes and 9000 edges. Of the 120 instances it was tested on, the algorithm solved 91 out of 96 instances with $500 \leq |V| \leq 2000$ and $813 \leq |E| \leq 6129$ in less than two hours on average, and eight out of 24 instances with $|V| = 3000$ and $4986 \leq |E| \leq 9085$. For the unsolved instances, the average gap obtained was less than 0.2%. Tests were run on an Intel Pentium IV 2.80GHz and 2GB RAM with a time limit of 10 hours. When compared with the results obtained by the same algorithm in MCPP instances of a similar size, the results obtained are slightly worse, which seems to imply that the WPP is harder than the MCPP.

A completely different approach to the WPP solution was proposed by Yan and Thompson [40]. The authors slightly modify the WPP formulation to get an equivalent set covering problem. Taking into account this transformation, an exact algorithm and a heuristic algorithm producing good results for instances with up to 90 nodes and 180 edges are proposed. Both methods are based on a branch-and-bound procedure previously designed for the solution of set covering, partitioning, and packing problems.

4.5 ▪ Related problems

The MCPP can be generalized by providing, for each arc and edge e, two integers u_e, $l_e \geq 0$, and requiring that e is used at least l_e and at most u_e times. This problem, called the *Bounded Mixed Postman Problem*, was suggested by Edmonds and Johnson [13] and studied by Zaragoza [42]. Another related problem mentioned in [42] is the *Restricted Mixed Postman Problem*, where given a subset $R \subseteq E \cup A$ of restricted edges and arcs, the problem consists of finding a restricted tour, i.e., a tour using each restricted edge and arc exactly once. The special cases when $R = A$ and $R = E$ are called the *Edges Postman Problem* and the *Arcs Postman Problem*, respectively, and have been studied by Veerasamy [37] and Zaragoza [42].

Given a mixed graph G, a cycle cover of G is a family $C = \{C_1, \dots, C_k\}$ of cycles of G such that each edge and arc of G belongs to at least one cycle in C. The weight of C is $\sum_{i=1}^{k} |C_i|$. The *Minimum Cycle Cover Problem* (MCC) is as follows: Given a strongly connected mixed graph G without bridges, find a cycle cover of G with weight as small as possible. In Lee and Wakabayashi [25] it is proved that this problem is NP-hard if G is a planar graph. However, the MCC (and the MCPP) defined on mixed graphs with bounded treewidth can be solved in polynomial time (Fernandes, Lee, and Wakabayashi [15]).

A problem related to the WPP on Eulerian graphs is the *Minimum Cost Eulerian Orientation Problem* (MCEOP). Let $G = (V, E)$ be an undirected Eulerian graph with two costs c_{ij} and c_{ji} associated with each edge $e = (i, j) \in E$, where c_{ij} and c_{ji} are the costs of orienting edge e as (i, j) or (j, i), respectively. An orientation of G is obtained by replacing edge e with arc (i, j) or arc (j, i). The MCEOP consists of finding the minimum cost Eulerian orientation of G. Note that the MCEOP is a WPP with an additional restriction: that we should traverse each edge e only once (either from i to j or from j to i). Guan and Pulleyblank [22] proposed several polynomial-time algorithms for this problem. Another polynomial-time algorithm for the MCEOP, similar to Win's algorithm for Eulerian graphs described in Section 4.4.1, as well as a complete description of its associated polytope, can be found in Win [38, 39].

Another problem related to the WPP is the *Plowing with Precedence Problem* (PPP) proposed by Dussault et al. [12]. This is a variant of the WPP motivated by the fact that deadhead travel over streets that have already been plowed is significantly faster than the time it takes to plow the street. The authors incorporate the observation that on some steep streets it is much more difficult, or even impossible, to plow uphill. The idea is to design routes that try to avoid plowing uphill on steep streets and take advantage of the faster traversal time on plowed streets. Note that this assumption requires the possibility of plowing against the direction of traffic and allows for the possibility of deadhead travel when only one side of the street has been plowed, regardless of the direction. For example, if a steep street has been plowed downhill, it is possible to traverse the street uphill without plowing as long as the vehicle eventually returns to clear the other side. If a street has not been plowed, however, then it is impossible to deadhead that street because the snow is too deep to traverse without plowing. Let $G = (V, E)$ be an undirected graph. Each edge $e = (i, j)$ has four costs: c^+_{ij}, the cost of plowing e from i to j; c^+_{ji}, the cost of plowing e from j to i; c^-_{ij}, the cost of traversing e from i to j after plowing; and c^-_{ji}, the cost of traversing e from j to i after plowing. If vertex j has a higher altitude than vertex i, then usually $c^+_{ij} >> c^+_{ji} >> c^-_{ij} \geq c^-_{ji}$. The PPP consists of finding the route that begins and ends at the depot (vertex 0), plows each edge twice (for each side of the street), and minimizes total cost. The order of a route is important since the cost of traversing an edge depends on whether or not the edge has been traversed previously; i.e., it is not possible to deadhead an edge $e=(i,j)$ from i to j, for instance, at cost c_{ij}^- if we have not previously plowed e at cost c_{ij}^+ or c_{ji}^+. Therefore, precedence is important in the PPP. Dussault et al. [12] formulate the PPP as an integer program and propose a local search procedure producing very good solutions. The algorithm optimally solves a relaxation of the PPP in which precedence is ignored. This relaxed problem is similar to the WPP. The optimal solution to this problem, called a partial solution by the authors, gives the number of times to plow and deadhead each edge in each direction and is used to construct a route for the PPP. Starting with this partial solution, an initial route using an alternative to Fleury's algorithm is constructed. This initial route is optimal if precedence is satisfied; otherwise, a local search heuristic is applied that modifies the route while taking precedence into account.

Acknowledgments

The authors wish to thank the Spanish Ministerio of Economía y Competitividad (project MTM2012-36163-C06-02) and the Generalitat Valenciana (project GVPROMETEO2013-049) for their support.

Bibliography

[1] E. BENAVENT, Á. CORBERÁN, AND J.M. SANCHIS, *Linear programming based methods for solving arc routing problems*, in Arc Routing: Theory, Solutions and Applications, M. Dror, ed., Kluwer, Boston, 2000, pp. 231–275.

[2] P. BRUCKER, *The Chinese postman problem for mixed graphs*, in Proceedings of the International Workshop, Bad Honnef, 1980, Lecture Notes in Computer Science 100, Springer, Berlin, 1981, pp. 354–366.

[3] A. CAPRARA, M. FISCHETTI, AND A.N. LETCHFORD, *On the separation of maximally violated mod-k cuts*, Mathematical Programming, 87 (2000), pp. 37–56.

[4] N. CHRISTOFIDES, *The optimum traversal of a graph*, Omega, 1 (1973), pp. 719–732.

[5] N. CHRISTOFIDES, E. BENAVENT, V. CAMPOS, Á. CORBERÁN, AND E. MOTA, *An optimal method for the mixed postman problem*, in System Modelling and Optimization, P. Thoft-Christensen, ed., Lecture Notes in Control and Information Sciences 59, Springer, Berlin, 1984, pp. 641–649.

[6] Á. CORBERÁN, R. MARTÍ, AND J.M. SANCHIS, *A GRASP heuristic for the mixed Chinese postman problem*, European Journal of Operational Research, 142 (2002), pp. 70–80.

[7] Á. CORBERÁN, G. MEJÍA, AND J.M. SANCHIS, *New results on the mixed general routing problem*, Operations Research, 53 (2005), pp. 363–376.

[8] Á. CORBERÁN, E. MOTA, AND J.M. SANCHIS, *A comparison of two different formulations for arc routing problems on mixed graphs*, Computers & Operations Research, 33 (2006), pp. 3384–3402.

[9] Á. CORBERÁN, M. OSWALD, I. PLANA, G. REINELT, AND J.M. SANCHIS, *New results on the windy postman problem*, Mathematical Programming, 132 (2012), pp. 309–332.

[10] Á. CORBERÁN, I. PLANA, AND J.M. SANCHIS, *Zigzag inequalities: A new class of facet-inducing inequalities for arc routing problems*, Mathematical Programming, 108 (2006), pp. 79–96.

[11] Á. CORBERÁN, A. ROMERO, AND J.M. SANCHIS, *The mixed general routing polyhedron*, Mathematical Programming, 96 (2003), pp. 103–137.

[12] B. DUSSAULT, B. GOLDEN, C. GRÖER, AND E. WASIL, *Plowing with precedence: A variant of the windy postman problem*, Computers & Operations Research, 40 (2013), pp. 1047–1059.

[13] J. EDMONDS AND E.L. JOHNSON, *Matching, Euler tours and the Chinese postman problem*, Mathematical Programming, 5 (1973), pp. 88–124.

[14] H.A. EISELT, M. GENDREAU, AND G. LAPORTE, *Arc routing problems, part 1: The Chinese postman problem*, Operations Research, 43 (1995), pp. 231–242.

[15] C.G. FERNANDES, O. LEE, AND Y. WAKABAYASHI, *Minimum cycle cover and Chinese postman problems on mixed graphs with bounded tree-width*, Discrete Applied Mathematics, 157 (2009), pp. 272–279.

[16] L.R. FORD AND D.R. FULKERSON, *Flows in Networks*, Princeton University Press, Princeton, NJ, 1962.

[17] G.N. FREDERICKSON, *Approximation algorithms for some postman problems*, Journal of the Association for Computing Machinery, 26 (1979), pp. 538–554.

[18] M. GRÖTSCHEL AND Z. WIN, *On the windy postman polyhedron*, Technical report 75, University of Augsburg, Augsburg, Germany, 1988.

[19] ——, *A cutting plane algorithm for the windy postman problem*, Mathematical Programming, 55 (1992), pp. 339–358.

[20] M.G. GUAN, *Graphic programming using odd or even points*, Chinese Mathematics, 1 (1962), pp. 273–277.

[21] ——, *On the windy postman problem*, Discrete Applied Mathematics, 9 (1984), pp. 41–46.

[22] M. GUAN AND W. PULLEYBLANK, *Eulerian orientations and circulations*, SIAM Journal on Algebraic Discrete Methods, 6 (1985), pp. 657–664.

[23] A. HERTZ AND M. MITTAZ, *Heuristic algorithms*, in Arc Routing: Theory, Solutions and Applications, M. Dror, ed., Kluwer, Boston, 2000, pp. 327–386.

[24] C.H. KAPPAUF AND G.J. KOEHLER, *The mixed postman problem*, Discrete Applied Mathematics, 1 (1979), pp. 89–103.

[25] O. LEE AND Y. WAKABAYASHI, *On the circuit cover problem for mixed graphs*, Combinatorics, Probability and Computing, 11 (2002), pp. 43–59.

[26] T.M. LIEBLING, *Graphentheorie in Planungs- und Tourenproblemen*, Lecture Notes in Operations Research and Mathematical Systems 21, Springer, Berlin, 1970.

[27] E. MINIEKA, *Optimization Algorithms for Networks and Graphs*, Marcel Dekker, New York, 1979.

[28] ——, *The Chinese postman problem for mixed networks*, Management Science, 25 (1979), pp. 643–648.

[29] D. NADDEF AND G. RINALDI, *The symmetric traveling salesman polytope and its graphical relaxation: Composition of valid inequalities*, Mathematical Programming, 51 (1991), pp. 359–400.

[30] Y. NOBERT AND J.-C. PICARD, *An optimal algorithm for the mixed Chinese postman problem*, Networks, 27 (1996), pp. 95–108.

[31] M. W. PADBERG AND M. R. RAO, *Odd minimum cut-sets and b-matchings*, Mathematics of Operations Research, 7 (1982), pp. 67–80.

[32] C.H. PAPADIMITRIOU, *On the complexity of edge traversing*, Journal of the Association for Computing Machinery, 23 (1976), pp. 544–554.

[33] W.L. PEARN AND J.B. CHOU, *Improved solutions for the Chinese postman problem on mixed networks*, Computers & Operations Research, 26 (1999), pp. 819–827.

[34] B. RAGHAVACHARI AND J. VEERASAMY, *A 3/2-approximation algorithm for the mixed postman problem*, SIAM Journal on Discrete Mathematics, 12 (1999), pp. 425–433.

[35] ——, *A 3/2-approximation algorithm for the windy postman problem*, Technical report, University of Texas, Dallas, TX, 2001.

[36] T.K. RALPHS, *On the mixed Chinese postman problem*, Operations Research Letters, 14 (1993), pp. 123–127.

[37] J. VEERASAMY, *Approximation Algorithms for Postman Problems*, Ph.D. thesis, University of Texas, Dallas, TX, 1999.

[38] Z. WIN, *Contributions to Routing Problems*, Ph.D. thesis, University of Augsburg, Augsburg, Germany, 1987.

[39] ——, *On the windy postman problem on Eulerian graphs*, Mathematical Programming, 44 (1989), pp. 97–112.

[40] H. YAN AND G. THOMPSON, *Finding postal carrier walk paths in mixed graphs*, Computational Optimization and Applications, 9 (1998), pp. 229–247.

[41] K. YAOYUENYONG, P. CHARNSETHIKUL, AND V. CHANKONG, *A heuristic algorithm for the mixed Chinese postman problem*, Optimization and Engineering, 3 (2002), pp. 157–187.

[42] F. J. ZARAGOZA, *Postman Problems on Mixed Graphs*, Ph.D. thesis, University of Waterloo, Waterloo, Ontario, Canada, 2003.

[43] ——, *Series-parallel graphs are windy postman perfect*, Discrete Mathematics, 308 (2008), pp. 1366–1374.

Chapter 5
The Undirected Rural Postman Problem

Gianpaolo Ghiani
Gilbert Laporte

5.1 ▪ Introduction

The undirected *Rural Postman Problem* (RPP) is defined on an undirected graph $G = (V,E)$, where V is the vertex set and E is the edge set. A subset E_R of the edges are *required*, which means they must be part of the solution. With each edge $e = (i,j)$ is associated a nonnegative cost c_{ij}. The RPP consists of designing a least cost tour traversing each edge of E_R at least once. Equivalently, the problem consists of determining a least cost set of deadheaded edges which, together with the edges of E_R, will yield a Eulerian graph.

The RPP was introduced by Orloff [32] and was proved to be NP-hard by Lenstra and Rinnooy Kan [27]. However, when the graph induced by the edges of E_R is connected, the RPP reduces to an undirected *Chinese Postman Problem* (CPP) (see Chapter 3) and can therefore be solved in polynomial time. Applications of the RPP arise in contexts where the edges of a graph must be serviced by an uncapacitated vehicle. There exist applications of the problem related to the control of plotting or drilling machines [23] and to the optimization of laser-plotter beam movements [19]. In addition, in the solution of a *Capacitated Arc Routing Problem* (see Chapter 7), every route can be optimized individually by means of an RPP algorithm.

A problem closely related to the RPP is the *General Routing Problem* (GRP) defined in Chapter 6. Since the GRP is a generalization of the RPP, many results obtained for the GRP can be applied to the RPP.

The remainder of this chapter is organized as follows. In Section 5.2 we recall some properties of RPP solutions. Mathematical formulations are presented in Section 5.3, followed by exact algorithms in Section 5.4. Constructive and improvement heuristics are described in Section 5.5. Finally, some variants and extensions of the RPP are introduced in Section 5.6.

5.2 ▪ Properties

An RPP optimal solution satisfies a number of properties. Let C_i $(i = 1,\dots,p)$ be the ith connected component of the subgraph induced by E_R and V_i $(i = 1,\dots,p)$ its set of vertices. Moreover, let $V_R = \cup_{i=1}^{p} V_i$.

Christofides et al. [5] and a number of subsequent authors have worked on a transformed graph in which only the vertices of V_R appear: First, they add to $G_R = (V_R, E_R)$ an edge between every pair of vertices of V_R having a cost equal to the shortest path length on G; they then simplify the graph by deleting one of the two parallel edges if they have the same cost, as well as all edges $e = (i,j) \notin E_R$ such that $c_{ij} = c_{ik} + c_{kj}$ for some $k \in V$. In the remainder of the chapter, we will work also on the transformed graph satisfying $V_R = V$.

Dominance Relation 1. *In every optimal solution, the maximum number of additional copies of edge e that must be added to G to make it Eulerian is equal to 1 if $e \in E_R$, or to 2 otherwise [5].*

Dominance Relation 2. *Let $e = (i,j)$ be an edge such that i and j belong to some connected component C_i. Then, in every optimal solution, the number of additional copies of e is equal to 1 [9].*

For the edges $e \in E_R$, these two dominance relations are equivalent.

Dominance Relation 3. *Let $e^{(1)}, e^{(2)}, \dots, e^{(l)}$ be the edges having exactly one vertex in a given component C_i and one vertex in another given component C_j $(i,j \in \{1,\dots,p\}, i \neq j)$. In any optimal solution, only an edge $e^{(r)}$ such that $c(e^{(r)}) = min\{c(e^{(1)}), c(e^{(2)}), \dots, c(e^{(l)})\}$ can be deadheaded twice [21].*

Consequently, no more than $p(p-1)/2$ edges need to appear twice in an optimal solution. Tighter properties can be derived as follows.

Dominance Relation 4. *There exists an optimal solution to the RPP in which at most $p-1$ edges are deadheaded twice [21].*

Dominance Relation 5. *Let $G_C = (V_C, E_C)$ be an auxiliary graph having a vertex i' for every component C_i and, for any pair of components C_i and C_j, an edge $e = (i', j')$ corresponding to a least cost edge between C_i and C_j. The only edges that can be deadheaded twice are those belonging to a minimum spanning tree on G_C [21].*

5.3 ▪ Mathematical formulations

The first formulation of the RPP is due to Christofides et al. [5]. Given a set of vertices $S \subset V$, let $\delta(S)$ be the set of edges of E with one extremity in S and one extremity in $V \backslash S$. If $S = \{i\}$, we simply write $\delta(i)$ instead of $\delta(\{i\})$. In the formulation proposed by Christofides et al., variable x_e represents the number of additional copies of edge e that must be added to G to make it Eulerian, and $2z_i$ represents the degree of vertex i:

$$\text{(5.1)} \qquad (\text{RPP}^{(\text{CHR})}) \qquad \text{minimize} \sum_{e \in E_R} c_e(1 + x_e) + \sum_{e \in E \backslash E_R} c_e x_e$$

$$\text{(5.2)} \quad \text{s.t.} \quad \sum_{e \in \delta(i) \cap E_R} (1 + x_e) + \sum_{e \in \delta(i) \cap E \setminus E_R} x_e = 2z_i, \qquad i \in V,$$

$$\text{(5.3)} \quad \sum_{e \in \delta(S)} x_e \geq 2, \qquad S = \cup_{k \in P} V_k, \; P \subset \{1, \ldots, p\}, P \neq \emptyset,$$

$$\text{(5.4)} \quad x_e \geq 0 \text{ and integer}, \qquad e \in E,$$

$$\text{(5.5)} \quad z_i \geq 0 \text{ and integer}, \qquad i \in V.$$

In this formulation, constraints (5.2), (5.4), and (5.5) state that each vertex of V_R must have an even degree, while constraints (5.3) force the solution graph to be connected. The resulting graph will therefore be Eulerian.

In the formulation of Corberán and Sanchis [9], a vertex set $S \in V$ is said to be R-odd (R-even) if an odd (even) number of required edges are incident to S and, again, x_e represents the number of deadheadings of edge e:

$$\text{(5.6)} \qquad (\text{RPP}^{(\text{CORB})}) \qquad \text{minimize} \sum_{e \in E} c_e x_e$$

$$\text{(5.7)} \quad \text{s.t.} \quad \sum_{e \in \delta(i)} x_e = 0 \bmod 2 \qquad \text{if } i \in V \text{ is } R\text{-even},$$

$$\text{(5.8)} \quad \sum_{e \in \delta(i)} x_e = 1 \bmod 2 \qquad \text{if } i \in V \text{ is } R\text{-odd},$$

$$\text{(5.9)} \quad \sum_{e \in \delta(S)} x_e \geq 2, \qquad S = \cup_{k \in P} V_k, P \subset \{1, \ldots, p\}, P \neq \emptyset,$$

$$\text{(5.10)} \quad x_e \geq 0 \text{ and integer}, \qquad e \in E.$$

As in the previous formulation, the constraints force the degree of each vertex to be even and the graph to be connected. It is worth noting that this formulation is nonlinear because of the modulo operations in contraints (5.7) and (5.8).

Ghiani and Laporte [21] take advantage of Dominance Relation 5 to simplify this formulation. Let E_{012} (E_{01}) be the set of edges that can be deadheaded at most twice (once) in an optimal solution. These authors reformulate the RPP as a pure binary integer program substituting for each edge $e \in E_{012}$ the integer variable x_e with two binary variables x'_e and x''_e:

$$\text{(5.11)} \qquad x_e = x'_e + x''_e,$$

$$\text{(5.12)} \qquad x'_e, x''_e = 0 \quad \text{or} \quad 1.$$

This is equivalent to replacing each edge $e \in E_{012}$ with a pair of parallel edges e' and e''. Let E' (E'') be the set of edges e' (e''), and let $\overline{E} = E_{01} \cup E' \cup E''$. Substituting (5.11) and (5.12) in ($\text{RPP}^{(\text{CORB})}$), Ghiani and Laporte [21] obtain the following formulation:

$$\text{(5.13)} \qquad (\text{RPP}^{(\text{GL1})}) \qquad \text{minimize} \sum_{e \in \overline{E}} c_e x_e$$

$$\text{s.t.} \quad \sum_{e\in\delta(i)} x_e = 0 \bmod 2 \qquad \text{if } i \in V \text{ is } R\text{-even}, \tag{5.14}$$

$$\sum_{e\in\delta(i)} x_e = 1 \bmod 2 \qquad \text{if } i \in V \text{ is } R\text{-odd}, \tag{5.15}$$

$$\sum_{e\in\delta(S)} x_e \geq 2, \qquad S = \cup_{k\in P} V_k, P \subset \{1,\ldots,p\}, P \neq \emptyset, \tag{5.16}$$

$$x_e \in \{0,1\}, \qquad e \in \overline{E}. \tag{5.17}$$

The modulo relations are then removed:

$$(\text{RPP}^{(\text{GL2})}) \qquad \text{minimize} \sum_{e\in\overline{E}} c_e x_e \tag{5.18}$$

$$\text{s.t.} \quad \sum_{e\in\delta(i)\setminus F} x_e \geq \sum_{e\in F} x_e - |F| + 1 \Big(i \in V_R, F \supseteq \delta(i), \; |F| \text{ is odd if } i \text{ is R-even}, |F| \text{ is even if } i \text{ is } R\text{-odd}\Big), \tag{5.19}$$

$$\sum_{e\in\delta(S)} x_e \geq 2, \qquad S = \cup_{k\in P} V_k, P \subset \{1,\ldots,p\}, P \neq \emptyset, \tag{5.20}$$

$$x_e \in \{0,1\}, \qquad e \in \overline{E}. \tag{5.21}$$

Constraints (5.19) are referred to as cocircuit inequalities by Barahona and Grötschel [3]. They state that an even (odd) number of edges are incident to each R-even (R-odd) vertex. Alternatively, constraints (5.19) impose that if an odd (even) number of edges $e \in F$ are incident to a R-even (R-odd) vertex $i \in V$, then at least another edge has to be incident to i. An in-depth analysis of the facets of the polytope of formulation RPP$^{(\text{GL1})}$ is provided in Reinelt and Theis [34] and Reinelt and Theis [35].

A quite different RRP formulation was proposed by Garfinkel and Webb [18]. These authors construct an augmented graph $G_A(V_A, E_A)$ in which V_A contains all vertices in V, plus a copy i' of each R-even vertex i. The edge set E_A contains an edge for every pair of vertices in V_A except (1) pairs of vertices belonging to the same component and (2) pairs of vertices of which at least one is R-even. The cost $\overline{c}_e$ of an edge $e = (i,j) \in E_A$ is equal to that of a shortest path between vertices i and j in G. Any RPP feasible solution corresponds to a perfect matching in G_A. The connectivity of the solution is ensured by the use of additional real-valued flow variables.

Let $S_1, \ldots, S_p$ be the vertex sets obtained by adding to each set $V_1, \ldots, V_p$ the corresponding copies. Moreover, given a set of vertices $S \subset V_A$, let $\delta_A(S)$ be the set of edges of E_A with one extremity in S and one extremity in $V \setminus S$. Binary variables x_{ij} indicate whether vertex i is matched to vertex j while real-valued y_{ij} variables represent the flow from a vertex i to a vertex j in a different set S_t. The formulation by Garfinkel and Webb [18] is as follows:

$$(\text{RPP}^{(\text{GW})}) \qquad \text{minimize} \sum_{e\in E_A} \overline{c}_e x_e \tag{5.22}$$

$$\text{s.t.} \quad \sum_{e \in \delta_A(i)} x_e = 1, \quad i \in V_A, \tag{5.23}$$

$$\sum_{i \in S_1} \sum_{j \notin S_1} y_{ij} = p - 1, \tag{5.24}$$

$$\sum_{i \notin S_t} \sum_{j \in S_t} (y_{ij} - y_{ji}) = 1, \quad t = 2, \dots, p, \tag{5.25}$$

$$y_{ij} \le (p-1) x_e, \quad i, j \in V_A, i \ne j, e \in E_A : e \in \delta_A(i) \cup \delta_A(i), \tag{5.26}$$

$$x_e \in \{0, 1\}, \quad e \in E_A, \tag{5.27}$$

$$y_{ij} \ge 0, \quad i, j \in V_A : (i,j) \in E_A, \; i \in S_t \text{ for a given } t \in \{1, \dots, p\} \text{ and } j \notin S_t. \tag{5.28}$$

In this model, the objective function and constraints (5.23) and (5.27) define a minimum perfect matching in V_A. Constraint (5.24) imposes that S_1 generates $p-1$ units of flow, while constraint (5.25) states that each set of vertices S_t ($t \ne 1$) absorbs one unit of flow. Finally, constraints (5.26) link the matching and flow variables.

Fernández et al. [14] introduced a variant of (RPP$^{(GW)}$) which reduces both the number of flow variables and the number of constraints. Let $w_{tt'}$ be the flow between set S_t ($t = 1, \dots, p$) and set $S_{t'}$ ($t' = 2, \dots, p$, $t \ne t'$). The new formulation is

$$(\text{RPP}^{(\text{FMGO})}) \quad \text{minimize} \sum_{e \in E_A} \overline{c}_e x_e \tag{5.29}$$

$$\text{s.t.} \quad \sum_{e \in \delta_A(i)} x_e = 1, \quad i \in V_A, \tag{5.30}$$

$$\sum_{j=2,\dots p} w_{1j} = p - 1, \tag{5.31}$$

$$\sum_{t'=1,\dots,p, t' \ne t} (w_{t't} - w_{tt'}) = 1, \quad t = 2, \dots, p, \tag{5.32}$$

$$w_{tt'} \le (p-1) \sum_{e=(i,j) \in E_A : i \in S_t, j \in S_{t'}} x_e, \quad t = 1, \dots, p, t' = 2, \dots, p, t \ne t', \tag{5.33}$$

$$x_e \in \{0, 1\}, \quad e \in E_A, \tag{5.34}$$

$$w_{tt'} \ge 0, \quad t = 1, \dots, p, t' = 2, \dots, p, t \ne t'. \tag{5.35}$$

As in (RPP$^{(GW)}$), the objective function and constraints (5.30) define a minimum perfect matching in V_A, constraint (5.31) imposes that S_1 generates $p-1$ units of flow, constraints (5.32) state that each set of vertices S_t ($t \ne 1$) absorbs one unit of flow, while constraints (5.33) link the matching and flow variables.

5.4 ▪ Exact algorithms

Christofides et al. [5] developed a lower bound by incorporating constraints (5.2) in the objective function in a Lagrangian fashion. Given a set of multipliers, the relaxation defined by the x variables is equivalent to a shortest spanning tree (SST) problem over $G_C = (V_C, E_C)$. These authors also devised an upper bound by matching the odd-degree vertices in the subgraph induced by the SST and the required edges. These bounding procedures were then inserted in a branch-and-bound scheme. Twenty-four randomly generated instances, with $|V| \leq 84$, $|A| \leq 184$, and $p \leq 8$, were solved to optimality.

Corberán and Sanchis [9] identified some families of facet-inducing inequalities of formulation (RPP$^{(\mathrm{CORB})}$), including the K-C inequalities (see Section 6.2.2.2) and the following R-odd cut inequalities:

$$\sum_{e \in \delta(S)} x_e \geq 1, \qquad S \subset V, S \text{ is } R\text{-odd}. \tag{5.36}$$

They embedded these inequalities within a cutting plane algorithm in which the separation problems were solved by visual inspection. They were able to solve 23 out of the 24 instances introduced by Christofides et al. [5], as well as two additional instances derived from the street network of Albaida (Spain). It is worth mentioning that polynomial exact separation routines are known for both the connectivity inequalities (5.9) and the R-odd cut constraints (5.36).

Letchford [28] introduced a generalization of the K-C inequalities, called the path-bridge inequalities. The author gives a polynomial-time exact separation routine for a simple subset of the path-bridge inequalities but reports no computational results.

Corberán, Letchford, and Sanchis [6] proposed a cutting plane algorithm for the GRP using connectivity constraints, R-odd cut inequalities, K-C inequalities, path-bridge inequalities, and honeycomb inequalities (see Section 6.2), which can also be applied to the RPP. Since it is based on facet-inducing inequalities, it can yield very strong lower bounds. The algorithm was tested on 118 RPP instances with $16 \leq |V| \leq 116$ and $24 \leq |E| \leq 200$. In 116 cases, optimality was reached with the cuts alone, while the remaining two instances required branching.

Ghiani and Laporte [21] presented an extension,

$$\sum_{e \in \delta(S) \setminus F} x_e \geq \sum_{e \in F} x_e - |F| + 1 \Big(S \subset V_R, F \supseteq \delta(S), \\ |F| \text{ is odd if } S \text{ is } R\text{-even}, |F| \text{ is even if } S \text{ is } R\text{-odd}\Big), \tag{5.37}$$

of the cocircuit inequalities (5.19) to a vertex subset S. They also implemented a branch-and-cut algorithm based on the connectivity inequalities (5.20), the R-odd cut inequalities (5.36), and a subset of (5.37),

$$\sum_{e \in \delta(S) \setminus \{f\}} x_e \geq x_f, \qquad S \subset V_R, f \in \delta(S), S \text{ is } R\text{-even}, \tag{5.38}$$

called R-even inequalities. The separation problem for the connectivity inequalities is solved by means of a fast heuristic proposed by Fischetti, Salazar, and Toth [16]. To separate the R-odd cut and R-even cut inequalities, Ghiani and Laporte [21] developed two constructive heuristics. Their algorithm was able to solve to optimality, within moderate solution times, instances with up to 350 randomly generated vertices.

Fernández et al. [14] developed a cutting plane algorithm for their improvement of the formulation of Garfinkel and Webb [18]. Lower bounds are obtained by iteratively solving the LP relaxation of the formulation (RPP$^{(\mathrm{FMGO})}$) enhanced with a set of previously violated connectivity, matching, and K-C inequalities. Violated K-C inequalities are added only when violated inequalities of the other two types have not been found. Upper bounds are provided by a variation of the Frederickson [17] heuristic. The algorithm was able to solve to optimality in 160 out of 173 instances, including the two Albaida instances. For the unsolved instances, the optimality gap was always below 1%.

Corberán and Prins [8] observed that these cutting-plane and branch-and-cut algorithms would certainly be capable of solving larger instances on today's computers because the main limitation was the memory at the time of the experimentations. They also noted that these methods could easily be combined to yield better exact algorithms. Some preliminary results along this direction are reported in Theis [36].

At present the largest Undirected RPP instances ($|V| = 1000$, $|A| \leq 3080$, and $p \leq 205$) were solved by the branch-and-cut procedure for the Windy General Routing Problem presented in Corberán, Plana, and Sanchis [7].

5.5 ▪ Heuristics

One of the most commonly used heuristics for the RPP is the constructive method of Frederickson [17]. Alternative procedures have since been proposed by a number of authors, including Christofides et al. [5], Pearn and Wu [33], Fernández de Córdoba, Garcia Raffi, and Sanchis [15], Hertz, Laporte, and Nanchen Hugo [26], Groves and van Vuuren [25], and Ghiani, Laganà, and Musmanno [20]. In what follows we will briefly describe the first and last four of these contributions, which present the most interest.

5.5.1 ▪ Frederickson's heuristic

Given an undirected graph with p mutually disjoint connected components, Frederickson's heuristic first links these components together by means of an SST. It identifies the odd-degree vertices on the resulting graph and connects them by solving a matching problem, as is done for the undirected CPP [12]. The resulting graph is Eulerian and yields a feasible RPP solution. The steps of the algorithm are as follows:

Step 1. Construct an SST over an auxiliary graph H containing one vertex for each connected component of required edges of G. The cost between any two vertices of H is equal to that of a shortest path between them on G. Let T be the edge set of the spanning tree.

Step 2. Solve a minimum cost matching problem on the set of odd-degree vertices in the graph induced by $E_R \cup T$, where the costs are those of shortest paths on G. Let M be the set of edges induced by the matching.

Step 3. Determine an Eulerian cycle in the graph induced by $E_R \cup T \cup M$.

If E_R is connected, then this algorithm is exact. Otherwise, if the costs c_{ij} satisfy the triangle inequality, it has a worst-case performance ratio of 3/2. This can be proved by using the same arguments as for the Christofides [4] heuristic for the *Traveling Salesman Problem* (TSP).

Frederickson's heuristic does not always yield an optimal solution because it is not always advantageous to connect the required edges before solving the matching problem (see, for example, Figure 5.1).

5.5.2 ▪ The Monte Carlo method of Fernández de Córdoba, Garcia Raffi, and Sanchis

Fernández de Córdoba, Garcia Raffi, and Sanchis [15] have proposed a rather simple yet effective heuristic based on a probabilistic scheme. Starting at an arbitrary vertex, a tour containing all required edges is iteratively constructed by selecting at each step an edge to

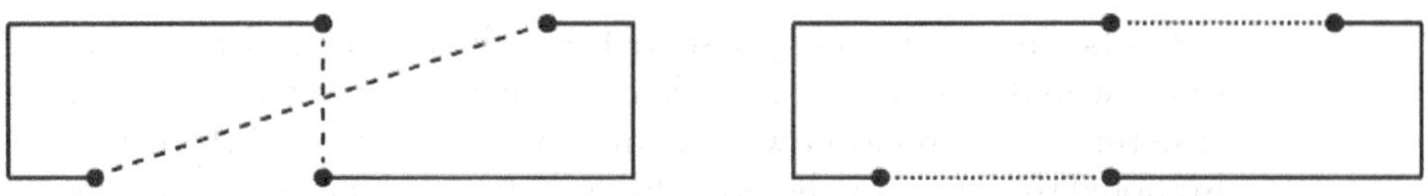

(a) Solution found by Frederickson's heuristic

(b) Optimal solution

Figure 5.1. *Example for which Frederickson's heuristic does not always yield an optimal solution.*

traverse in a probabilistic manner. When all required edges have been traversed, the tour is closed by returning to the starting vertex through a shortest path.

The probability of selecting an edge depends on its nature and on its cost relative to those of the other edges available for traversal. The authors consider three types of edges: (1) required edges; (2) nonrequired edges incident to a yet untraversed required edge; (3) other nonrequired edges, with decreasing probabilities of selecting edges of type 1, 2, and 3, respectively. Similarly, the probability of selecting an edge is inversely related to its cost. The solution generated by this randomized procedure is then improved by applying three simplification operations: (1) Remove pairs of traversal of the same edge when there are at least three traversals; (2) if a nonrequired edge is traversed twice, remove the two traversals if this does not disconnect the graph; (3) consider a path $(e_1,\ldots,e_k)$ of edges traversed twice between two vertices i and j; then delete one copy of each edge and add one copy of each edge of another path from i to j (e.g., a shortest path).

This heuristic can be repeated a large number of times (1000 times in the authors' implementation). On 26 small and medium-size instances with $7 \leq |V| \leq 102$, $10 \leq |E| \leq 184$, and $4 \leq |E_R| \leq 99$, it identified 19 optimal solutions.

5.5.3 ▪ The improvement heuristics of Hertz, Laporte, and Nanchen Hugo

Hertz, Laporte, and Nanchen Hugo [26] provided a set of improvement heuristics which are applicable to any feasible RPP solution.

The first of these heuristics, called SHORTEN, transforms a solution into an equivalent one starting with a long chain of unrequired edges, which is then shortened by applying a shortest path algorithm. It scans the edges of a solution and applies one of two operations to each edge. If edge (i,j) is required but appears later in the solution as nonrequired, then the operator POSTPONE changes the status of these two edges. If an edge (i,j) is required and does not appear later as nonrequired, then the operator REVERSE attempts to identify a path $(i,j,\ldots,k,i)$ in which edge (k,i) is nonrequired, and it reverses this path. These two procedures are applied as long as it is possible to do so. The longest chain $(i,\ldots,j)$ at the start of the resulting solution is then replaced with a shortest path from i to j.

To illustrate, consider the RPP solution depicted in Figure 5.2(a), where the thick edges are required and the others are not. The statuses of the first and last edges are first reversed. In Figure 5.2(b), the first required edge is (3,4) and this edge does not appear later in the solution. However, the chain (3,4,2,3) can be reversed to yield the solution of Figure 5.2(c). This solution can no longer be transformed by applying POSTPONE or REVERSE. However, it can be shortened by replacing the initial chain (1,2,3,2,4) with the shorter chain (1,2,4), as shown in Figure 5.2(d).

The next two heuristics, called DROP and ADD, are very simple and technical. They are meant to be used as building blocks in other heuristics. DROP removes a required edge (i,j) from the solution. This is achieved by replacing it with a shortest path between i

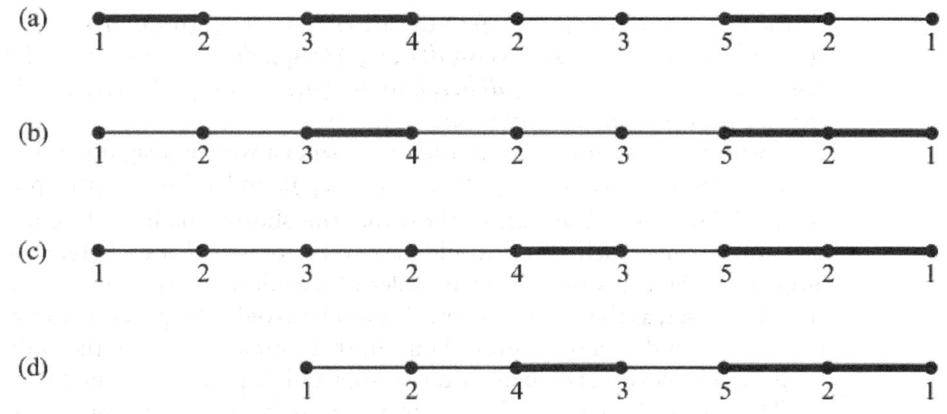

Figure 5.2. *RPP solution.*

and j. ADD inserts a new edge (i,j) into an existing solution by linking i and j to the solution by means of shortest paths.

Using SHORTEN, DROP, and ADD, Hertz, Laporte, and Nanchen Hugo [26] have created the arc routing equivalent of the classical r-opt postoptimization procedures for the TSP [30]. However, the implementation of r-opt in an arc routing context can be rather intricate because required edges are sometimes removed in the process. To illustrate, we restrict our attention to 2-opt. This procedure repeatedly removes two edges from the solution and reconnects the two remaining chains by means of the shortest chains. SHORTEN is then called. If any required edge is missing from the solution because it has been removed, it is reinserted in the solution by means of ADD. The complexity of this heuristic, computed by Groves and van Vuuren [25], is $O(|E|^5)$ per iteration, which makes it unsuitable for large instances.

Another postoptimization procedure is called DROP-ADD: It removes a required edge from the solution, calls SHORTEN, and reinserts the edge by means of ADD. This process is repeated as long as improvements can be obtained.

The algorithms 2-opt and DROP-ADD were tested on several classes of RPP instances. Overall, 2-opt performed better than DROP-ADD but was more time consuming. On some instance classes for which the optimum was known, the latter heuristic yielded an average optimality gap of around 3%, whereas applying 2-opt as a postoptimizer closed this gap to around 0%.

5.5.4 ▪ The heuristics of Groves and van Vuuren

Groves and van Vuuren [25] have designed rather efficient 2-opt and 3-opt heuristics for the undirected RPP. Both operate on a vector $((v_{s_1}, v_{t_1}), \ldots, (v_{s_n}, v_{t_n}))$ whose components are the required edges. A 2-opt (resp., 3-opt) move consists of removing and relocating two (resp., three) components from this vector. A directed layered auxiliary graph $\mathscr{L}$ with vertex set $V(\mathscr{L}) = \{b, s_1t_1, t_1s_1, \ldots, s_nt_n, t_ns_n, e\}$ is then constructed. In this set, b and e are the vertices at which the tour must begin and end. The elements s_it_i and t_is_i of $V(\mathscr{L})$ represent the two traversal directions of edge (v_{s_i}, v_{t_i}). For each $i (1 \leq i \leq n-1)$, an arc of $\mathscr{L}$ is directed from s_it_i to $s_{i+1}t_{i+1}$ and $t_{i+1}s_{i+1}$, and from t_is_i to $s_{i+1}t_{i+1}$ and $t_{i+1}s_{i+1}$. Arcs are also created from b to s_1t_1 and to t_1s_1, and from s_nt_n and t_ns_n to e. Let $d(i,j)$ be the length of a shortest path from vertex i to vertex j in G. Then assign a cost $d(t_i, s_{i+1})$ (resp., $d(s_i, t_{i+1})$) to the arc $(s_i t_i, s_{i+1} t_{i+1})$ (resp., $(t_i s_i, t_{i+1} s_{i+1})$) for every $i (1 \le i \le n-1)$. Also assign a cost $d(t_i, t_{i+1})$ (resp., $d(s_i, s_{i+1})$) for every $i (1 \le i \le n-1)$. Assign a cost $d(b, s_1)$ (resp., $d(b, t_1)$) to arc $(b, s_1 t_1)$ (resp., $(b, t_1 s_1)$) and a cost $d(t_n, e)$ (resp., $d(s_n, e)$) to arc $(s_n t_n, e)$ (resp., $(t_n s_n, e)$).

Each path from b to e in $\mathcal{L}$ then represents a way of assigning traversal directions to the edges of the vector $((v_{s_1}, v_{t_1}), \dots, (v_{s_n}, v_{t_n}))$, and a shortest path provides the best traversal directions. The authors show that this shortest path can be computed in $O(n)$ operations. Since such a path would have to be computed at each iteration of a 2-opt or 3-opt algorithm, this would add an order of magnitude to the complexity of each iteration. However, as the authors show, this can be avoided by properly exploiting available information and data structures. Thus, in their implementation, the authors manage to keep the complexity of each iteration of 2-opt and 3-opt to $O(n^2)$ and $O(n^3)$, respectively.

The authors have tested their 2-opt and 3-opt heuristics on instances with Euclidean distances and $507 \le |V| \le 979$, $2528 \le |E| \le 3499$, $304 \le |E_R| \le 350$, and on instances with random distances and $536 \le |V| \le 976$, $2641 \le |E| \le 3451$, $257 \le |E_R| \le 345$. Their results are summarized in Table 5.1.

Table 5.1. *Deviation of the upper bound over the lower bound.*

Data set	Frederickson	2-opt	3-opt
Euclidean	5.9%	3.8%	3.0%
Random	4.1%	2.5%	2.0%

On Euclidean instances with $3130 \le |V| \le 4927$, $20660 \le |E| \le 29835$, $2066 \le |E_R| \le 2984$, the 2-opt heuristic consistently outperformed the Frederickson heuristic within slightly larger computation times.

5.5.5 ▪ The heuristic of Ghiani, Laganà, and Musmanno

Ghiani, Laganà, and Musmanno [20] described a fast heuristic whose use is recommended when a high-quality solution is needed within a short amount of time (e.g., in laser plotter applications). At each iteration, the procedure inserts a connected component of the required edges. A 2-opt local search is then performed in an attempt to reduce the cost. Computational results on a set of benchmark instances with up to 350 vertices show that the algorithm is competitive with the classical Frederickson procedure.

5.6 ▪ Variants

A number of variants and extensions of the undirected RPP, typically related with real-life applications, have been investigated. Two of these, the *Prize-Collecting RPP* [2, 1] and the *Profitable Arc Tour Problem* [13], are presented in Chapter 12. In what follows, we describe variants related to the RPP with multiple edge services, the RPP with deadline classes, and the dynamic RPP.

5.6.1 ▪ The RPP with multiple edge services

We summarize two papers in which edges may have to be serviced several times in a solution. The first, by Ghiani et al. [22], introduces the periodic arc routing problem, in which each required edge $e \in E_R$ must be serviced n_e times over an m-period horizon, and the periods in which the edges are serviced must be equally spaced. This implies of course that m must be divisible by n_e for all e. This is more restrictive than the conditions imposed in the *Periodic Vehicle Routing Problem* (see, e.g., [11]) in which customers specify allowed combinations of service periods, without requiring equal spacing. Applications of the *Periodic RPP* arise naturally in garbage collection and in street sweeping. The authors have proposed a heuristic in which all edges e having the same service frequency n_e are initially assigned the same service day combination, and a local search is then performed to yield cost reductions. Three procedures are applied: path transfers, cycle transfers, and component transfers.

The first two routines attempt to remove deadheaded edges by suitably moving paths of serviced edges from a service combination to another having the same frequency. The third subroutine verifies, for any service frequency n, whether a cost saving can be obtained by assigning a different service day combination to a connected component induced by the edges with frequency n. In order to assess this heuristic, the authors have generated a number of instances whose optimal solution is known a priori: First, an Eulerian random graph was created; then frequencies were assigned to the edges in such a way that no deadheading would be required. The authors succeeded in solving instances with $|V| = 100, 150, 200, 250$ and $m = 6$. The proposed heuristic yielded optimality gaps varying between 0.58% and 2.92%, depending on the instance classes.

Groves, le Roux, and van Vuuren [24] also solved a variant in which edges may be serviced several times, and a suitable spacing must be achieved between two consecutive services of the same edge. Applications of this problem are encountered in the maintenance of transportation networks and in garbage collection or snow plowing operations. The two goals of cost minimization and of service spacing are somewhat conflictual. They were handled simultaneously through a bicriteria objective. The authors have considered two ways of modeling the spacing constraint: through the number of edge traversals between consecutive services of the same edge, and through a temporal deviation within a specified time window. The second of these options proved to be the most satisfactory.

The authors developed a simple construction heuristic which links circuit segments through several copies of the graph. Local search moves, such as those described in Groves and van Vuuren [25], are then applied. Instances defined on six types of graphs, with $20 \leq |E| \leq 298$, were solved. Since no good lower bounds are available for this problem, it is difficult to assess the accuracy of this heuristic. It was observed that the average improvement yielded by the local search phase over the construction phase was equal to 13%.

5.6.2 ▪ The RPP with deadline classes

In the RPP with deadline classes [29], the set E_R of required edges is partitioned into $\{E_R^1, \ldots, E_R^p\}$ and service for each class k of edges ($k = 1, \ldots, p$) must be completed by time T^k. The authors have developed an integer linear programming model for this problem. They have also proposed several classes of valid inequalities which exploit the structure of the problem, but they have not attempted to identify conditions under which these induce facets. The valid inequalities are called strong cumulative inequalities, strong cumulative connectivity inequalities, strong cumulative parity inequalities, and strong cumulative path-bridge inequalities. Some of these inequalities are natural counterparts of similar inequalities derived for the RPP [9, 10, 6, 28]. The authors have also provided lifted versions of these inequalities. The connectivity and parity inequalities can be separated in polynomial time, while heuristics are used for the strong cumulative path-bridge inequalities. The problem was solved by branch-and-cut. Instances with $17 \leq |V| \leq 50$, $35 \leq |E| \leq 110$, $22 \leq |E_R| \leq 67$, and $p = 1$ or 2 were solved to optimality.

5.6.3 ▪ The dynamic RPP

The dynamic RPP investigated by Moreira et al. [31] is motivated by an industrial application in which pieces of different shapes have to be cut out of a circular area by means of an electrified copper string. In a first phase, the pieces are nested (or packed) in the circular area (Figure 5.3). In a second phase, which is relevant as far as the RPP is concerned, the pieces are cut out by the copper string and fall out of the circle when their entire perimeter has been cut. Any movement of the cutting tool generates an effective cut, unless of course it is performed over a region already cut out. Like the laser-beam application described by Ghiani and Improta [19], this problem can be modeled as an RPP. However, because pieces are sometimes cut out, the graph on which the RPP is solved changes in a dynamic fashion (Figure 5.4). Also, because some of the pieces may share common edges, other practical considerations complicate the problem; namely, there may not remain uncut pieces in areas that have been cut out.

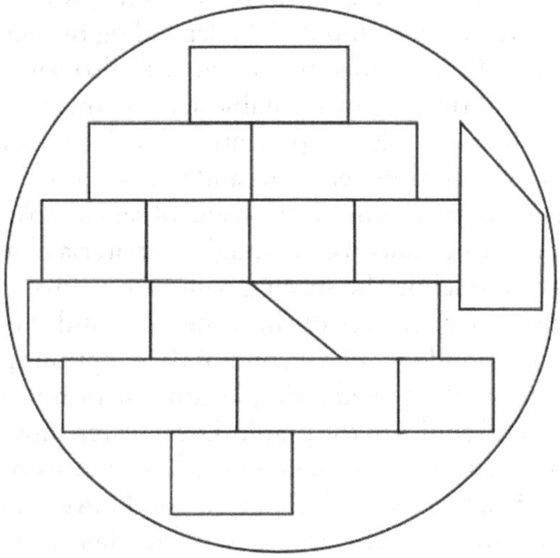

Figure 5.3. *Pieces nested in a circular area.*

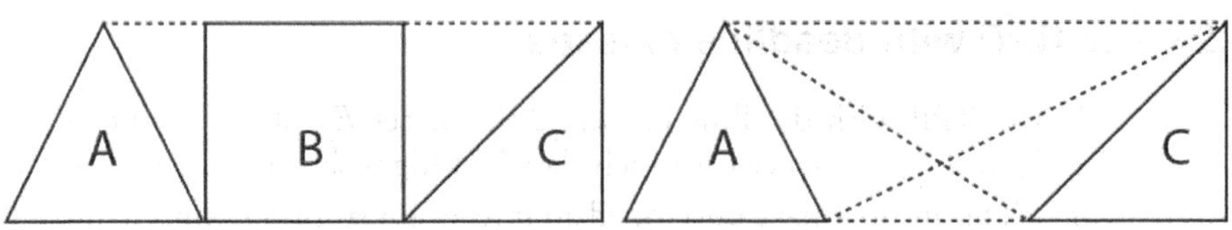

(a) Before B has been cut out (b) After B has been cut out

Figure 5.4. *RPP graph before and after piece B has been cut out.*

The authors have devised two constructive heuristics for this problem. The first, called "higher up vertex" (HUV): Given the current vertex, is a higher up vertex among the vertices reachable by uncut edges, under some side conditions? The second, called "minimum empty path" (MEP) is a variant of HUV which uses different criteria for the execution of unrequired movements. The two heuristics were tested and compared on 10 industrial instances containing between 19 and 52 pieces, and between 183 and 639 vertices. Deviations from a lower bound were computed. On average, HUV yielded an optimality gap of 17.5%, whereas this gap was equal to 15.2% for MEP.

Bibliography

[1] J. ARÁOZ, E. FERNÁNDEZ, AND O. MEZA, *Solving the prize-collecting rural postman problem*, European Journal of Operational Research, 196 (2009), pp. 886–896.

[2] J. ARÁOZ, E. FERNÁNDEZ, AND C. ZOLTAN, *Privatized rural postman problems*, European Journal of Operational Research, 33 (2006), pp. 3432–3449.

[3] F. BARAHONA AND M. GRÖTSCHEL, *On the cycle polytope of a binary matroid*, Journal of Combinatorial Theory, 40 (1986), pp. 40–62.

[4] N. CHRISTOFIDES, *Worst-case analysis of a new heuristic for the travelling salesman problem*, Tech. Report 388, School of Industrial Administration, Carnegie-Mellon University, Pittsburgh, PA, 1976.

[5] N. CHRISTOFIDES, V. CAMPOS, Á. CORBERÁN, AND E. MOTA, *An algorithm for the rural postman problem*, Tech. Report IC.O.R.81.5, Imperial College, London, 1981.

[6] Á. CORBERÁN, A. N. LETCHFORD, AND J. M. SANCHIS, *A cutting plane algorithm for the general routing problem*, Mathematical Programming, Series A, 90 (2001), pp. 291–316.

[7] Á. CORBERÁN, I. PLANA, AND J. M. SANCHIS, *A branch & cut algorithm for the windy general routing problem and special cases*, Networks, 49 (2007), pp. 245–257.

[8] Á. CORBERÁN AND C. PRINS, *Recent results on arc routing problems: An annotated bibliography*, Networks, 56 (2010), pp. 50–69.

[9] Á. CORBERÁN AND J. M. SANCHIS, *A polyhedral approach for the rural postman problem*, European Journal of Operational Research, 79 (1994), pp. 95–114.

[10] ——, *The general routing problem polyhedron: Facets from the RPP and GTSP polyhedra*, European Journal of Operational Research, 108 (1998), pp. 538–550.

[11] J.-F. CORDEAU, M. GENDREAU, AND G. LAPORTE, *A tabu search heuristic for periodic and multi-depot vehicle routing problems*, Networks, 30 (1997), pp. 105–119.

[12] J. EDMONDS AND E. L. JOHNSON, *Matching, Euler tours and the Chinese postman problem*, Mathematical Programming, 5 (1973), pp. 88–124.

[13] D. FEILLET, P. DEJAX, AND M. GENDREAU, *The profitable arc tour problem: Solution with a branch and price algorithm*, Transportation Science, 39 (2005), pp. 539–552.

[14] E. FERNÁNDEZ, O. MEZA, R. GARFINKEL, AND M. ORTEGA, *On the undirected rural postman problem: Tight bounds based on a new formulation*, Operations Research, 51 (2003), pp. 281–291.

[15] P. Fernández de Córdoba, L. M. Garcia Raffi, and J. M. Sanchis, *A heuristic algorithm based on Monte Carlo methods for the rural postman problem*, Computers & Operations Research, 25 (1998), pp. 1097–1105.

[16] M. Fischetti, J. J. Salazar, and P. Toth, *A branch-and-cut algorithm for the symmetric generalized travelling salesman problem*, Operations Research, 45 (1997), pp. 378–393.

[17] G. N. Frederickson, M. S. Hecht, and C. E. Kim, *Approximation algorithms for some routing problems*, SIAM Journal on Computing, 7 (1978), pp. 178–193.

[18] R. Garfinkel and I. Webb, *On crossings, the crossing postman problem and the rural postman problem*, Networks, 34 (1999), pp. 173–180.

[19] G. Ghiani and G. Improta, *The laser-plotter beam routing problem*, The Journal of the Operational Research Society, 52 (2001), pp. 887–903.

[20] G. Ghiani, D. Laganà, and R. Musmanno, *A constructive heuristic for the undirected rural postman problem*, Computers & Operations Research, 33 (2006), pp. 3450–3457.

[21] G. Ghiani and G. Laporte, *A branch-and-cut algorithm for the undirected rural postman problem*, Mathematical Programming, 87 (2000), pp. 467–481.

[22] G. Ghiani, R. Musmanno, G. Paletta, and C. Triki, *A heuristic for the periodic rural postman problem*, Computers & Operations Research, 32 (2005), pp. 219–228.

[23] M. Grötschel, M. Jünger, and G. Reinelt, *Optimal control of plotting and drilling machines: A case study*, Zeitschrift für Operations Research, 35 (1991), pp. 61–84.

[24] G. W. Groves, J. le Roux, and J. H. van Vuuren, *On a routing and scheduling problem concerning multiple edge traversals in graphs*, Networks, 46 (2005), pp. 69–81.

[25] G. W. Groves and J. H. van Vuuren, *Efficient heuristics for the rural postman problem*, ORiON, 21 (2005), pp. 33–51.

[26] A. Hertz, G. Laporte, and P. Nanchen Hugo, *Improvement procedures for the undirected rural postman problem*, INFORMS Journal on Computing, 11 (1999), pp. 53–62.

[27] J. K. Lenstra and A. H. G. Rinnooy Kan, *On general routing problems*, Networks, 6 (1976), pp. 273–280.

[28] A. N. Letchford, *New inequalities for the general routing problem*, European Journal of Operational Research, 96 (1997), pp. 317–322.

[29] A. N. Letchford and R. W. Eglese, *The rural postman problem with deadline classes*, European Journal of Operational Research, 105 (1998), pp. 390–400.

[30] S. Lin, *Computer solutions of the traveling salesman problem*, Bell System Technical Journal, 44 (1965), pp. 2245–2269.

[31] L. M. MOREIRA, J. F. OLIVEIRA, A. M. GOMEZ, AND J. S. FERREIRA, *Heuristics for a dynamic rural postman problem*, Computers & Operations Research, 34 (2007), pp. 3281–3294.

[32] C. S. ORLOFF, *A fundamental theorem in vehicle routing*, Networks, 27 (1974), pp. 35–64.

[33] W. L. PEARN AND T. C. WU, *Algorithms for the rural postman problem*, Computers & Operations Research, 22 (1995), pp. 819–828.

[34] G. REINELT AND D. O. THEIS, *Transformation of facets of the general routing problem polytope*, SIAM Journal on Optimization, 16 (2005), pp. 220–234.

[35] ——, *On the general routing polytope*, Discrete Applied Mathematics, 156 (2008), pp. 368–384.

[36] D. O. THEIS, *Das General Routing Problem mit binären Variablen*, Ph.D. thesis, Diplomarbeit, Ruprecht-Karls-Universität Heidelberg, Institut für Informatik, Heidelberg, Germany, 2001.

Chapter 6

The Rural Postman Problem on Directed, Mixed, and Windy Graphs

Ángel Corberán
Isaac Plana
José María Sanchis

6.1 ▪ Introduction

The previous chapter provided an overview of the *Rural Postman Problem* (RPP) on an undirected graph. Recall that given a graph and a subset of required edges, this problem consists of finding a minimum cost tour (a closed walk) traversing every required edge at least once. This problem was proposed by Orloff [49] and shown to be NP-hard by Lenstra and Rinnooy Kan [45]. This chapter deals with its directed, mixed, and windy versions. For simplicity, and given that here we will work with directed, mixed, and windy graphs, we will use the term "link" to refer to (undirected) edges and (directed) arcs indistinctly.

A problem closely related to the RPP is the *General Routing Problem* (GRP). Given a graph with nonnegative costs associated with its links, a subset of required links, and a subset of required vertices, the GRP consists of finding a minimum cost tour traversing each required link at least once and visiting each required vertex at least once. The GRP is a generalization of the RPP, so many of the results obtained on the GRP can be applied to the RPP. Since it is not as widely known as the RPP and it is not a pure arc routing problem, the GRP will not be addressed in this chapter, but the results found in several studies on the GRP defined on directed, mixed, and windy graphs will be referenced and applied to the corresponding version of the RPP.

The *Mixed Rural Postman Problem* (MRPP), defined over a mixed graph (containing arcs and edges), is an NP-hard problem (Corberán, Romero, and Sanchis [27]), since it contains the RPP as a particular case when there are no arcs in the graph. In the MRPP, we have a subset of required edges and a subset of required arcs. It can be assumed that all the vertices of the graph are incident with at least one required link. If this is not the case, the instance can easily be transformed into an equivalent one in which this property holds (see e.g. [27]).

When the RPP is defined on a directed graph (containing only arcs), we have the *Di-*

rected Rural Postman Problem (DRPP), which is also NP-hard (Lenstra and Rinnooy Kan [45]). This problem will be studied at the end of this chapter.

The *Windy Rural Postman Problem* (WRPP) is defined on an undirected graph where each edge has two associated costs, one for each direction of traversal. This problem also contains the RPP as a special case and therefore is NP-hard too (Benavent et al. [9] and [8]). As with the MRPP, we can assume that all the vertices in the graph are incident with at least one required edge. The study of the WRPP is of special interest for two reasons. First, it generalizes the RPP on undirected, directed, and mixed graphs. Secondly, while the MRPP has obvious applications in the context of garbage collection, mail delivery, etc., there are several other specific real-life applications that can be modeled using windy graphs only. Two of these applications are described in what follows.

The periodical inspection of complex structures, such as bridges or skyscrapers, requires checking the integrity of the surface of the beams in their skeleton (see Figure 6.1). Since this is a risky task, it could be carried out by remotely controlled robots equipped with a camera, such as the one in Figure 6.2 (Balaguer et al. [7]). The beam structure can be represented by means of an undirected graph where some edges representing the surfaces of the beams (dashed lines in the graph in Figure 6.1) are required, while some others representing the movements needed to move from one surface to another (bold and dotted lines in the graph in Figure 6.1) are not required. Moreover, each edge has two associated costs corresponding to the amount of energy required to perform this movement in each possible direction. Note that these two costs can be different, given that gravity affects energy consumption. Therefore, the problem of designing a route that traverses all the required edges at minimum cost is a WRPP (Benavent et al. [8]).

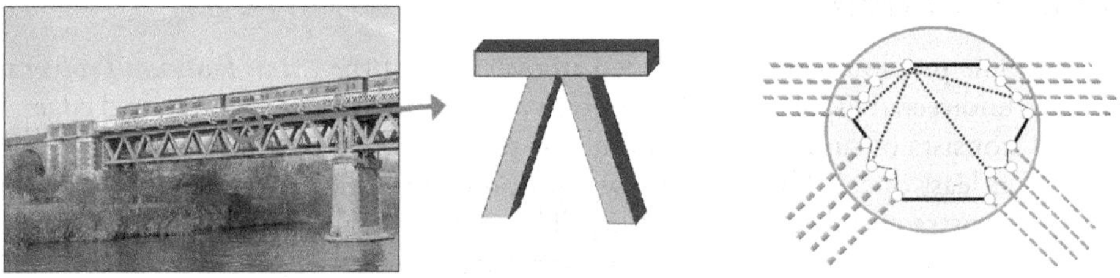

Figure 6.1. *Bridge, beam structure, and graph modeling [8].*

A further application of the WRPP is related to the design of the movements of cutting plotters, such as that in Figure 6.3. These machines are used to print and cut signs, stickers, etc. To this end, the machine is equipped with a razor that can be moved in two directions. The objective is to design the movements of the paper and the razor in order to complete the cutting of all the designs in the minimum possible time. Moreover, the cost of moving the paper in one direction is different from that of moving it in the opposite one, and therefore we may have nonsymmetric costs for the edges. The movements that represent the cutting of the paper are modeled as required edges, while the movements of the paper and the razor that do not require cutting correspond to nonrequired edges.

6.2 ▪ The mixed Rural Postman Problem

In this section, a formal definition of the MRPP is given, some notation is introduced, and heuristic and exact algorithms for this problem are presented.

Consider a strongly connected mixed graph $G = (V, E, A)$ with a vertex set V, an edge set E, and an arc set A, and a nonnegative cost c_e for each $e \in E \cup A$. Given a subset

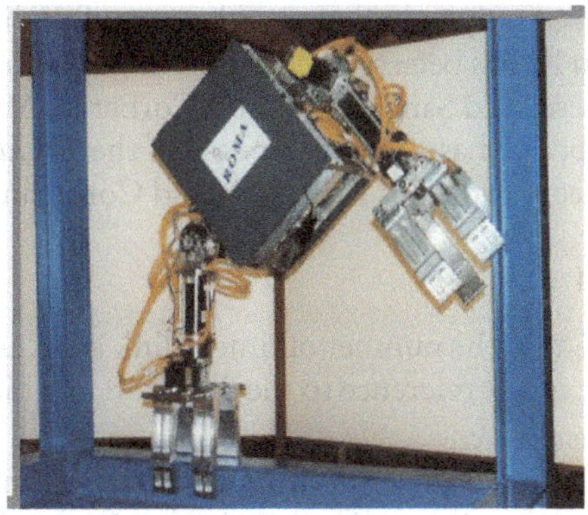

Figure 6.2. *A climber Robot.*

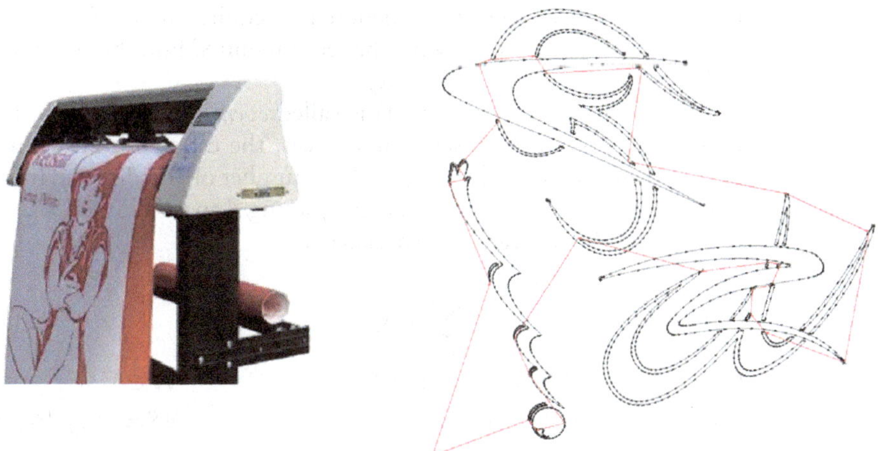

Figure 6.3. *Cutting plotter and design fragment (provided by S. Defever).*

of required edges $E_R \subseteq E$ and a subset of required arcs $A_R \subseteq A$, the MRPP consists of finding a minimum cost tour traversing all the required edges and all the required arcs at least once.

In what follows, it will be assumed that all the edges in the graph are required, i.e., $E_R = E$, since the original graph can easily be transformed into another one satisfying this property (see e.g. Christofides et al. [16]) by replacing each nonrequired edge with two opposite nonrequired arcs of the same cost. Note that the graph induced by the required links $G(V, E \cup A_R)$ is in general nonconnected. The vertex sets of its connected components, denoted by $V_1, \dots, V_p$, are called R-sets.

6.2.1 ▪ Problem formulations

The MRPP, just as the *Mixed Chinese Postman Problem* (MCPP) and other ARPs defined on mixed graphs, can be formulated in two ways by using two or just one variable per edge of the graph. Both formulations have been widely used in the literature and compared in Corberán, Mota, and Sanchis [22] (see also Chapter 4). However, these formulations and their associated polyhedra have not been studied directly in the literature for the MRPP. Formulation F1 has been studied for the *Mixed General Routing Problem* (MGRP), in Corberán, Romero, and Sanchis [27] and Corberán, Mejía, and Sanchis [21], while formulation F2 is obtained as a particular case of the *Windy General Routing Problem* (WGRP), which has been studied in Plana [50] and Corberán, Plana, and Sanchis [25, 26].

6.2.1.1 ▪ Formulation F1

Let us define x_e as the number of times a link $e \in E \cup A$ appears in the tour. If it is necessary to make an explicit reference to the direction in which a required link is traversed, x_{ij} will be used instead of x_e.

For $S_1, S_2 \subseteq V$, $(S_1 : S_2)$ denotes the set of links with one end-point in S_1 and the other in S_2, $A(S_1 : S_2) = \{(i,j) \in A : i \in S_1, j \in S_2\}$, and $E(S_1 : S_2)$ denotes the set of edges with one endpoint in S_1 and the other one in S_2. Moreover, $\delta^+(S) = A(S : V \setminus S)$, $\delta^-(S) = A(V \setminus S : S)$, and $\delta(S) = E(S : V \setminus S)$, while $E(S)$ and $A(S)$ denote the sets of edges and arcs, respectively, with both endpoints in S. Furthermore, $\delta^*(S) = \delta(S) \cup \delta^+(S) \cup \delta^-(S) = (S : V \setminus S)$. In order to denote the required links of a given set, subindex R will be used (e.g., $(S_1 : S_2)_R$ represents the set of required links between S_1 and S_2). Finally, for a subset of links F, $x(F) = \sum_{e \in F} x_e$.

A mixed graph $G = (V, E, A)$ is called *even* if all its vertices have even degree and *balanced* if, given any subset S of vertices, the difference between the number of arcs directed from S to $V \setminus S$, $|\delta^+(S)|$, and the number of arcs directed from $V \setminus S$ to S, $|\delta^-(S)|$, is no greater than the number of edges joining S and $V \setminus S$, $|\delta(S)|$.

Thus, the MRPP can be formulated as

$$(6.1) \quad \text{(MRPP-F1)} \quad \min \sum_{e \in E \cup A} c_e x_e$$

$$(6.2) \quad \text{s.t.} \quad x(\delta^*(i)) \equiv 0 \bmod 2 \qquad \forall i \in V,$$

$$(6.3) \quad x(\delta^+(S)) \geq 1 \qquad \forall S = \cup_{k \in Q} V_k, \; Q \subset \{1, \ldots, p\},$$

$$(6.4) \quad x(\delta^+(S)) - x(\delta^-(S)) \leq x(\delta(S)) \qquad \forall S \subset V,$$

$$(6.5) \quad x_e \geq 1 \text{ and integer} \qquad \forall e \in E \cup A_R,$$

$$(6.6) \quad x_e \geq 0 \text{ and integer} \qquad \forall e \in A \setminus A_R.$$

The above formulation is stated in terms of tours, where each variable represents the number of times the associated link is traversed, while the one given in Corberán, Romero, and Sanchis [27] was presented using what the authors called semitours, which are the result of removing one copy of each required link from a tour. In what follows, the inequalities will be expressed considering the solutions as tours.

As with the MCPP (see Chapter 4), if $K_1, \ldots, K_q$ denote the sets of vertices of the connected components of the graph (V, E), all the MRPP solutions satisfy the following q system equations:

$$(6.7) \quad x(\delta^+(K_i)) = x(\delta^-(K_i)), \quad i = 1, \ldots, q,$$

of which $q - 1$ are linearly independent.

6.2.1.2 ▪ MRPP polyhedron associated with formulation F1

Let $\text{MRPP}_{F1}(G)$ be the convex hull of all the tours $x \in \mathbb{R}^{E \cup A}$ satisfying (6.2) to (6.6). $\text{MRPP}_{F1}(G)$ is an unbounded polyhedron with $\dim(\text{MRPP}_{F1}(G)) = |E \cup A| - q + 1$ and the following inequalities included in the formulation of the MRPP are facet-inducing under mild conditions (Corberán, Romero, and Sanchis [27]):

- Trivial inequalities: $x_e \geq 1 \; \forall e \in E \cup A_R$ and $x_e \geq 0 \; \forall e \in A \setminus A_R$.
- Connectivity inequalities (6.3).
- Balanced-set inequalities (6.4).

In what follows, other inequalities defining facets of $\mathrm{MRPP}_{F1}(G)$ are presented.

R-odd cut inequalities

A cutset $\delta^*(S)$ is called R-odd if it contains an odd number of required links. The following R-odd cut inequalities were shown to be valid and facet-inducing for $\mathrm{MRPP}_{F1}(G)$ in Corberán, Romero, and Sanchis [27]:

$$x(\delta^*(S)) \geq |\delta^*_R(S)| + 1, \; \forall \delta^*(S) \;\; R\text{-odd cutset of } G. \tag{6.8}$$

K-C and K-C_{02} inequalities

K-C inequalities were proposed for the undirected RPP (Corberán and Sanchis [28]) and generalized to the MRPP in Corberán, Romero, and Sanchis [27]. Let K be an integer with $K \geq 3$. Let us consider a partition of V into $K+1$ subsets $\{M_0, M_1, \ldots, M_K\}$ such that each R-set V_i, $1 \leq i \leq p$, is contained in exactly one of the node sets $M_0 \cup M_K, M_1, \ldots, M_{K-1}$ and $(M_0 : M_K)$ contains a positive and even number of required links. Given a link $(i,j) \in E \cup A$ with $i \in S$, $j \in T$, the coefficient corresponding to its associated variable is $c(S,T)$, where $c(M_0, M_K) = K-2$ and $c(M_i, M_j) = |i-j|$, $\forall i,j : \{i,j\} \neq \{0,K\}$ (see Figure 6.4(a)).

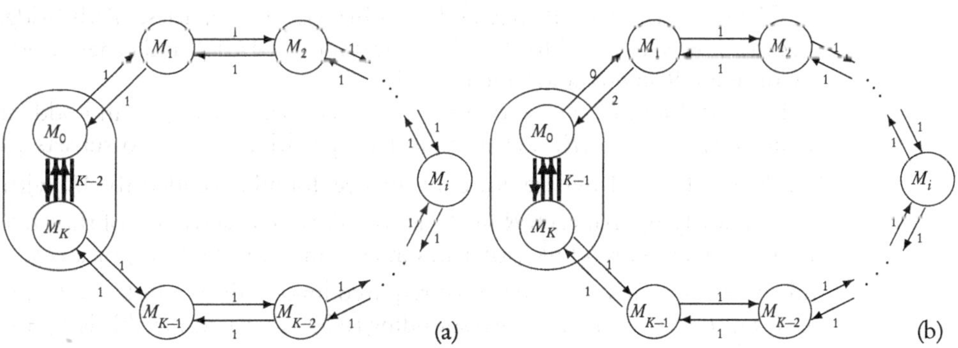

Figure 6.4. *K-C and K-C_{02} configurations.*

The coefficients associated with variables corresponding to links joining vertices of the same subset M_i are zero. The K-C inequality is

$$(K-2)\, x\,(M_0 : M_K) + \sum_{\substack{0 \leq i < j \leq K \\ (i,j) \neq (0,K)}} |i-j|\, x(M_i : M_j) \geq 2(K-1) + (K-2)|(M_0 : M_K)_R|. \tag{6.9}$$

These inequalities are valid for the MRPP, and they are facet-inducing for $\text{MRPP}_{F1}(G)$ if the induced subgraphs $G(M_i)$, $i = 0,1,\dots,K$, are strongly connected, the sets $A(M_i : M_{i+1})$ and $A(M_{i+1} : M_i)$ are nonempty, and

$$|E(M_0 : M_K)| \geq \big||A_R(M_K : M_0)| - |A_R(M_0 : M_K)|\big|,$$

i.e., if there are enough edges in $(M_0 : M_K)$ to balance this cutset.

K-C inequalities have symmetric coefficients, i.e., $c(S,T) = c(T,S)$, because they were obtained from those proposed for the undirected RPP. Therefore, they ignore the directed nature of the MRPP, where we can have opposite arcs with different costs. The following inequalities, called K-C_{02}, try to exploit this fact by using different coefficients for some opposite arcs. They are based on the same partition of V, with the only difference being that now K can be 2. Figure 6.4(b) shows the coefficients of the links joining consecutive sets. Arcs in $A(M_i : M_j)$ not represented in the figure have a coefficient equal to the length of the shortest path from M_i to M_j. Note that the links in $(M_0 : M_K)$ now have a coefficient $K-1$, while the opposite arcs in $(M_0 : M_1)$ now have coefficients 0 and 2. In fact, these 0-2 coefficients could be placed between any pair of consecutive subsets M_i, M_{i+1}, but the corresponding inequalities would induce the same facets. If we denote the coefficient associated with each link e by α_e, the K-C_{02} inequality is

$$\sum_{e \in E \cup A} \alpha_e x_e \geq 2(K-1) + (K-1)|(M_0 : M_K)_R|. \tag{6.10}$$

In Corberán, Romero, and Sanchis [27] it is proved that these inequalities are valid and facet-inducing for $\text{MRPP}_{F1}(G)$ under certain conditions.

Path-Bridge and Path-Bridge$_{02}$ inequalities

In the K-C and K-C_{02} inequalities there is a single path of sets $M_1,\dots,M_{K-1}$ joining sets M_0 and M_K. The generalizations of these inequalities in which any number of paths is considered are the Path-Bridge and Path-Bridge$_{02}$ inequalities. Path-Bridge inequalities were introduced by Letchford [46] for the undirected GRP and generalized to the MGRP in Corberán, Romero, and Sanchis [27].

Let $P \geq 1$ and $B \geq 0$ be two integers such that $P + B \geq 3$ and odd, and let $n_i \geq 2$ be integers, $i = 1,\dots,P$. Let us consider a partition of V into subsets $\{A, Z, M_j^i : i = 1,\dots P,\ j = 1,\dots n_i\}$ (where, for convenience, for all i we identify M_0^i with A and $M_{n_i+1}^i$ with Z) satisfying that each R-set V_i is contained in exactly one of the node sets $A \cup Z$ or M_j^i (i.e., each required link either lies in certain $E(M_j^i) \cup A(M_j^i)$ or crosses from A to Z) and $(A : Z)$ contains a number B of required links. Given a link $e = (i,j) \in E \cup A$ with $i \in S$, $j \in T$, the coefficient corresponding to its associated variable is $\alpha_e = c(S,T)$, where

- $c(A,Z) = c(Z,A) = 1$,
- $c(M_j^i, M_q^i) = \dfrac{|j-q|}{n_i - 1} \qquad \forall j,q \in \{0,1,\dots,n_i+1\}, 0 < |j-q| < n_i + 1,$
- $c(M_j^i, M_q^r) = \dfrac{1}{n_i - 1} + \dfrac{1}{n_r - 1} + \left|\dfrac{j-1}{n_i - 1} - \dfrac{q-1}{n_r - 1}\right|$
  $\forall i,r \in \{1,\dots,P\},\ i \neq r,\ j \in \{1,\dots,n_i\},\ q \in \{1,\dots,n_r\}.$

Again, coefficients associated with variables corresponding to links joining vertices in the same set M_j^i are zero. Some of these coefficients are shown in Figure 6.5. The Path-Bridge inequality is then

$$\sum_{e \in E \cup A} \alpha_e x_e \geq B + 1 + \sum_{i=1}^{P} \frac{n_i + 1}{n_i - 1} + |(A:Z)_R|. \tag{6.11}$$

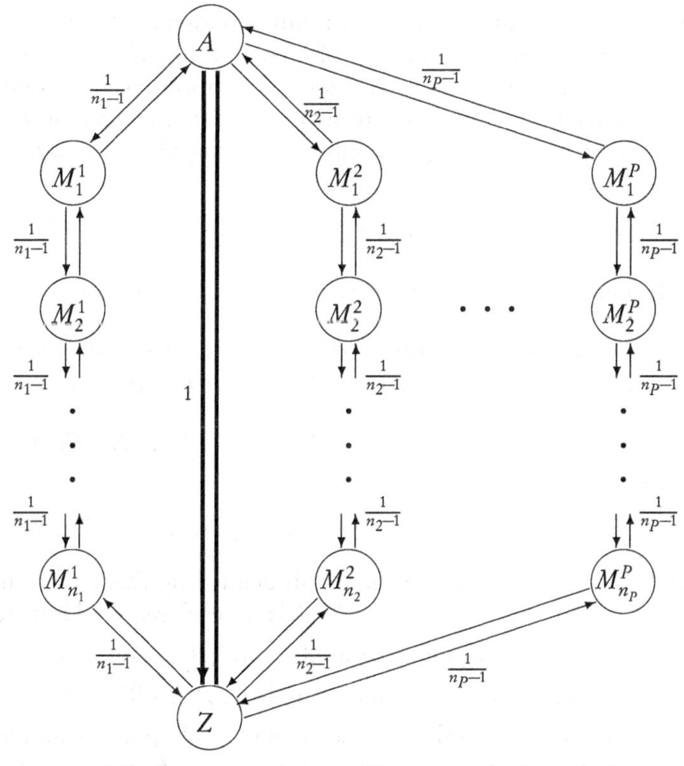

Figure 6.5. *Coefficients of a Path-Bridge inequality.*

In the special case in which n_i is the same number n for each path, these inequalities are called regular Path-Bridge inequalities and can be rewritten in a simpler form where all the coefficients are integer by multiplying them by $n-1$.

Path-Bridge inequalities are valid for $\text{MRPP}_{F1}(G)$. Provided that all the induced graphs $G(M_i)$ are strongly connected and all the sets $A(M_j^i : M_{j+1}^i)$ and $A(M_{j+1}^i : M_j^i)$, $i = 1, \ldots, P$, $j = 0, 1, \ldots, n_i$, are nonempty, they are also facet-inducing for $\text{MRPP}_{F1}(G)$ under the following conditions:

$$|E(A:Z)| \geq \big||A_R(A:Z)| - |A_R(Z:A)|\big| \text{ if } P \text{ is odd, or}$$

$$|E(A:Z)| \geq \big||A_R(A:Z)| - |A_R(Z:A)|\big| + 1 \text{ if } P \text{ is even.}$$

As happens with K-C inequalities, there is an asymmetric version of the Path-Bridge inequalities, called Path-Bridge$_{02}$, that are valid and facet-inducing for $\text{MRPP}_{F1}(G)$ (see Corberán, Romero, and Sanchis [27] for details).

Honeycomb and Honeycomb$_{02}$ inequalities

The Honeycomb inequalities were introduced in Corberán and Sanchis [29] for the undirected GRP and in Corberán, Mejía, and Sanchis [21] for the MRPP. Like Path-Bridge inequalities, they are also a generalization of K-C inequalities. However, the generalization is in a different direction and neither class contains the other. In the partition associated with a K-C inequality, an R-set (or a cluster of R-sets) is divided into two parts (M_0 and M_K). Now this partition is generalized simultaneously both in the number of parts an R-set is divided into and in the number of divided R-sets.

Consider a partition of the set of vertices V into K vertex sets $\{A_1,\ldots,A_L,A_{L+1},\ldots,A_K\}$, $K \geq 3$, in such a way that each R-set is contained in exactly one A_i and the induced subgraphs $G(A_i)$, $i = 1,\ldots,K$, are strongly connected. Suppose each set A_i, $i = 1,\ldots,L$, can be partitioned into $\gamma_i \geq 2$ subsets, $A_i = B_i^1 \cup B_i^2 \cup \ldots \cup B_i^{\gamma_i}$, satisfying the following conditions:

1. $|\delta_R^*(B_i^p)|$ is even for $p = 1,\ldots,\gamma_i$.
2. The induced subgraphs $G(B_i^p)$, $p = 1,\ldots,\gamma_i$, are strongly connected.
3. The graph with vertex set $B_i^1,\ldots,B_i^{\gamma_i}$ and having an edge (B_i^p,B_i^q) for every pair of vertices $B_i^p \neq B_i^q$, such that $(B_i^p : B_i^q)_R \neq \emptyset$ in the original graph, is connected.

For notational convenience, $B_i^0 = A_i$, $i = L+1,\ldots,K$. We therefore have the following partition of V:

$$\mathscr{B} = \{B_1^1,\ldots,B_1^{\gamma_1},B_2^1,\ldots,B_2^{\gamma_2}, \ldots ,B_L^1,\ldots,B_L^{\gamma_L}, B_{L+1}^0,\ldots,B_K^0\}.$$

Let $G_{\mathscr{C}} = (\mathscr{B},\mathscr{E} \cup \mathscr{A})$ be the graph defined by this partition, where the set of links $\mathscr{E} \cup \mathscr{A}$ contains a required link (B_i^p,B_j^q) for each required link between sets B_i^p and B_j^q in the original graph and a nonrequired arc (B_i^p,B_j^q) between each couple of vertices B_i^p and B_j^q such that there is a nonrequired arc from B_i^p to B_j^q in the original graph.

Let us suppose that there is a set T of pairs of opposite nonrequired arcs in $G_{\mathscr{C}}$ joining vertices corresponding to different A_i, $i = 1,\ldots K$, such that the undirected graph with vertex set $\mathscr{B}$ and having an edge (B_i^p,B_j^q) for each pair of opposite arcs (B_i^p,B_j^q), (B_j^q,B_i^p) in T is a spanning tree. For each pair of vertices B_i^p, B_j^q in $\mathscr{B}$, let $d(B_i^p,B_j^q)$ denote the number of arcs in the unique path in $(\mathscr{B},T)$ joining B_i^p to B_j^q. Assume that the following two conditions are also satisfied:

4. $d(B_i^p,B_i^q) \geq 3 \quad \forall\, i = 1,\ldots,L$ and $\forall\, p \neq q$.
5. The degree of every vertex B_i^p, $p \neq 0$, in $(\mathscr{B},T)$ is equal to 1.

The graph $(\mathscr{B},T)$ can be seen in Figure 6.6, where the arcs in T are represented by thin lines and the required links by bold lines and it looks like a honeycomb where each cell is a K-C partition defined by a pair of vertices B_i^p, B_i^q (related to the same A_i) and by the unique path in T joining B_i^p and B_i^q.

The coefficients of the variables corresponding to the links in the original graph in a Honeycomb inequality are as follows (see Figure 6.6):

- For the links $e \in (B_i^p : B_i^q)$ with $p,q \neq 0$ (the nodes obtained by splitting the sets A_i, $i = 1,\ldots,L$), $\alpha_e = d(B_i^p,B_i^q) - 2$.

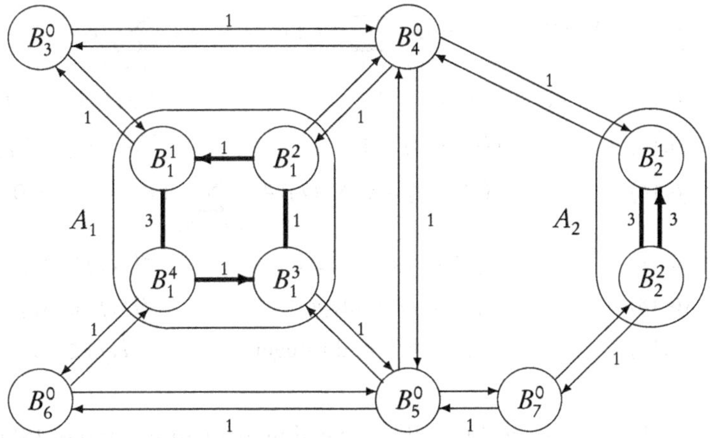

Figure 6.6. *Honeycomb partition.*

- For the arcs $e \in A(B_i^p : B_j^q)$, $i \neq j$, $\alpha_e = d(B_i^p, B_j^q)$.

- For the remaining links, $\alpha_e = 0$.

Then the following inequality is called a Honeycomb inequality and is valid for the MRPP and is facet-inducing under certain conditions:

$$\sum_{e \in E \cup A} \alpha_e x_e \geq 2(K-1) + \sum_{e \in \mathcal{R}} \alpha_e, \tag{6.12}$$

where $\mathcal{R}$ is the set of required links joining two distinct sets B_i^p and B_i^q with $p, q \neq 0$. For example, the right-hand side of the Honeycomb inequality represented in Figure 6.6 is $2 \times (7-1) + 3 + 1 + 1 + 1 + 3 + 3 = 24$.

In Corberán, Mejía, and Sanchis [21], a more general version of the Honeycomb inequalities (6.12) is presented by ignoring condition 5, but in this case some of the coefficients in the inequality have to be calculated by means of a sequential lifting procedure.

As happens with K-C and Path-Bridge inequalities, there is an asymmetric version of the Honeycomb inequalities. They are called Honeycomb$_{02}$, but the only case which has been studied is the one where one R-set is partitioned. These inequalities are valid and facet-inducing for MRPP$_{F1}(G)$ (see [21] for details).

6.2.1.3 ▪ Formulation F2

Instead of using just one variable representing each edge, the MRPP can also be formulated using two variables x_{ij} and x_{ji} associated with each edge $e = (i, j)$, representing the number of times edge e is traversed in the corresponding direction. Then the MRPP can be formulated as

$$\text{(6.13) (MRPP-F2)} \quad \min \sum_{e=(i,j)\in E} c_e(x_{ij}+x_{ji}) + \sum_{(i,j)\in A} c_{ij}x_{ij}$$

$$\text{(6.14)} \quad \text{s.t.} \quad x_{ij}+x_{ji} \geq 1 \qquad \forall e=(i,j)\in E,$$

$$\text{(6.15)} \quad x(\delta^+(S)) \geq 1 \qquad \forall S=\cup_{k\in Q} V_k, \ Q\subset\{1,\dots,p\},$$

$$\text{(6.16)} \quad x(\delta^+(i)) - x(\delta^-(i)) + \sum_{(i,j)\in\delta(i)} (x_{ij}-x_{ji}) = 0 \qquad \forall i\in V,$$

$$\text{(6.17)} \quad x_{ij} \geq 0 \text{ and integer} \qquad \forall (i,j)\in A\setminus A_R,$$

$$\text{(6.18)} \quad x_{ij} \geq 1 \text{ and integer} \qquad \forall (i,j)\in A_R,$$

$$\text{(6.19)} \quad x_{ij}, x_{ji} \geq 0 \text{ and integer} \qquad \forall (i,j)\in E.$$

Inequalities (6.14), called traversing inequalities, and (6.18) imply that all the required edges and arcs, respectively, are traversed. Connectivity inequalities (6.15) ensure that the solution is connected, while symmetry equations (6.16) force it to be symmetric.

6.2.1.4 ▪ MRPP polyhedron associated with formulation F2

Let $\text{MRPP}_{F2}(G)$ be the convex hull of all the vectors in $\mathbb{R}^{2|E|+|A|}$ satisfying (6.14) to (6.19). Although this polyhedron has not been studied directly, it can be seen as a particular case of the WGRP and WRPP polyhedra, studied in Plana [50] and Corberán, Plana, and Sanchis [25, 26]. In this section several results regarding the polyhedra of the WGRP and the WRPP are adapted to the MRPP.

The dimension of $\text{MRPP}_{F2}(G)$ is $2|E|+|A|-|V|+1$, and the following inequalities from the formulation are, under mild conditions, facet-defining for $\text{MRPP}_{F2}(G)$:

- Inequalities (6.14) and (6.15).
- Trivial inequalities $x_{ij} \geq 0 \ \forall (i,j)\in A\setminus A_R$, $x_{ij}\geq 1 \ \forall (i,j)\in A_R$, and $x_{ij}, x_{ji} \geq 0$ $\forall (i,j)\in E$.

If integrality conditions are satisfied, symmetry inequalities (6.16) imply that all the vertices are even. However, once the integrality conditions have been removed from F2, symmetry conditions alone no longer guarantee that the vertices are of even degree. However, given an R-odd cutset $\delta^*(S)$, the following R-odd cut constraints must also be satisfied by any solution to F2:

$$\text{(6.20)} \quad x(\delta^*(S)) \geq |\delta^*_R(S)|+1 \ \forall \delta^*(S) \ R\text{-odd cutset.}$$

These R-odd constraints are also facet-inducing for $\text{MRPP}_{F2}(G)$ if graphs $G(S)$ and $G(V\setminus S)$ are strongly connected and $|\delta(S)| > ||\delta^+_R(S)|-|\delta^-_R(S)||$.

Furthermore, any valid inequality for the WRPP is also a valid inequality for the MRPP. However, the facet-inducing inequalities for WRPP(G) described in Section 6.3 are also facet-inducing for $\text{MRPP}_{F2}(G)$ only if the arcs in G are of the "appropriate direction," i.e., if the removed variables do not avoid the existence of all the tours needed to define a facet. The following inequalities, which can be easily adapted from the WRPP, are valid and facet-defining for $\text{MRPP}_{F2}(G)$ under some conditions (see Corberán, Plana, and Sanchis [26]):

- K-C and K-C_{02} inequalities.
- Path-Bridge and Path-Bridge_{02} inequalities.

- Honeycomb and Honeycomb$_{02}$ inequalities.
- Even and Odd zigzag inequalities.
- Even-even and Odd-odd zigzag inequalities.

6.2.1.5 ▪ Formulations comparison

Formulations F1 and F2 are compared in Corberán, Mota, and Sanchis [22] (see also Chapter 4 in this book). It is easy to see that both (integer) formulations are equivalent. However, the lower bounds obtained from their LP-relaxations can be different.

Consider the following initial LP relaxation, $LP0_{F1}$, corresponding to formulation F1, containing

- constraints $x_e \geq 1\ \forall e \in E \cup A_R$ and $x_e \geq 0\ \forall e \in A \setminus A_R$,
- one connectivity inequality (6.3) for each R-set,
- one balanced-set inequality (6.4) for each "unbalanced" vertex,
- the system equations (6.7), and
- one R-odd cut inequality (6.8) for each R-odd "balanced" vertex,

and $LP0_{F2}$, associated with formulation F2, as the LP containing

- constraints $x_{ij} \geq 0\ \forall (i,j) \in A \setminus A_R$ $x_{ij} \geq 1\ \forall (i,j) \in A_R$, and $x_{ij}, x_{ji} \geq 0\ \forall (i,j) \in E$,
- constraints (6.14),
- one connectivity inequality (6.15) for each R-set,
- one symmetry equation (6.16) for each vertex, and
- one R-odd cut inequality (6.20) for each R-odd "balanced" vertex.

It is shown in Corberán, Mota, and Sanchis [22] that $LP0_{F2}$ is stronger than $LP0_{F1}$. Now consider the above formulations after adding all the families of inequalities that are known to be separable in polynomial time, which are connectivity, balanced-set, and R-odd cut inequalities. In [22] it is also shown that both extended formulations produce the same bound, while the computational effort to solve them is, at least in theory, comparable.

6.2.2 ▪ Exact methods

In Blais and Laporte [12] it is shown how the MRPP can be solved exactly by transforming it into an equivalent vertex routing problem for which an exact solution procedure is available. To transform an MRPP on a mixed graph $G = (V,E,A)$ into an equivalent vertex routing problem on a directed graph $\widetilde{G} = (\widetilde{V},\widetilde{A})$, each required arc $(i,j) \in A_R$ is replaced by a required vertex v_{ij}, and each required edge $e = (i,j) \in E_R$ is first replaced by two opposite arcs (i,j) and (j,i) of cost $\widetilde{c}_{ij} = \widetilde{c}_{ji} = c_e$ and then substituted for with two vertices v_{ij} and v_{ji}, only one of which is required. The vertex set $\widetilde{V}$ is then made up of all the vertices just defined. Each vertex pair (v_{ki}, v_{lj}) in the transformed problem defines an arc of $\widetilde{A}$ having a cost $\widetilde{c}_{ij} = s_{il} + c_{lj}$, where s_{il} represents the cost of the shortest path in G from vertex i to vertex l and $\widetilde{c}_{ii} = 0$. This transformation is depicted in Figure 6.7, taken from Blais and Laporte [12].

If $E_R \neq \emptyset$, the resulting problem on $\widetilde{G}$ is not a TSP, but a *Generalized TSP* (GTSP), since only one of the two vertices v_{ij} and v_{ji} corresponding to a required edge $e = (i,j)$

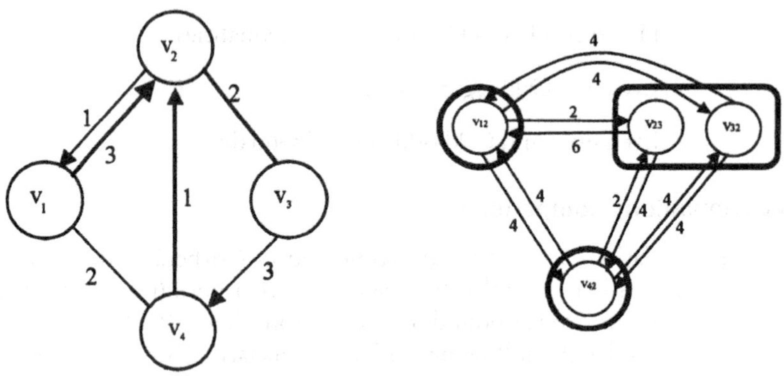

Figure 6.7. *Graph transformation for a mixed graph [12].*

must appear in the solution. To obtain the appropriate GTSP, the vertex set $\widetilde{V}$ is partitioned into clusters, with a cluster $\{v_{ij}, v_{ji}\}$ for each required edge $e = (i, j)$. The GTSP then consists of determining a minimum cost Hamiltonian circuit on $\widetilde{G}$ passing through each cluster exactly once. Solving this asymmetric GTSP on $\widetilde{G}$ provides a solution for the original MRPP on G. The asymmetric GTSP is solved using two methods. The direct approach applies the Lagrangian-based branch-and-bound algorithm proposed in Noon and Bean [47]. The undirected one first transforms the asymmetric GTSP into an equivalent *Asymmetric TSP* (ATSP) and then solves it by means of the exact method described in Carpaneto, Dell'Amico, and Toth [13]. The transformation procedure was assessed on randomly generated mixed graphs with $|V| = 500$, $|E| = |A| = 1000$, and with up to $|A_R| = 200$ and $|E_R| = 10$ or $|A_R| = 10$ and $|E_R| = 25$. The authors point out that the difficulty of the problem is clesely related to the size of $|E_R|$, which determines the number of clusters in the GTSP. These exact procedures were able to solve instances involving up to 25 required edges in G and up to 230 vertices in $\widetilde{G}$.

Corberán, Romero, and Sanchis [27] implement a preliminary cutting-plane algorithm for the MRPP using formulation F1 and including separation procedures only for connectivity, R-odd cut, and balanced-set inequalities. Later on, Corberán, Mejía, and Sanchis [21] improve it by adding separation procedures for the K-C, Honeycomb (with $L=1$), and regular Path-Bridge inequalities. The initial LP relaxation used is the one denoted above as LP0_{F1}. As mentioned before, the separation of connectivity, R-odd cut, and balanced-set inequalities is done in polynomial time, while the authors apply heuristic algorithms for the separation of K-C, regular Path-Bridge, and Honeycomb inequalities. When the separation procedures do not find violated inequalities and the current LP solution x^* is not a tour for the MRPP, a branch-and-bound procedure is applied. If the resulting integer solution is a tour, it is optimal. Otherwise, the procedure terminates with a tight lower bound, but no feasible solution. This algorithm was tested on 100 randomly generated MRPP instances with the following features: $3 \le p \le 12$, $20 \le |V| \le 100$, $15 \le |E| \le 220$, $5 \le |E_R| \le 150$, $50 \le |A| \le 350$, $5 \le |A_R| \le 200$. All of these instances were solved to optimality.

In Corberán, Plana, and Sanchis [25], a branch-and-cut algorithm for the WRPP is presented. This algorithm is based on a formulation with two variables per edge similar to formulation F2 for the MRPP and will be described in detail in Section 6.3.2. It was tested on 117 MRPP instances with up to 1000 vertices, 2000 edges, and 2000 arcs, solving all of them to optimality within an average time of half an hour.

6.2.3 ▪ Heuristic methods

A constructive heuristic and a tabu search algorithm for the MRPP are described in Corberán, Martí, and Romero [20]. The heuristic algorithm consists of four stages. In the first stage, a subset of arcs is chosen so that the graph including the required links of G and this subset of arcs is connected. In the second stage, the connectivity of this new graph is improved by solving a minimum cost flow problem in a directed graph, as in the heuristic algorithm proposed in Christofides et al. [15]. A feasible solution is obtained in the third stage. First, another minimum cost flow is computed in order to construct a balanced mixed graph, which is then made even by solving a minimum cost matching on the odd-degree vertices, and the added edges are oriented to get a symmetric directed graph. Finally, stage four tries to improve the MRPP solution.

The tabu search procedure starts from an MRPP tour obtained with the heuristic algorithm described above. In order to move from one solution to another one, the tabu search procedure uses the following movement: Let (i, j) be an arc connecting two R-sets in the current solution, and let $[k, i]$ and $[j, s]$ be paths from k to i and from j to s, respectively, in the augmentation of the graph induced by the required links. Then the movement consists of deleting path $[[k, i], (i, j), [j, s]]$ from the current solution and adding the shortest path from k to s. Note that the resulting graph may be disconnected, and thus it would not be a feasible MRPP solution. The tabu search algorithm is divided into two phases: intensification and diversification. At each iteration of the intensification phase, an arc is chosen with a probability proportional to its cost and the corresponding movement described above is applied. The removed arc is marked as tabu so that it is not added again in the intensification phase. This phase ends when no arc is found for which a movement can be applied. The diversification phase is applied for a fixed number of iterations. At each iteration a non-tabu arc connecting R-sets is chosen at random as in the intensification phase. If the corresponding movement can be applied, it is performed; otherwise, two alternative movements are considered. If the current solution is connected, the shortest path to be added in the movement is computed in the graph obtained by deleting the tabu arcs. If the current solution is nonconnected, a connected component is randomly selected and all the arcs connecting this component for which the associated shortest path does not include a tabu arc are considered. The lowest cost one, considering the arc plus the shortest return path costs, is selected. If all the above paths include a tabu arc, we ignore the tabu status. Both phases are alternated until the best-known solution for a fixed number of consecutive iterations is not improved. Additionally, a long-term diversification phase is implemented to complement the diversification phase in the basic procedure. This phase consists of a restarting mechanism in which information generated by the search history is used to guide the construction process.

Both the constructive heuristic and the tabu search algorithm were tested on a set of 100 randomly generated instances. These instances have up to 12 R-sets, 100 vertices, 220 edges, 150 required edges, 350 arcs, and 200 required arcs. The constructive method presented an average deviation of 1.87% with respect to a lower bound obtained by the cutting-plane algorithm described in Corberán, Romero, and Sanchis [27]. In three of the 100 instances, the solution provided matched the lower bound and was thus proved to be optimal. The tabu search procedure was tested using different sets of parameters. For all the combinations tried, the average deviation was less than or equal to 0.53%, with the best one obtaining an average deviation of 0.48%. The average time used by the different versions of the tabu search ranges from 57.8 to 326.3 seconds on a Pentium-150 PC, while the time consumed by the constructive algorithm was negligible.

Another heuristic procedure for the MRPP is the one proposed by Chyu [17]. The author applies this algorithm to the special case in which for each arc (i,j) in the graph there is an opposite arc (j,i) and both arcs have the same cost. The heuristic includes four subheuristics, with each one being designed for some particular structure of the required arcs set. The best result produced among the four subheuristics is then selected. This heuristic has a worst-case performance ratio of between 3/2 and 15/8, depending on the ratio of the cost of the required edges to that of the required arcs in the problem.

6.3 ▪ The windy Rural Postman Problem

In this section we define the Windy Rural Postman Problem (WRPP) and describe its heuristic and exact solution. Consider a windy graph $G=(V,E)$ with a vertex set V and an edge set E, and two nonnegative costs c_{ij}, c_{ji} for each $(i,j)\in E$, representing the costs of traversing it in each direction. Given a subset of required edges $E_R \subseteq E$, the WRPP consists of finding a minimum cost tour traversing all the required edges at least once.

6.3.1 ▪ Problem formulation

The formulation for the WRPP was first proposed in Benavent et al. [8]. In order to present it, we need some notation. Given $S \subseteq V$, $\delta(S)$ denotes the set of edges with an endpoint in S and the other in $V \setminus S$. Given $S_1, S_2 \subseteq V$, $(S_1 : S_2)$ represents the set of edges with one endpoint in S_1 and the other in S_2, while $\delta_R(S)$ and $(S_1 : S_2)_R$ denote the previous sets referring only to the required edges. A vertex is R-even (R-odd) if it is incident with an even (odd) number of required edges, and a subset of vertices $S \subseteq V$ is called R-even (R-odd) if it contains an even (odd) number of R-odd vertices. As in the MRPP, the connected components of the graph induced by the required edges, $G_R = (V, E_R)$, denoted by $V_1, \ldots, V_p$, are called R-sets.

The formulation uses two variables x_{ij} and x_{ji} associated with each edge (i,j), representing the number of times edge (i,j) is traversed from i to j and from j to i, respectively. Given $F \subseteq E$, $x(F)$ denotes the sum of all the variables associated with the edges in F, $x(F)=\sum_{(i,j)\in F}(x_{ij}+x_{ji})$, and given $S_1, S_2 \subset V$, $x(S_1, S_2)$ denotes the sum of the variables associated with the traversal of the edges in $(S_1 : S_2)$ from S_1 to S_2, that is $x(S_1 : S_2) = x(S_1, S_2) + x(S_2, S_1)$. The WRPP formulation is

$$\text{(WRPP)} \quad \min \sum_{(i,j)\in E} (c_{ij}x_{ij} + c_{ji}x_{ji}) \tag{6.21}$$

$$\text{s.t.} \quad x_{ij} + x_{ji} \geq 1 \qquad \forall (i,j)\in E_R, \tag{6.22}$$

$$x(S, V\backslash S) \geq 1 \qquad \forall S = \cup_{k\in Q} V_k, \ Q \subset \{1,\ldots,p\}, \tag{6.23}$$

$$\sum_{(i,j)\in\delta(i)} (x_{ij} - x_{ji}) = 0 \qquad \forall i \in V, \tag{6.24}$$

$$x_{ij}, x_{ji} \geq 0 \qquad \forall (i,j)\in E, \tag{6.25}$$

$$x_{ij}, x_{ji} \text{ integer} \qquad \forall (i,j)\in E. \tag{6.26}$$

Inequalities (6.22) imply that each required edge is traversed. Connectivity inequalities (6.23) and symmetry equations (6.24) ensure that the tours are connected and symmetric, respectively.

6.3.1.1 ▪ The WRPP polyhedron

Let WRPP(G) denote the convex hull of all the tours for the WRPP on graph G, i.e., all the integer vectors $x \in \mathbb{R}^{2|E|}$ satisfying (6.22) to (6.25). In Plana [50] and Corberán, Plana, and Sanchis [26] it is shown that WRPP(G) is an unbounded polyhedron of dimension $2|E|-|V|+1$ and that inequalities (6.22), (6.23), and (6.25) are facet-inducing for WRPP(G) under mild conditions.

In Benavent et al. [8], it is also shown how to obtain valid inequalities for an asymmetric Arc Routing Problem from inequalities for its symmetric version. Let $\mathscr{P}$ be an Arc Routing Problem defined on an undirected graph $G=(V,E)$ and assume that it is formulated with a variable y_e associated with each edge $e=(i,j)\in E$, representing the number of times the edge is traversed in any direction. Let $\mathscr{P}'$ be the "windy" version of this Arc Routing Problem, i.e., the problem defined on the same graph $G=(V,E)$, but with two nonnegative costs associated with each edge. Assume that this problem is formulated with two variables x_{ij} and x_{ji} associated with each edge $(i,j)\in E$, representing the number of times the edge is traversed in each direction. Then the following result holds:

Let $\sum_{e\in E}\alpha_e y_e \geq \alpha_0$ be a valid inequality for $\mathscr{P}$. The corresponding inequality

$$\sum_{e=(i,j)\in E} \alpha_e(x_{ij}+x_{ji}) \geq \alpha_0$$

is valid for $\mathscr{P}'$ if, for every feasible solution $x\in\mathbb{Z}^{2|E|}$ for problem $\mathscr{P}'$, the vector $y\in\mathbb{Z}^{|E|}$, defined as $y_e = x_{ij}+x_{ji}$ for all $e=(i,j)\in E$, is a feasible solution for problem $\mathscr{P}$.

Since problems defined on directed or mixed graphs can be considered as windy problems where some variables have been set to zero, the above result implies that

- All the valid inequalities for the (Undirected) Chinese Postman Problem produce valid inequalities for the Directed and Mixed Chinese Postman Problems and for the Windy Postman Problem, and

- From the (Undirected) Rural Postman Problem valid inequalities for their Directed, Mixed, and Windy versions can be obtained.

Moreover, the authors conjecture that, if the condition of the above result holds, the following result is also true:

Given a facet-inducing inequality $\sum_{e\in E}\alpha_e y_e \geq \alpha_0$ for the polyhedron associated with problem $\mathscr{P}$, the corresponding inequality $\sum_{e=(i,j)\in E}\alpha_e(x_{ij}+x_{ji}) \geq \alpha_0$ is facet-inducing for the polyhedron associated with problem $\mathscr{P}'$.

Although, as far as we know, this conjecture has not been proved yet, the following families of inequalities, some of them derived from the undirected RPP, are known to be facet-inducing for WRPP(G).

R-odd cut inequalities

Given that any tour must cross any given edge cutset an even number of times, the following R-odd cut inequalities:

(6.27) $$x(\delta(S)) \geq |\delta_R(S)|+1 \quad \forall S\subset V \quad \text{such that } \delta(S) \text{ is } R\text{-odd},$$

are valid for WRPP(G). They are facet-inducing for WRPP(G) if subgraphs $G(S)$ and $G(V\setminus S)$ are connected (Corberán, Plana, and Sanchis [26]).

K-C and K-C$_{02}$ inequalities

K-C and K-C$_{02}$ inequalities have already been described for the MRPP in this chapter. For the WRPP, these inequalities are based on the same partition of V and have the same right-hand side and the same coefficients of the corresponding inequalities for the MRPP (see Figure 6.4). In particular, given a variable associated with the traversal of an edge in a given direction, its coefficient is the same as the corresponding arc in a K-C or K-C$_{02}$ inequality for the MRPP. Both K-C and K-C$_{02}$ inequalities are valid and facet-inducing for WRPP(G) (Corberán, Plana, and Sanchis [26]).

Path-Bridge, Path-Bridge$_{02}$, Honeycomb, and Honeycomb$_{02}$ inequalities

Again, the partition of V in which these inequalities are based, the coefficients of the variables, and their right-hand sides are the same as those of the corresponding inequalities for the MRPP (see Section 6.2.1.2). All of them are valid and facet-inducing for WRPP(G) (Corberán, Plana, and Sanchis [26]).

Zigzag inequalities

When solving the WRPP, it is not unusual to find fractional solutions containing closed walks resembling a zigzag whose edges satisfy $x_{ij}+x_{ji}=1.5$ if $(i,j)\in E_R$ and $x_{ij}+x_{ji}=0.5$ otherwise. Consider, for example, the vector represented in Figure 6.8(a), where arcs drawn in solid lines correspond to variables of value 1, while arcs in dotted lines correspond to variables of value 0.5. This vector satisfies all the previously described inequalities but can be cut by means of zigzag inequalities. There are four families of zigzag inequalities. The first two, Odd and Even zigzag inequalities, were introduced in Corberán, Plana, and Sanchis [24], while the other two, Odd-odd and Even-even zigzag inequalities, were proposed in Corberán et al. [23] for the *Windy Postman Problem* (WPP) and extended to the WRPP in Corberán, Plana, and Sanchis [26].

Consider a partition of set V into four subsets, M^1, M^2, M^3, and M^4. Let α_{ij} denote the number of required edges in $(M^i : M^j)$, and suppose one of the following conditions is satisfied:

Even case: M^1, M^2, M^3, and M^4 are R-even, and $\alpha_{12}=\alpha_{34}=\alpha_{14}=\alpha_{23}=0$.

Odd (simple) case: M^1, M^2, M^3, and M^4 are R-odd, and $\alpha_{12}+\alpha_{34}=\alpha_{14}+\alpha_{23}$.

The Even and Odd (simple) zigzag inequalities are

$$x(\delta(M^1\cup M^2))+2x(M^2,M^1)+2x(M^4,M^3)\geq \alpha_{13}+\alpha_{24}+\alpha_{14}+\alpha_{23}+2. \tag{6.28}$$

Examples of these inequalities are represented in Figures 6.8(b) and 6.8(c), where the required edges are represented in bold lines and each number represents the coefficient of the variable associated with the traversal of the edge from the nearest vertex to the farthest one.

In Corberán, Plana, and Sanchis [24] it is shown that inequalities (6.28) are valid and facet-inducing for WRPP(G) if the following conditions hold: if $(M^i : M^j)\neq\emptyset, \forall i,j$, and $\alpha_{13},\alpha_{24}\geq 2$ (even case) or $\alpha_{13}\geq|\alpha_{12}-\alpha_{14}|+1$ and $\alpha_{24}\geq|\alpha_{12}-\alpha_{23}|+1$ (odd case). More

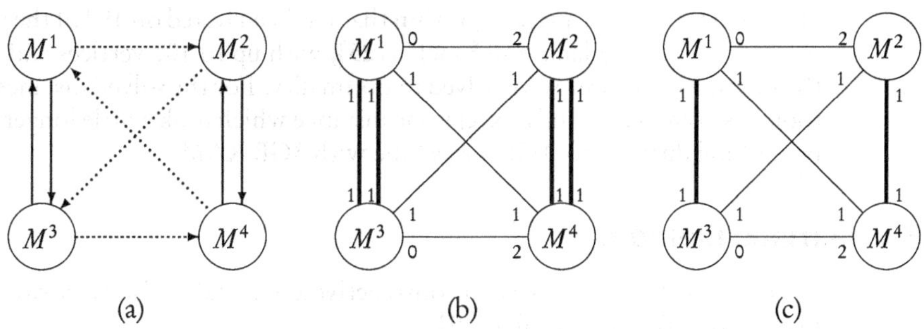

Figure 6.8. *Fractional solution and Even and Odd zigzag inequalities.*

general Odd zigzag inequalities are also proposed in the same paper (see also Chapter 4 in this book).

The two other families of inequalities, the Even-even and the Odd-odd zigzag inequalities, are defined by a partition of the set of vertices into four subsets, just as with the Odd and Even zigzag inequalities. However, in this case not all of these subsets have to be incident with an odd, respectively even, number of required edges. As mentioned above, they were introduced for the WPP in Corberán et al. [23] (see also Chapter 4) and extended to the WRPP in Corberán, Plana, and Sanchis [26], where it is proved that they are valid and facet-inducing under certain conditions.

6.3.2 ▪ Exact methods

In Corberán, Plana, and Sanchis [25], a branch-and-cut algorithm to solve the WRPP is presented. This procedure starts with an initial LP including constraints (6.22), (6.24), and (6.25), as well as one connectivity constraint (6.23) associated with each R-set. Furthermore, R-odd cutset inequalities (6.27) associated with R-odd vertices and with the connected components S of the subgraph of G_R induced by the R-odd vertices, if $|S|$ is odd, are also included.

At each iteration of the cutting-plane algorithm, exact and heuristic separation algorithms are used for identifying violated R-odd cut and connectivity inequalities. Moreover, heuristic algorithms for separating K-C, K-C_{02}, HC, HC_{02}, PB, Even, and Odd zigzag inequalities are used. The cutting-plane procedure is applied at each node of the tree until no new violated inequalities are found or until a stopping criterion, called tailing-off, is satisfied.

The constructive algorithms H1 and H2 described in Section 6.3.3 are used to obtain an initial upper bound. If the gap at the root node with respect to this upper bound is greater than 1.5%, an artificial upper bound with a value 1.005 times the lower bound is used. If the procedure ends without finding an optimal solution, it is restarted using a greater artificial upper bound (1.010 times the lower bound). If, again, the algorithm ends unsuccessfully, it is restarted for the last time using the true upper bound.

This branch-and-cut algorithm was tested on 522 WRPP instances with up to 1000 vertices, 4000 edges, and 200 R-sets, obtaining 510 optimal solutions. The time limit was set to 10 hours, but the average time used by the algorithm for solving the hardest instances (those with around 1000 vertices) to optimality was less than 1500 seconds on a Pentium IV at 2.8 GHz machine with 1GB RAM.

Another exact procedure for the WRPP is described in Afsar, Jozefowiez, and Lopez

[1]. It is a branch-and-price algorithm that has been tested on 198 of the smallest instances used in Corberán, Plana, and Sanchis [25], with up to 100 vertices and 184 edges. All of the instances but three were solved to optimality. For the solved instances, this procedure took less than five seconds, except one instance which took a little longer than 200 seconds on an Intel dual core 3.0 GHz machine with 3GB RAM.

6.3.3 ▪ Heuristic methods

Here we briefly describe several constructive and metaheuristic algorithms for the WRPP presented in Benavent et al. [8, 9].

6.3.3.1 ▪ Constructive algorithms

In Benavent et al. [8], three heuristic algorithms for the solution of the WRPP are presented. The first algorithm, WRPP1, consists of three phases. First, the components of graph G_R are joined to obtain a connected graph by adding the edges corresponding to a shortest spanning tree. Then, a minimum cost matching problem on the odd-degree vertices of the resulting graph is solved, obtaining a connected and even windy graph. Win's exact algorithm for the WPP defined on Eulerian graphs (Win [55]), described in Chapter 4, is then applied to obtain a feasible solution for the WRPP.

The second heuristic, WRPP2, begins by assigning the direction associated with its minimal traversing cost to each edge in G_R, in order to obtain a directed graph G_R^d. Then a balanced mixed graph G^1 is obtained by solving a minimum cost flow on a special graph. If G^1 has odd-degree vertices, a minimum cost matching problem is solved to obtain an even mixed graph G^2. Again, the edges in G^2 are oriented in the direction of their minimal traversing cost and a new network flow problem is solved to obtain a symmetric digraph G^3. Finally, as graph G^3 may be disconnected, a last phase consisting of the solution of a shortest spanning tree connecting its components is executed.

The third constructive heuristic, WRPP3, is simpler and consists of first computing a shortest spanning tree connecting the components of G_R, then orienting the edges according to the minimum traversal cost, and finally solving a transportation problem to get a connected and symmetric directed graph that defines a feasible WRPP solution.

The solutions generated by all the algorithms are improved by means of two simple procedures described in the same paper. The first one removes redundant cycles from the solution graph, while the second one consists of first constructing an Eulerian tour traversing the solution graph and then substituting any path of nonrequired arcs connecting two consecutive required edges with a shortest path.

The above algorithms have no finite worst-case bound (Benavent et al. [8]). In fact, contrary to what happens with the WPP, it is not known if the WRPP admits a polynomial time heuristic with a constant worst-case bound. As far as we know, this is an open question.

In Benavent et al. [9], two new constructive heuristics, several multistart algorithms, and a Scatter Search procedure for the WRPP are presented. The constructive heuristics, called H1 and H2, are obtained by modifying the previous heuristics WRPP1 and WRPP2, respectively, in a similar way Raghavachari and Veerasamy [51] do to the Frederickson heuristic for the Mixed Chinese Postman Problem (Frederickson [34]). Local search based on 2- and Or-interchanges (Or [48]) is applied, as well as another method, based on the one proposed in Lacomme, Prins, and Ramdane-Chérif [43]. Given an order of traversal of the required edges, this method, called the reversing direction procedure, consists of finding the optimal direction in which each required edge has to be traversed.

The multistart algorithms are based on introducing some random elements into the constructive heuristics described above. The one producing the best results uses the heuristic WRPP2 with randomized costs in the computation of the matching and the shortest spanning tree. The improvement phase consists of applying first the Or-interchange, then the 2-interchange, and finally the reversing direction procedure.

The Scatter Search procedure starts with the construction of an initial reference set of solutions (RefSet) from a large set P of diverse solutions. This initial population P contains the five solutions provided by the constructive algorithms described earlier and is completed with solutions generated with the construction phase of the best multistart algorithm. The k best solutions in P are included in RefSet. Another l solutions are added in such a way that the diversity of RefSet is as large as possible. To combine the solutions in the reference set, a version of the classical order crossover similar to that used in Lacomme, Prins, and Ramdane-Chérif [42] is applied. A local search phase consisting of the Or-interchange, the 2-interchange, and the reversing direction procedures is applied to the resulting solution. The procedure stops when, after combining all the solutions in RefSet, no new solutions are incorporated.

Table 6.1 (taken from Benavent et al. [9]) summarizes the results obtained with the best constructive algorithm (H2), the best multistart (MS), and the Scatter Search (SS) on four sets of randomly generated instances. Their sizes and characteristics are similar to those of the 117 instances mentioned at the end of the previous section. For each set of instances, the average gap, number of optimal solutions found, and average time in seconds is reported. The gap is computed with respect to a lower bound obtained with the cutting-plane procedure described in Benavent et al. [8].

Table 6.1. *Results obtained with heuristic algorithms for the WRPP.*

		H2	MS	SS
Albaida	Average gap (%)	2.21	0.45	0.29
(72 instances)	# of optima	8	22	21
	Time (s)	0.11	150.46	4.06
Madrigueras	Average gap (%)	2.23	0.61	0.44
(72 instances)	# of optima	5	14	12
	Time (s)	0.43	162.53	11.86
500 vertices	Average gap (%)	1.38	1.38	1.06
(27 instances)	# of optima	0	0	0
	Time (s)	2.95	200	145.96
1000 vertices	Average gap (%)	0.72	0.72	0.63
(27 instances)	# of optima	1	1	1
	Time (s)	26.88	200	1543.606

6.4 ▪ Related problems

In this section we present the results obtained for some problems closely related to the MRPP and WRPP. Among them, the DRPP is addressed, as well as the *Stacker Crane Problem* (SCP), which, although it can be considered a special case of the MRPP and DRPP, has some special features that recommend its separate presentation. Other variants described here are more related to real-life applications, like those with turn penalties and forbidden tours or zigzag servicing, which allows the servicing of the two sides of a street segment simultaneously.

6.4.1 ▪ The directed Rural Postman Problem

As with the MRPP, the DRPP can be solved exactly by transforming it into an ATSP (Blais and Laporte [12]). To transform a DRPP on a directed graph $G = (V,A)$ into an equivalent vertex routing problem on a directed graph $\widetilde{G} = (\widetilde{V},\widetilde{A})$, each required arc $(i,j) \in A_R$ is replaced by a required vertex v_{ij}. These vertices define the vertex set $\widetilde{V}$. Associated with each pair of vertices (v_{ki}, v_{lj}) there is an arc of $\widetilde{A}$ with cost $\widetilde{c}_{ij} = s_{il} + c_{lj}$, where s_{il} represents the cost of the shortest path in G from vertex i to vertex l, and c_{lj} is the cost of the arc (l,j) in G. The resulting problem on graph $\widetilde{G}$ is an ATSP, whose solution solves the original DRPP on G. The exact method used in Blais and Laporte [12] for solving the ATSP is the one described in Carpaneto, Dell'Amico, and Toth [13]. With this methodology, the authors were able to solve randomly generated directed graphs with $|V| = 5000$ and $|A| = 50000$, and with up to $|A_R| = 3500$.

A direct algorithm for the solution of the *Directed General Routing Problem* (DGRP), and the DRPP as special case, is proposed in Ávila et al. [5] and Ávila [3]. A formulation of the problem, similar to formulation MRPP-F2 presented in Section 6.2.1.3, is used to define the DGRP polyhedron. Several families of inequalities are proved to be facet-inducing of this polyhedron: trivial, connectivity, K-C, Path-Bridge, and a new family, the Asymmetric 2PB inequalities. Based on this polyhedral study, the authors implement a branch-and-cut procedure that is capable of solving to optimality 36 randomly generated DRPP instances with 1000 vertices and up to 3200 arcs and 553 R-sets. Moreover, 70 instances generated with the same characteristics as those used in Blais and Laporte [12] have been optimally solved in short computing times.

6.4.2 ▪ The stacker crane problem

The SCP basically consists of finding a tour that starts and ends at a given vertex and traverses the set of arcs (representing jobs) of a mixed graph at minimum cost. Its name refers to the practical problem of operating a crane. The crane must start from an initial position, perform a set of movements, and return to the initial position. The objective is to schedule the movements of the crane so as to minimize the total tour cost. This problem can be considered a special case of both the MRPP and the DRPP. The SCP was proposed by Frederickson, Hecht, and Kim [35]. They defined the SCP as an arc routing problem on a mixed graph where a given vertex represents the depot, i.e., the initial and final position of the crane. Each arc and edge of the graph has an associated nonnegative cost. The goal is to find a minimum cost tour, starting and finishing at the depot, which traverses all the arcs in the graph.

It is shown in Frederickson, Hecht, and Kim [35] that the SCP is NP-hard, and a heuristic algorithm with a worst-case performance ratio of 9/5 and $O(n^3)$ complexity is developed. This procedure consists of two algorithms: The first one computes a minimum-cost matching and then a minimum-cost spanning tree. The second one transforms the SCP instance into a TSP one and then uses Christofides' algorithm for the TSP [14]. Both procedures need the costs of the instances to satisfy the triangle inequality. The first procedure works better if the cost of the arcs is large compared to the cost of the edges of an optimal solution, while the second one works better in the opposite situation.

In Berbeglia et al. [11], the SCP is presented as a pickup and delivery problem. This is an important class of vehicle routing problems in which commodities or people have to be transported from origins to destinations. They have been the object of study in recent years because of their many applications in logistics, ambulatory services, and robotics. The SCP can be considered as a pick-up and delivery problem where single objects have to be transported from their origins to their destinations using a unit capacity vehicle.

Other papers dealing with the SCP can be found in the literature. Eiselt, Gendreau, and Laporte [33] present a survey on the RPP and devote a section to the SCP. Zhang [56] proposes a simplification of Frederickson's algorithm with a worst-case ratio of 2 and $O(n^2)$ complexity. Zhang and Zheng [57] transform the SCP into an ATSP and solve it using existing algorithms for the ATSP. Hassin and Khuller [38] propose a $\frac{1}{2}$ z-approximation algorithm (a polynomial-time algorithm that produces a solution whose distance from the optimal one is at most $\frac{1}{2}$ times the distance between the optimal solution and the worst possible solution) for the ATSP that can also be used for the SCP. Srour and van de Velde [54] study the difficulty of the SCP, presenting a statistical study that compares the difficulty of the solution of SCP instances with that of general ATSP instances. The motivation for this study is the claim by some authors that although the SCP and the ATSP are known to be NP-hard, the SCP can be easily solved when transformed into an ATSP (Laporte [44]). Their conclusion is that SCP instances are not necessarily easier than other ATSP instances, but a special subset of SCP instances, called drayage problems, are more readily solved. The study by Cirasella et al. [18] of ATSP heuristics on 11 types of instances contains one type corresponding to the SCP.

In Ávila et al. [4], a metaheuristic algorithm based on the combination of a Multistart and an Iterated Local Search algorithm is presented. This algorithm was tested on several SCP instances generated from 92 instances for the RPP proposed in Hertz, Laporte, and Nanchen-Hugo [39], providing good solutions in reasonable computing times.

More recently, the SCP has been studied as a special case of the DGRP in Ávila et al. [5] and Ávila [3]. The branch-and-cut algorithm proposed in these works has been tested on instances randomly generated on grids of size 50×50 and 100×100 with a number of jobs varying from 100 to 3000. All the instances defined on a 50×50 grid have been solved to optimality in very short computing times. Also, 27 out of 40 instances generated on a 100×100 grid have been solved to optimality within the time limit of one hour. For those instances that could not be solved optimally, the average gap between the final lower bound and the best solution found was less than 0.33%.

6.4.3 ▪ The DRPP and MRPP with turn penalties

When a real-life problem is modeled and solved as an MRPP, or as a particular case of it, it is commonly assumed that all turns are allowed and that they are not time consuming. While some turns can be considered irrelevant, others are more time consuming and/or dangerous in their execution, and most U-turns may even be forbidden. Therefore, the MRPP solution could be unfeasible if the traffic signals are respected, especially inside a city.

To model these situations, Benavent and Soler [10] and Corberán et al. [19] introduce, respectively, the DRPP and MRPP with Turn Penalties, which take into account turn penalties and forbidden turns. Both papers provide theoretical results on the complexity of these problems and propose a transformation into an ATSP that is then solved using already known algorithms. In Soler, Martínez, and Micó [53], a more general problem is considered: the MGRP with Turn Penalties and forbidden turns. In this problem, the route must visit all the required vertices and traverse all the required arcs and edges of the graph, without making forbidden turns. Again, the authors propose a transformation of this problem into an ATSP. In this paper, however, as the ATSP can be considered a particular case of the MGRP, the authors use the exact algorithm for the MGRP described in Corberán, Plana, and Sanchis [25] instead of the known ATSP algorithms. Computational results on a set of 128 instances with up to 200 vertices, 440 arcs, 176 required arcs, 26 required vertices, and 40 required edges are presented. 120 of these 128 instances were solved to optimality in less than two hours of computing time.

6.4.4 ▪ The directed clustered RPP

The *Directed Clustered Rural Postman Problem* (DCRPP) is a restricted version of the DRPP in which each connected component of required arcs has to be completely serviced before starting the service of another component. The use of arcs from different connected components for deadheading is allowed. This problem is introduced by Dror and Langevin [32]. The motivation for this problem stems from the observation that, in the case of postal delivery, for example, the mail is presorted to facilitate the speed of mail distribution and is usually assigned in connected components, forcing the completion of a traversal of a given subset of arcs before starting the traversal of a new subset. Other applications appear in the context of torch cutting and the servicing of parking meters. In Dror and Langevin [32], the DCRPP is first transformed into an Asymmetric Generalized TSP (AGTSP) and then solved exactly by a Lagrangian-based method by Noon and Bean [47]. The authors succeeded in solving random instances of up to 81 vertices, 247 arcs (including 88 required arcs), and 15 connected components.

6.4.5 ▪ The generalized DRPP

The *Generalized Directed Rural Postman Problem* (GDRPP), also known as the *Close-Enough Arc Routing Problem*, is an arc routing problem with some interesting real-life applications, such as routing for meter reading. Let $G=(V,A)$ be a directed graph with set of vertices V and set of arcs A, and let $c_{ij} \geq 0$ be the cost associated with the traversal of arc $(i,j) \in A$. Given a family of arc subsets $\mathbb{H}=\{H_1,\dots,H_L\}$, the GDRPP consists of finding a minimum cost tour starting and ending at the depot and traversing at least one arc from each subset H_m, $m=1,\dots,L$. These subsets H_m do not need to be disjoint nor induce connected subgraphs.

This problem, without considering the presence of a depot, was introduced by Drexl [30]. He noted that the GDRPP is NP-hard since, when each set H_m consists of a single arc, the GDRPP reduces to the DRPP, and proposed a formulation and a branch-and-cut algorithm producing good computational results. A more recent version of his work was published in [31].

However, it was Shuttleworth et al. [52] who presented this problem in the context of constructing routes for meter reading, calling it the *Close-Enough Traveling Salesman Problem*. In this application, a vehicle with a receiver travels through a series of neighborhoods. If the vehicle gets closer than a certain distance to a meter (customer), the receiver is able to record the gas, water, or electricity consumption. Therefore, the vehicle does not need to traverse every street, but only a few, in order to get close enough to each meter. The set of streets from which a certain meter m can be read defines the set H_m. In their work, Shuttleworth et al. proposed four heuristics to solve eight real-life instances with an average of 900 customers and 9000 streets each. The procedures have essentially two phases: First a subset of streets to be traversed is selected, and then a route starting and ending at the depot and traversing all these streets is found. The first phase is obviously related to the resolution of a Set Covering Problem (SCP), while the second consists of solving a DRPP (or a Directed General Routing Problem if the depot is not incident with a selected street).

More recently, Hà et al. [37] studied this problem, which they called the Close-Enough Arc Routing Problem, and proposed a new formulation, which they compared with a previous one introduced by the same authors [36] and the one by Drexl [30, 31]. Moreover, they proposed a branch-and-cut algorithm providing very good computational results on large-size instances.

In Ávila et al. [6] a slightly different formulation is presented and its associated polyhedron is studied. The authors describe several new families of valid inequalities that are used in a branch-and-cut algorithm, whose performance improves that of the branch-and-cut by Hà et al. [37].

6.4.6 ▪ The WRPP with zigzag service

The *WRPP with zigzag service* (WRPPZ) is a variant of the WRPP proposed by Irnich [40]. In this problem, the streets can be divided into four classes. First, street segments with houses on one side of the street only, which must be serviced exactly once. Second, street segments with houses on both sides, where the sides must be serviced separately and where, consequently, the segments must be traversed at least twice. Third, street segments with houses on both sides with the option of servicing both sides simultaneously by a single zigzag traversal (passing through the street segment once) or of servicing the two sides separately by two traversals. Fourth, the so-called nonrequired street segments may be used to get from one point to another. All street segments may be traversed in both directions for deadheading. Again, different directions of traversal of a segment may incur different costs. Moreover, the different traversal modes (one-side service, zigzag service, deadheading) may incur different costs.

In Irnich [40], an integer programming formulation for the WRPPZ is presented. The author also proposes a procedure to transform the WRPPZ into an ATSP, which is later transformed into a TSP and solved using Concorde (Applegate et al. [2]). From a theoretical point of view, it is clear that the additional flexibility provided by the zigzagging option enables better feasible solutions to be build. The computational results obtained indicate that the extent of improvement depends on the cost of services and on the distribution of the types of service edges.

The same author presents an extended version of the WRPPZ, called EXT-WRPP, in Irnich [41]. In addition to the WRPPZ, the new problem considers several relevant practical aspects such as turn restrictions, clustered street segments which must be serviced consecutively, public transport for reaching districts served on foot, and open tours. The author points out that the EXT-WRPP is of great practical importance for mail delivery, since in Germany there are more than 50,000 districts to be serviced every day, each district corresponding to an EXT-WRPP. For solving this problem, a transformation into an asymmetric or symmetric TSP is proposed. Since the computational experiments show that the solution of the resulting TSP instances with standard algorithms does not work satisfactorily, a new heuristic method is proposed in the paper that outperforms standard TSP heuristics with respect to solution quality and computing time.

Acknowledgments

The authors would like to thank the Spanish Ministerio of Economía y Competitividad (project MTM2012-36163-C06-02) and the Generalitat Valenciana (project GVPROMETEO2013-049) for their support.

Bibliography

[1] H.M. AFSAR, N. JOZEFOWIEZ, AND P. LOPEZ, *A branch-and-price algorithm for the windy rural postman problem*, RAIRO Operations Research, 45 (2011), pp. 353–364.

[2] D. APPLEGATE, R.E. BIXBY, V. CHVÁTAL, AND W. COOK, *Concorde*, 1999. Available at http://www.math.princeton.edu/tsp/concorde.html.

[3] T. ÁVILA, *Algunos problemas de rutas por arcos*, Ph.D. thesis, University of Valencia, Valencia, Spain, 2014.

[4] T. ÁVILA, Á. CORBERÁN, I. PLANA, AND J.M. SANCHIS, *An ILS-based metaheuristic for the stacker crane problem*, in Evolutionary Computation in Combinatorial Optimization, J.K. Hao and M. Middendorf, eds., Lecture Notes in Computer Science 7245, Springer, Berlin, 2012, pp. 25–36.

[5] ——, *The stacker crane problem and the directed general routing problem*, Technical report, Department of Statistics and Ops. Res., University of Valencia, Valencia, Spain, 2013.

[6] ——, *A new branch-and-cut algorithm for the generalized directed rural postman problem*, Technical report, Department of Statistics and Ops. Res., University of Valencia, Valencia, Spain, 2013.

[7] C. BALAGUER, A. GIMÉNEZ, J.M. PASTOR, V.M. PADRÓN, AND M. ABDERRAHIM, *A climbing autonomous robot for inspection applications in 3D complex environments*, Robotica, 18 (2000), pp. 287–297.

[8] E. BENAVENT, A. CARROTTA, Á. CORBERÁN, J.M. SANCHIS, AND D. VIGO, *Lower bounds and heuristics for the windy rural postman problem*, European Journal of Operational Research, 176 (2007), pp. 855–869.

[9] E. BENAVENT, Á. CORBERÁN, E. PIÑANA, I. PLANA, AND J.M. SANCHIS, *New heuristic algorithms for the windy rural postman problem*, Computers & Operations Research, 32 (2005), pp. 3111–3128.

[10] E. BENAVENT AND D. SOLER, *The directed rural postman problem with turn penalties*, Transportation Science, 33 (1999), pp. 408–418.

[11] G. BERBEGLIA, J.-F. CORDEAU, I. GRIBKOVSKAIA, AND G. LAPORTE, *Static pickup and delivery problems: A classification scheme and survey*, TOP, 15 (2007), pp. 1–31.

[12] M. BLAIS AND G. LAPORTE, *Exact solution of the generalized routing problem through graph transformations*, The Journal of the Operational Research Society, 54 (2003), pp. 906–910.

[13] G. CARPANETO, M. DELL'AMICO, AND P. TOTH, *Exact solution of large scale, asymmetric traveling salesman problems*, ACM Transactions on Mathematial Software, 21 (1995), pp. 395–409.

[14] N. CHRISTOFIDES, *Worst-case analysis of a new heuristic for the traveling salesman problem*, Technical report, Graduate School of Industrial Administration, Carnegie-Mellon University, Pittsburgh, PA, 1976.

[15] N. Christofides, E. Benavent, V. Campos, Á. Corberán, and E. Mota, *An optimal method for the mixed postman problem*, in System Modelling and Optimization, P. Thoft-Christensen, ed., Lecture Notes in Control and Information Sciences 59, Springer, Berlin, 1984, pp. 641–649.

[16] N. Christofides, V. Campos, Á. Corberán, and E. Mota, *An algorithm for the rural postman problem on a directed graph*, Mathematical Programming Study, 26 (1986), pp. 155–166.

[17] C.C. Chyu, *A mixed strategy heuristic for the mixed arc routing problem*, Journal of the Chinese Institute of Industrial Engineers, 18 (2001), pp. 68–76.

[18] J. Cirasella, D.S. Johnson, L.A. McGeoch, and W. Zhang, *Traveling salesman problem: Algorithms, instance generators, and tests*, in ALENEX 2001 Proceedings, A. L. Buchsbaum and J. Snoeyink, eds., Lecture Notes in Computer Science 2153, Springer, Berlin, 2001, pp. 32–59.

[19] Á. Corberán, R. Martí, E. Martínez, and D. Soler, *The rural postman problem on mixed graphs with turn penalties*, Computers & Operations Research, 29 (2002), pp. 887–903.

[20] Á. Corberán, R. Martí, and A. Romero, *Heuristics for the mixed rural postman problem*, Computers & Operations Research, 27 (2000), pp. 183–203.

[21] Á. Corberán, G. Mejía, and J.M. Sanchis, *New results on the mixed general routing problem*, Operations Research, 53 (2005), pp. 363–376.

[22] Á. Corberán, E. Mota, and J.M. Sanchis, *A comparison of two different formulations for arc routing problems on mixed graphs*, Computers & Operations Research, 33 (2006), pp. 3384–3402.

[23] Á. Corberán, M. Oswald, I. Plana, G. Reinelt, and J.M. Sanchis, *New results on the windy postman problem*, Mathematical Programming, 132 (2012), pp. 309–332.

[24] Á. Corberán, I. Plana, and J.M. Sanchis, *Zigzag inequalities: A new class of facet-inducing inequalities for arc routing problems*, Mathematical Programming, 108 (2006), pp. 79–96.

[25] ——, *A branch & cut algorithm for the windy general routing problem and special cases*, Networks, 49 (2007), pp. 245–257.

[26] ——, *The windy general routing polyhedron: A global view of many known arc routing polyhedra*, SIAM Journal on Discrete Mathematics, 22 (2008), pp. 606–628.

[27] Á. Corberán, A. Romero, and J.M. Sanchis, *The mixed general routing polyhedron*, Mathematical Programming, 96 (2003), pp. 103–137.

[28] Á. Corberán and J.M. Sanchis, *A polyhedral approach to the rural postman problem*, European Journal of Operational Research, 79 (1994), pp. 95–114.

[29] ——, *The general routing problem polyhedron: Facets from the RPP and GTSP polyhedra*, European Journal of Operational Research, 108 (1998), pp. 538–550.

[30] M. Drexl, *On Some Generalized Routing Problems*, Ph.D. thesis, Aachen University, Aachen, Germany, 2007.

[31] ——, *On the generalized directed rural postman problem*, The Journal of the Operational Research Society, 65 (2014), pp. 1143–1154.

[32] M. DROR AND A. LANGEVIN, *A generalized traveling salesman problem approach to the directed clustered rural postman problem*, Transportation Science, 31 (1997), pp. 187–192.

[33] H.A. EISELT, M. GENDREAU, AND G. LAPORTE, *Arc routing problems, part 2: The rural postman problem*, Operations Research, 43 (1995), pp. 399–414.

[34] G.N. FREDERICKSON, *Approximation algorithms for some postman problems*, Journal of the ACM, 26 (1979), pp. 538–554.

[35] G.N. FREDERICKSON, M.S. HECHT, AND C.E. KIM, *Approximation algorithms for some routing problems*, SIAM Journal on Computing, 7 (1978), pp. 178–193.

[36] M-H. HÀ, N. BOSTEL, A. LANGEVIN, AND L-M. ROUSSEAU, *An exact algorithm for close enough traveling salesman problem*, in Proceedings of the 1st International Conference on Operations Research and Enterprise Systems, 2012, pp. 233–238.

[37] ——, *Solving the close enough arc routing problem*, Networks, 63 (2014), pp. 107–118.

[38] R. HASSIN AND S. KHULLER, *z-approximations*, Journal of Algorithms, 41 (2001), pp. 429–442.

[39] A. HERTZ, G. LAPORTE, AND P. NANCHEN-HUGO, *Improvement procedures for the undirected rural postman problem*, INFORMS Journal on Computing, 11 (1999), pp. 53–62.

[40] S. IRNICH, *A note on postman problems with zigzag service*, INFOR, 43 (2005), pp. 33–39.

[41] ——, *Solution of real-world postman problems*, European Journal of Operational Research, 190 (2008), pp. 52–67.

[42] P. LACOMME, C. PRINS, AND W. RAMDANE-CHÉRIF, *A genetic algorithm for the capacitated arc routing problem and its extensions*, in Applications of Evolutionary Computing, E.J.W. Boers, ed., Lecture Notes In Computer Science 2037, Springer-Verlag, Berlin, 2001, pp. 473–483.

[43] ——, *Fast algorithms for general arc routing problems*, in IFORS 2002, Edinburgh, UK, 2002.

[44] G. LAPORTE, *Modeling and solving several classes of arc routing problems as traveling salesman problems*, Computers & Operations Research, 24 (1997), pp. 1057–1061.

[45] J.K. LENSTRA AND A. RINNOOY KAN, *On the general routing problem*, Networks, 6 (1976), pp. 273–280.

[46] A.N. LETCHFORD, *New inequalities for the general routing problem*, European Journal of Operational Research, 96 (1997), pp. 317–322.

[47] C.E. NOON AND J.C. BEAN, *A Lagrangian based approach for the asymmetric generalized traveling salesman problem*, Operations Research, 39 (1991), pp. 623–632.

[48] I. OR, *Traveling Salesman-Type Combinatorial Problems and Their Relation to the Logistics of Regional Blood Banking*, Ph.D. thesis, Northwestern University, Evanston, IL, 1976.

[49] C.S. ORLOFF, *A fundamental problem in vehicle routing*, Networks, 4 (1974), pp. 35–64.

[50] I. PLANA, *The Windy General Routing Problem*, Ph.D. thesis, University of Valencia, Valencia, Spain, 2005.

[51] B. RAGHAVACHARI AND J. VEERASAMY, *A 3/2 approximation algorithm for the mixed postman problem*, SIAM Journal on Discrete Mathematics, 12 (1999), pp. 425–433.

[52] R. SHUTTLEWORTH, B.L. GOLDEN, S. SMITH, AND E. WASIL, *Advances in meter reading: Heuristic solution of the close enough traveling salesman problem over a street network*, in The Vehicle Routing Problem: Lastest Advances and New Challenges, B.L. Golden, S. Raghavan, and E. Wasil, eds., Springer, Berlin, 2008, pp. 487–501.

[53] D. SOLER, E. MARTÍNEZ, AND J. MICÓ, *A transformation for the mixed general routing problem with turn penalties*, The Journal of the Operational Research Society, 59 (2008), pp. 540–547.

[54] J. SROUR AND S. VAN DE VELDE, *Are stacker crane problems easy? A statistical study*, Computers & Operations Research, 40 (2013), pp. 674–690.

[55] Z. WIN, *On the windy postman problem on Eulerian graphs*, Mathematical Programming, 44 (1989), pp. 97–112.

[56] L. ZHANG, *Polynomial algorithms for the k-Chinese postman problem*, in Proceedings of the IFIP 12th World Computer Congress. Volume 1: Algorithms, Software, Architecture, J. Van Leeuwen, ed., Elsevier, Amsterdam, 1992, pp. 430–435.

[57] L. ZHANG AND W. ZHENG, *Genetic coding for solving both the stacker crane problem and its k-variant*, in Proceedings of the IEEE International Conference on Systems, Man and Cybernetics, IEEE, Piscataway, NJ, 1995, pp. 1061–1066.

Part II

Arc Routing Problems with Several Vehicles

Chapter 7

The Capacitated Arc Routing Problem: Heuristics

Christian Prins

7.1 ▪ Introduction

Whereas all previous chapters have covered arc routing problems (ARPs) with a single vehicle, this chapter opens the second part of the book, devoted to problems involving several vehicles with limited capacity. More precisely, the chapter addresses the seminal problem in this category, the *Capacitated Arc Routing Problem* (CARP). This important problem plays the same role in arc routing as the *Capacitated Vehicle Routing Problem* (CVRP) in node routing: Even if recent extensions get closer to the real world by handling more constraints, the CARP is already hard enough to constitute a good laboratory to develop and test new ideas in arc routing. The CARP is obviously more difficult than uncapacitated ARPs and its study is also more recent, explaining why many results concern constructive heuristics and metaheuristics. This chapter focuses on such approaches, while Chapters 8, 9, 10, respectively, survey lower bounds, exact algorithms, and more general problems.

The CARP is usually defined on an undirected connected graph $G = (V, E)$. To demarcate from more recent versions on directed or mixed graphs (Chapter 10), we call it *Undirected CARP* (UCARP) in what follows. The set V of n nodes includes one depot (node 1) with identical vehicles of capacity Q. The set E of m edges contains a subset E_R of t required edges or *tasks*, to be serviced by a vehicle. A traversal cost c_e is associated with each edge e in E. Each task has a nonnegative demand q_e. The goal is to find a minimum cost set of routes to process all tasks, the number of routes being a decision variable. A route (or trip) is a closed walk containing the depot and a set of traversed edges. Some of these edges are serviced by the route, while the others (*deadheading edges*) are traversed but not serviced. The total demand handled in a route is limited to Q. Like in the CVRP, split services are prohibited: Each task must be processed by a single vehicle.

The UCARP was defined in 1981 by Golden and Wong [24], who proved its NP-hardness and showed that it is even NP-hard to find a 3/2 approximation when the triangle inequality holds. In fact, Christofides [14] introduced in 1973 the case where all edges are required, the *Capacitated Chinese Postman Problem* (CCCP).

The UCARP has obvious applications such as urban refuse collection, winter gritting, meter reading, etc. In waste collection, for instance, G represents the street network and each edge e models a street segment of length c_e, with an amount of waste q_e which can be removed in any direction. The goal is to determine a set of routes to collect all the waste and minimize the total distance traveled. If priority is given to the total duration, processing times or service costs c_e^{serv} are often introduced, c_e denoting deadheading times. In this example, required edges are defined by $q_e \neq 0$. In some applications like meter reading, no commodity is distributed or collected and the demand of an edge is the time c_e^{serv} to inspect it. In such cases vehicle capacity is often replaced by a maximum working time.

The goal of this chapter is to recall classical constructive heuristics and metaheuristics for the UCARP, before focusing on recent developments. The classical period is the one covered in the book *Arc Routing: Theory, Solutions and Applications*, edited in 2000 by Dror [18]. Many new works on arc routing since Dror's book are discussed in a survey by Wohlk [52] for the period 2000 to 2007 and in an annotated bibliography by Corberán and Prins [16] for 2000 to 2010. These surveys are completed here by the most recent references.

In the algorithms of this chapter, E_R is the set of edges with nonzero demands. An $n \times n$ distance matrix $D = (d_{ij})$ is precomputed. The length d_{ij} of a shortest path SP_{ij} between nodes i and j can be found in $O(mn \log n)$ using Dijkstra's algorithm [17]. G is coded as a symmetric digraph with two arcs per edge. Arcs (i, j) with $i < j$ are indexed from 1 to m and the opposite arcs from $m + 1$ to $2m$. Arc number u is defined by its extremities $beg(u)$ and $end(u)$, and by the cost c_u and the demand q_u from the edge of origin. The opposite arc has index $m + u$. We call *required arcs* the arcs coding required edges. A route R_k is coded as a list of required arcs, connected by implicit shortest paths and with implicit copies of the depot at the extremities. Its cost and load are denoted by $cost(k)$ and $load(k)$.

The constructive heuristics are presented in section 7.2 for the classical period, followed by the novelties in section 7.3. The classical and new metaheuristics are described in sections 7.4 and 7.5. Performance indicators are delayed until section 7.6 that provides a numerical comparison on three sets of standard instances.

7.2 ▪ Classical constructive heuristics for the CARP

This section deals with heuristics that build a single solution for the UCARP, contrary to metaheuristics that generate a sequence of solutions or work on a population of solutions. Arbitrarily, we decided to focus on four heuristics: *Construct-Strike*, *Path-Scanning*, *Augment-Merge* and *Ulusoy's method*. Construct-Strike deserves to be recalled as the earliest UCARP heuristic [14]. The three others have the merit to have survived over time, due to their efficiency and ability to handle additional constraints. Other heuristics are available, but they will be only briefly mentioned.

7.2.1 ▪ Construct-strike

Chapter 3 presents a classical technique to solve the *Undirected Chinese Postman Problem* (UCPP). The principle is to make the given graph G Eulerian, by adding copies of existing edges at minimal total cost. The set V_{odd} of odd-degree nodes in G is determined first. A complete graph G_{odd} is then defined on V_{odd}, where the cost of each edge $[i, j]$ is the cost d_{ij} of a shortest path SP_{ij} between nodes i and j in G. A minimum cost perfect matching in G_{odd} can be computed in $O(n^3)$. Finally, for each edge $[i, j]$ in the resulting matching, a copy of each edge of SP_{ij} is added to G. It is then easy to extract an Eulerian cycle.

The idea of *Construct-Strike* proposed in 1973 by Christofides is similar. The graph is made Eulerian but in general the tasks cannot be included in a single route, due to capacity constraints. Hence, simple cycles (feasible routes) are successively extracted and their edges erased, while keeping the graph connected (ignoring isolated nodes, which can appear). When this is no longer possible, new edges are added to make all node-degrees even and to avoid an isolated depot. In fact, to delay the application of the time-consuming matching algorithm, the heuristic attempts at the beginning to extract routes even if the graph is not Eulerian. The matching algorithm is called at the first failure.

The heuristic is sketched in Algorithm 7.1. Variable k counts the number of unsuccessful attempts to extract a route R in step 6; it is reset in step 8 whenever a route is discovered. In step 13, two shortest paths are added between the depot and its closest node if $k = 1$, between the depot and its second closest node if $k = 2$, and so on. In step 16, the costs for the matching problem correspond to the shortest paths SP_{ij} in G, not in $\overline{G}$.

Christofides applied his algorithm to one graphical example, without saying how to build R. Pearn [40] proposed one method in an improved version, described in subsection 7.2.5. Another possibility is the route construction approach used in the next heuristic, Path-Scanning. The preservation of connectivity if one edge $[i, j]$ is removed can be checked by a graph search from node i. According to Pearn [40], Construct-Strike can be implemented in $O(mn^3)$.

Algorithm 7.1 – Construct-Strike

1. Take a copy $\overline{G}$ of G
2. **loop**
3. $k \leftarrow 0$
4.   **loop**
5. $k \leftarrow k + 1$
6.   build in $\overline{G}$ a simple cycle R with the depot and at least one task, such that $\overline{G}$ stays connected (except isolated nodes) if edges of R are erased
7.   **exit when** R is not found
8.   remove the edges of R from $\overline{G}$ and reset k to 0
9.   **endloop**
10.   **exit when** all required edges are treated
11.   remove all edges added previously by steps 13 and 15 (if any)
12.   **if** $\overline{G}$ has all even or zero node degrees and depot is not isolated **then**
13.   add to $\overline{G}$ two copies of the shortest path in G from the depot to its kth nearest node
14.   **else** //odd-degree nodes exist and/or depot is isolated
15.   if depot is isolated, add a copy $n+1$ and edge $[1, n+1]$ with $c_{1,n+1} = \infty$
16.   make $\overline{G}$ Eulerian solving a minimum cost matching problem
17.   if a copy of the depot has been added, shrink nodes 1 and $n+1$
18.   **endif**
19. **endloop**

7.2.2 ▪ Path-scanning

Golden, DeArmon, and Baker [23] proposed in 1983 the *Path-Scanning* (PS) heuristic for the UCPP. The method is explained as follows in their paper. Routes are built one by one. Starting at the depot, the emerging route ending at node i is progressively extended by adding at the end one required edge $[i,j]$, not yet serviced and compatible with vehicle capacity. As several required edges are often incident to i, five rules are used to break ties and select the next edge $[i,j]$:

1: Maximize the ratio c_{ij}/r_{ij}, r_{ij} being the remaining demand once $[i,j]$ is treated.
2: Minimize the ratio c_{ij}/r_{ij}.
3: Maximize the return cost from j to the depot.
4: Minimize this return cost.
5: If the vehicle is less than half-full, then apply rule 3, otherwise rule 4.

The construction of the current route stops when all required edges incident to node i are already served or do not fit the vehicle residual capacity. In that case, the vehicle returns to the depot via a shortest path. Successive routes are built in this way, until all required edges are covered. The heuristic is executed successively with the five rules, and the best of the resulting five solutions is returned.

In 1992, Evans and Minieka [21] replaced rules 1 and 2 by the minimization and the maximization of c_{ij}/q_{ij}, respectively. It is easy to show that these new rules select the same edges as the original rules. Moreover, contrary to Golden et al., who finish the emerging route when no edge incident to i can be added, they propose to proceed to the nearest vertex having incident edges which can be served. The route returns to the depot if no such vertex exists. This version has been adopted later by most authors; see for instance [31], [7], [43], and [45].

Algorithm 7.2 describes more formally this version, using our notations and coding conventions. The loop (3–22) builds successive routes R_k. The loop (5–20) appends required arcs at the end of the emerging trip (recall that we code each edge as two arcs), and the trip is finally closed (line 21). Node i is the last node of R_k, $\bar{d}$ the minimum distance between i and unserviced arcs in the list *free*, and $\bar{u}$ the arc selected. Priority is given to the arc nearest to i (line 8) and ties are broken by a function $better(u,\bar{u},rule)$ returning $true$ if and only if arc u is better than $\bar{u}$ for the selected rule (line 11). To implement rule 3 for instance, the function $better$ consists in returning *true* if and only if $d_{end(u),1} > d_{end(\bar{u}),1}$.

PS has clearly an $O(t^2)$ complexity. It is efficient in practice because a rule with poor results on one instance is often compensated by another rule. For instance, rule 3 drives the vehicle away from the depot, which works well on towns with radial streets, whereas rule 4 keeps the vehicle close to the depot, which favors cities with concentric avenues. Moreover, the method is flexible. Windy edges can be tackled using distinct costs on the two arcs coding each edge, while durations can be handled by replacing deadheading costs by service costs in rules 1-2 and line 18. Forbidden turns are easily handled, as explained in [31]. Finally, the heuristic can be adapted to more complex networks like a mixed graph [7].

7.2.3 ▪ Augment-merge

Augment-Merge (AM), just sketched in 1981 by Golden and Wong [24], was detailed in 1983 by Golden, DeArmon, and Baker [23]. It is inspired by the *Clarke and Wright*

Heuristic (CWH) for the CVRP [15]. The heuristic builds first one direct trip for each required edge. Then, a *Merge* phase is executed, in which each iteration considers each pair of routes and evaluates the saving if the two routes are merged (concatenated). There are four ways of merging two given routes, as each route can be reversed or not. The merger with the largest saving is executed, and this process is repeated until no merger is possible without violating vehicle capacity.

Algorithm 7.2 – Path-Scanning for one priority rule

```
k ← 0
copy all required arcs in a list free
repeat
  k ← k+1; R_k ← ∅; load(k), cost(k) ← 0; i ← 1
  repeat
    d̄ ← ∞
    for each u ∈ free | load(k) + q_u ≤ Q do
      if d_{i,beg(u)} < d̄ then
        d̄ ← d_{i,beg(u)}
        ū ← u
      else if (d_{i,beg(u)} = d̄) and better(u, ū, rule)
        ū ← u
      endif
    endfor
    add ū at the end of route R_k
    remove arc ū and its opposite ū + m from free
    load(k) ← load(k) + q_ū
    cost(k) ← cost(k) + d̄ + c_ū
    i ← end(ū)
  until (free = ∅) or (d̄ = ∞)
  cost(k) ← cost(k) + d_{i1}
until free = ∅
```

Compared with CWH, a preliminary phase called *Augment* is added. The initial routes are sorted in nonincreasing order of costs. Recall that we represent a route as a list of required arcs (see introduction). Starting from the longest route, each route $R_k = ((i,j)), k = 1,2,\dots,t-1$, is compared with each shortest route $R_p, p = k+1, k+2, \dots, t$ such that the sum of their loads fits vehicle capacity. If the unique edge serviced by R_p lies on SP_{1i} or SP_{j1}, it can be transferred in R_k and R_p can be replaced by an empty trip. The cost of R_k does not change, but a saving equal to the cost of R_p is incurred. The Augment phase can strongly reduce the total cost and the number of routes before starting the Merge phase.

This nontrivial heuristic is detailed in Algorithm 7.3, where F_k and L_k denote the first and last node of route R_k. The Augment phase can be implemented in $O(nt^2)$, while the Merge phase is dominated by the sort line 24 in $O(t^2 \log t)$. The whole complexity is then $O(t^2(n + \log t))$, or $O(m^2 n)$ if all edges are required.

7.2.4 ▪ Ulusoy's route-first cluster-second method

In 1983, Beasley [3] gave a route-first cluster-second heuristic for the CVRP. A first phase ignores vehicle capacity and runs any TSP algorithm to get a giant tour T covering all customers. A second phase *Split*, based on an auxiliary graph H, cuts T into feasible CVRP routes. The nodes of H include one dummy node 0 and one node per customer. Each subsequence of customers $(T_i, T_{i+1}, \dots, T_j)$ is modeled in H as one arc $(i-1, j)$, weighted by the cost of the corresponding trip. The optimal splitting (subject to the sequence) is given by a least-cost path from node 0 to node n in H. Any TSP heuristic can be used in the first phase because an optimal giant tour does not necessarily lead to an optimal CVRP solution after splitting.

Algorithm 7.3 - Augment-Merge

//Initial routes
$k \leftarrow 0$
for $u \leftarrow 1$ **to** m **such that** $q_u \neq 0$ **do**
    $k \leftarrow k+1;\ R_k \leftarrow (beg(u), end(u));\ load(k) \leftarrow q_u$
    $cost(k) \leftarrow d_{1,beg(u)} + c_u + d_{end(u),1}$
endfor
//Augment phase
for $k \leftarrow 1$ **to** $t-1$ **such that** $R_k \neq \emptyset$ **do**
    assume that $R_k = ((x,y))$: expand the route by inserting all arcs of SP_{1x}
    before (x,y) and all arcs of SP_{y1} after (x,y)
    for $p \leftarrow k+1$ **to** t **such that** $R_p \neq \emptyset$ and $load(k) + load(p) \leq Q$ **do**
        let (v,w) be the unique required arc in R_p
        if (v,w) or (w,v) lies on R_k **then**
            switch the status of this arc to "serviced" in R_k
            $load(k) \leftarrow load(k) + q_{vw};\ R_p \leftarrow \emptyset$
        endif
    endfor
    remove nonrequired arcs and unserviced required arcs from R_k
endfor
//Merge phase
prepare an empty list Λ
for each pair of distinct trips $\{R_k, R_p\}$ **do**
    add to Λ the four arcs $(L_k, F_p), (F_k, F_p), (F_k, L_p)$, and (L_k, L_p)
endfor
sort the arcs (i,j) in Λ in nonincreasing order of savings $s_{ij} = d_{i1} + d_{1,j} - d_{ij}$
for each arc $(i,j) \in \Lambda$ **do**
    if i, j are still the end-nodes of two distinct trips R_k and R_p
    and $load(k) + load(p) \leq Q$ **then** //Merge the two trips
        if i is at the beginning of R_k **then** invert route R_k **endif**
        if j is at the end of R_p **then** invert route R_p **endif**
        copy the required arcs of R_p at the end of R_k
        $load(k) \leftarrow load(k) + load(p);\ cost(k) \leftarrow cost(k) + s_{ij}; R_p \leftarrow \emptyset$
    endif
endfor

In 1985, Ulusoy [48] transposed this heuristic to the UCARP by replacing T by a sequence of t required arcs, one per required edge. Algorithm 7.4 provides a compact implementation of this version, proposed by Lacomme, Prins, and Ramdane-Chérif [31]. It is based on Bellman's algorithm for directed acyclic graphs [17], but the auxiliary graph H is not generated explicitly. For each node i in H, V_i is the cost of a shortest path from node 0 to node i in H, also called *label* of i.

Roughly speaking, the algorithm performs two nested loops to inspect each subsequence $(T_i, T_{i+1}, \ldots, T_j)$, compatible with vehicle capacity, and computes the cost of the associated trip, corresponding to arc $(i-1, j)$ in H. If $V_{i-1} + cost < V_j$, this means that a better path to node j has been found via node $i-1$ and label V_j can be immediately updated. The cost of the shortest path can be read at the end in label V_t. P_j denotes the predecessor of node j along the shortest path: It can be used to extract the successive trips after splitting.

Ulusoy designed his heuristic for the CCPP and solves a UCPP to get an optimal giant tour, using the matching technique of Chapter 3. When not all edges are required, the computation of an optimal giant tour becomes an NP-hard undirected rural postman problem and the giant tour can be determined using one of the heuristics of Chapter 5. An alternative is to set vehicle capacity Q to a large value and apply a UCARP heuristic like Construct-Strike, Path-Scanning, or Augment-Merge. In the last decade, the *Split* procedure found applications in several metaheuristics (see section 7.5).

Split runs in $O(t^2)$ because the two loops in Algorithm 7.4 check in $O(1)$ each subsequence $(T_i, T_{i+1}, \ldots, T_j)$. The trick is to deduce in $O(1)$ the load (line 5) and cost (line 9) for $(T_i, T_{i+1}, \ldots, T_j, T_{j+1})$ from those for $(T_i, T_{i+1}, \ldots, T_j)$, instead of running a loop to recompute these values. In practice, many subsequences are ignored due to excessive load. Let b denote the average number of tasks per feasible trip. As b is also the average node degree in H, the auxiliary graph has $O(bt)$ arcs and we get a more accurate complexity in $O(bt)$. For instance, if all demands exceed $Q/3$, a trip treats at most two tasks and *Split* runs in $O(t)$.

Algorithm 7.4 – Split (phase 2 of Ulusoy's algorithm)

```
set V_0 and P_0 to 0, and V_1, V_2, ..., V_n to +∞
for i ← 1 to t do
  load, cost ← 0; j ← i
  repeat
    load ← load + q(T_j)
    if j = i then
      cost ← d(1, beg(T_i)) + c(T_i) + d(end(T_i), 1)
    else
      cost − d(end(T_{j−1}), 1) + d(end(T_{j−1}), beg(T_j)) + c(T_j) + d(end(T_j), 1)
    endif
    if load ≤ Q then
      if V_{i−1} + cost < V_j then
        V_j ← V_{i−1} + cost
        P_j ← i − 1
      endif
      j ← j + 1
    endif
  until (j > t) or (load > Q)
endfor
```

7.2.5 ▪ Other classical heuristics

While Path-Scanning, Augment-Merge, and Ulusoy's heuristics have still been used in the last decade by several authors to design improved versions or serve as components in metaheuristics, the following classical heuristics, presented in chronological order, have mainly a historical interest.

Chapleau et al. [13] developed a *Parallel-Insert* heuristic for a limited fleet size K, inspired by insertion heuristics for the TSP. In a first phase, starting with an empty set of routes, the heuristic tries to insert the required edge $[i,j]$ farthest from the depot in the routes with enough capacity. In case of failure, a new route is created with $[i,j]$. This process is iterated with the remaining task most remote from the depot, until all tasks are treated or K routes are created. In a second phase, each iteration consists in determining the least-loaded route and inserting the unserviced task that induces the smallest detour. An optional improvement procedure performing edge exchanges can be added. If after the two phases some tasks are still uncovered, K is incremented and phases 1 and 2 are repeated.

The *Modified-Construct-Strike heuristic* from Pearn [40] no longer requires a connected graph when a route is removed and specifies how each route is built, contrary to the original version [14]. For an emerging route ending at node i, the heuristic determines the path with maximum demand from i to the depot, such that the load of the emerging route plus the demand on the path fits vehicle capacity. The first edge $[i,j]$ of this path is added at the end of the route. According to Pearn [40], these improvements increase the complexity up to $O(mn^4)$.

In the same paper, Pearn described a randomized version of Path-Scanning. Instead of building one complete solution for each rule, a rule is chosen randomly for each edge added to the emerging route. The resulting *Random-Path-Scanning* heuristic can be executed several times to return the best solution.

Pearn designed two years after [41] a heuristic called *Augment-Insert*. The routes are built one by one. Each new route R_k is initialized with the free task $[i,j]$ such that $\delta_{ij} = d_{1i} + d_{j1}$ is maximal, and then grown like in the Augment phase of Augment-Merge [23]. After these obvious insertions which do not increase the cost of R_k, the heuristic considers the remaining tasks in decreasing order of δ_{ij}: If the insertion of the incumbent task at the beginning or at the end of R_k is feasible and induces a cost variation below a given threshold, the task is inserted; otherwise, it is ignored. The insertions in the current route stop when no candidate exists or if the first and last tasks of R_k are incident to the depot.

Benavent et al. [8] studied a cluster-first route-second algorithm, *Cycle-Assignment*, for a fleet of K vehicles. The clustering phase determines the set of tasks (cluster) C_k assigned to each vehicle k. If necessary, artificial edges are added to the partial graph induced by required edges, to make it connected and even. Then K well-spread nodes called *centers* are selected in this graph. Complete cycles in the graph are successively assigned to each center, so that the set of cycles assigned to a center is connected and its total load respects vehicle capacity. If some tasks are left unassigned, a generalized assignment problem is solved, where assignment costs approximate the insertion cost of each remaining task into each route. The routing phase builds a route in each cluster C_k. If the tasks assigned to C_k form a connected graph, this problem reduces to an undirected Chinese postman problem that can be solved in polynomial time [19]. Otherwise, the resulting NP-hard rural postman problem is solved via a heuristic proposed by Frederickson [22].

7.3 ▪ Recent constructive heuristics

The previous heuristics were published either without numerical results or tested on the only set of benchmark problems available at that time, the *gdb* instances (see section 7.6). A few new constructive heuristics have been proposed in the last decade, and some historical heuristics have led to improved versions.

7.3.1 ▪ Wøhlk's heuristics

In her 2005 thesis [51], Wøhlk described four constructive heuristics: *Modified Path-Scanning* (MPS), *Double Outer Scan*, *Node Duplication Heuristic*, and *A-ALG*.

In the original Path-Scanning [23], the route arrived at a node i returns to the depot if all required edges incident to i are already served or have a too large demand. In contrast, MPS follows a shortest path to the closest task $[k,l]$ which can be added to the route, serves this task, then sets $i = l$ to continue building the route with the selected rule. In fact, MPS is very similar to the version amended by Evans and Minieka [21] and formalized in Algorithm 7.2. The only difference is that Algorithm 7.2 applies the current rule when several edges are at the same minimum distance to i, while MPS ignores this rule.

For Double Outer Scan, Wøhlk defines the *parallel distance* of a required edge $[i,j]$ to a node k as $d_{ki} + d_{jk}$. Each new route is initialized as a chain, reduced to the unserviced edge with the largest parallel distance to the depot. This chain is repeatedly extended by adding the free task incident to one of its end-nodes and maximizing the parallel distance to the depot. When extension becomes impossible, the two end-nodes of the resulting chain are connected to the depot via shortest paths; then, like in the first phase of Augment-Merge, unserviced tasks along these paths are treated by the emerging route, as long as vehicle capacity allows it. Successive routes are built in this way until all tasks are treated. Finally, the second phase of Augment-Merge is executed to try merging pairs of routes.

Jansen designed a heuristic called *Shortest Optimal Tour Partitioning* (SOTP) with approximation factor $7/2-3/Q$ for the capacitated general routing problem [29]. When simplified for the UCARP, SOTP computes a giant tour T using Frederickson's 3/2-approximation heuristic for the rural postman problem [22], then partitions T into feasible trips using Ulusoy's splitting algorithm of subsection 7.2.4. A-ALG is an improved version described in Wøhlk's thesis [51] and published later in a journal [53]. It includes a new splitting procedure to determine the best traversal direction of each required edge in T. Jansen's performance ratio is not improved, but the results of A-ALG are always at least as good as the ones of SOTP.

In fact, the splitting procedure in A-ALG gives the same results as *Split with Flips* described in 7.3.3 but relies on a different algorithm (dynamic programming). This idea of improving a route by flipping its required edges was also used by Muyldermans [39], in his *dir-opt* move used by in a local search for the UCARP.

7.3.2 ▪ Improvements to augment-merge and path-scanning

In a paper on the mixed CARP, Belenguer et al. [7] evaluate Path-Scanning and Augment-Merge on five sets of instances, including the *gdb*, *val*, and *egl* UCARP instances presented in section 7.6. They show that these heuristics degrade rapidly with instance size, especially when not all edges are required. This drawback was already noticed in 1983 by Golden et al. [23]. For Augment-Merge, the tests in [7] indicate that the Augment phase brings little improvement on average, and is even prejudicial when some edges are not required. The results are improved by a version *Merge* without the Augment phase, and even further by a variant *Improved Merge*: When two pairs of routes can be merged with the same maximal saving, the pair with the largest discrepancy in load is preferred. This situation is frequent in standard UCARP instances because edge costs are small integers.

Two versions of Path-Scanning (PS) are also studied in [7]. One called *PS with Random Criterion* (PSRC) is identical to Pearn's Random-Path-Scanning [40] while the other *PS with Random Task* (PSRT) determines at each iteration the set F of feasible unserved edges that are nearest to the last node of the emerging route, then draws one edge at random among these candidates. Note that the five rules of PS are no longer used in this version. The main interest of these randomizations is to allow multiple executions to get better solutions. The tests show that 4 to 11 executions are enough to do better than PS and its five runs. PSRC is better than PSRT on average, but the difference becomes negligible beyond 100 runs.

In 2009, Santos et al. [45] improved PSRT with what they call an *ellipse rule*. When a vehicle is near the end of its route, this rule forces the vehicle to service only edges close to the shortest path to return to the depot. More precisely, let Q_{tot} be the total demand, Q_{rem} the total demand not yet serviced, C_{tot} the total cost of required edges, i the last node of the emerging trip, and β a given positive factor. Like in PSRT, the set F of candidates closest to i is determined and one edge of F is selected at random. However, if $Q_{rem} \leq \beta \times Q_{tot}/t$, F is filtered to keep only the edges $[p,j]$ such that $d_{ip}+c_{pj}+d_{j1} \leq C_{tot}/t+d_{i1}$. The resulting heuristic, called *RSE-ER*, returns excellent results when executed 10,000 times. We will see in Section 7.6 that some metaheuristics can do better in comparable running times.

7.3.3 ▪ Improved tour splitting heuristics

In 2009, Prins, Labadi, and Reghioui [43] derived improved algorithms from Ulusoy's heuristic, by randomizing the giant tour construction (three methods) or playing on the splitting procedure (four methods).

The first two methods to get various giant tours, named URC and URT, consist in calling, respectively, the heuristics PSRC and PSRT (see previous subsection). The vehicle capacity is set to the total demand to return a single route. For an emerging route waiting at node i, the third method called URTF determines the set F of candidate edges $[k,l]$ like in URT/PSRT, but then F is partitioned into F_1, the tasks which increase the distance to the depot ($d_{l1} > d_{i1}$), and F_2, the other tasks which reduce this distance ($d_{l1} \leq d_{i1}$). If $load \mod Q < Q/2$, URTF draws the next edge of the tour in F_1, and in F_2 otherwise. In other words, the giant tour tends to go away from the depot if its load is in one interval $[k\,Q,(k+0.5)\cdot Q]$ for some integer $k \geq 0$; otherwise, it gets closer to the depot. Hence, the probability of splitting the tour when it gets close to the depot is higher. The name RTF with F like "flower" comes from the flower-shaped resulting tour.

The first splitting procedure is the *Basic Split* (BS): The route for a subsequence $(T_i, T_{i+1}, \ldots, T_j)$ is $(1, T_i, T_{i+1}, \ldots, T_j, 1)$, i.e., it begins and ends at the depot. The version *Split with Shifts* (SS) allows circular shifts like $(1, T_{i+1}, \ldots, T_j, T_i, 1)$. The arc $(i-1, j)$ that models the subsequence in the auxiliary graph corresponds to the route with the best shift. This technique can be viewed also as a best insertion of the depot in the subsequence, considered as a circular list. The version *Split with Flips* (SF) allows edge reversals. Like the basic Split, the SS and SF variants are optimal (subject to the giant tour T) and can be implemented with an $O(bt)$ complexity, b denoting the average cardinal of feasible subsequences. The last version, *Split with Shifts and Flips (SSF)*, combines circular shifts of subsequences and edge reversals but displays a higher complexity in $O(b^2 t)$.

The three giant tour construction methods and the four splitting procedures give 12 randomized heuristics, e.g., URT-SS means a giant tour computed by URT, followed by *Split with Shifts*. For 20 executions, all these versions outperform PSRC and PSRT on average. For a given version of Split, the best results are obtained from giant tours built with URTF, and, for a given giant tour construction heuristic, the SF algorithm yields the best solutions but is time consuming. A few performance indicators are provided in section 7.6 for a good compromise, URTF-SS.

7.4 ▪ Classical metaheuristics for the CARP

Very few metaheuristics were designed in the classical period ending in 2000. These are either simulated annealing procedures or tabu search algorithms, reflecting the dominant positions of these techniques at that time.

7.4.1 ▪ Simulated annealing

Eglese introduced in 1994 the first metaheuristic dedicated to the UCARP, a *Simulated Annealing* (SA) algorithm. The goal is to solve winter gritting problems in a rural environment. The network is made even and decomposed into elementary cycles called *cyclenodes* according to a chessboard pattern. A route is defined as a tree rooted at the depot and spanning a subset of cycles. The SA procedure works on a solution encoded as a set of trees, which is randomly modified using elementary moves affecting at most two trees.

7.4.2 ▪ Tabu search

The first *Tabu Search* (TS) for the UCARP was proposed in 1996 by Eglese and Li [20]. The year after, Belenguer, Benavent, and Cognata [6] designed the first tabu search able to reach deviations to lower bounds below 1% on two classical sets of benchmark problems presented in section 7.6, the *gdb* and *val* instances. Solutions violating vehicle capacity are accepted but penalized. The only move considered consists in changing the route servicing one edge e, provided the new route traverses one of the end-nodes of e. After the determination of the best non-tabu move at each iteration, a heuristic for the rural postman problem is applied to each route to try to improve it. In spite of its good results, this metaheuristic is unjustly ignored, probably because it was presented at a conference.

In 2000, Hertz, Laporte, and Mittaz published CARPET [26], a tabu search in which each route is encoded as a list of all edges traversed, serviced or not. This TS involves three main improvement procedures (Shorten, Drop, and Add), initially designed for the rural postman problem [27]. Simplifying, Shorten exploits the multiple edge traversals which are frequent in arc routing. It delays the service of each edge until its last traversal, enabling the replacement of the edges leading to the first serviced edge by a shortest chain. Drop switches the status of an edge from serviced to nonserviced in a route, whereas Add inserts a required edge into a route. Both procedures then call Shorten. The neighborhood considered is defined by Drop-Add moves. Compared to the route encoding mentioned in introduction, the fact that CARPET stores all the edges traversed increase the complexity of neighborhood explorations. A tabu list prevents the service of an edge from coming back to its route of origin during a given number of iterations.

CARPET includes several refinements. Solutions violating vehicle capacity are accepted, but penalized. Diversification is achieved by concatenating the routes into a giant tour, calling Shorten, and splitting the tour into feasible routes. Periodic intensification is realized using additional moves. It outperforms the simpler TS in [6] on the *gdb* instances, but the latter is better on the *val* instances.

7.5 ▪ Recent metaheuristics

The metaheuristics published in the last decade reflect three trends observed in vehicle routing in general: (i) fewer publications on earlier metaheuristics like simulated annealing and tabu search, (ii) development of population-based heuristics (like genetic algorithms and scatter search) and multiagent methods (like ant colonies), and (iii) an increase in "lighter" algorithms offering a good compromise between solution quality and running time (like GRASP and VNS). Except for two methods that code each route as a list of traversed edges (serviced or not) like in CARPET [26], all the other algorithms use the more natural encoding explained in the introduction: a list of required arcs connected by implicit shortest paths. This section describes local-search-based metaheuristics (VND, VNS, GRASP, Tabu Search, and GLS) and then population-based algorithms (memetic algorithms, scatter search, and ant colonies).

7.5.1 ▪ Simulated annealing

Wøhlk designed in her thesis [51] a simulated annealing algorithm called DYPSA, working on giant tours. Starting from a random ordering of the tasks, the basic iteration consists in swapping two randomly selected tasks in this tour and applying the same splitting procedure as in A-ALG (see 7.3.1). Recall that this dynamic algorithm both splits the tour into feasible UCARP trips and determines the best traversal direction of each edge.

7.5.2 ▪ Variable neighborhood search

Variable Neighborhood Descent (VND) is a conceptually simple metaheuristic involving p neighborhoods $N_1, N_2, \ldots, N_p$. Starting from an initial solution and $i = 1$, each iteration consists in exploring N_i to try finding a better solution. In case of success, the search moves to the new solution and i is reset to one; otherwise, i is incremented to browse the next neighborhood.

Variable Neighborhood Search (VNS) uses p neighborhoods, too, but requires a local search, based on other neighborhoods. Starting with $i = 1$, each iteration randomly draws one neighbor of the incumbent solution in N_i (step called *shaking*) and applies local search to it. If the result outperforms the incumbent solution, it replaces it and i is reset to 1. Otherwise, the shaking step and the local search are repeated for the next value of i. Both VND and VNS stop when the last neighborhood N_p brings no improvement.

In 2001, Hertz and Mittaz [28] presented a VND for the UCARP, based on the same route encoding and improvement procedures as in CARPET. N_1 contains solutions obtained by moving a task to another route using the Drop and Add procedures. For $i > 1$, N_i consists in selecting i routes, merging them into a single tour, improving the tour, and finally splitting it to go back to a UCARP solution. The maximum value p considered for i is $\lceil q_{tot}/Q \rceil$, q_{tot} denoting the total demand. To reduce the search space for $i > 1$, only 2500 neighbors are randomly sampled and infeasible solutions are not considered. This VND is very close to CARPET in terms of solution quality but runs in general faster.

Polacek et al. [42] published in 2008 a remarkably simple but efficient VNS, involving only two kinds of moves. Designed for the CARP with intermediate facilities, it has been tested on UCARP instances, too. The initial solution is computed by splitting a giant tour using Ulusoy's heuristic [48]. The shaking step relies on a random cross-exchange: Two subsequences of edges are selected in two routes R_k and R_p and exchanged while keeping all edge directions. The case $k = p$ is allowed, and R_p can even be a new empty route. The successive neighborhoods correspond to a growing length of subsequences exchanged. The local search considers each individual route and attempts to find improving reversals of subsequences (the edge directions concerned are also reversed). The objective function allows capacity violations, but strongly penalized. The solution returned by the local search is accepted using a threshold accepting criterion. This VNS was tested with 10 runs on the *val* and *egl* instances and obtained 3 and 17 new best solutions, respectively, while all other best known solutions were reproduced.

7.5.3 ▪ GRASP algorithms

The *Greedy Randomized Adaptive Search Procedure* (GRASP) is a simple metaheuristic whose each iteration builds one trial solution, using a greedy randomized heuristic, and improves it using a local search procedure. Prins and Wolfler Calvo designed in 2005 the first GRASP for the UCARP [44]. Trial solutions are constructed using a randomized version of the Merge heuristic (subsection 7.3.2) and improved by the same local search as in the Lacomme et al. memetic algorithm (subsection 7.5.6). In Merge, the pairs of routes are sorted in nonincreasing order of saving: The randomization consists in merging the current pair with a given probability α. This basic GRASP can be completed by a reactive mechanism that tests several values of α to favor the one giving the best solutions on average, and/or by a path relinking applied as a postoptimization procedure to a small pool of good and diverse solutions. No infeasible solution is considered in all these versions.

For two given UCARP solutions S_{beg} and S_{end}, the path relinking builds a path between the giant tours T_{beg} and T_{end} obtained by concatenating the routes of these solutions and removing copies of the depot node. Intermediate giant tours are generated by progressively repairing the pairs of required edges that are adjacent in T_{end} but separated in T_{beg}. These tours are split like in Ulusoy's heuristic [48] and immediately improved by calling the local search.

The tests on the *gdb*, *val*, and *egl* instances with a single run show that the basic GRASP takes place between the tabu search CARPET [26] and the Lacomme et al. memetic algorithm [31] in terms of solution quality, while being at least six times faster. The versions using the reactive mechanism, the path relinking, or both get closer to the MA results but save less running time.

Very recently, Usberti et al. introduced a *GRASP with Evolutionary Path Relinking* (GRASP-EVPR) [49], a recent variant of GRASP which has given excellent results on the truck and trailer routing problem [50]. The constructive phase is inspired by the PSRT heuristic described in subsection 7.3.2. At each iteration, the minimum and maximum distances d_{min} and d_{max} between unserviced edges and the last node i of the emerging route are determined.

The *Restricted Candidate List* (RCL) is initialized with the edges $[p,j]$ such that $d_{ip} \leq d_{min} + \alpha(d_{max} - d_{min})$ and reduced using the ellipse rule of Santos et al. recalled in subsection 7.3.2. The next edge added to the route is randomly selected in the resulting RCL. The local search tests relocations of one or two adjacent required edges, and exchanges of two required edges. These moves can invert the reinserted edges and are applied to one or two routes. The local search may return solutions infeasible with respect to vehicle capacities.

The GRASP is made reactive by allowing several values for α and the parameter β used by the ellipse rule (see 7.3.2). The path relinking process uses the same broken pairs distance as in [44] but works on complete solutions which may violate vehicle capacities. It is said to be evolutionary because it is embedded in the GRASP: At each GRASP iteration, the five best elite solutions are relinked with the constructed solution and the best solution found so far. Moreover, the search is intensified every 100 iterations by relinking each solution from the pool with the five best solutions from the same pool. This GRASP is one of the three heuristic algorithms able to retrieve all optimal solutions to *gdb* instances, if the best of 15 runs is considered, but it is also more time consuming than most other UCARP metaheuristics.

7.5.4 ▪ Tabu search

Brandão and Eglese published in 2008 a very effective tabu search [12]. Like CARPET, the list coding each route contains all edges traversed (required or not) and capacity violations are accepted. However, the new algorithm uses simpler moves and is fully deterministic (CARPET involves some random decisions). The moves consider two distinct routes and include the relocation of one or two tasks and the exchange of two tasks, with a test of the two directions for each inserted edge. At each iteration, once the best non-tabu move has been executed, Frederickson's heuristic for the rural postman problem [22] is applied to the two individual routes that have been changed to try to reduce their costs further.

A simple version TSA-1, initialized with the Path-Scanning heuristic [24], is first described. A more complex version TSA-2 performs five runs using five different initial heuristics, plus one run from the best resulting solution. These two phases are repeated with different parameters, and a final run is executed from the best of the two resulting solutions. Hence, TSA-2 should be considered as a metaheuristic with multiple executions. Brandão and Eglese presented a comprehensive comparison with CARPET [26] and the MA of [31], using the *gdb*, *val*, and *egl* instances. TSA-1 outperforms CARPET but not the MA, while TSA-2 is the best on *val* and *egl* instances. Moreover, TSA-1 is still today the fastest UCARP metaheuristic.

Mei, Tang, and Yao [38] proposed in 2009 a *Global Repair Operator* (GRO) to exploit infeasible solutions in the previous tabu search. Brandão and Eglese associate with each edge traversed in a route a binary variable equal to one iff the edge is serviced by this route. As vehicle capacity violations are penalized in the objective function, the local search is often able to repair infeasible routes by moving some of its edges to another route. Mei et al. observe that an infeasible solution can be repaired in a simpler way, by flipping binary variables without changing the sequence of edges traversed by each route.

The GRO solves a bin packing problem where (i) the routes are considered as bins with limited capacity Q, and (ii) the required edges are items with weights q_e, to be assigned to these bins. The problem is solved using a kind of best- fit heuristic, followed to a short tabu search limited to t iterations, in which the moves consist in changing the bin of a required edge. The GRO is simply applied at each iteration of TSA-1 to the best non-tabu solution found in the neighborhood of the incumbent solution. The tests show that the results are strongly improved: TSA-1 with GRO (called RTS*) is even better and faster than TSA-2.

7.5.5 ▪ Guided local search

An effective *Guided Local Search* (GLS) was designed in 2003 by Muyldermans in his Ph.D. thesis [39] and published the same year in a journal article [10]. GLS is an interesting local-search-based metaheuristic that modifies the costs of the components of a solution to escape local optima, contrary to some methods that randomly perturb solution structure. For the UCARP, the longest deadheading path present in the routes is penalized.

Muyldermans and, independently, Lacomme et al. [31] were the first researchers to represent the network as a symmetric digraph and the routes as lists of required arcs. The moves used in their local search procedures are also similar, but the GLS allows capacity violations and involves an innovative move called *dir-opt*. For a given subsequence of k tasks, dir-opt selects the traversal direction of each task to minimize the total cost of the subsequence. This can be computed in $O(k)$ by solving a shortest path problem in a layered network. This technique has been also used by Wøhlk [51] for her heuristic A-ALG and her simulated annealing algorithm DYPSA, and by Prins, Labadi, and Reghioui [43] in tour splitting procedures. The GLS is accelerated using various ingredients such as lists of neighbors and edge marking techniques.

In [10] the method is appraised with 10^5 iterations on *gdb* files, with 10^5 and 5×10^5 iterations on *val* files, and on 100 new instances obtained by partitioning the road network of Flanders (Belgium). The thesis [39] adds tests with 10^5, 5×10^5, and 10^6 iterations on larger problems, the *egl* files. GLS was the first algorithm to find definitive solution values for the *gdb* instances. Moreover, GLS is known to be the fastest UCARP heuristic with TSA-1 and TSA-2, described in subsection 7.5.4.

7.5.6 ▪ Memetic algorithms

Lacomme, Prins, and Ramdane-Chérif [30] proposed in 2001 a *Memetic Algorithm* (MA), i.e., a genetic algorithm in which children-solutions generated by the crossover operator are improved via a local search procedure. The key-point in this MA is to use chromosomes encoded as giant tours (RPP tours) and to decode them optimally (subject to the sequence) using Ulusoy's splitting procedure [48]. The fitness of a chromosome is nothing but the cost of the resulting UCARP solution. As there exists at least one optimal chromosome, i.e., one giving an optimal UCARP solution after splitting, the MA can search the smaller space of RPP tours, without loss of information.

Each offspring solution undergoes a first-improvement local search, called with a small probability p_{LS} and working on the individual routes instead of the giant tour. The moves considered include the relocation of one or two consecutive edges, the exchange of two required edges, and 2-opt moves. All moves may involve one or two routes, and the difference with node routing is that the two directions of an edge are tested in reinsertions. The resulting routes are finally concatenated to form a new chromosome that replaces an existing one in the worst half of the population.

The same authors presented in 2004 a more efficient version [31] of this MA. The average running time is on average halved by performing short restarts with a partial renewal of the population and an increased local search rate p_{LS}. The two versions of the MA were with Muyldermans' GLS [39] the first metaheuristics able to outperform the tabu search CARPET [26].

In 2009, Mei, Tang, and Yao [37] proposed to modify the local search in the MA from Lacomme et al. Starting from a given feasible solution obtained by splitting a child-chromosome using Ulusoy's algorithm, the moves are now allowed to violate vehicle capacity, at the expense of a penalty. However, the solution returned at the end is the best feasible solution met during the search, even if a better but infeasible solution is obtained last. With 30 runs, the resulting MA called ILMA2 improves 14 times the results of Lacomme et al. on the large *egl* instances. However, as the original MA was tested with a single run, it is difficult to know whether the gain comes from the proposed technique or from the multiple executions.

The same authors [47] introduced a *Memetic Algorithm with Extended Neighborhood Search (MAENS)*, in which the Split procedure is used as a large neighborhood move. Solutions are represented by chromosomes obtained by concatenating the required arcs of each route, separated by copies of the depot node. Offspring solutions are generated by applying the *Sequence Based Crossover* (SBX) initially designed for the CVRP. The local search involves the same moves as in [31] but ends with a new move called *Merge-Split* (MS). MS selects a subset of routes at random, merges the customers into an unordered list, and applies the Path-Scanning UCARP heuristic to get five solutions. The trips of each solution are concatenated to get a giant tour, and the basic Split is applied. The best resulting solution is returned.

The results of this metaheuristic, averaged over 30 executions, are very close to the solution costs obtained by the previously best algorithm, the tabu search TSA-2 from Brandão and Eglese [12]. The best solution of the 30 runs is even better, and all optimal solution values are retrieved for *gdb* instances. However, MAENS is one of the slowest metaheuristics for the UCARP.

The year 2011 saw the publication of a *Biased Random-Key GA* (BRKGA) by Martinez et al. [36]. Random-key GAs, designed for sequencing problems, use chromosomes represented as strings of randomly generated numbers, which allow one to define crossover and mutation operators that are independent of the problem treated. BRKGA is a variant in which parents are selected in a different way for mating. The BRKGA for the UCARP can be considered as a memetic algorithm since it employs a local search procedure, inspired by the one used in Muyldermans' GLS [39]. Numerical results are reported only on a few instances extracted from the classical sets, which makes difficult a comparison with other metaheuristics.

Just before sending this chapter to the publisher, we detected two new memetic algorithms. The first one, by Liu, Jiang, and Geng [34], is derived from the MA from Lacomme et al. [31]: same chromosomes encoded as giant tours, same splitting procedure to decode them, same OX crossover, and same local search. However, the new MA is reinforced by (i) allowing infeasible solutions during the search, (ii) using a second crossover which keeps the longest common subsequence contained in the two parents, and (iii) running a short iterated local search, after the local search, if the cost of the child is close enough to that of the best parent. The resulting MA is evaluated on the *gdb*, *val*, and *egl* instances, and the quality of its solutions is very close to two state-of-the-art metaheuristics, the Ant Colony Optimization approach by Santos et al. [46] and the GRASP by Usberti et al. [49].

The second MA, designed by Liu, Singh, and Ray [33], belongs to the family of generational genetic algorithms. It is proposed with a random generator to mimic the networks encountered in various practical applications. The population composed of N_p complete UCARP solutions is partitioned into pairs, selected to have the maximum number of common vehicle trips. Three recombination operators are applied to these pairs of solutions but return giant tours which must be evaluated using a splitting procedure. The children are added to the population, which is then sorted in ascending cost order and truncated to keep the N_p best individuals. The current generation ends by applying a local search to the current best solution. It is difficult to appraise the performance of this MA in general, because it is tested only on three randomly generated instances with 167, 647, and 999 edges.

7.5.7 ▪ Scatter search

Greistorfer [25] published in 2003 an original *Tabu Scatter Search* metaheuristic (TSCS). Starting from a pool containing one good heuristic solution and randomly generated solutions, a tabu search phase is executed. Each time a new best solution is discovered, it is inserted in the pool. Then, a scatter search phase combines up to five solutions randomly chosen in the pool, using a sophisticated recombination operator that solves a transportation problem based on edge-to-route assignment frequencies. A new tabu search phase is then launched from the resulting solution, etc. TSCS is evaluated on random graphs and on the 23 *gdb* instances: Compared to CARPET, it improves one best-known solution and converges twice as fast.

7.5.8 ▪ Ant colony optimization

An *Ant Colony Optimization* (ACO) algorithm was elaborated in 2010 by Santos et al. [46]. Three heuristics are called to prepare a population of distinct UCARP solutions. The first two are the PSRT and RSE-ER algorithms derived from Path-Scanning and described in subsection 7.3.2. The third one consists in applying to the result of RSE-ER the same local search as in the ACO iterations.

These solutions are used to determine the initial pheromone levels. At each main iteration, each ant builds one giant tour (vehicle capacity is relaxed) by selecting required arcs according to pheromone intensity. Each giant tour is disaggregate into several feasible UCARP routes using Ulusoy's splitting heuristic [48], and the UCARP solutions obtained undergo a local search procedure with 12 move types. The resulting solutions are tested one by one and replace the worst solution in the population in case of improvement. The current iteration ends by updating pheromone levels.

The algorithm has been tested using a population of 10 solutions, 30 ants, and 500 iterations on seven sets of instances, with very good results. For instance, 6 new best solutions to *egl* instances were discovered by this method. It is also the fastest method among the recent metaheuristics tested with multiple executions.

7.6 ▪ Comparison on standard instances

This chapter is a good opportunity to provide an updated comparison of classical and recent heuristics on three sets of standard UCARP benchmarks. The 23 *gdb* instances [23] have 7 to 27 nodes and 11 to 55 edges, all required. They are numbered from 1 to 25, but nobody uses instances 8 and 9 due to inconsistencies (disconnected graphs). The 34 *val* instances [9] contain 24 to 50 nodes and 34 to 97 edges, all required. Finally, the 24 larger *egl* instances [32] have 77 to 140 nodes, 98 to 190 edges, and 51 to 190 required edges.

All demands and edge costs are integers. Service costs are defined only in *val* problems. Note that *gdb* and *val* problems are in fact CCPP instances.The three sets can be downloaded on the Internet: `http://www.uv.es/~belengue/carp.html`.

Two tables are provided for each set. The first table concerns general performance indicators of heuristics, while the second one reports the current best lower bound and upper bound for each instance.

The first table gathers the following data: the name of each heuristic, the section of this chapter where the method is described (*Where*), the reference for solution values (*Reference*), the average and maximum deviation in percent to the best known lower bound or to the optimum when known (D_{avg} and D_{max}), the number of proven optima retrieved (*Opt*), and the average running time in seconds (*Time*). Empty cells correspond to missing information in the reference.

The metaheuristics with multiple executions are clearly identified by their number of runs. For such methods, the cost used for each instance to compute the overall deviations is the best cost of the different runs, but the running time is given for one run. The randomized constructive heuristics have also multiple executions, but they are so fast that we report the total duration of their runs. For all algorithms, we considered the solution values achieved using the standard setting of parameters, and not the best results that are sometimes reported using several parameter sets.

The second table gives the current status of each instance, with the number of nodes n, the number of edges m, the number of required edges (tasks) t, the best lower bound or proven optimum found by an exact algorithm (*LB*) and its reference, and the best upper bound found by a heuristic algorithm (*UB*), with the reference. Asterisks denote optimal solution costs. Each reference corresponds to the first paper mentioning the reported value, even if it has been obtained using several runs with different parameters.

Like most authors of metaheuristics, we measure the cost of a solution as the sum of the c_e over all edges traversed in the solution, ignoring service costs. Muyldermans [39] uses the total deadheading cost, which is equal to our cost minus the constant sum $S_1 = \sum_{e \in E_R} c_e$ of traversal costs for required edges. Finally, authors of exact methods prefer the real cost, i.e., the sum of the c_e for deadheading traversals plus the constant sum $S_2 = \sum_{e \in E_R} c_e^{serv}$ of service costs. Note that the first and last conventions give different values only on *val* instances. We provide in the second table the values of S_1 for all files and those of S_2 for *val* files, to help the reader check the literature. For instance, our optimal cost for instance val8C is 521. As $S_1 = 347$ and $S_2 = 483$, this cost becomes $521 - 347 + 483 = 657$ in the Bode and Irnich exact algorithm [11], and $521 - 347 = 174$ in Muyldermans [39].

The heuristics considered in the first table are the ones published with results for each instance. Indeed, we had to recompute all deviations using the most recent lower bounds, which explains small discrepancies with the original papers. The results for Path-Scanning concern the most frequent version [21, 31, 45], described in Algorithm 7.2. Augment-Merge and its version Improved Merge without Augment correspond to the implementations of Belenguer et al. [7]. Reference [43] gives indicators for URTF-SF with 50 runs only, but we re-executed our code with 10,000 runs to provide a comparison with the RSE-ER heuristic [45].

The running times displayed have been scaled for a 2.4 GHz Pentium IV PC, considering that they are inversely proportional to the clock speed. Our experience shows that the resulting times are not less precise than the scaling factors found in the literature, which are valid only for programs using exactly the same language, the same operating system, the same compiler, etc. This is even truer for the three sets of UCARP instances, which do not involve floating point computations. We proceed in fact like in most articles surveyed in this chapter [12, 47, 46, 49].

Tables 7.1 and 7.2 concern the smallest problems, the *gdb* instances. The horizontal lines in Table 7.1 demarcate constructive heuristics from randomized versions and metaheuristics. The three last open instances were closed in 2006 by an exact method [35], but the optima were already reached by different heuristics in 2001. Four metaheuristics are able to find all optima: GLS [39], MAENS [47], GRASP [49], and RTS* [38]. However, only GLS achieves such results in one execution only.

Table 7.1. *Performance on the 23 gdb instances* ($n = 7$–27, $t = 11$–55).

Heuristic	Section	Reference	D_{avg}	D_{max}	Opt	Time
Double Outer-Scan	7.3.1	Wøhlk [51]	23.82	46.00	0	
Construct-Strike (CS)	7.2.1	Golden et al. [23]	14.13	43.01	2	
Path-Scanning (PS)	7.2.2	Lacomme et al. [31]	10.23	33.05	1	< 1 ms
Node Duplication Heur	7.3.1	Wøhlk [51]	9.15	19.21	0	
A-ALG	7.3.1	Wøhlk [51]	8.31	19.65	0	
Modified-PS	7.3.1	Wøhlk [51]	7.22	22.11	3	
Augment-Merge	7.2.3	Belenguer et al. [7]	7.19	24.24	2	< 1 ms
Merge	7.3.2	Belenguer et al. [7]	6.38	14.55	0	< 1 ms
Improved Merge	7.3.2	Belenguer et al. [7]	5.45	14.55	1	< 1 ms
Modified-CS	7.2.5	Pearn [40]	4.12	40.83	11	
Random-PS (best 30)	7.2.5	Pearn [40]	4.55	23.58	5	
URT-B (best 50)	7.3.3	Prins et al. [43]	3.45	12.87	4	1 ms
URTF-S (best 50)	7.3.3	Prins et al. [43]	2.10	13.86	10	1 ms
RSE-ER (best 10,000)	7.3.2	Santos et al. [45]	1.13			1.30
URTF-S (best 10,000)	7.3.3	Prins et al. [43]	0.88	7.59	16	0.19
DYPSA (best 5)	7.5.1	Wøhlk [51]	0.46	2.78	18	
TSA-1	7.5.4	Brandão & Eglese [12]	0.46	1.77	15	0.12
TS	7.4.2	Belenguer et al. [6]	0.44	5.28	20	
CARPET	7.4.2	Hertz et al. [26]	0.35	4.62	19	3.75
GRASP (mean 15)	7.5.3	Usberti et al. [49]	0.11	1.23	19	6.37
ACO (mean 15)	7.5.8	Santos et al. [46]	0.10	1.49	21	1.42
TSA-2	7.5.4	Brandão & Eglese [12]	0.07	0.86	21	1.46
ACO (best 15)	7.5.8	Santos et al. [46]	0.04	0.86	22	1.42
MAENS (mean 30)	7.5.6	Tang et al. [47]	0.04	0.75	21	5.25
MA	7.5.6	Lacomme et al. [31]	0.02	0.57	22	2.20
GLS (10^5 iter)	7.5.5	Muyldermans [39]	0.00	0.00	23	0.36
MAENS (best 30)	7.5.6	Tang et al. [47]	0.00	0.00	23	5.25
GRASP (best 15)	7.5.3	Usberti [49]	0.00	0.00	23	6.37
RTS*	7.5.4	Mei et al. [38]	0.00	0.00	23	0.50

On a 2.4 GHz PC, all deterministic constructive heuristics like Path-Scanning need less than 1 ms on average. Randomized methods need from 1 ms (URTF-SF, 50 runs) to 1.3 (RSE-ER with 10,000 runs). The fastest metaheuristics are TSA-1 and GLS (less than 0.5), followed by other methods using a single run (RTS*, TSA-2, MA, CARPET). Although ACO looks fast (1.42 s per run), all recent algorithms (ACO, GRASP, MAENS) are slower than earlier methods if one considers the number of runs they require. In fact, GLS is as good as MAENS and GRASP and better than ACO, while being much faster. The authors of GRASP, ACO, and MAENS also give the mean solution cost of the different runs. Even if we admit that a single run of these metaheuristics would yield a cost close to the reported mean of several runs, GLS published in 2003 is still faster.

Table 7.2. *Current status of gdb instances.*

File	n	m	t	S_1	LB	Reference	UB	Reference
gdb1	12	22	22	252	316*	2003 [5]	316*	1983 [23]
gdb2	12	26	26	291	339*	1992 [9]	339*	1997 [6]
gdb3	12	22	22	233	275*	1992 [9]	275*	1989 [40]
gdb4	11	19	19	238	287*	2003 [5]	287*	1989 [40]
gdb5	13	26	26	316	377*	2003 [5]	377*	1997 [6]
gdb6	12	22	22	260	298*	2003 [5]	298*	1997 [6]
gdb7	12	22	22	262	325*	2003 [5]	325*	1983 [23]
gdb10	27	46	46	210	348*	2006 [35]	348*	2000 [26]
gdb11	27	51	51	219	303*	2003 [5]	303*	2001 [30]
gdb12	12	25	25	252	275*	1992 [9]	275*	1989 [40]
gdb13	22	45	45	256	395*	2003 [5]	395*	1997 [6]
gdb14	13	23	23	236	458*	2006 [35]	458*	1997 [6]
gdb15	10	28	28	509	536*	2006 [35]	536*	2001 [30]
gdb16	7	21	21	96	100*	1992 [9]	100*	1989 [40]
gdb17	7	21	21	56	58*	1992 [9]	58*	1983 [23]
gdb18	8	28	28	119	127*	1992 [9]	127*	1989 [40]
gdb19	8	28	28	84	91*	1992 [9]	91*	1983 [23]
gdb20	9	36	36	158	164*	1992 [9]	164*	1989 [40]
gdb21	8	11	11	45	55*	1992 [9]	55*	1989 [40]
gdb22	11	22	22	105	121*	1992 [9]	121*	1997 [6]
gdb23	11	33	33	149	156*	1992 [9]	156*	1989 [40]
gdb24	11	44	44	191	200*	1992 [9]	200*	1989 [40]
gdb25	11	55	55	223	233*	1992 [9]	233*	1989 [40]

The results for *val* instances are in Tables 7.3 and 7.4. These problems were recently closed by an exact algorithm designed by Bode and Irnich [11]. The heuristic RSE-ER is not included because reference [45] provides only an average gap over all instances, which cannot be corrected using the lower bounds improved in [11]. All optimal solutions were already reached in 2003 by different metaheuristics, except three found only in 2008 by the VNS from Polacek et al. [42]. No metaheuristic is able to find all optimal solutions values, but the VNS retrieves 32 optima out of 36. All deterministic constructive heuristics last at most 2 ms on average on a 2.4 GHz PC. Randomized versions require less than 1. Except for TSA-1, which is still very fast (1.2 s), the metaheuristics with a single execution last between 6 and 27 s. The others (ACO, MAENS, GRASP, VNS) require 11, 57, 77, and 90 s per run, respectively, to be multiplied by at least 10 runs.

Finally, Tables 7.5 and 7.6 concern the *egl* instances. No constructive heuristic, even randomized, is able to find optimal solutions. The Improved Merge heuristic performs remarkably well in spite of its simplicity: The comparison with Augment-Merge shows how prejudicial the Augment phase can be. Contrary to the two previous instance sets, Double Outer Scan falls here under 10%. The multiple execution of randomized heuristics, which worked well on *gdb* and *val* instances, reaches here its limits: Even with 10,000 runs, the improvement is marginal. The average duration per run of metaheuristics is now less spread, and, in spite of their multiple executions, ACO, VNS, MAENS, and ILMA2 bring a reduced improvement and cannot find more than 9 optima. Although important advances have been realized recently by two exact methods [2, 11], 13 problems are still open.

Table 7.3. *Performance on the 34 val instances* ($n = 24$–50, $t = 34$–97).

Heuristic	Section	Reference	D_{avg}	D_{max}	Opt	Time
Double Outer-Scan	7.3.1	Wøhlk [51]	34.44	50.59	0	
Path-Scanning (PS)	7.2.2	Lacomme et al. [31]	16.27	35.10	0	1 ms
Modified-Path-Scanning	7.3.1	Wøhlk [51]	14.78	29.04	0	
A-ALG	7.3.1	Wøhlk [51]	12.31	27.65	0	
Node duplication heur	7.3.1	Wøhlk [51]	12.14	28.52	0	
Merge	7.3.2	Belenguer et al. [7]	12.06	20.25	0	2 ms
Augment-Merge	7.2.3	Belenguer et al. [7]	10.24	17.03	0	2 ms
Improved Merge	7.3.2	Belenguer et al. [7]	5.54	10.13	1	2 ms
URT-B (best 50)	7.3.3	Prins et al. [43]	9.16	17.22	0	3 ms
URTF-S (best 50)	7.3.3	Prins et al. [43]	5.91	12.63	0	4 ms
URTF-S (best 10,000)	7.3.3	Prins et al. [43]	2.81	9.85	6	0.68
DYPSA (best 5)	7.5.1	Wøhlk [51]	8.74	17.39	0	
CARPET	7.4.2	Hertz et al. [26]	1.51	8.10	17	26.61
TS	7.4.2	Belenguer et al. [6]	0.83	5.41	24	
TSA-1	7.5.4	Brandão & Eglese [12]	0.75	4.61	20	1.23
MA	7.5.6	Lacomme et al. [31]	0.18	2.08	29	15.98
MAENS (mean 30)	7.5.6	Tang et al. [47]	0.18	1.64	27	56.75
GRASP (mean 15)	7.5.3	Usberti et al. [49]	0.17	1.65	26	76.87
TSA-2	7.5.4	Brandão & Eglese [12]	0.14	1.54	30	11.78
RTS*	7.5.4	Mei et al. [38]	0.13	1.71	30	5.67
ACO (mean 15)	7.5.8	Santos et al. [46]	0.11	1.39	30	10.54
VNS (mean 10)	7.5.2	Polacek et al. [42]	0.08	0.83	26	[1]90.00
MAENS (best 30)	7.5.6	Tang et al. [47]	0.07	1.14	31	56.75
GRASP (best 15)	7.5.3	Usberti et al. [49]	0.07	1.04	30	76.87
GLS (5×10^5 iter)	7.5.5	Muyldermans [39]	0.05	0.77	31	16.94
ACO (best 15)	7.5.8	Santos et al. [46]	0.04	0.77	31	10.54
VNS (best 10)	7.5.2	Polacek et al. [42]	0.01	0.26	32	[1]90.00

[1]Corresponding to a fixed time of 60 s allocated to each run, on a 3.6 GHz PC.

7.7 ▪ Conclusion

The heuristic algorithms for the UCARP covered in this chapter display a wide variety of techniques and almost all kinds of metaheuristics are represented, except particle swarm optimization. Constructive heuristics are often inspired by similar methods designed for node routing problems. Some tricks introduced initially for better solving UCPP instances become useless and even prejudicial when not all edges are required, like the Augment phase in Augment-Merge. The randomization of simple heuristics leads to surprisingly good results if multiple executions are allowed. However, this technique degrades rapidly on the largest instances, for which metaheuristics can get much better solutions in comparable time scales.

The metaheuristics include many ingredients for intensification and diversification, for instance, path relinking and large neighborhoods. A clear tendency is to use the increasing power of computers to do multiple runs in randomized metaheuristics. The performance of some older algorithms doing one run only is still remarkable in these conditions. For instance, our tables with running times scaled for a 2.4 GHz PC show that TSA-2 [12] achieves in 2008 an average solution gap of 1.04% on the *egl* instances, while five years later the GRASP designed by Usberti et al. [49] reaches 0.43%, but using 15 runs of 998 s each. The GLS of Muyldermans published in 2003 [39] is another good example: It is almost as good as the best method on *val* instances, while running 50 times faster.

It is interesting to see that the giant tour approach has lead to efficient metaheuristics for the UCARP in the last decade: Giant tours are used in memetic algorithms to encode

Table 7.4. *Status of val instances.*

File	n	m	t	S_1	S_2	LB	Reference	UB	Reference
val1a	24	39	39	146	220	173*	1992 [9]	173*	1997 [6]
val1b	24	39	39	146	220	173*	1992 [9]	173*	1997 [6]
val1c	24	39	39	146	220	245*	2006 [35][1]	245*	1997 [6]
val2a	24	34	34	185	256	227*	1998 [4]	227*	1997 [6]
val2b	24	34	34	185	256	259*	2003 [5]	259*	1997 [6]
val2c	24	34	34	185	256	457*	2006 [35][1]	457*	2000 [26]
val3a	24	35	35	65	89	81*	1998 [4]	81*	1997 [6]
val3b	24	35	35	65	89	87*	1998 [4]	87*	1997 [6]
val3c	24	35	35	65	89	138*	2006 [35]	138*	2000 [26]
val4a	41	69	69	343	365	400*	1998 [4]	400*	1997 [6]
val4b	41	69	69	343	365	412*	1998 [4]	412*	1997 [6]
val4c	41	69	69	343	365	428*	2003 [5]	428*	2001 [30]
val4d	41	69	69	343	365	530*	2012 [11]	530*	2001 [30]
val5a	34	65	65	367	510	423*	1998 [4]	423*	1997 [6]
val5b	34	65	65	367	510	446*	2003 [5]	446*	1997 [6]
val5c	34	65	65	367	510	474*	2012 [11]	474*	1997 [6]
val5d	34	65	65	367	510	575*	2012 [11]	575*	2008 [42]
val6a	31	50	50	190	297	223*	1992 [9]	223*	1997 [6]
val6b	31	50	50	190	297	233*	2006 [1]	233*	1997 [6]
val6c	31	50	50	190	297	317*	2006 [1]	317*	2000 [26]
val7a	40	66	66	249	352	279*	1992 [9]	279*	1997 [6]
val7b	40	66	66	249	352	283*	1998 [4]	283*	1997 [6]
val7c	40	66	66	249	352	334*	2006 [35][1]	334*	1997 [6]
val8a	30	63	63	347	483	386*	1992 [9]	386*	1997 [6]
val8b	30	63	63	347	483	395*	1998 [4]	395*	1997 [6]
val8c	30	63	63	347	483	521*	2012 [11]	521*	2003 [39]
val9a	50	92	92	278	405	323*	1992 [9]	323*	1997 [6]
val9b	50	92	92	278	405	326*	1992 [9]	326*	1997 [6]
val9c	50	92	92	278	405	332*	2003 [5]	332*	1997 [6]
val9d	50	92	92	278	405	388*	2012 [11]	388*	2008 [42]
val10a	50	97	97	376	585	428*	1992 [9]	428*	1997 [6]
val10b	50	97	97	376	585	436*	1998 [4]	436*	1997 [6]
val10c	50	97	97	376	585	446*	1998 [4]	446*	1997 [6]
val10d	50	97	97	376	585	525*	2006 [35]	525*	2008 [42]

chromosomes [30, 31, 37], in local search procedures as large neighborhood operators [47], in a VNS to provide good initial solutions [42], and even by ants [46] to generate new solutions!

Concerning instances, the *gdb* and *val* problems are all solved to optimality, but they can be useful for the preliminary testing of heuristics. The *egl* instances look much harder since nearly one half are still open. An interesting observation is that three nonoptimal best-known solutions have been found by exact methods, but no metaheuristic is able to reproduce them, an unusual situation for the CVRP. This indicates that the existing metaheuristics are perhaps not so powerful. Some sets of instances which have still many open problems, like the ones designed by Muyldermans [39], should be used to better separate the existing metaheuristics.

In our opinion, two research directions should be investigated in priority. The first one consists in developing metaheuristics able to tackle in reasonable running times the huge instances met in real applications like urban refuse collection (10,000 edges or more): Indeed, only simple classical constructive heuristics like Path-Scanning are currently fast enough on such networks. The second direction already followed by the node routing community is to exploit parallel or multicore computers which are now widespread and relatively cheap. Beyond a simple adaptation of existing metaheuristics to multiprocessor

Table 7.5. *Performance on the 24 egl instances* ($n = 77$–140, $t = 51$–190).

Heuristic	Section	Reference	D_{avg}	D_{max}	Opt	Time
Path-Scanning (PS)	7.2.2	Lacomme et al. [31]	24.69	35.11	0	3 ms
Modified-PS	7.3.1	Wøhlk [51]	23.97	30.47	0	
Augment-Merge	7.2.3	Belenguer et al. [7]	21.48	30.41	0	5 ms
Node duplication heur	7.3.1	Wøhlk [51]	13.66	17.02	0	
A-ALG	7.3.1	Wøhlk [51]	13.22	15.90	0	
Double Outer-Scan	7.3.1	Wøhlk [51]	9.97	24.41	0	
Merge	7.3.2	Belenguer et al. [7]	5.36	10.17	0	10 ms
Improved Merge	7.3.2	Belenguer et al. [7]	4.49	8.19	0	10 ms
URT-B (best 50)	7.3.3	Prins et al. [43]	12.37	16.48	0	7 ms
URTF-S (best 50)	7.3.3	Prins et al. [43]	11.24	15.50	0	8 ms
URTF-S (best 10,000)	7.3.3	Prins et al. [43]	8.47	12.44	0	1.48
DYPSA (best 5)	7.5.1	Wøhlk [51]	13.97	28.21	0	
TSA-1	7.5.4	Brandão & Eglese [12]	1.82	3.72	3	26.37
MA	7.5.6	Lacomme et al. [31]	1.12	2.97	7	219.58
GLS (10^6 iter)	7.5.5	Muyldermans [39]	1.11	3.04	6	213.31
TSA-2	7.5.4	Brandão & Eglese [12]	1.04	2.86	4	169.98
MAENS (mean 30)	7.5.6	Tang et al. [47]	0.88	2.22	2	585.08
ACO (median 15)	7.5.8	Santos et al. [46]	0.87	2.47	8	209.58
VNS (mean 10)	7.5.2	Polacek et al. [42]	0.85	2.40	4	[1]270.00
RTS*	7.5.4	Mei et al. [38]	0.78	2.52	8	104.83
GRASP (median 15)	7.5.3	Usberti et al. [49]	0.74	2.20	7	998.25
VNS (best 10)	7.5.2	Polacek et al. [42]	0.57	1.86	9	[1]270.00
ACO (best 15)	7.5.8	Santos et al. [46]	0.53	2.01	9	209.58
ILMA2 (best 30)	7.5.6	Mei et al. [37]	0.52	1.64	9	256.42
MAENS (best 30)	7.5.6	Tang et al. [47]	0.44	1.58	9	585.08
GRASP (best 15)	7.5.3	Usberti et al. [49]	0.43	1.42	9	998.25

[1]Corresponding to a fixed time of 180 s allocated to each run, on a 3.6 GHz PC.

Table 7.6. *Current status of egl instances.*

File	n	m	t	S_1	LB	Ref.	UB	Ref.
egl-e1-A	77	98	51	1468	*3548	[35]	*3548	[31]
egl-e1-B	77	98	51	1468	*4498	[2]	*4498	[31]
egl-e1-C	77	98	51	1468	*5595	[2]	*5595	[31]
egl-e2-A	77	98	72	1879	*5018	[1]	*5018	[31]
egl-e2-B	77	98	72	1879	6301	[11]	6317	[12]
egl-e2-C	77	98	72	1879	*8335	[2]	*8335	[12]
egl-e3-A	77	98	87	2188	*5898	[35]	*5898	[31]
egl-e3-B	77	98	87	2188	7728	[11]	7775	[42]
egl-e3-C	77	98	87	2188	10244	[2]	10292	[42]
egl-e4-A	77	98	98	2453	6408	[11]	6444	[46]
egl-e4-B	77	98	98	2453	8935	[2]	[1]8983	[46]
egl-e4-C	77	98	98	2453	11493	[2]	11561	[47]
egl-s1-A	140	190	75	1394	*5018	[2]	*5018	[31]
egl-s1-B	140	190	75	1394	*6388	[2]	*6388	[12]
egl-s1-C	140	190	75	1394	*8518	[2]	*8518	[31]
egl-s2-A	140	190	147	3174	9825	[2]	9884	[46]
egl-s2-B	140	190	147	3174	13017	[2]	13100	[12]
egl-s2-C	140	190	147	3174	*16425	[2]	*16425	[12]
egl-s3-A	140	190	159	3379	10160	[11]	10220	[46]
egl-s3-B	140	190	159	3379	13648	[2]	13682	[42]
egl-s3-C	140	190	159	3379	*17188	[2]	[1]17207	[47]
egl-s4-A	140	190	190	4186	12149	[11]	12268	[46]
egl-s4-B	140	190	190	4186	16105	[11]	16321	[42]
egl-s4-C	140	190	190	4186	20430	[2]	[1]20517	[46]

[1]Better UBs found by an exact method [2]: 8961, *17188, and 20481.

environments, new families of algorithms should be designed to fully exploit multiple processors.

Bibliography

[1] R. BALDACCI AND V. MANIEZZO, *Exact methods based on node-routing formulations for undirected arc routing problems*, Networks, 47 (2006), pp. 52–60.

[2] E. BARTOLINI, J.F. CORDEAU, AND G. LAPORTE, *Improved lower bounds and exact algorithm for the capacitated arc routing problem*, Mathematical Programming, 137 (2013), pp. 409–452.

[3] J.E. BEASLEY, *Route-first cluster-second methods for vehicle routing*, Omega, 11 (1983), pp. 403–408.

[4] J.M. BELENGUER AND E. BENAVENT, *The capacitated arc routing problem: Valid inequalities and facets*, Computational Optimization and Applications, 10 (1998), pp. 165–187.

[5] ——, *A cutting plane algorithm for the capacitated arc routing problem*, Computers & Operations Research, 30 (2003), pp. 705–728.

[6] J.M. BELENGUER, E. BENAVENT, AND F. COGNATA, *Un metaheuristico para el problema de rutas por arcos con capacidades*, in Proceedings of the 23rd national SEIO meeting, Valencia, Spain, 1997.

[7] J.M. BELENGUER, E. BENAVENT, P. LACOMME, AND C. PRINS, *Lower and upper bounds for the mixed capacitated arc routing problem*, Computers & Operations Research, 33 (2006), pp. 3363–3383.

[8] E. BENAVENT, V. CAMPOS, A. CORBERÁN, AND E. MOTA, *The capacitated arc routing problem: A heuristic algorithm*, Qüestiió, 14 (1990), pp. 107–122.

[9] E. BENAVENT, V. CAMPOS, A. CORBERÁN, AND E. MOTA, *The capacitated Chinese postman problem: Lower bounds*, Networks, 22 (1992), pp. 669–690.

[10] P. BEULLENS, L. MUYLDERMANS, D. CATTRYSSE, AND D. VAN OUDHEUSDEN, *A guided local search heuristic for the capacitated arc routing problem*, European Journal of Operational Research, 147 (2003), pp. 629–643.

[11] C. BODE AND S. IRNICH, *Cut-first branch-and-price-second for the capacitated arc-routing problem*, Operations Research, 60 (2012), pp. 1167–1182.

[12] J. BRANDÃO AND R. EGLESE, *A deterministic tabu search algorithm for the capacitated arc routing problem*, Computers & Operations Research, 35 (2008), pp. 1112–1126.

[13] L. CHAPLEAU, J.A. FERLAND, G. LAPALME, AND J.M. ROUSSEAU, *A parallel insert method for the capacitated arc routing problem*, Operations Research Letters, 3 (1984), pp. 95–99.

[14] N. CHRISTOFIDES, *The optimum traversal of a graph*, Omega, 1 (1973), pp. 719–732.

[15] G. CLARKE AND J.W. WRIGHT, *Scheduling of vehicles from a central depot to a number of delivery points*, Operations Research, 12 (1964), pp. 568–581.

[16] Á. CORBERÁN AND C. PRINS, *Recent results on arc routing problems: An annotated bibliography*, Networks, 56 (2010), pp. 50–69.

[17] T.H. CORMEN, C.E. LEISERSON, AND R.L. RIVEST, *Introduction to Algorithms*, MIT Press, Cambridge, MA, 1990.

[18] M. DROR, *Arc Routing : Theory, Solutions and Applications*, Kluwer Academic Publishers, Boston, MA, 2000.

[19] J. EDMONDS AND E.L. JOHNSON, *Matching, Euler tours and the Chinese postman problem*, Mathematical Programming, 5 (1973), pp. 88–124.

[20] R.W. EGLESE AND L.Y.O LI, *A tabu search based heuristic for arc routing with a capacity constraint and time deadline*, in Meta-heuristics: Theory and Applications, I.H. Osman and J.P. Kelly, eds., Kluwer Academic Publishers, Dordrecht, The Netherlands, 1996, pp. 633–649.

[21] J.R. EVANS AND E. MINIEKA, *Optimization Algorithms for Networks and Graphs*, Marcel Dekker, New York, 1992.

[22] G.N. FREDERICKSON, *Approximation algorithms for some postman problems*, Journal of the ACM, 26 (1979), pp. 538–554.

[23] B.L. GOLDEN, J.S. DEARMON, AND E.K. BAKER, *Computational experiments with algorithms for a class of routing problems*, Computers & Operations Research, 10 (1983), pp. 47–59.

[24] B.L. GOLDEN AND R.T. WONG, *Capacitated arc routing problems*, Networks, 11 (1981), pp. 305–315.

[25] P. GREISTORFER, *A tabu scatter search metaheuristic for the arc routing problem*, Computers and Industrial Engineering, 44 (2003), pp. 249–266.

[26] A. HERTZ, G. LAPORTE, AND M. MITTAZ, *A tabu search heuristic for the capacitated arc routing problem*, Operations Research, 48 (2000), pp. 129–135.

[27] A. HERTZ, G. LAPORTE, AND P. NANCHEN-HUGO, *Improvement procedures for the undirected rural postman problem*, INFORMS Journal on Computing, 11 (1999), pp. 53–62.

[28] A. HERTZ AND M. MITTAZ, *A variable neighborhood descent algorithm for the undirected capacitated arc routing problem*, Transportation Science, 35 (2001), pp. 425–434.

[29] K. JANSEN, *Bounds for the general capacitated routing problem*, Networks, 23 (1993), pp. 165–173.

[30] P. LACOMME, C. PRINS, AND W. RAMDANE-CHÉRIF, *A genetic algorithm for the capacitated arc routing problem and its extensions*, in Applications of Evolutionary Computing, E.J.W. Boers, ed., Lecture Notes In Computer Science 2037, Springer-Verlag, Berlin, 2001, pp. 473–483.

[31] ——, *Competitive memetic algorithms for arc routing problems*, Annals of Operations Research, 131 (2004), pp. 159–185.

[32] L.Y.O. Li and R.W. Eglese, *An iterative algorithm for vehicle routeing for winter-gritting*, The Journal of the Operational Research Society, 47 (1996), pp. 217–228.

[33] M. Liu, H.K. Singh, and T. Ray, *Application specific instance generator and a memetic algorithm for capacitated arc routing*, Transportation Research C, 43 (2014), pp. 249–266.

[34] T. Liu, Z. Jiang, and N. Geng, *A memetic algorithm with iterated local search for the capacitated arc routing problem*, International Journal of Production Research, 51 (2013), pp. 3075–3084.

[35] H. Longo, M. Poggi de Aragão, and E. Uchoa, *Solving capacitated arc routing problems using a transformation to the CVRP*, Computers & Operations Research, 33 (2006), pp. 1823–1837.

[36] C. Martinez, I. Loiseau, M.G.C. Resende, and S. Rodriguez, *BRKGA algorithm for the capacitated arc routing problem*, Electronic Notes in Theoretical Computer Science, 281 (2011), pp. 69–83.

[37] Y. Mei, K. Tang, and X. Yao, *Improved memetic algorithm for capacitated arc routing problem*, in IEEE Congress on Computing and Processing (CEC'09), IEEE, Washington, DC, 2009, pp. 1699–1706.

[38] ——, *A global repair operator for capacitated arc routing problem*, IEEE Transactions on Systems, Man, and Cybernetics—Part B, 39 (2009), pp. 723–734.

[39] L. Muyldermans, *Routing, Districting and Location for arc traversal problems*, Ph.D. dissertation, Catholic University of Leuven, Leuven, Belgium, 2003.

[40] W.L. Pearn, *Approximate solutions for the capacitated arc routing problem*, Computers & Operations Research, 16 (1989), pp. 589–600.

[41] ——, *Augment-insert algorithms for the capacitated arc routing problem*, Computers & Operations Research, 18 (1991), pp. 189–198.

[42] M. Polacek, K.F. Doerner, R.F. Hartl, and V. Maniezzo, *A variable neighborhood search for the capacitated arc routing problem with intermediate facilities*, Journal of Heuristics, 14 (2008), pp. 405–423.

[43] C. Prins, N. Labadi, and M. Reghioui, *Tour splitting algorithms for vehicle routing problems*, International Journal of Production Research, 47 (2009), pp. 507–536.

[44] C. Prins and R. Wolfler Calvo, *A fast GRASP with path relinking for the capacitated arc routing problem*, in 3rd International Network Optimization Conference (INOC 2005), L. Gouveia and C. Mourão, eds., University of Lisbon, Lisbon, Portugal, 2005, pp. 289–295.

[45] L. Santos, J. Coutinho-Rodrigues, and J.R. Current, *An improved heuristic for the capacitated arc routing problem*, Computers & Operations Research, 36 (2009), pp. 2632–2637.

[46] ——, *An improved ant colony optimisation based algorithm for the capacitated arc routing problem*, Transportation Research B, 44 (2010), pp. 246–266.

[47] K. Tang, Y. Mei, and X. Yao, *Memetic algorithm with extended neighborhood search for capacitated arc routing problems*, IEEE Transactions on Evolutionay Computation, 13 (2009), pp. 1151–1166.

[48] G. Ulusoy, *The fleet size and mix problem for capacitated arc routing*, European Journal of Operational Research, 22 (1985), pp. 329–337.

[49] F.L. Usberti, P.M. França, and A.L.M. França, *GRASP with evolutionary path-relinking for the capacitated arc routing problem*, Computers & Operations Research, 40 (2013), pp. 3206–3217.

[50] J.G. Villegas, C. Prins, C. Prodhon, A.L. Medaglia, and N. Velasco, *A GRASP with evolutionary path relinking for the truck and trailer routing problem*, Computers & Operations Research, 38 (2011), pp. 1319–1334.

[51] S. Wøhlk, *Contributions to Arc Routing*, Ph.D. dissertation, University of Southern Denmark, Aarhus, Denmark, 2005.

[52] ——, *A decade of capacitated arc routing*, in The Vehicle Routing Problem: Latest Advances and New Challenges, B. Golden, S. Raghavan, and E. Wasil, eds., Springer, New York, 2008, pp. 29–48.

[53] ——, *An approximation algorithm for the capacitated arc routing problem*, The Open Operational Research Journal, 2 (2008), pp. 8–12.

Chapter 8

The Capacitated Arc Routing Problem: Combinatorial Lower Bounds

Dino Ahr
Gerhard Reinelt

8.1 ▪ Introduction

In this chapter we will discuss algorithms for obtaining good lower bounds on the value of an optimal solution of capacitated arc routing problems. In contrast to primal heuristics attempting to find good feasible solutions, such algorithms are sometimes called *dual heuristics*. On the one hand, their availability is important for assessing the quality of solutions obtained by primal heuristics. On the other hand, lower bounds are indispensable for devising a branch-and-bound scheme for solving problem instances to proven optimality.

In the following we assume that a connected (undirected) graph $G = (V,E)$, $|V| = n$, is given. Every edge $e \in E$ has a nonnegative *length* $w(e)$ and a nonnegative *demand* $d(e)$. The edges $E_R = \{e \in E \mid d(e) > 0\}$ with positive demand are also called *required edges*.

For $U \subseteq V$ the set of edges with one endnode in U and one endnode in $V \setminus U$ is denoted $\delta(U)$ (or $\delta_G(U)$ to emphasize for which graph the node degree is computed). It is clear that $\delta(U) = \delta(V \setminus U)$. The set of edges with endnode v is denoted $\delta(v)$. A node $v \in V$ is called *even (odd)* if $|\delta(v)|$ is even (odd). We say that a node is *R-odd* if it is incident to an odd number of required edges. Obviously, there is an even number of R-odd nodes.

The required edges have to be serviced by vehicles which are located at the *depot node* $v_1 \in V$. Each vehicle has a *capacity* $Q \in \mathbb{N}$. We are interested in *capacitated vehicle tours* given by a set of tours C_p, $p = 1,\ldots,K$, each starting and ending at the depot, such that all required edges are serviced by exactly one tour and such that the sum of the demands of the edges serviced in each tour is at most Q. Note that required edges are serviced only once, but may be used by other vehicles. Of course, in a capacitated vehicle tour all nodes are even (allowing for multiple edges) and the depot has degree $2K$.

We speak of a *K-vehicle* tour if we want to emphasize that K vehicles are used. The *Capacitated Arc Routing Problem* (CARP) consists of finding a shortest capacitated vehicle tour with respect to the edge lengths w.

The number of vehicles K is not treated as a variable; we set $K = \lceil \sum_{e \in E} \frac{d(e)}{Q} \rceil$.

Note that actually this K is only a lower bound for the least number K^* of vehicles required for a capacitated vehicle tour in G; i.e., it could happen that K vehicles are not enough. However, for all instances in our computational experiments we have $K^* = K$. Hence, in the following K will be considered as an implicit input value.

In this chapter we give a survey of dual heuristics for the CARP where we restrict ourselves to so-called combinatorial algorithms which do not make use of linear relaxations of integer programming models. We will present the ideas of various approaches and discuss their implementation. Some dominance relations between the bounds are pointed out, and the chapter is completed by giving computational results for a set of benchmark problems from the literature.

8.2 ▪ Combinatorial lower bounds

There has been active research on combinatorial lower bounds and their efficient computation for decades starting with the work of Christofides [10] dating back to 1973. Partial but not comprehensive overviews of combinatorial lower bounds are given in Benavent et al. [7], Eiselt, Gendreau, and Laporte [14], Assad and Golden [3], and Breslin and Keane [9]. A comprehensive overview is Ahr [1]. More recent but rather compact surveys can be found in Wøhlk [29] and Corberán and Prins [13]. In the past decade research on lower bounds has focused on LP-based methods, e.g., [20], [5], [8], and [24]. In fact, the latest pure combinatorial approach has been published by Wøhlk in 2006 [28]. The reason, as we will also conclude at the end of this chapter, is that combinatorial approaches seem to have fully exploited their possibilites. Further improvements can only be achieved by applying LP-based methods.

8.2.1 ▪ The bound of Christofides

This idea for obtaining a lower bound is proposed by Christofides [10] and starts with the computation of an optimal 1-vehicle tour C^*, i.e., the computation of an optimal solution of the classical *Chinese Postman Problem* (CPP).

Now, assume that C^* contains l edges incident to the depot. A feasible (and hence an optimal) tour must have at least $2K - l$ further edges incident to the depot. Every such edge is, of course, at least as long as the shortest edge with endnode v_1.

CHRISTOFIDES LOWER BOUND

(1) Let C^* be an optimal 1-vehicle tour in G, and let l be the number of edges of C^* incident to the depot. Let e^* be an edge such that $w(e^*) = \min_{e \in \delta(v_1)} w(e)$.

(2) Set $\text{CLB} = w(C^*) + (2K - l)w(e^*)$.

The running time for computing the bound of Christofides (CLB) is dominated by the solution of the CPP in time $O(|V|^3)$.

Although the reasoning seems to be intuitively appealing, this bound is not correct, as was pointed out by Golden and Wong [16]. Consider the graph G depicted on the left-hand side of Figure 8.1, where edge labels show edge lengths. All demands are equal to 1; furthermore, $Q = 3$ and $K = 2$.

The middle part of Figure 8.1 shows G with the edges added for the computation of CLB, namely, $\{v_2, v_4\}$ for obtaining the shortest 1-vehicle tour C^* with length 22 and $\{v_1, v_2\}$ and $\{v_1, v_3\}$ as cheapest edges incident to the depot. Hence $\text{CLB} = 22 + 6 = 28$.

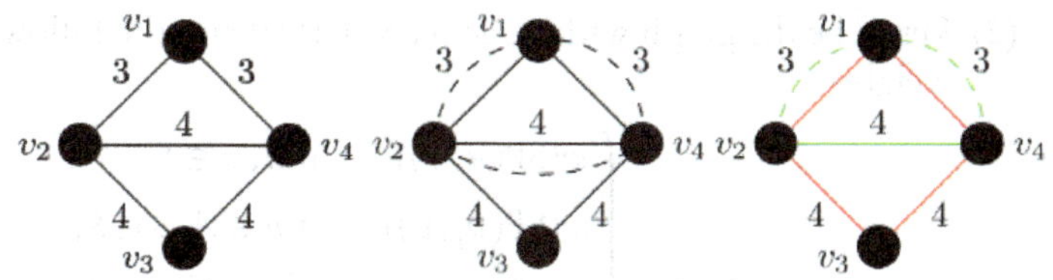

Figure 8.1. *A counterexample for the CLB [1].*

However, on the right-hand side of Figure 8.1 we have an optimal solution with weight 24 composed of the two tours (v_1, v_2, v_4, v_1) and $(v_1, v_2, v_3, v_4, v_1)$ with length $10+14 = 24$.

Clearly, the incorrectness of CLB is due to the fact that adding edges to the depot node in order to obtain degree $2K$ leads to a change of parity of the other nodes. This must be taken into account when adding edges to achieve even parity of all nodes.

8.2.2 ▪ The matching bound

The *Matching Lower Bound* (MLB) is given by Golden and Wong [16]. It is initialized with the sum of the edge weights $w(E_R)$ of the required edges, which is clearly a lower bound because E_R is a subset of every feasible tour. The matching bound now looks for the cheapest way to augment E_R such that all R-odd nodes become even and the depot receives degree $2K$.

The best way to connect two R-odd nodes or to connect the depot to an R-odd node is to use a shortest path between them. Let $\text{SP}(u, v)$ denote the shortest path between two nodes u and v and $w(\text{SP}(u, v))$ its length.

Furthermore, there must be at least $2K$ edges incident to the depot node. Hence, in total, we can add $2K - |\delta_R(v_1)|$ edges to the depot node. These edges can either be edges of a shortest path connecting the depot to an R-odd node or edges of a shortest path connecting the depot to the nearest required edge (and also including it). Let $\text{SP}(v_1, E_R)$ denote this path with length $w(\text{SP}(v_1, E_R))$. This improvement was first mentioned in Pearn [25].

For combining these ideas we construct the graph $\overline{G}$ on the R-odd nodes and $2l$ copies of the depot node, where $l = 2K - |\delta_R(v_1)|$ if $|\delta_R(v_1)|$ is even and $l = 2K - |\delta_R(v_1)| - 1$ otherwise. Note that we need an even number of copies of the depot node for the correct construction. A perfect matching in $\overline{G}$ added to E_R gives a graph with the requested property of the node degrees of the depot and the R-odd nodes. Adding the weight of a minimum perfect matching to $w(E_R)$ therefore yields a lower bound.

MATCHING LOWER BOUND

(1) Let $S = \{v \in V \mid v \text{ is } R\text{-odd}\}$ and l be as defined above. Let the node sets $A = \{a_1, \ldots, a_l\}$ and $B = \{b_1, \ldots, b_l\}$ each consist of l (distinguishable) copies of the depot node.

(2) Let $\overline{G}$ be the graph with node set $S \cup A \cup B$ and the following edges $e = \{u, v\}$ with weights

$$\overline{w}(e) = \begin{cases} w(\mathrm{SP}(u,v)) & \text{if } u, v \in S, \\ w(\mathrm{SP}(v_1,v)) & \text{if } u \in A, v \in S, \\ w(\mathrm{SP}(v_1,u)) & \text{if } u \in S, v \in A, \\ w(\mathrm{SP}(v_1,E_R)) & \text{if } u = a_i \text{ and } v = b_i, i = 1,\ldots,l, \\ 0 & \text{if } u, v \in B. \end{cases}$$

(3) Compute a minimum weight perfect matching M in $\overline{G}$.

(4) Set $\mathrm{MLB} = w(E_R) + \overline{w}(M)$.

The time complexity is dominated by the computation of the perfect matching in step (3). Since $\overline{G}$ contains $O(|V|)$ nodes, we have an overall time complexity $O(|V|^3)$.

Assad, Pearn, and Golden [4] show that MLB is a nondecreasing function of the number K of vehicles, or more precisely that $\mathrm{MLB}(K) \leq \mathrm{MLB}(K+1)$, for $K > |\delta(v_1)|$.

By adding paths $\mathrm{SP}(v_1, E_R)$ to the depot, some end nodes of the paths may become odd. This allows for further improvements, namely, by adding a cheapest incident edge, making this node even again.

8.2.3 ▪ The node scanning bound

The *Node Scanning Lower Bound* (NSLB) is due to Assad, Pearn, and Golden [4]. This approach augments E_R by adding further edges or paths to the depot. However, the parity of the nodes is not taken into account.

A single vehicle tour starts and ends at the depot. The first part of the tour consists of a path of unserviced edges until the first serviced edge is encountered, and the last part connects the last serviced edge to the depot by a path of unserviced edges. Since both paths can be empty, we have either 0, 1, or 2 such paths (which are called *d-paths* in Benavent et al. [7]).

If we consider the set of all such paths in a feasible solution, then, for a fixed node $v \neq v_1$, clearly the number of paths having v as an endnode cannot exceed $|\delta_R(v)|$. In order to obtain degree $2K$ at the depot node, we add $2K - |\delta_R(v_1)|$ least-cost paths while observing that no more than $|\delta_R(v)|$ paths from a node v are available.

NODE SCANNING LOWER BOUND

(1) Sort the nodes such that $w(\mathrm{SP}(v_1,v_2)) \leq w(\mathrm{SP}(v_1,v_3)) \leq \cdots \leq w(\mathrm{SP}(v_1,v_n))$.

(2) Let j be the smallest integer such that $|\delta_R(v_2)| + |\delta_R(v_3)| + \cdots + |\delta_R(v_j)| \geq 2K - |\delta_R(v_1)|$.

(3) Define $\deg(i) = |\delta_R(v_i)|$ for $i = 2,\ldots,j-1$, and $\deg(j) = 2K - |\delta_R(v_1)| - \sum_{i=2}^{j-1} \deg(i)$.

(4) Set $\mathrm{NSLB} = w(E_R) + \sum_{i=2}^{j} \deg(i) w(\mathrm{SP}(v_1,v_i))$.

The time complexity is dominated by the shortest path computation required for performing step (1). If all demands are greater than $Q/2$, then K can be set to $|E|$ and all the shortest paths added to the depot are needed since each tour can only service one edge. Computational results reported in Assad, Pearn, and Golden [4] lead to the conclusion that, compared to the matching bound, this bound only gives good results for sparse graphs with large demands.

8.2.4 ▪ The bound of Pearn

The *Pearn Lower Bound* (PLB) is presented in Pearn [25]. It combines the ideas of both bounds MLB and NSLB and aims at adding $2K - |\delta_R(v_1)|$ edges or paths to the depot node. Let again S be the set of the R-odd nodes of G. The bounding approach proceeds iteratively through two stages for every $p = 0, 2, \ldots, |S|$.

In the first stage the goal is to obtain even parity of the R-odd nodes while also taking into account that odd nodes could be connected to the depot. This is realized by computing a minimum weight perfect matching $M(p)$ in the graph $\overline{G}(p)$ which consists of the R-odd nodes S and p copies $a_1, \ldots, a_p$ of the depot node. All nodes are connected by edges weighted with shortest path distances. The only exceptions are the nodes $a_1, \ldots, a_p$, which are not connected with each other, since the intention is to match each copy of the depot with a node from S. By adding edges represented by the matching edges $M(p)$ to E_R, the degree of the depot node is increased by at least p, since each node $a_i, i = 1, \ldots, p$, is matched. Further increase can occur if a shortest path connection represented by a matching edge passes through the depot node. A special case can happen if $|\delta_R(v_1)|$ is odd, i.e., $v_1 \in S$. Then it is likely that one of the nodes $a_i, i = 1, \ldots, p$, is matched to the depot node v_1. Clearly, in that case, this edge will not be counted for, increasing the degree of the depot node. More precisely, if $\hat{G}(p)$ is the multigraph with edges $E_R \cup M(p)$, then the new degree of the depot node is $|\delta_{\hat{G}(p)}(v_1)|$ and there are $L = 2K - |\delta_{\hat{G}(p)}(v_1)|$ possibilities left to add further edges to the depot node.

In the second stage, essentially the procedure NSLB is applied to add L further edges and paths, respectively, to the depot node and finally arrive at the degree $2K$. The intermediate lower bound PLB(p) is computed as the sum of the weights of the required edges $w(E_R)$, the weight of the matching edges $w(M(p))$, and the weights of the edges added by the NSLB routine.

For every p we obtain an intermediate lower bound PLB(p), and the minimum of these values is a valid lower bound for G.

PEARN LOWER BOUND

(1) Sort the nodes such that $w(\mathrm{SP}(v_1, v_2)) \leq w(\mathrm{SP}(v_1, v_3)) < \cdots \leq w(\mathrm{SP}(v_1, v_n))$.

(2) For $p = 0, 2, \ldots, |S|$,

(2.1) Let $A(p) = \{a_1, \ldots, a_p\}$ denote the set of p copies of the depot node. Let $\overline{G}(p)$ be the graph with node set $\overline{V}(p) = S \cup A(p)$ and edges $e = \{u, v\}$ with weights

$$\overline{w}(e) = \begin{cases} w(\mathrm{SP}(u, v)) & \text{if } u, v \in S, \\ w(\mathrm{SP}(v_1, v)) & \text{if } u \in A(p),\ v \in S, \\ w(\mathrm{SP}(v_1, u)) & \text{if } u \in S,\ v \in A(p). \end{cases}$$

Observe that nodes in $A(p)$ are not connected with each other.

(2.2) Compute a minimum weight perfect matching $M(p)$ in $\overline{G}(p)$.

(2.3) Let $\hat{G}(p)$ be the multigraph with edge set $E_R \cup M(p)$, and let $|\delta_{\hat{G}(p)}(v_1)|$ be the degree of the depot node in $\hat{G}(p)$. Let $L = 2K - |\delta_{\hat{G}(p)}(v_1)|$.

(2.4) Let j be the smallest integer with $|\delta_R(v_2)| + |\delta_R(v_3)| + \cdots + |\delta_R(v_j)| \geq L$, $\deg(i) = |\delta_R(v_i)|$ for $i = 2, \ldots j - 1$, and $\deg(j) = L - \sum_{i=2}^{j-1} \deg(i)$.

(2.5) Compute $\mathrm{PLB}(p) = w(E_R) + \overline{w}(M(p)) + \sum_{i=2}^{j} \deg(i) w(\mathrm{SP}(v_1, v_i))$.

(3) Set $\mathrm{PLB} = \min\{\mathrm{PLB}(p) \mid p = 0, 2, \ldots, |S|\}$.

The loop of step (2) is dominated by the computation of the minimum weight perfect matching in step (2.2). Since $\overline{G}(p)$ contains $O(|V|)$ nodes and $|S| = O(V)$, we have the overall time complexity $O(|V|^4)$.

8.2.5 ▪ The node duplication bound

Saruwatari, Hirabayashi, and Nishida [26] define the *Node Duplication Lower Bound* (NDLB) in the context of the development of a branch-and-bound scheme for solving the CARP to optimality.

The basic idea is similar to the matching bound. A special graph $\overline{G}$ is constructed, and a matching is computed for it in order to determine a least-cost augmentation which makes all nodes of the original graph even and let the depot node have degree $2K$. The improvement of NDLB compared to MLB is that the endnodes of edges or paths added to the depot node in order to obtain degree $2K$ will be forced to have even degree.

A nice feature of NDLB is that we can easily exclude an edge from the augmentation (by setting its cost to infinity in $\overline{G}$) if we know that this edge cannot be part of a feasible or optimal solution (which is very interesting for branch-and-bound algorithms).

An improved version NDLB$^+$ of NDLB is possible where connections between two required edges e and f are forbidden if $d(e) + d(f) > Q$. The additional steps to be performed for NDLB$^+$ are marked by "+" in the algorithm. For the pure NDLB they are omitted.

NODE DUPLICATION LOWER BOUND

(1) Create the complete graph $\overline{G} = (\overline{V}, \overline{E})$ with weights $\overline{w} : \overline{E} \to \mathbb{R}_0^+$ as follows:

- For each $v_i \in V_R \setminus \{v_1\}$, let $\overline{V}_i$ consist of $|\delta_R(v_i)|$ copies of the node v_i.
- Let $l = \max\{2K - |\delta_R(v_1)|, 0\}$ if $|\delta_R(v_1)|$ is even and $l = \max\{2K - |\delta_R(v_1)|, 1\}$ if $|\delta_R(v_1)|$ is odd. Let $\overline{V}_1$ consist of l copies of the node v_1.

\+ Augment the node set $\overline{V}_1$ with $|\delta_R(v_1)|$ copies of the depot node.

- Set
$$\overline{V} = \overline{V}_1 \cup \bigcup_{v_i \in V_R \setminus \{v_1\}} \overline{V}_i.$$
- For each required edge $e = \{v_i, v_j\} \in E_R \setminus \delta_R(v_1)$, which is not incident to the depot node, select nodes $\overline{v}_{i_p} \in \overline{V}_i$ and $\overline{v}_{j_q} \in \overline{V}_j$ and add $\overline{e} = \{\overline{v}_{i_p}, \overline{v}_{j_q}\}$ to $\overline{E}$ with $d(\overline{e}) = d(e)$. The selection of nodes should be such that each node from $\bigcup_{v_i \in V_R} \overline{V}_i$ is only incident to exactly one required edge $\overline{e}$. Let $\overline{E}_R \subset \overline{E}$ be these new edges (which are counterparts to the required edges of G).

\+ Do the same as in the previous step for all required edges e incident to the depot node.

- Add remaining edges (with zero demand) to $\overline{E}$ to make $\overline{G}$ complete.

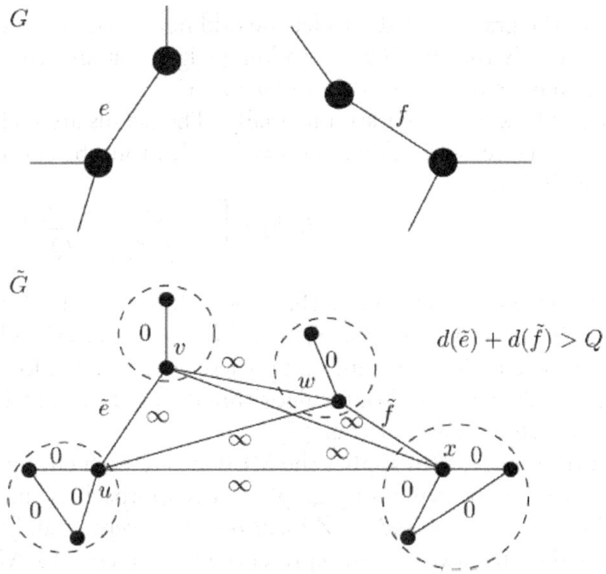

Figure 8.2. *Illustration of the improvement step of NDLB+ [1].*

- For every $\overline{e} = \{\overline{u}, \overline{v}\} \in \overline{E}$, assign the weight

$$\overline{w}(\overline{e}) = \begin{cases} \infty & \text{if } \overline{e} \text{ is required}, \\ w(\mathrm{SP}(v_i, v_j)) & \text{if } \overline{u} \in \overline{V}_i, \overline{v} \in \overline{V}_j, i \neq j, \\ 0 & \text{if } \overline{u}, \overline{v} \in \overline{V}_i, v_i \in V_R \setminus \{v_1\}, \\ \infty & \text{if } \overline{u}, \overline{v} \in \overline{V}_1. \end{cases}$$

+ For every pair of edges $\overline{e} = \{u, v\}, \overline{f} = \{w, x\} \in \overline{E}_R$ with $d(\overline{e}) + d(\overline{f}) > Q$, set $\overline{w}(\{u, w\}) = \overline{w}(\{u, x\}) = \overline{w}(\{v, w\}) = \overline{w}(\{v, x\}) = \infty$ (see Figure 8.2).
+ Remove those $|\delta_R(v_1)|$ copies of the depot node from $\overline{V}$ which are incident to required edges.

  (Note that the required edges have only been added in order to be able to take them into account in the previous step.)

(2) Compute a minimum weight perfect matching M in $\overline{G}$.

(3) Set $\text{NDLB} = w(E_R) + \overline{w}(M)$.

Again, the time complexity is dominated by computation of the minimum weighted perfect matching in step (2). But now $\overline{G}$ contains two nodes for each edge $e \in E$. Hence, we obtain the time complexity $O(|E|^3)$.

8.2.6 ▪ The bounds of Win

Win [27] presents a powerful new idea for improving lower bounds which is used in almost all of the more recent approaches. The strategies discussed so far were based on adding edges or paths to the depot node in order to reach degree $2K$ and/or on adding edges to the graph in order to let the odd nodes have even degree. Win proposes to consider not only the cut $\delta(v_1)$ for adding edges, but also cuts $\delta(U)$, $v_1 \in U \subseteq V$, with U successively being extended from $\{v_1\}$ to V.

We will explain the idea informally. The details are included in the algorithms based on this idea discussed below. Let $S = V \setminus U$. From the cut separating V into U and S, we can conclude that at least

$$K(S) = \left\lceil \sum_{e \in E_R(S) \cup \delta_R(S)} \frac{d(e)}{Q} \right\rceil$$

vehicles are needed to service the subgraph induced by S and the cut edges $\delta_R(S)$. Let $k(S) = 2K(S) - |\delta_R(S)|$. If $k(S) > 0$, then we can clearly add $k(S)$ copies of the cheapest edge to the cut $\delta(S)$. Summing up these edge weights for each cut and adding them to $w(E_R)$ yields the lower bound algorithm denoted by 8.2.5 in Win [27], which is denoted by ZAW1LB in Benavent et al. [7].

Furthermore, Win applies the MLB idea to each cut by treating U as the depot node and $V \setminus U$ as the remaining graph. This approach is summarized as algorithm 8.2.10 in Win [27] and denoted by *ZAW2LB* in Benavent et al. [7]. BCCM2LB which will be discussed in the next section represents a refinement of ZAW2LB.

8.2.7 ▪ The bounds of Benavent et al.

Benavent et. al. [7] define four further lower bounds named LB1, LB2, LB3, and LB4. We will call them BCCM1LB, BCCM2LB, BCCM3LB, and BCCM4LB, respectively.

The idea of the first lower bound BCCM1LB is similar to PLB. But while PLB considers the matching of the odd nodes separately from the augmentation with shortest d-paths, BCCM1LB takes the interaction between both aspects into account.

BCCM1 LOWER BOUND

(1) Sort the nodes such that $w(\mathrm{SP}(v_1, v_2)) \le w(\mathrm{SP}(v_1, v_3)) \le \cdots \le w(\mathrm{SP}(v_1, v_n))$.

(2) Define node sets A, B, and S' as follows:

- Let $l = \max\{2K - |\delta_R(v_1)|, 0\}$, and let A consist of l copies of the depot.
- Let i' be the smallest integer with $|\delta_R(v_2)| + |\delta_R(v_3)| + \cdots + |\delta_R(v_{i'})| \ge 2K - |\delta_R(v_1)|$. Let B contain $\deg(i) = |\delta_R(v_i)|$ copies of v_i for $i = 2, \ldots, i'$. (Note that we do not reset the value of $\deg(i')$.)
- Let $S' = \{v \in V \mid v \text{ is } R\text{-odd}\} \setminus \{B \cup v_1\}$.

(3) Let $\overline{G}$ be the complete graph with node set $A \cup B \cup S'$ and edges $e = \{u, v\}$ with weights

$$\overline{w}(e) = \begin{cases} \infty & \text{if } u, v \in A, \\ w(\mathrm{SP}(u, v)) & \text{otherwise.} \end{cases}$$

(3) Compute a minimum weight perfect matching M in $\overline{G}$.

(4) Set BCCM1LB $= w(E_R) + \overline{w}(M)$.

It is difficult to estimate how many nodes will be contained in B. In the worst case B could contain copies for each node $v \in V$, hence leading to the same complexity as NDLB. However, usually the size of B will be rather small and therefore BCCM1LB will be faster than NDLB.

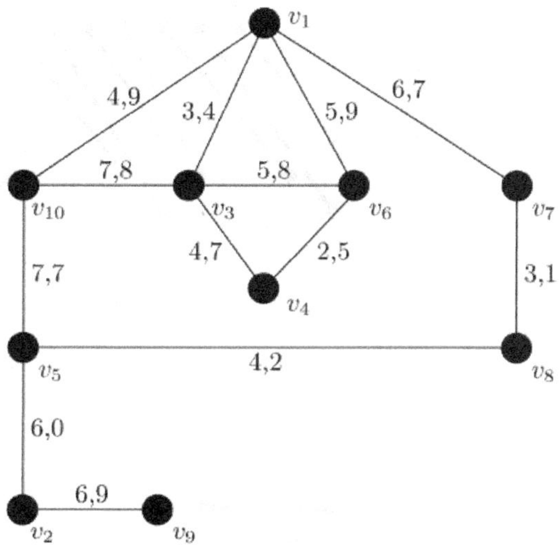

Figure 8.3. *The input graph G for algorithms NDLB and BCCM1LB [1].*

The bound BCCM1LB is identical to NDLB if we do not set the weights of the required edges to infinity in $\overline{G}$ for NDLB, although less nodes are duplicated for BCCM1LB (essentially only those nodes are duplicated which could be matched to the copies of the depot node). However, setting the weights of the required edges to infinity in $\overline{G}$ leads to better lower bounds for NDLB in some cases as the following example shows.

Consider the graph G depicted in Figure 8.3. Each edge e is labeled with $w(e)$ and $d(e)$. Furthermore, we have $K = 6$ and $Q = 13$. For the computation of BCCM1LB we determine the sequence v_3, v_{10}, v_6, v_7,... in step (1). Then in step (2) we get $l = 2 \cdot 6 - 4 = 8$. Hence A consists of eight copies of the depot node. The set B contains four copies of node v_3, three copies of node v_{10}, and three copies of node v_6. Finally, the set S' consists only of nodes v_2 and v_9 since the other R odd nodes are already contained in B. The resulting graph $\overline{G}$ with the matching edges M computed in step (3) is depicted in Figure 8.4. Copies of the same node are surrounded by a dashed circle and labeled with the corresponding node of G. Since $w(E_R) = 56$ and $\overline{w}(M) = 35$, we obtain the lower bound 91.

In step (1) of NDLB we create node sets $\overline{V}_i$ consisting of $|\delta_R(v_i)|$ nodes for each node v_i from $V \setminus \{v_1\}$. Thus, we create one copy of v_2 and v_9, two copies of v_4, v_5, v_7 and v_8, three copies of v_6 and v_{10}, and four copies of v_3. As for BCCM1LB we create eight copies of the depot node. The nodes in $\overline{V}_i$ connected with each other with zero weight edges except for $\overline{V}_1$ where internal weights are set to ∞. Required edges also receive an infinte weight. Figure 8.5 shows the resulting matching M on $\overline{G}$ (matching edges are bold). Note that the matching is identical to the one computed for BCCM1LB except for the edge connecting nodes v_2 and v_9. For NDLB both nodes are matched each with a copy of v_5, whereas for BCCM1LB v_2 and v_9 are matched with each other. This slight difference of the matching computed by NDLB increases its weight by 12, and hence we obtain the lower bound 103.

The second lower bound BCCM2LB generalizes the idea of BCCM1LB to successive cut sets according to the bound ZAW2LB of Win [27].

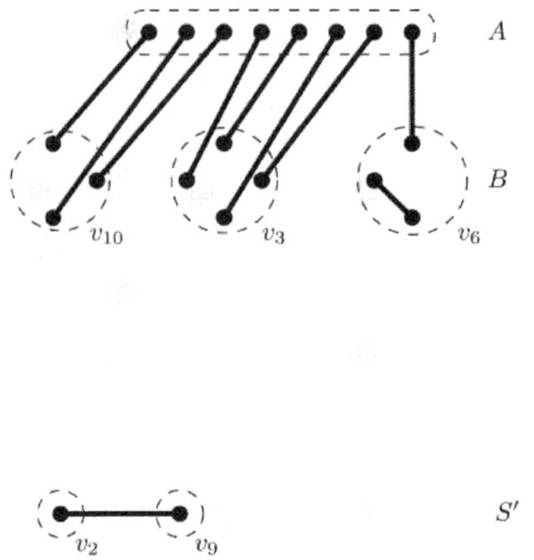

Figure 8.4. *The matching for $\overline{G}$ computed by BCCM1LB [1].*

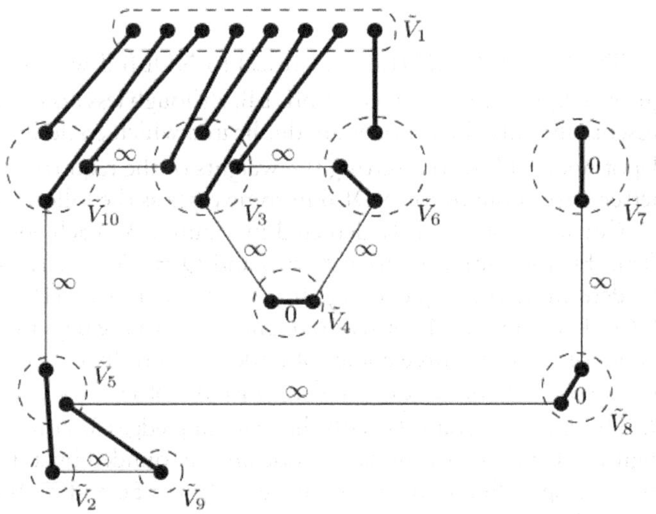

Figure 8.5. *The matching for $\overline{G}$ computed by NDLB [1].*

BCCM2 LOWER BOUND

(1) Let $U = \{v_1\}$ and $L1 = L2 = 0$.

(2) While $U \neq V$:

(2.1) Let $G'_1 = (V'_1, E'_1), \ldots, G'_t = (V'_t, E'_t)$ be the connected components of the graph induced by the node set $V \setminus U$.

(2.2) For $j = 1, \ldots, t$:

(2.2.1) Set $k_j = K(V_j') = \lceil \sum_{e \in E_j' \cup \delta(V_j')} d(e)/Q \rceil$ (minimum number of vehicles required to service edges in E_j' and $\delta(V_j')$), $q_j = |\delta_R(V_j')|$ (number of required edges in the cut $\delta(V_j')$), $r_j = \max\{0, 2k_j - q_j\}$ (number of additional edges which must be traversed to service edges in E_j' and $\delta(V_j')$), and let e_j^* be the cheapest edge in the cut $\delta(V_j')$.

(2.2.2) Let $S_j' = \{v \in V \mid v \text{ is } R\text{-odd}\} \cap V_j'$ be the set of R-odd nodes contained in V_j'.

If $S_j' \neq \emptyset$ or $r_j > 0$, then construct the complete graph $\overline{G}_j = (\overline{V}_j, \overline{E}_j)$ with weights $\overline{w} : \overline{E}_j \rightarrow \mathbb{R}_0^+$ as follows:

- Rename the nodes in V_j' such that $w(\text{SP}(U, v_{j_1})) \leq w(\text{SP}(U, v_{j_2})) \leq \ldots$.
- Let A_j consist of r_j copies of a "supernode" representing the node set U.
- Let $j_{i'}$ be the smallest integer such that $|\delta_R(v_{j_1})| + |\delta_R(v_{j_2})| + \cdots + |\delta_R(v_{j_{i'}})| \geq r_j$. Let $d_j(i) = |\delta_R(v_{j_i})|$ for $i = 1, \ldots, j_{i'}$. Let B_j contain $d_j(i)$ copies of v_{j_i}, $i = 1, \ldots, j_{i'}$.
- Let $S_j'' = S_j' \setminus B_j$ be the set of R-odd nodes of V_j' without nodes already contained in B_j and C_j be a set of $\max\{0, |S_j'| - r_j\}$ nodes representing potential matching nodes in U.
- Let $\overline{V}_j = A_j \cup B_j \cup S_j'' \cup C_j$ and $\overline{E}_j$ consist of edges $e = \{u, v\}$ with weights defined by

$$\overline{w}_j(e) = \begin{cases} w(\text{SP}(u,v)) & \text{if } u, v \in S_j'' \cup B_j, \\ w(\text{SP}(U,v)) & \text{if } u \in A_j \cup C_j, v \in S_j'' \cup B_j, \\ w(\text{SP}(u,U)) & \text{if } u \in S_j'' \cup B_j, v \in A_j \cup C_j, \\ 0 & \text{if } u, v \in C_j, \\ \infty & \text{if } u, v \in A_j. \end{cases}$$

- Compute a minimum weight perfect matching M_j in $\overline{G}_j$.

(2.3) $L2 = \max\{L2, w(E_R) + \sum_{j=1}^{t} \overline{w}_j(M_j) + L1\}$.

(2.4) $L1 = L1 + \sum_{j=1}^{t} r_j w(e_j^*)$.

(2.5) Let $W = \{v \in V \setminus U \mid v \text{ is adjacent to a node in } U\}$ and $U = U \cup W$.

(3) Set BCCM2LB $= L2$.

Step (2.2.2) is in fact the application of BCCM1LB to the component G_j'. In the worst case loop (2) is executed $O(|V|)$ times, and hence we have the overall time complexity $O(|V|)$ times the BCCM1LB time complexity.

The idea of the third lower bound BCCM3LB is similar to that of the NDLB$^+$. A different idea (compared to the improvement step of NDLB$^+$) is used to forbid some of the edges in the constructed graph. Assume that K vehicles are used. Then we know that each vehicle must service at least the demand

$$Q_{\min} = \max\left\{0, \sum_{e \in E} d(e) - (K-1)Q\right\}$$

because this is the remaining demand when $K-1$ vehicles are fully loaded. When considering the nodes of the graph in NSLB order, i.e., sorted in nondecreasing order according to their distance to the depot node, then we know that a node u with $d(\mathrm{SP}(v_1,u)) < Q_{\min}$ will not be matched with $|\delta_R(u)|$ edges but with $|\delta_R(u)|-1$ edges. Hence we can forbid one connection between node u and the depot by setting the weights of the edges connecting one copy of u with each of the depot node copies to infinity.

The idea of the fourth lower bound BCCM4LB is completely different from the ideas discussed so far. It is based on a dynamic programming approach for determining lower bounds for single routes by determining so-called q-routes and was used by Christofides, Mingozzi, and Toth [12, 11] for developing a branch-and-bound algorithm for the Capacitated Vehicle Routing Problem.

We did not implement BCCM3LB and BCCM4LB because computational experiments carried out in Benavent et al. [7] showed that BCCM2LB performed best.

8.2.8 ▪ The multiple cuts node duplication bound

The *Multiple Cuts Node Duplication Lower Bound* (MCNDLB) is proposed by Wøhlk [28]. It is based on the framework of BCCM2LB but uses a different construction of $\overline{G}_j$ in step (2.2.2). Here $\overline{G}_j$ is constructed according to the NDLB or the NDLB$^+$ scheme instead of the BCCM1LB scheme. Analogously to NDLB, we can distinguish between the pure MCNDLB and the improved version MCNDLB$^+$.

MULTIPLE CUTS NODE DUPLICATION LOWER BOUND

(1)–(2.2.1) See algorithm BCCM2 LOWER BOUND.

(2.2.2) Generate the complete graph $\overline{G}_j=(\overline{V}_j,\overline{E}_j)$ with weights $\overline{w}:\overline{E}_j\to\mathbb{R}_0^+$ as follows:

- For each node $v_i\in V_j'\cap V_R$ let $\overline{V}_{j_i}$ consist of $|\delta_R(v_i)|$ copies of the node v_i.
- Let A_j consist of r_j copies of a "supernode" representing the node set U.
+ Add further q_j copies of a "supernode" representing the node set U to A_j.
- Let
$$\overline{V}_j=A_j\cup\bigcup_{v_i\in V_j'\cap V_R}\overline{V}_{j_i}.$$
- For each required edge $e=\{v_i,v_l\}\in E_j'\cap E_R$ select nodes $\overline{v}_{i_p}\in\overline{V}_{j_i}$ and $\overline{v}_{l_q}\in\overline{V}_{jl}$ and add $\overline{e}=\{\overline{v}_{i_p},\overline{v}_{l_q}\}$ to $\overline{E}_j$ with $d(\overline{e})=d(e)$. The selection of nodes should be in such a way that each node from $\bigcup_{v_i\in V_j'\cap V_R}\overline{V}_{j_i}$ is only incident to exactly one required edge $\overline{e}$.

  We denote these new edges (which are counterparts to the required edges of G_j') by $\overline{E}_{j_R}\subset\overline{E}_j$.
+ As in the previous step for required edges $e=\{v_i,v_l\}\in\delta(V_j')\cap E_R$, with say $v_i\in A_j$, we select nodes $\overline{v}_{i_p}\in A_{j_i}$ and $\overline{v}_{l_q}\in\overline{V}_{jl}$ and add $\overline{e}=\{\overline{v}_{i_p},\overline{v}_{l_q}\}$ to $\overline{E}_j$ with $d(\overline{e})=d(e)$.
- Add remaining edges (with zero demand) to $\overline{E}_j$ to make $\overline{G}_j$ complete.

- For $\overline{e} = \{\overline{u}, \overline{v}\} \in \overline{E}_j$ assign the weight

$$\overline{w}(\overline{e}) = \begin{cases} \infty & \text{if } \overline{e} \text{ is required}, \\ w(\mathrm{SP}(v_i, v_l)) & \text{if } \overline{u} \in \overline{V}_{j_i}, \overline{v} \in \overline{V}_{j_l}, i \neq l, \\ 0 & \text{if } \overline{u}, \overline{v} \in \overline{V}_{j_i}, v_i \in V_j' \cap V_R, \\ w(\mathrm{SP}(U, v_l)) & \text{if } \overline{u} \in A_j, \overline{v} \in \overline{V}_{j_l}, \\ w(\mathrm{SP}(v_i, U)) & \text{if } \overline{u} \in \overline{V}_{j_i}, \overline{v} \in A_j, \\ \infty & \text{if } \overline{u}, \overline{v} \in A_j. \end{cases}$$

+ For every pair of required edges $\overline{e} = \{u, v\}, \overline{f} = \{w, x\} \in \overline{E}_{j_R}$, set $\overline{w}(\{u, w\}) = \overline{w}(\{u, x\}) = \overline{w}(\{v, w\}) = \overline{w}(\{v, x\}) = \infty$ if $d(\overline{e}) + d(\overline{f}) > Q$.

+ Remove those q_j nodes from A_j which are incident to required edges from $E_j' \cap E_R$.

- Finally, let C_j be a set of $\max\{0, |\overline{V}_j \setminus A_j| - |A_j|\}$ nodes if $|\overline{V}_j \setminus A_j| - |A_j|$ is even and $\max\{1, |\overline{V}_j \setminus A_j| - |A_j|\}$ nodes otherwise. Let $\overline{V}_j = \overline{V}_j \cup C_j$, and add remaining edges connecting C_j to $\overline{E}_j$ to make $\overline{G}_j$ complete. The new edges $\overline{e} = \{\overline{u}, \overline{v}\}$ have the weights

$$\overline{w}(\overline{e}) = \begin{cases} 0 & \text{if } \overline{u}, \overline{v} \in C_j, \\ \infty & \text{if } \overline{u} \in A_j \text{ and } \overline{v} \in C_j \text{ or if } \overline{u} \in C_j \text{ and } \overline{v} \in A_j, \\ w(\mathrm{SP}(U, v_l)) & \text{if } \overline{u} \in C_j, \overline{v} \in \overline{V}_{j_l}, \\ w(\mathrm{SP}(v_i, U)) & \text{if } \overline{u} \in \overline{V}_{j_i}, \overline{v} \in C_j. \end{cases}$$

  Note that the node set C_j is added to $\overline{G}_j$ in order to provide (together with A_j) enough nodes, namely $|\overline{V}_j \setminus A_j|$, which simulate the case that a node from $\overline{V}_j \setminus A_j$ will be matched with a node from U.

- Compute a minimum weight perfect matching M_j in $\overline{G}_j$.

(2.3)–(2.5) See algorithm BCCM2 LOWER BOUND.

(3) Set MCNDLB $= L2$.

Analogously to the analysis of BCCM2LB, we have the overall time complexity $O(|V|)$ times the complexity of NDLB.

8.2.9 ▪ The hierarchical relaxations bound

The *Hierarchical Relaxations Lower Bound* (HRLB) was presented by Amberg and Voss [2]. In principle, this approach is based on an LP formulation with so-called *aggregate parity constraints* and *aggregate capacity constraints*. In each iteration a cut $\delta(U)$ (starting with $U = \{v_1\}$) is considered. For each such cut two phases are performed. In the first phase the aim is to fulfill the corresponding capacity constraint. This is done by adding enough copies of the cut edge of minimum weight such that the constraint is satisfied. Then, this minimum weight is considered as shadow price for the constraint and the other edges of the cut are assigned opportunity costs by subtracting the shadow price from their weight. In the second phase a minimum cost perfect matching on the odd nodes according to the opportunity costs is computed. For each iteration a lower bound is computed as the sum of the edge weights added in order to fulfill the capacity constraints and the weight of the matching edges. The overall lower bound is computed as the maximum lower bound from the single iterations plus $w(E_R)$.

While adding copies of the cut edge of minimum weight, we must also consider the fact that the reduced value of the edge (in case the opportunity cost is not equal to the actual cost) can only be used a certain number of times (which is equal to the minimum of the number of times edges were added having *this* edge in the cut set). Also, after each iteration, we extend the cut set $\delta(U)$ by adding the node incident to the cut edge of minimum weight.

As is pointed out in Amberg and Voß [2], the computation of HRLB can also be interpreted as solving a linear programming relaxation based on a certain subset of cuts in a combinatorial way which could also be characterized as a dual ascent approach.

8.3 ▪ Improvements

In the following we describe some improvements of lower bounding procedures proposed by Breslin and Keane [9] and Ahr [1]. In order to capture the core idea of these improving modifications, we briefly review the main aspects and the evolution of the lower bounding algorithms.

Basically, the following three properties of a capacitated vehicle tour are exploited for determining lower bounds.

- The depot node must have degree $2K$.
- Each R-odd node must have even degree.
- For successive cuts $\delta(S)$ we have to ensure that at least $K(S)$ vehicles cross the cut. Note that the first property is the special case for the cut $\delta(v_1)$, i.e., $K = K(\{v_1\})$.

The first generation of approaches (MLB, NSLB, PLB, NDLB, and NDLB$^+$ as well as BCCM1LB) considers only the first and the second properties. The second generation (BCCM2LB, BCCM3LB, MCNDLB, MCNDLB$^+$, and HRLB) takes all of the above properties into account.

A general improvement for methods exploiting the third property can be made by computing improved lower bounds or even optimal solutions $K^*(S)$ for the number of vehicles required for servicing a given node set S instead of using the lower bound $K(S)$ given in section 8.2.6. Since this problem is a *Bin Packing Problem* (BPP), the widely known algorithms of Martello and Toth [22, 23] can be used to compute $K^*(S)$. For example, this enhancement was applied in Hertz, Laporte, and Mittaz [17] and Hertz and Mittaz [18].

However, a more serious limitation of the existing procedures is that they only consider a series of nested cuts, and thus the vast majority of possible cuts is ignored. Hence, improved lower bounds can be expected if a larger set of cuts will be considered.

Breslin and Keane [9] suggest an improvement following this line for BCCM2LB. Instead of adding all nodes of W at once to U in step (2.5), we add only one node at a time. The order in which nodes of W are selected can be arbitrary, but Breslin and Keane [9] find out that it is most promising to add them in increasing order of their node degree. Clearly, we must not update *L1* until all nodes of W have been added to U. Since all the cuts of the ordinary BCCM2LB computation are included, this modification cannot obtain a worse result than BCCM2LB. Ahr [1] proposes to apply this approach to

MCNDLB$^+$ because it also uses the framework of BCCM2LB. We denote the procedures including this improvement by BCCM2LBMOD and MCNDLBMOD. Computational results show that these modified procedures obtain improved lower bounds for some instances.

Obviously, the above modification considers more cuts than before but we are still far away from considering all possible cuts. Another deficiency of BCCM2LB and MCNDLB$^+$ is that a potential matching of odd nodes in the node set U is not taken into account when computing *L2* in step (2.3). This is due to the fact that it is very complicated to do the bookkeeping of the correct parities of nodes in U which are dependent on the edge copies which account for *L1* as well as the matching of the nodes chosen for the shortest path connections to the supernode U.

8.4 ▪ Dominance relations between the bounds

Figure 8.6 shows the relations between the CARP lower bounds. An arrow connecting two bounds means that the bound to which the arrow points is dominated by the other bound. Unless obvious from the discussion of this chapter, arrow labels give the reference where the proof can be found.

Wøhlk [28] gives an example showing that the bounds BCCM2LB and NDLB$^+$ are incomparable. Furthermore, Assad, Pearn, and Golden [4] show that NSLB and MLB are incomparable.

We found no proof where to put HRLB in this hierarchy. Computational results which will be discussed in the next section exhibit that HRLB is equal or superior to all other lower bounds.

8.5 ▪ Computational experiments and conclusions

We now give a survey on the performance of the combinatorial lower bounds. As benchmarks we have chosen instances from the well-established benchmark sets from Benavent at el. [7] (denoted carpBCCM92), Golden, DeArmon, and Baker [15] (denoted carpGDB83), Kiuchi et al. [19] (denoted carpKSHS95), and Li and Eglese [21] (denoted carpLE96).

Tables 8.1–8.4 give the specifications of the instances with the name of the instance, the numbers $|V|$ and $|E|$ of nodes and edges, the number $|E_R|$ of required edges, the vehicle capacity Q, and the number of vehicles K.

Tables 8.5–8.7 present the lower bounds of this chaper as well as results for branch-and-cut-based lower bounds in order to assess the quality of the combinatorial lower bounds. Optimal solution values are printed in bold, and best lower bound values are underlined. Each table gives for every instance

- the combinatorial bounds NDLB (N), NDLB (N+), BCCM1LB ($B1$), BCCM2LB ($B2$), BCCM2LBMOD ($B2m$), MCNDLB ($M1$), MCNDLB$^+$ (M+), MCNDLBMOD ($M+m$), and HRLB,
- the bounds obtained by the cutting plane algorithm of Belenguer and Benavent [6] *without* separating disjoint path inequalities (CPA1) and *with* separating disjoint path inequalities (CPA2),
- the bound obtained with exact separation of aggregate capacity constraints as described by Ahr [1] (EAC), and
- the value of the best heuristic solution from Belenguer and Benavent [6] (UB).

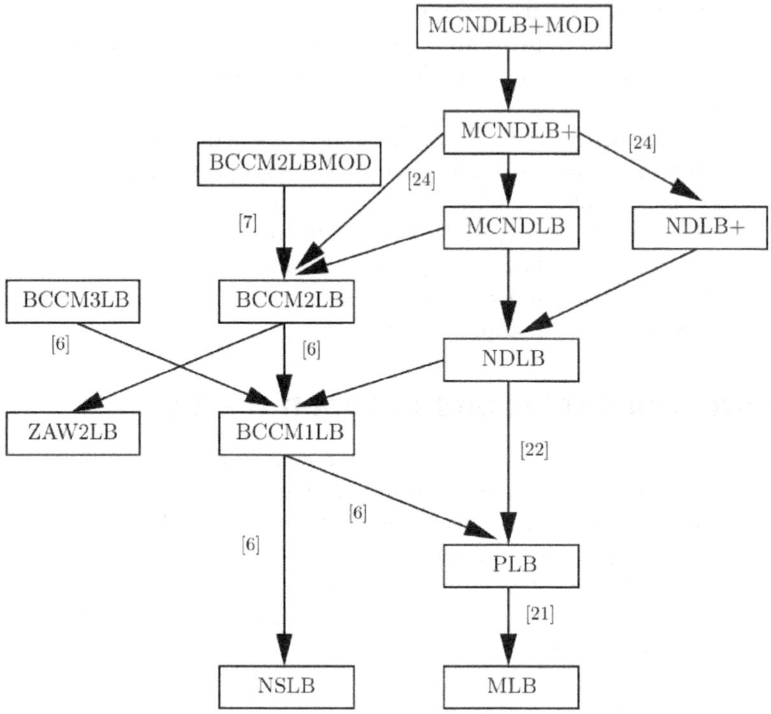

Figure 8.6. *Dominance relations between combinatorial lower bounds [1].*

For the set carpBCCM92 we observe that all algorithms yield the same results except for instances `1C`, `2B`, `2C`, `3B`, `3C`, `4D`, `5B`, and `7C`. For each of these instances NDLB, NDLB$^+$ and BCCM1LB obtain the same result and the same holds for BCCM2LB, BCCM2LBMOD, MCNDLB, MCNDLB$^+$, and MCNDLBMOD, but the latter results are better. A further exception is instance `4C`, where all algorithms obtain the bound 524; however, algorithms BCCM2LBMOD and MCNDLBMOD achieve a better bound of 525.

For the sets carpKSHS95 and carpGDB83 all algorithms perform equally well except for instance `gdb8` from carpGDB83, where NDLB, NDLB$^+$, and BCCM1LB are inferior.

The most interesting results have been obtained for the set carpLE96. The fact that NDLB dominates BCCM1LB is confirmed by instances `egl-e1-A`, `egl-e1-B`, `egl-e3-A`, `egl-e3-B`, and `egl-e3-C`. Furthermore, the domination of MCNDLB over BCCM2LB is confirmed by instances `egl-e1-A`, `egl-e2-A`, `egl-e3-A`, `egl-e3-B`, and `egl-e3-C`. For instances `egl-e1-A`, `egl-e2-A`, and `egl-e3-A` the improving modification achieved the best results.

We have to mention that for the instance set carpLE96 we have obtained worse results for our implementation of BCCM2LB than those reported in Belenguer and Benavent [6]. The reason for this is not clear; perhaps the implementation of BCCM2LB used in the scope of Belenguer and Benavent [6] applied an improved bound $K^*(S)$. Nevertheless, for instances `egl-e1-A`, `egl-e2-A`, `egl-e3-A`, `egl-e3-B`, and `egl-e3-C` we obtained improved combinatorial lower bounds compared to Belenguer and Benavent [6].

For none of the instances NDLB$^+$ dominated NDLB or MCNDLB$^+$ dominated

Table 8.1. *Benchmark set* carpBCCM92 *[1].*

Inst.	$\|V\|$	$\|E\|$	$\|E_R\|$	Q	K	Inst.	$\|V\|$	$\|E\|$	$\|E_R\|$	Q	K
1A	24	39	39	200	2	1B	24	39	39	120	3
1C	24	39	39	45	8	2A	24	34	34	180	2
2B	24	34	34	120	3	2C	24	34	34	40	8
3A	24	35	35	80	2	3B	24	35	35	50	3
3C	24	35	35	20	7	4A	41	69	69	225	3
4B	41	69	69	170	4	4C	41	69	69	130	5
4D	41	69	69	75	9	5A	34	65	65	220	3
5B	34	65	65	165	4	5C	34	65	65	130	5
5D	34	65	65	75	9	6A	31	50	50	170	3
6B	31	50	50	120	4	6C	31	50	50	50	10
7A	40	66	66	200	3	7B	40	66	66	150	4
7C	40	66	66	65	9	8A	30	63	63	200	3
8B	30	63	63	150	4	8C	30	63	63	65	9
9A	50	92	92	235	3	9B	50	92	92	175	4
9C	50	92	92	140	5	9D	50	92	92	70	10
10A	50	97	97	250	3	10B	50	97	97	190	4
10C	50	97	97	150	5	10D	50	97	97	75	10

Table 8.2. *Benchmark set* carpGDB83 *[1].*

Inst.	$\|V\|$	$\|E\|$	$\|E_R\|$	Q	K	Inst.	$\|V\|$	$\|E\|$	$\|E_R\|$	Q	K
gdb1	12	22	22	5	5	gdb2	12	26	26	5	6
gdb3	12	22	22	5	5	gdb4	11	19	19	5	4
gdb5	13	26	26	5	6	gdb6	12	22	22	5	5
gdb7	12	22	22	5	5	gdb8	27	46	46	27	10
gdb9	27	51	51	27	10	gdb10	12	25	25	10	4
gdb11	22	45	45	50	5	gdb12	13	23	23	35	7
gdb13	10	28	28	41	6	gdb14	7	21	21	21	5
gdb15	7	21	21	37	4	gdb16	8	28	28	21	5
gdb17	8	28	28	41	5	gdb18	9	36	36	37	5
gdb19	8	11	11	27	3	gdb20	11	22	22	27	4
gdb21	11	33	33	27	6	gdb22	11	44	44	27	8
gdb23	11	55	55	27	10						

MCNDLB. Clearly, this case can only occur when demands are large compared to the vehicle capacity, which was not the case for our instance sets.

When comparing HRLB to the other combinatorial lower bounds, we observe that HRLB is equal or in many cases superior to NDLB, NDLB, BCCM1LB, BCCM2LB, BCCM2LBMOD, MCNDLB, MCNDLB$^+$, and MCNDLBMOD for each instance (except for instance gdb17, which might be caused by an error in the implementation).

Turning to cutting-plane-based algorithms, we see that for smaller instance sets HRLB can compete with CPA1, CPA2, and EAC. For instance, set carpKSHS95 HRLB is equal and for carpGDB83 HRLB is only inferior in 3 of 23 instances. Moving to instance set carpBCCM92, we see that HRLB can only compete in 17 of 34 instances, and finally for carpLE96 HRLB is inferior for all instances.

Table 8.3. *Benchmark set* carpKSHS95 *[1].*

Inst.	$\|V\|$	$\|E\|$	$\|E_R\|$	Q	K	Inst.	$\|V\|$	$\|E\|$	$\|E_R\|$	Q	K
k1	8	15	15	150	4	k2	10	15	15	150	4
k3	6	15	15	150	4	k4	8	15	15	150	4
k5	8	15	15	150	3	k6	9	15	15	150	3

Table 8.4. *Benchmark set* carpLE96 *[1].*

Inst.	$\|V\|$	$\|E\|$	$\|E_R\|$	Q	K	Inst.	$\|V\|$	$\|E\|$	$\|E_R\|$	Q	K
egl-e1-A	77	98	51	305	5	egl-e1-B	77	98	51	220	7
egl-e1-C	77	98	51	160	10	egl-e2-A	77	98	72	280	7
egl-e2-B	77	98	72	200	10	egl-e2-C	77	98	72	140	14
egl-e3-A	77	98	87	280	8	egl-e3-B	77	98	87	190	12
egl-e3-C	77	98	87	135	17	egl-e4-A	77	98	98	280	9
egl-e4-B	77	98	98	180	14	egl-e4-C	77	98	98	130	19
egl-s1-A	140	190	75	210	7	egl-s1-B	140	190	75	150	10
egl-s1-C	140	190	75	103	14	egl-s2-A	140	190	147	235	14
egl-s2-B	140	190	147	160	20	egl-s2-C	140	190	147	120	27
egl-s3-A	140	190	159	240	15	egl-s3-B	140	190	159	160	22
egl-s3-C	140	190	159	120	29	egl-s4-A	140	190	190	230	19
egl-s4-B	140	190	190	160	27	egl-s4-C	140	190	190	120	35

We observe that the quality of HRLB lies between pure combinatorial lower bounds on the one side and pure cutting-plane-based lower bounds on the other side. This is due to the fact that HRLB realizes a kind of hybrid approach between the two in that it essentially solves linear programming relaxations based on a certain subset of cuts in a combinatorial way.

As a conclusion we see that the evolution of the lower bounding algorithms for the CARP has led to sophisticated and successful approaches. The latest advances could be achieved by considering a larger set of cuts. This is the crucial point but also the limiting factor of the combinatorial algorithms. Since there are exponentially many cuts, not all of them can be considered. As will be described in Chapter 9, LP-based methods will overcome this problem.

Table 8.5. *Results for instances* carpBCCM92 *[1]*.

	N	N+	B1	B2	B2m	M	M+	M+m	HRLB	CPA1	EAC	CPA2	UB
1A	247	247	247	247	247	247	247	247	247	247	247	247	247
1B	247	247	247	247	247	247	247	247	247	247	247	247	247
1C	279	279	279	280	280	280	280	280	293	309	309	309	319
2A	296	296	296	296	296	296	296	296	298	298	298	298	298
2B	305	305	305	318	318	318	318	318	328	328	328	330	330
2C	386	386	386	482	482	482	482	482	502	526	526	526	528
3A	103	103	103	103	103	103	103	103	103	105	105	105	105
3B	105	105	105	108	108	108	108	108	109	111	111	111	111
3C	123	123	123	140	140	140	140	140	153	159	159	161	162
4A	514	514	514	514	514	514	514	514	522	522	522	522	522
4B	518	518	518	518	518	518	518	518	534	534	534	534	534
4C	524	524	524	524	525	524	524	525	548	550	550	550	550
4D	558	558	558	565	565	565	565	565	623	640	642	644	652
5A	562	562	562	562	562	562	562	562	566	566	566	566	566
5B	566	566	566	567	567	567	567	567	584	586	586	589	589
5C	582	582	582	582	582	582	582	582	606	610	610	612	617
5D	656	656	656	656	656	656	656	656	712	714	714	714	724
6A	330	330	330	330	330	330	330	330	330	330	330	330	330
6B	334	334	334	334	334	334	334	334	336	336	336	338	340
6C	372	372	372	372	372	372	372	372	392	414	414	418	424
7A	382	382	382	382	382	382	382	382	382	382	382	382	382
7B	382	382	382	382	382	382	382	382	382	386	386	386	386
7C	402	402	402	403	403	403	403	403	412	430	430	436	437
8A	522	522	522	522	522	522	522	522	522	522	522	522	522
8B	528	528	528	528	528	528	528	528	531	531	531	531	531
8C	587	587	587	587	587	587	587	587	633	645	645	653	663
9A	450	450	450	450	450	450	450	450	450	450	450	450	450
9B	453	453	453	453	453	453	453	453	453	453	453	453	453
9C	459	459	459	459	459	459	459	459	459	459	459	459	459
9D	493	493	493	493	493	493	493	493	501	505	505	509	518
10A	637	637	637	637	637	637	637	637	637	637	637	637	637
10B	641	641	641	641	641	641	641	641	645	645	645	645	645
10C	649	649	649	649	649	649	649	649	653	655	655	655	655
10D	697	697	697	697	697	697	697	697	724	731	731	732	739

Table 8.6. *Results for instances* carpGDB83 *[1].*

	N	N+	B1	B2	B2m	M	M+	M+m	HRLB	CPA1	EAC	CPA2	UB
gdb1	310	310	310	310	310	310	310	310	**316**	**316**	**316**	**316**	**316**
gdb2	**339**	**339**	**339**	**339**	**339**	**339**	**339**	**339**	**339**	**339**	**339**	**339**	**339**
gdb3	**275**	**275**	**275**	**275**	**275**	**275**	**275**	**275**	**275**	**275**	**275**	**275**	**275**
gdb4	274	274	274	274	274	274	274	274	**287**	**287**	**287**	**287**	**287**
gdb5	376	376	376	376	376	376	376	376	**377**	**377**	**377**	**377**	**377**
gdb6	295	295	295	295	295	295	295	295	**298**	**298**	**298**	**298**	**298**
gdb7	312	312	312	312	312	312	312	312	**325**	**325**	**325**	**325**	**325**
gdb8	294	294	294	325	325	325	325	325	328	344	344	344	348
gdb9	277	277	277	277	277	277	277	277	297	**303**	**303**	**303**	**303**
gdb10	**275**	**275**	**275**	**275**	**275**	**275**	**275**	**275**	**275**	**275**	**275**	**275**	**275**
gdb11	**395**	**395**	**395**	**395**	**395**	**395**	**395**	**395**	**395**	**395**	**395**	**395**	**395**
gdb12	428	428	428	428	428	428	428	428	444	450	450	450	458
gdb13	536	536	536	536	536	536	536	536	536	536	536	536	538
gdb14	100	100	100	100	100	100	100	100	100	100	100	100	100
gdb15	**58**	**58**	**58**	**58**	**58**	**58**	**58**	**58**	**58**	**58**	**58**	**58**	**58**
gdb16	**127**	**127**	**127**	**127**	**127**	**127**	**127**	**127**	**127**	**127**	**127**	**127**	**127**
gdb17	**91**	**91**	**91**	**91**	**91**	**91**	**91**	**91**	87	**91**	**91**	**91**	**91**
gdb18	**164**	**164**	**164**	**164**	**164**	**164**	**164**	**164**	**164**	**164**	**164**	**164**	**164**
gdb19	**55**	**55**	**55**	**55**	**55**	**55**	**55**	**55**	**55**	**55**	**55**	**55**	**55**
gdb20	**121**	**121**	**121**	**121**	**121**	**121**	**121**	**121**	**121**	**121**	**121**	**121**	**121**
gdb21	**156**	**156**	**156**	**156**	**156**	**156**	**156**	**156**	**156**	**156**	**156**	**156**	**156**
gdb22	200	200	200	200	200	200	200	200	200	200	200	200	200
gdb23	**233**	**233**	**233**	**233**	**233**	**233**	**233**	**233**	**233**	**233**	**233**	**233**	**233**

Table 8.7. *Results for instances* carpKSHS95.

	N	N+	B1	B2	B2m	M	M+	M+m	HRLE	CPA1	EAC	CPA2	UB
k1	14039	14039	14039	14039	14039	14039	14039	14039	**14661**	**14661**	**14661**	**14661**	**14661**
k2	9275	9275	9275	9275	9275	9275	9275	9275	**9863**	**9863**	**9863**	**9863**	**9863**
k3	**9320**	**9320**	**9320**	**9320**	**9320**	**9320**	**9320**	**9320**	**9320**	**9320**	**9320**	**9320**	**9320**
k4	10774	10774	10774	10774	10774	10774	10774	10774	11098	11098	11098	11098	11498
k5	**10957**	**10957**	**10957**	**10957**	**10957**	**10957**	**10957**	**10957**	**10957**	**10957**	**10957**	**10957**	**10957**
k6	**10197**	**10197**	**10197**	**10197**	**10197**	**10197**	**10197**	**10197**	**10197**	**10197**	**10197**	**10197**	**10197**

Acknowledgment

We would like to thank Sidharth C. Pardeshi for implementing the hierarchical lower bound algorithm.

Bibliography

[1] D. AHR, *Multiple Postmen Problems: Fundamentals and New Algorithms*, VDM Verlag Dr. Müller, Saarbrücken, Germany, 2007.

[2] A. AMBERG AND S. VOSS, *A hierarchical relaxations lower bound for the capacitated arc routing problem*, in Proceedings of the 35th Annual Hawaii International Conference on System Sciences, Vol. 3, R.H. Sprague, ed., IEEE, Washington, DC, 2002, pp. 1–10.

[3] A.A. ASSAD AND B.L. GOLDEN, *Arc routing methods and applications*, in Network Routing, M.G. Ball, T.L. Magnanti, C.L. Monma, and G.L. Nemhauser, eds., Handbooks in Operations Research and Management Science 8, Elsevier, New York, 1995, pp. 375–483.

[4] A.A. ASSAD, W.L. PEARN, AND B.L. GOLDEN, *The capacitated Chinese postman problem: Lower bounds and solvable cases*, American Journal of Mathematics and Management Science, 7 (1987), pp. 63–88.

[5] E. BARTOLINI, J.-F. CORDEAU, AND G. LAPORTE, *Improved lower bound and exact algorithm for the Capacitated Arc Routing Problem*, Mathematical Programming Series A, 137 (2013), pp. 409–452.

[6] J.M. BELENGUER AND E. BENAVENT, *A cutting plane algorithm for the capacitated arc routing problem*, Computers & Operations Research, 30 (2003), pp. 705–728.

[7] E. BENAVENT, V. CAMPOS, A. CORBERÁN, AND E. MOTA, *The capacitated arc routing problem: Lower bounds*, Networks, 22 (1992), pp. 669–690.

[8] C. BODE AND S. IRNICH, *Cut-First Branch-and-Price-Second for the Capacitated Arc-Routing Problem*, Operations Research, 60 (2012), pp. 1167–1182.

[9] P. BRESLIN AND A. KEANE, *The Capacitated Arc Routing Problem: Lower Bounds*, Master's thesis, Department of Management Information Systems, University College Dublin, Dublin, Ireland, 1997.

[10] N. CHRISTOFIDES, *The optimum traversal of a graph*, Omega, 1 (1973), pp. 719–732.

[11] N. CHRISTOFIDES, A. MINGOZZI, AND P. TOTH, *State-space relaxation procedures for the computation of bounds to routing problems*, Networks, 11 (1981), pp. 145–164.

[12] ——, *Exact algorithms for the vehicle routing problem based on spanning tree and shortest path relaxations*, Mathematical Programming, 20 (1981), pp. 255–282.

[13] A. CORBERÁN AND C. PRINS, *Recent results on arc routing problems: An annotated bibliography*, Networks, 56 (2010), pp. 50–69.

[14] H.A. EISELT, M. GENDREAU, AND G. LAPORTE, *Arc routing problems II: The rural postman problem.*, Operations Research, 43 (1995), pp. 399–414.

[15] B.L. GOLDEN, J.S. DEARMON, AND E.K. BAKER, *Computational experiments with algorithms for a class of routing problems*, Computers & Operations Research, 10 (1983), pp. 47–59.

[16] B.L. GOLDEN AND R.T. WONG, *Capacitated arc routing problems*, Networks, 11 (1981), pp. 305–315.

[17] A. HERTZ, G. LAPORTE, AND M. MITTAZ, *A tabu search heuristic for the capacitated arc routing problem*, Operations Research, 48 (2000), pp. 129–135.

[18] A. HERTZ AND M. MITTAZ, *A variable neighborhood descent algorithm for the undirected capacitated arc routing problem*, Transportation Science, 35 (2001), pp. 425–434.

[19] M. KIUCHI, Y. SHINANO, R. HIRABAYASHI, AND Y. SARUWATARI, *An exact algorithm for the capacitated arc routing problem using parallel branch and bound method*, in Abstracts of the 1995 Spring National Conference of the Operational Research Society of Japan, 1995, pp. 28–29.

[20] A. N. LETCHFORD AND A. OUKIL, *Exploiting sparsity in pricing routines for the capacitated arc routing problem*, Computers & Operations Research, 36 (2009), pp. 2320–2327.

[21] L.Y.O. LI AND R.W. EGLESE, *An interactive algorithm for vehicle routing for winter-gritting*, The Journal of the Operational Research Society, 47 (1996), pp. 217–228.

[22] S. MARTELLO AND P. TOTH, *Knapsack problems: Algorithms and computer implementations*, Wiley Interscience Series in Discrete Mathematics and Optimization, Wiley, New York, 1990.

[23] ——, *Lower bounds and reduction procedures for the bin packing problem*, Discrete Applied Mathematics, 28 (1990), pp. 59–70.

[24] R. MARTINELLI, M. POGGI, AND A. SUBRAMANIAN, *Improved bounds for large scale capacitated arc routing problem*, Computers & Operations Research, 40 (2013), pp. 2145–2160.

[25] W.L. PEARN, *New lower bounds for the capacitated arc routing problem*, Networks, 18 (1988), pp. 181–191.

[26] Y. SARUWATARI, R. HIRABAYASHI, AND N. NISHIDA, *Node duplication lower bounds for the capacitated arc routing problem*, Journal of the Operations Research Society of Japan, 35 (1992), pp. 119–133.

[27] Z. WIN, *Contributions to Routing Problems*, Ph.D. thesis, University of Augsburg, Augsburg, Germany, 1987.

[28] S. WØHLK, *New lower bound for the capacitated arc routing problem*, Computers & Operations Research, 33 (2006), pp. 3458–3472.

[29] ——, *A decade of capacitated arc routing*, in The Vehicle Routing Problem: Latest Advances and New Challenges, B.L. Golden, S. Raghavan, and E. Wasil, eds., Operations Research/Computer Science Interfaces Series 43, Springer, New York, 2008.

Chapter 9
The Capacitated Arc Routing Problem: Exact Algorithms

José Manuel Belenguer
Enrique Benavent
Stefan Irnich

9.1 ▪ Introduction

The *Capacitated Arc Routing Problem* (CARP) was formally introduced by Golden and Wong [46] in 1981. Roughly speaking, the CARP consists of finding a set of minimum cost routes for vehicles with limited capacity that service some street segments with known demand, represented by a subset of edges of a graph. The literature contains numerous real-world applications of the CARP, for instance, discussed in the survey by Assad and Golden [7] and Part III of this book. Most applications are raised by operations on street networks, e.g., urban waste collection, street sweeping, snow plowing, etc. One of the most important applications comes from refuse collection activities, where an amount of waste to be collected at each street segment defines the demand, and edge costs correspond to distances or travel times. Although the CARP is the reference model for these applications, many additional characteristics and constraints occur in practice, giving rise to several CARP variants. In this chapter, we survey the methods that have been proposed to solve the CARP to proved optimality. We have classified these methods into three groups: branch-and-bound based on combinatorial lower bounds, cutting-plane and branch-and-cut, and column-generation and branch-and-price methods. Furthermore, some authors have solved the CARP by previously transforming it into a node routing problem. In the following sections, we will give a brief historical perspective of the development of each class of methods focussing on the most recent works. The reader can find more detailed descriptions of the older approaches in the original papers as well as in the excellent arc-routing surveys by Assad and Golden [7], Eiselt, Gendreau, and Laporte [40], Dror [38], Corberán and Prins [33], and Wøhlk [75].

The CARP is defined over an undirected graph $G = (V,E)$, where V is the set of nodes and E the set of edges. Each edge $e \in E$ has an associated nonnegative integer demand $q_e \geq 0$. Edges with positive demand form the subset $E_R \subseteq E$ of required edges. These have to be serviced exactly once, while all edges $e \in E$ can be traversed without providing service (this is called deadheading). The CARP requires the determination of a set of feasible vehicle tours that service all required edges exactly once at minimum cost. Herein, the tour costs consist of two components, that is, service costs c_e^{serv} for servicing required edges e and deadheading costs c_e for all edges e deadheaded. A tour is feasible if it starts and ends at the depot $d \in V$ and respects the vehicle capacity Q. Some authors define the CARP for an unlimited fleet of vehicles (Bartolini, Cordeau, and Laporte [12]), others fix the number of vehicles (Bode and Irnich [24]), or assume that a fleet K of $|K|$ (homogeneous) vehicles each with capacity Q is stationed at the depot d and not all the vehicles have to be used (Belenguer and Benavent [16]). By solving a standard bin-packing problem with items $e \in E_R$ of weight q_e and bins of capacity Q, one can determine the minimum fleet size.

The CARP generalizes several other combinatorial problems such as the bin packing problem (Martello and Toth [63]), the rural postman problem (Lenstra and Rinnooy Kan [55]) (when $Q \geq \sum_{e \in E} q_e$), and the capacitated Chinese postman problem (Christofides [31]) (when $E_R = E$), and is therefore also $\mathcal{NP}$-hard.

Throughout this chapter we use the following standard notation: Given a subset $S \subseteq V$, the cut set $\delta(S)$ is the set of edges with exactly one endpoint in S, and $E(S)$ is the set of edges with both endpoints in S. The subscript $_R$ indicates the restriction to subsets of required edges, i.e, $\delta_R(S) = \delta(S) \cap E_R$ and $E_R(S) = E(S) \cap E_R$. For singleton sets $S = \{i\}$ (with a node $i \in V$), the abbreviation $\delta(i)$ is used instead of $\delta(\{i\})$ (also $\delta_R(i)$). For any subset $F \subseteq E$ and any parameter or variable y, the term $y(F)$ is defined as $\sum_{e \in F} y_e$.

In principle, exact approaches for solving the CARP can be classified into two groups: (1) methods that determine a lower bound (LB) and that are complemented with (meta)heuristics to provide an upper bound (UB), hopefully closing the gap ($UB = LB$); (2) fully integrated exact methods that use an enumeration approach in order to close the gap and to produce a feasible integral solution. The first group (1) applies either combinatorial arguments based on bounds resulting from matching and similar models (see Chapter 8 in this book) or bounds resulting from a CARP relaxation solved using an MIP model (later on we will see that aggregate compact formulations are models of that kind). Approaches from the second group (2) comprise variants of branch-and-bound. The literature contains approaches that either use branch-and-bound in combination with combinatorial lower bounds, with cutting planes (branch-and-cut), or with models derived from a Dantzig–Wolfe decomposition (branch-and-price).

We describe in this chapter only those methods that were computationally tested. Six sets of benchmark instances have been used in the literature for this purpose. They are referred to as *gdb* (Golden, DeArmon, and Baker [45]), *kshs* (Kiuchi et al. [54]), *val* (Benavent et al. [21]), *egl* (Li and Eglese [58]), *bmcv* (Beullens et al. [23]), and *egl-g* (Brandão and Eglese [29]). In the first three, *gdb*, *kshs*, and *val*, all edges are required. The set *gdb* contains 23 instances with between 7 and 27 nodes and 11 to 55 edges, the set *kshs* has 6 instances with 6 to 10 nodes and 15 edges, and finally, set *val* has 34 instances with between 24 and 50 nodes and 34 to 97 edges. The benchmark set *egl*, is based on two graphs representing two road networks in Lancashire, England [58], with 77 nodes and 98 edges, and 140 nodes and 190 edges, respectively. The set *egl* contains 24 instances that were generated by changing the set of required edges and the capacities of the vehicles. The data set *bmcv* has 100 instances (partitioned into four classes) ranging from 26 to 97 nodes, 35 to 142 edges, and 28 to 121 required edges. The graphs were obtained by partitioning the intercity road network of Flanders, Belgium, into districts. Finally, the set *egl-g* contains 10 large-scale instances with 255 nodes, 375 edges, and 347 or 375 required edges. They were also created using the road network of Lancashire. The data of some of these sets of instances can be found at `http://www.uv.es/belengue/carp.html`.

We have decided to not include comprehensive tables showing computational results in the form of bounds, gaps, computation times, and other performance indicators for all instances (or benchmark sets) and the respective methods. The reason is first and foremost space limitations, but also the fact that due to intensive research in the field such tables shortly become outdated. Instead, we will maintain the website `http://logistik.bwl.uni-mainz.de/benchmarks.php`, and (try to) keep the information about new results up-to-date there.

The remainder of this chapter is structured as follows: Section 9.2 presents transformations of the CARP to the classical *Capacitated Vehicle Routing Problem*(CVRP) (see Toth and Vigo [72, 73]) and some VRP variants. These transformations allow the determination of tight lower bounds as well as the application of fully integrated exact approaches. Section 9.3 outlines some early branch-and-bound approaches that are based on combinatorial lower bounds. Section 9.4 then summarizes (mixed) integer programming formulations. To the compact formulations, cutting-plane methods can be applied as presented in Section 9.5. For those models that are valid formulations for the CARP (not just relaxations), embedding the corresponding linear relaxation into branch-and-bound also gives rise to fully integrated approaches. Column-generation approaches are presented in Section 9.6. Variants and extensions of the classical CARP are sketched in Section 9.7, which is followed by final conclusions and an outlook in Section 9.8.

9.2 ▪ Transformation into node routing problems

Three different transformations of the CARP into a node routing problem are presented now. In 1987, Pearn, Assad, and Golden [68] proposed the first transformation into a CVRP by replacing each required edge by three nodes. This transformation was improved in 2006 by Baldacci and Maniezzo [9] and Longo, Poggi de Aragão, and Uchoa [59]. In 2013, a different transformation into the *Generalized Routing Problem* (GVRP) was proposed by Bartolini, Cordeau, and Laporte [12]. In the GVRP, customers are grouped in clusters, and exactly one of the costumers of each cluster has to be visited by the set of vehicles. Some approaches (as also techniques explained in following subsections) are easier to present for simple graphs $G = (V, E)$ because we can then uniquely refer to an edge $e \in E$, and likewise to its endpoints, by writing $e = \{i, j\}$. The associated demand is $q_{ij} = q_e$. It should be noted that if we assume G to be simple, we do it only for the sake of convenience, since all approaches also work for graphs with parallel edges and loops.

Moreover, all transformations heavily rely on the property that optimal CARP tours perform deadheading along shortest paths. More precisely, if edge $e = \{i, j\} \in E_R$ is served in direction from i to j and edge $e' = \{k, l\} \in E_R$ is served next in direction from k to l, then deadheading from node j to node k uses a shortest j-k-path in G with respect to deadheading costs (c_e). We denote the unique shortest-path distance by ℓ_{jk}.

9.2.1 ▪ A transformation into a CVRP with $3|E_R| + 1$ nodes

The transformation, proposed by Pearn, Assad, and Golden [68], replaces each edge $\{i, j\} \in E_R$ by three new nodes s_{ij}, m_{ij}, and s_{ji}. A corresponding CVRP instance is defined on the complete undirected graph $H^1 = (N^1, A^1)$ with node set

$$N^1 = \bigcup_{\{i,j\} \in E_R} \{s_{ij}, s_{ji}, m_{ij}\} \cup \{0\},$$

where 0 denotes the depot. The edge costs $\hat{c}$ are defined as follows:

$$\hat{c}(s_{ij}, s_{kl}) = \tfrac{1}{4}(c_{ij}^{serv} + c_{kl}^{serv}) + \ell_{ik},$$

$$\hat{c}(0, s_{ij}) = \tfrac{1}{4}c_{ij}^{serv} + \ell_{di},$$

$$\hat{c}(m_{ij}, v) = \begin{cases} \tfrac{1}{4}c_{ij}^{serv} & \text{if } v = s_{ij} \text{ or } v = s_{ji}, \\ \infty & \text{otherwise.} \end{cases}$$

The costs of the edges incident with nodes m_{ij} have the purpose of forcing a CVRP route to pass through the three nodes associated with edge $\{i,j\}$ either in the sequence (s_{ij}, m_{ij}, s_{ji}) or (s_{ji}, m_{ij}, s_{ij}). The demands $q(v)$ for $v \in N^1$ are

$$q(s_{ij}) = q(m_{ij}) = q(s_{ji}) = \frac{1}{3}q_{ij}.$$

Figure 9.1(a) shows a graph $G = (V, E)$ that we will use to illustrate all transformations. For simplicity, the numbers beside the edges indicate both costs and demands, i.e., $c_e^{serv} = c_e = q_e$ for all $e \in E$. The capacity of the vehicle is $Q = 35$, and node 1 is the depot d. In Figures 9.1(b) and (c) a feasible solution with cost 80 is shown. Solid lines represent serviced edges, and dotted lines represent deadheaded edges.

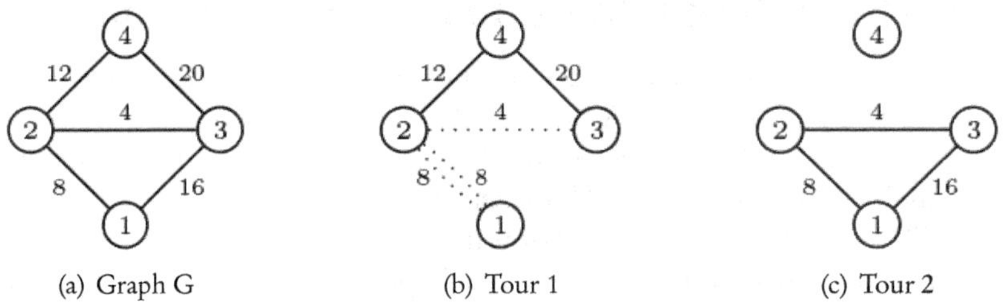

Figure 9.1. *Graph G and a feasible solution with two tours. Tour 1 is realized by route* $(1-2-3=4=2-1)$ *and tour 2 by route* $(1=2=3=1)$, *where "=" indicates a service and "–" a deadheading.*

Figure 9.2 shows the two tours of our example in the transformed graph $H^1 = (N^1, A^1)$, which has $3|E_R| + 1 = 16$ nodes and 120 edges.

9.2.2 ▪ Transformations into a CVRP with $2|E_R| + 1$ nodes

Baldacci and Maniezzo [9] and Longo, Poggi de Aragão, and Uchoa [59] proposed, independently, a transformation into a CVRP instance with an associated graph with only $2|E_R| + 1$ nodes. In both transformations, the graph is the same but the costs of the edges are different. An edge $\{i,j\} \in E_R$ is now represented by only two nodes s_{ij} and s_{ji}. The resulting CVRP instance is defined on the complete undirected graph $H^2 = (N^2, A^2)$ with node set

$$N^2 = \bigcup_{\{i,j\} \in E_R} \{s_{ij}, s_{ji}\} \cup \{0\}.$$

Again, node 0 is the depot. The demands $q(v)$ for $v \in N^2$ are

$$q(s_{ij}) = q(s_{ji}) = \frac{1}{2}q_{ij}.$$

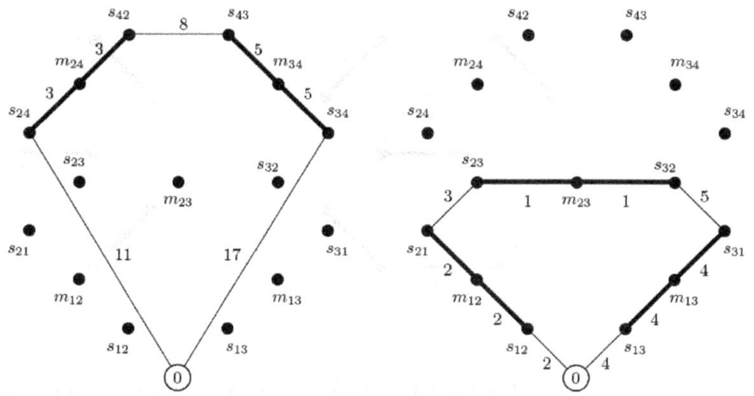

Figure 9.2. *The two routes in the transformed graph $H^1 = (N^1, A^1)$. The numbers beside the edges represent costs.*

In the transformation by Longo, Poggi de Aragão, and Uchoa [59], the edge costs $\hat{c}^L$ are

$$\hat{c}^L(s_{ij}, s_{kl}) = \begin{cases} c_{ij}^{serv} & \text{if } s_{kl} = s_{ji} \text{ (i.e., } i = l \text{ and } j = k), \\ \ell_{ik} & \text{if } s_{kl} \neq s_{ji}, \end{cases}$$

$$\hat{c}^L(0, s_{ij}) = \ell_{di},$$

whereas Baldacci and Maniezzo [9] define the edge costs $\hat{c}^B$ as follows:

$$\hat{c}^B(s_{ij}, s_{kl}) = \begin{cases} -M & \text{if } s_{kl} = s_{ji} \text{ (i.e., } i = l \text{ and } j = k), \\ \frac{1}{2}(c_{ij}^{serv} + c_{kl}^{serv}) + \ell_{ik} & \text{if } s_{kl} \neq s_{ji}, \end{cases}$$

$$\hat{c}^B(0, s_{ij}) = \frac{1}{2} c_{ij}^{serv} + \ell_{di},$$

where M is any upper bound of the CARP instance.

The idea of Longo, Poggi de Aragão, and Uchoa [59] to avoid the middle nodes m_{ij} is simply to fix the variables associated with the $|E_R|$ edges $\{s_{ij}, s_{ji}\}$ to one in the transformed CVRP instance. In contrast, Baldacci and Maniezzo [9] do not fix any edges and use the costs to produce the same result.

Let us compare the costs of the two routes of our example with the costs in both transformations. In the graph H^2, tour 1 is equivalent to the route $(0, s_{24}, s_{42}, s_{43}, s_{34}, 0)$ and tour 2 to $(0, s_{12}, s_{21}, s_{23}, s_{32}, s_{31}, s_{13}, 0)$. Using costs $\hat{c}^L$, the routes have costs 52 and 28, respectively. However, with respect to $\hat{c}^B$, the routes have costs $52 - 2M$ and $28 - 3M$, respectively. In general, feasible CARP and CVRP solutions have identical costs using the transformation of Longo, Poggi de Aragão, and Uchoa [59] (with costs $\hat{c}^L$), while the cost of the same transformed CVRP solution of Baldacci and Maniezzo [9] is by $M|E_R|$ smaller (using costs $\hat{c}^B$).

Figure 9.3 shows the transformed graph $H^2 = (N^2, A^2)$, which has $2|E_R| + 1 = 11$ nodes and 55 edges. Several edges of H^2 are omitted to allow a better visualization.

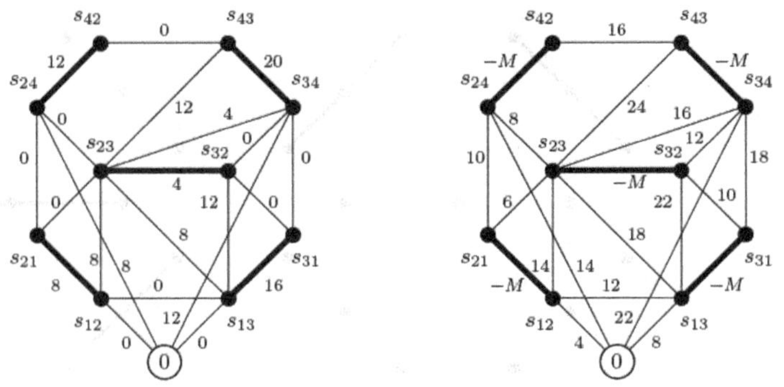

Figure 9.3. *Transformed graphs with both transformations.*

9.2.3 ▪ A transformation into an asymmetric generalized VRP (GVRP) with $2|E_R|+1$ nodes

Recently, Bartolini, Cordeau, and Laporte [12] proposed a transformation of the CARP into an asymmetric *Generalized VRP* (GVRP) on a directed graph $\overline{G}=(\overline{V},\overline{A})$, obtained from G as follows:

- Each edge $\{i,j\}\in E_R$ is again associated with two nodes, s_{ij} and s_{ji}. The node set is $\overline{V}=\bigcup_{\{i,j\}\in E_R}\{s_{ij},s_{ji}\}\cup\{0\}$, which is partitioned into $|E_R|+1$ clusters $\overline{V}_0,\overline{V}_{e_1},\dots,\overline{V}_{e_m}$, where $\overline{V}_0$ contains the depot 0 and $\overline{V}_e$, for $e=\{i,j\}\in E_R$, contains the nodes s_{ij} and s_{ji}. The arc set $\overline{A}$ contains (antiparallel) arcs (u,v) and (v,u) for every pair of nodes $u,v\in\overline{V}$ that are taken from two different clusters.

- The demand for each node $v=s_{ij}\in\overline{V}\setminus\{0\}$ is $\overline{q}_v=q_{ij}$. In addition, $\overline{q}_0=0$.

- For each node pair $u=s_{ij},v=s_{kl}\in\overline{V}$, the costs of arcs $\{u,v\}$ and $\{v,u\}$ are $\overline{c}_{uv}=\ell_{jk}+\frac{1}{2}(c^{serv}_{ij}+c^{serv}_{kl})$ and $\overline{c}_{uv}=\ell_{li}+\frac{1}{2}(c^{serv}_{ij}+c^{serv}_{kl})$, respectively. Moreover, for each $v=s_{ij}\in\overline{V}\setminus\{0\}$, the costs of the two arcs $(0,v)$ and $(v,0)$ are $\overline{c}_{0v}=\ell_{di}+\frac{1}{2}c^{serv}_{ij}$ and $\overline{c}_{0v}=\ell_{dj}+\frac{1}{2}c^{serv}_{ij}$, respectively.

Figure 9.4 shows two alternative solutions of the GVRP instance equivalent to our two tours of the CARP and with the same cost. Recall that visiting two nodes s_{ij} and s_{kl} consecutively is equivalent, in the original graph G, to service edge $\{i,j\}$ from node i to node j, to deadhead along a shortest path from node j to node k and, finally, to service the edge $\{k,l\}$ from k to l.

In Figure 9.4(*a*), tour 1 is realized by the route $(1-2-3=4=2-1)$ and tour 2 by the route $(1=2=3=1)$ in the original graph G. In Figure 9.4(*b*), the solution results from the reversal of the routes in the original graph. Note, however, that the reverse routes $(1-2=4=3-2-1)$ and $(1=3=2=1)$ do *not* correspond to reversed routes in the GVRP, since reverse routes incur in different costs. For example, while $(0,s_{12},s_{23},s_{31},0)$ with costs $4+6+10+8=28$ corresponds to $(1=2=3=1)$, the reversal $(0,s_{31},s_{23},s_{12},0)$ corresponds to $(1-2-3=1-2=3-2-1=2-1)$ with a higher cost of $8+4+16+8+4+4+8+8+8=68$.

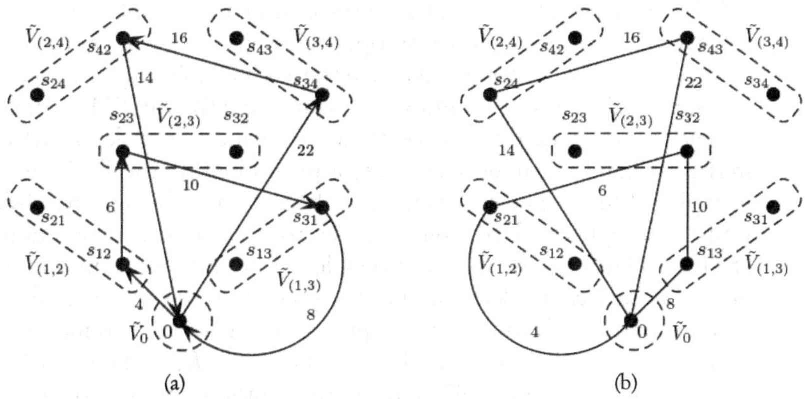

Figure 9.4. *Two feasible GVRP solutions equivalent to the two CARP routes.*

9.3 ▪ Branch-and-bound based on combinatorial lower bounds

Three branch-and-bound algorithms for the CARP have been described in the literature based on the *Node Duplication Lower Bound* (NDLB) of Saruwatari, Hirabayashi, and Nishida [70]. The NDLB is a lower bound for the CARP and results from computing a *Minimum Cost Perfect Matching* (MCPM) on an extended graph $G_1 = (V_1, E_1)$ that is constructed as follows: For each required edge $e \in E_R$, two nodes i_e and j_e are created and a demand edge $\{i_e, j_e\}$ with infinite cost is added. As a result, each original node i is substituted by $|\delta_R(i)|$ nodes in V_1. We denote this set of nodes by $fam(i)$. Furthermore, $2K_{min}$ copies of the depot, denoted by $fam(d)$, are included, where $K_{min} = \lceil \sum_{e \in E_R} q_e / Q \rceil$. Then, non-demand edges among each pair of nodes in V_1 not yet linked are added. The cost of the non-demand edges joining nodes included in the same set $fam(i)$, for $i \in V \setminus \{d\}$ are set to zero, while the cost of an edge $\{p, q\}$ among nodes $p \in fam(i)$ and $q \in fam(j)$, $i \neq j$ are equal to the shortest-path distance ℓ_{ij} between i and j in graph G. Finally, the cost of edges among copies of the depot is set to infinity.

A MCPM in G_1 consists of a set of non-demand edges, and it is easy to see that these edges, together with the required edges, form a subgraph of G_1 where each node is incident with one demand (required) edge and one non-demand edge, except for the copies of the depot that are incident with only one non-demand edge. This graph can be decomposed into a set of alternating paths or cycles of demand and non-demand edges. The alternating paths start and end in two different copies of the depot and correspond obviously to tours starting and ending at the depot. If it happens that there is no alternating cycle and all the alternating paths satisfy the capacity constraint, then the set of tours is an optimal CARP solution. Note also that, if a given non-demand edge appears in the solution, it implies that the two demand edges incident with its endpoints will be serviced by the same vehicle. Therefore, if the sum of the demands of these edges exceeds Q, the non-demand edge cannot be used. This is exploited by Saruwatari, Hirabayashi, and Nishida [70], who declare these kind of edges as prohibited and set their cost to infinity before computing the MCPM.

The first branch-and-bound algorithm for the CARP that uses the NDLB was presented by Hirabayashi, Saruwatari, and Nishida [49], and it is called the *Tour Construction Algorithm* (TCA). In this algorithm, each branch-and-bound node corresponds to a partial solution where some non-demand edges are fixed and some are prohibited. The branching decisions modify the costs of edges of graph G_1 so that the computation of the

MCPM provides a lower bound on the cost of any completion of the partial solution. If the MCPM does not produce an optimal solution in a node, the branching rule selects, from all the free non-demand edges in the MCPM solution, the edge whose exclusion provides the highest lower bound when computing the MCPM. If the chosen edge is $\{i,j\}$, two nodes are created: In the so-called right-hand-side node, the edge $\{i,j\}$ is fixed, while in the left-hand-side node, edge $\{i,j\}$ is prohibited. In the TCA, the right-hand-side node is explored first. The accumulation of fixed edges in right-hand-side nodes eventually leads to completely fixed alternating paths in G_1 that correspond to feasible tours. Every time a feasible tour is fixed, a new lower bound on the minimum number of vehicles needed to service the demand of the edges that do not belong to fixed tours is computed, and, if necessary, the number of copies of the depot in G_1 is increased. This is important because the initial lower bound on the fleet size, K_{min}, may not be tight. Hirabayashi, Saruwatari, and Nishida [49] report computational results for the TCA on graphs with randomly generated arc costs and demands where $|E_R|$ ranges from 15 to 50. Unfortunately, these instances are not available any more, and no further details are given by the authors on how these instances were generated. Beullens et al. [23] also implemented this branch-and-bound algorithm and could solve two previously unsolved instances, the *gdb12* with $|E_R| = 23$ and *kshs4* with $|E_R| = 15$. They also mentioned that this method is not applicable to larger instances.

Kiuchi et al. [54] implemented a parallel tour construction algorithm that is a parallelization of the TCA to be used in a cluster of workstations. A workstation acts as master and controls the branching, and each node of the branch-and-bound is solved on a slave workstation. They report computational results for the 6 *kshs* instances. This set of instances has been used subsequently by several authors, but the reader should be aware of the fact that for the instances *kshs2* and *kshs4* incorrect results were reported in [54].

Finally, Saruwatari, Hirabayashi, and Nishida [71] introduced a branch-and-bound algorithm for the CARP called the *Subtour Elimination Algorithm* (SEA) that is also based on the computation of the NDLB in each node. In this case, each non-fathomed node generates multiple descendant nodes. Let $(e_1, e_2, \ldots, e_{2p+1})$ be an alternating path that is included in the MCPM solution of a branch-and-bound node. If the demand of the path exceeds the vehicle capacity or if it is an alternating cycle disconnected from the depot (edges with odd indices in this path are non-demand edges), then $p+1$ descendant nodes are created: The rth node prohibits the edge e_{2r-1}, but fixes the preceding non-demand edges $\{e_1, e_3, \ldots, e_{2r-3}\}$ to one, for $r = 1, \ldots, p+1$. Saruwatari, Hirabayashi, and Nishida [71] report computational results for randomly generated instances with $|E_R|$ equal to 10 or 15. On these instances, the results are similar to the TCA of Hirabayashi, Saruwatari, and Nishida [49] in terms of CPU time, although the number of nodes generated tends to be higher in the SEA.

9.4 ▪ Integer programming formulations

Several IP and MIP formulations of the CARP exist in the literature (see for instance Eglese and Letchford [39]). We will only describe here those models that have been used to solve CARP instances to proved optimality.

The first formulation of the CARP was proposed by Golden and Wong [46] in 1981. This formulation employs an exponential number of variables and constraints, and, to the best of our knowledge, it has not been used to solve the CARP. Belenguer [14] proposed in 1990 the two-index formulation presented in Subsection 9.4.1 (see also Belenguer and Benavent [16]). This formulation is the basis of the branch-and-cut algorithm of Belenguer and Benavent [16] that is described in Subsection 9.5.2. In 1994, Welz [74] presented another two-index formulation and a branch-and-bound algorithm that are also briefly described in the above mentioned subsections. In the years 1997 and 1998, the one-index IP formulation described in Subsection 9.4.2 was independently proposed by Letchford [56] and Belenguer and Benavent [16]. This formulation is the basis of a cutting-plane algorithm devised by Belenguer and Benavent [17] in 2003 that was further developed in the works of Ahr [2] in 2004 and Martinelli, Poggi de Aragão, and Subramanian [65] in 2013.

We will start describing two-index formulations, followed by the one-index formulation and the set-covering formulation.

9.4.1 ▪ Two-index formulation

Belenguer and Benavent [16] use the decision variables $x_e^k \in \{0,1\}$ to indicate whether vehicle k for $k \in K$ serves edge $e \in E_R$, and decision variables y_e^k to count the number of times vehicle k deadheads through edge $e \in E$. With these variables the two-index formulation of the CARP is

$$\text{(9.1)} \qquad \min \sum_{k\in K}\sum_{e\in E_R} c_e^{serv} x_e^k + \sum_{k\in K}\sum_{e\in E} c_e y_e^k$$

$$\text{(9.2)} \qquad \text{s.t.} \sum_{k\in K} x_e^k = 1 \quad \forall\, e \in E_R,$$

$$\text{(9.3)} \qquad x^k(\delta_R(S)) + y^k(\delta(S)) \geq 2x_f^k \quad \forall\, S \subseteq V\setminus\{d\}, f \in E_R(S), k \in K,$$

$$\text{(9.4)} \qquad x^k(\delta_R(i)) + y^k(\delta(i)) = 2p_i^k \quad \forall\, i \in V, k \in K,$$

$$\text{(9.5)} \qquad q^\top x^k \leq Q \quad \forall\, k \in K,$$

$$\text{(9.6)} \qquad p^k \in \mathbb{Z}_+^{|V|}, x^k \in \{0,1\}^{|E_R|}, y^k \in \mathbb{Z}_+^{|E|} \quad \forall\, k \in K.$$

Constraints (9.2) and (9.5) ensure that each required edge will be serviced and the capacity of the vehicles is not exceeded. Constraints (9.3) (called connectivity constraints) state that, if route k services edge e, then it must traverse any edge cutset separating f from the depot. Finally, constraints (9.4) assure that each vehicle uses an even number of edges incident with each node. As a result, the graph induced by the edges traversed by the vehicles is Eulerian.

Belenguer and Benavent [16] introduce some families of valid inequalities to tighten the linear relaxation of this formulation. For instance, parity constraints (9.4) are not useful when we relax the integrality constraints and they are substituted by

$$\text{(9.7)} \qquad x^k(\delta_R(S)\setminus H) + y^k(\delta(S)) > x^k(H) - |H| + 1,$$

where $S \subseteq V\setminus\{d\}$, and $H \subseteq \delta_R(S)$ with $|H|$ odd. To see that they are valid, note that if all edges in H are serviced by vehicle k (that is $x^k(H) = |H|$), given that $|H|$ is odd, this vehicle will traverse at least once more the edge cutset $\delta(S)$, so $x^k(\delta_R(S)\setminus H) + y^k(\delta(S)) \geq 1$. These constraints are also known as cocircuit constraints in the literature (Barahona and Grötschel [11]).

For any $S \subseteq V\setminus\{d\}$, we denote by $D(S)$ the demand on all edges with one or two endpoints in S, given by $q(E_R(S)\cup\delta_R(S))$. Let $Q_{min} = \max\{q(E_R) - (|K|-1)Q, 0\}$. Q_{min} is the minimum demand that any vehicle must service in any feasible solution to the CARP. Then, the following inequalities are valid:

$$\text{(9.8)} \qquad x^k(\delta_R(S)) + y^k(\delta(S)) \geq 2 \quad \forall k \in K \text{ and } S \subseteq V\setminus\{d\}, q(E_R) - D(S) < Q_{min}.$$

Obviously, each vehicle capacity constraint (9.5) is a knapsack-like constraint. Then, any valid inequality for the binary knapsack polytope conv$\{z \in \{0,1\}^{|E_R|} : \sum_{e\in E_R} q_e z_e \leq Q\}$, generates $|K|$ valid inequalities for the CARP, one for each vehicle $k \in K$.

Other valid inequalities were proposed in Belenguer and Benavent [16] for this formulation, among them disaggregate versions of the capacity and odd-cut inequalities that are introduced in Subsection 9.4.2.

Welz [74] also proposed a two-index formulation for the CARP but based on directed variables. Each edge $\{i,j\}$ is substituted by two arcs (i,j) and (j,i). He uses binary variables x^k_{ij} and z^k_{ij}, for which the value 1 indicates that vehicle k services or traverses arc (i,j), respectively. It is easy to show that it is never necessary for any vehicle to traverse an edge more than once in a given direction. Apart from this difference, his formulation is quite similar to that of Belenguer and Benavent [16]: Instead of parity constraints he uses symmetry constraints to assure that each vehicle departs from each node as many times as it enters; Welz [74] includes a constraint stating that each required edge is serviced in one of the two possible directions and capacity constraints are similar to (9.5). He uses as connectivity constraints inequalities that are much weaker than (9.3). Finally, he proposes the following odd-cut inequalities to tighten his formulation:

$$\sum_{k\in K} z^k(\delta^+(S)) \geq \lceil |\delta_R(S)|/2\rceil \quad \forall S \subseteq V \setminus \{d\}, |\delta_R(S)| \text{ odd}, \tag{9.9}$$

where $\delta^+(S)$ denotes the set of arcs (i,j) with $i \in S$ and $j \notin S$. These constraints are derived from the fact that, if a cutset $\delta(S)$ has an odd number of required edges, then it will be deadheaded at least once. They are called odd-cut constraints and have also been used in the one-index formulation described in next subsection, although, in their nondirected version. More details on Welz's formulation can be found in Eglese and Letchford [39].

9.4.2 ▪ One-index formulation

The one-index formulation, first considered independently by Letchford [56] and Belenguer and Benavent [16], solely uses aggregate deadheading variables

$$y_e = \sum_{k\in K} y^k_e \in \mathbb{Z}_+, \tag{9.10}$$

one for each edge $e \in E$.

For any $S \subseteq V \setminus \{d\}$, let $K(S) = \lceil q(E_R(S) \cup \delta_R(S))/Q\rceil$. Obviously, $K(S)$ is a lower bound on the minimum number of vehicles needed to service the edges $E_R(S) \cup \delta_R(S)$.

The one-index formulation is

$$\min \quad c^\top y \tag{9.11}$$

$$\text{s.t.} \quad y(\delta(S)) \geq 2K(S) - |\delta_R(S)| \quad \forall\, \emptyset \neq S \subseteq V \setminus \{d\}, \tag{9.12}$$

$$y(\delta(S)) \geq 1 \quad \forall\, \emptyset \neq S \subseteq V, |\delta_R(S)| \text{ odd}, \tag{9.13}$$

$$y \in \mathbb{Z}^{|E|}_+. \tag{9.14}$$

The objective function (9.11) includes only the cost of the deadheading. This is correct because the cost of service $\sum_{e\in E_R} c^{serv}_e$ is a constant and, therefore, not relevant when minimizing the overall costs. Constraints (9.12) are called aggregate capacity constraints, and they express that at least $K(S)$ vehicles must traverse the edge cutset $\delta(S)$ to service the edges in $E_R(S) \cup \delta_R(S)$. Constraints (9.13) are the already mentioned odd-cut constraints, which are the nondirected version of constraints (9.9).

The number of vehicles is free in the one-index formulation. The advantage of the formulation is that it is compact and herewith allows fast LP-(re)optimization, since the number of variables is only $|E|$, which is much smaller than that of the two-index formulations with at least $K(|E|+|E_R|)$ variables. The main drawback of the one-index formulation is, however, that it constitutes only a CARP relaxation as it may contain integer solutions that are infeasible for the CARP. Furthermore, the integer solutions provided by this formulation only show the aggregation of all the routes. It is a difficult problem (in fact it is strongly $\mathcal{NP}$ -hard; see Eglese and Letchford [39, p. 219]) just to determine if a given aggregate solution can be separated in feasible individual tours. Despite this fact, Belenguer and Benavent [16, 17] used this formulation to compute tight lower bounds for the CARP with a cutting-plane algorithm. The formulation was tightened in [17] by introducing the so-called disjoint-path inequalities that are based on the following idea. Given a subset of nodes $S \subseteq V \setminus \{d\}$, at least $K(S)$ vehicles are needed to service the demand of the edges $E_R(S) \cup \delta_R(S)$. In addition to this demand, the current deadheading may force those vehicles to service some additional demand on the path between the depot and those edges and the path back to the depot. This demand has not been taken into account when computing $K(S)$. If the capacity $QK(S)$ of the $K(S)$ vehicles is exceeded, then either more vehicles are needed or additional deadheading must occur on the paths between the depot and the edges.

Disjoint-path inequalities are defined by node sets $S_0 \subset S_1 \subset \cdots \subset S_t \subseteq V \setminus \{d\}$ and a subset of edges $E' \subseteq E$. In general, they can be written as

$$2y(E')+\sum_{p=0}^{t} y(\delta(S_p)) \geq \sum_{p=0}^{t} \alpha(S_p)+2,$$

where $\alpha(S)$ is defined as follows. If $|\delta_R(S)|$ is even, $\alpha(S)$ is the maximum of $2K(S)-|\delta_R(S)|$ and 0, while if $|\delta_R(S)|$ is odd, $\alpha(S)$ is the maximum of $2K(S)-|\delta_R(S)|$ and 1.

A disjoint-path inequality is valid if it can be proved that the total demand that has to be serviced by $K(S_0)$ vehicles in the paths from the depot to S_0, when $\alpha(S_p)$ deadheadings are allowed for the edges of $\delta(S_p)$, for $p=0,\ldots,t$, and no deadhead is allowed in the edges of E', exceeds the remaining capacity of those vehicles, after serving $E_R(S_0) \cup \delta_R(S_0)$. A lower bound on this total demand can be computed by solving a minimum cost flow problem, or just by adding the least-demand edges of the edge cutsets $\delta_R(S_p)$ for $p = 1,\ldots,t$ (see Belenguer and Benavent [17] for the details).

Summarizing, all the inequalities used by Belenguer and Benavent [17] in the one-index formulation can be written as

$$\sum_{e \in E} d_{es} y_e \geq r_s, \quad s \in \mathcal{S}, \tag{9.15}$$

where s is the index referring to a particular inequality, $d_{es} \in \mathbb{Z}$ the coefficient of edge $e \in E$, r_s the RHS in the inequality, and $\mathcal{S}$ the set of all valid inequalities.

9.4.3 ▪ Set-partitioning and set-covering formulations

Many variants of vehicle routing and arc routing problems can be formalized as (extended) *Set-Partitioning* (SP) or *Set-Covering* (SC) problems, where each task (i.e., a node, edge, or arc to be serviced) induces one row in the SC/SP model that has to be covered once. Desaulniers et al. [35] provide with their unified framework a systematic scheme that can be used to properly derive such an (extended) SP or SC formulation from a compact formulation via Dantzig–Wolfe decomposition.

Even without this theoretic background, the derivation of the SP model is straightforward: Let $r \in \Omega$ the set of all feasible CARP tours. For such a tour $r \in \Omega$ and an edge $e \in E$, the numbers $\bar{x}_{er} \in \{0,1\}$ and $\bar{y}_{er} \in \mathbb{Z}_+$ indicate how many times tour r services and deadheads through edge e, respectively. The decision variables $\lambda_r \in \{0,1\}$ indicate whether or not the tour $r \in \Omega$ is in the solution. The SP model now reads as follows:

$$\min \sum_{r\in\Omega} c_r \lambda_r \tag{9.16}$$

$$\text{s.t.} \sum_{r\in\Omega} \bar{x}_{er} \lambda_r = 1 \quad \forall\, e \in E_R, \tag{9.17}$$

$$\lambda_r \in \{0,1\} \quad \forall\, r \in \Omega. \tag{9.18}$$

The objective (9.16) minimizes the sum of all tour costs, and (9.17) are the partitioning constraints. Whenever the cost for deadheading an edge does not exceed the cost for serving ($c_e \le c_e^{serv}$ for all $e \in E_R$), an equivalent model results from replacing the partitioning constraints (9.17) by covering constraints (≥ 1). Notice that the same type of SP or SC model results for other arc routing and node routing problems whenever $\bar{x}_{er}$ indicates whether or not a task e is covered by route r.

Restricting the fleet size can be accomplished by extending (9.16)–(9.18) with the constraint

$$\sum_{r\in\Omega} \lambda_r = |K|. \tag{9.19}$$

Gómez-Cabrero, Belenguer, and Benavent [47] were the first who suggested complementing the SC formulation with valid inequalities of the one-index formulation. Odd-cut constraints, capacity cuts, and disjoint-path inequalities all solely refer to the deadheading variables y_e (of a compact formulation). Recall from (9.15) that all such valid inequalities are indexed by $s \in \mathcal{S}$. Moreover, with the definition $d_{sr} = \sum_{e\in E} \bar{y}_{er} d_{es}$ they can be reformulated in the tour variables λ_r:

$$\sum_{r\in\Omega} d_{sr} \lambda_r \ge r_s \quad \forall\, s \in \mathcal{S}. \tag{9.20}$$

An interesting observation reported in Letchford and Oukil [57] is that the linear relaxation bound of the pure SP formulation (9.16)–(9.19) (without inequalities (9.20)) is most of the time inferior to the lower bound provided by the linear relaxation of the compact formulations (9.1)–(9.6) and (9.11)–(9.15). This contrasts with the situation in many node routing problems, where SP formulations provide the better lower bounds.

As the number $|\Omega|$ of feasible CARP tours $r \in \Omega$ is typically huge, a solution procedure based on the direct solution of the SP or SC model is practically impossible. However, with combined cut-and-column-generation techniques (see Section 9.6) it is possible to dynamically generate tours and inequalities as needed, and herewith to solve the LP-relaxation of (9.16)–(9.20).

9.5 ▪ Cutting-plane methods and branch-and-cut

In this section we summarize the exact methods based on the two-index and one-index formulations, as well as the branch-and-cut method published in 2006 by Baldacci and Maniezzo [9] that solves the CARP after transforming it into a CVRP.

9.5.1 ▪ Branch-and-cut for CVRP

Baldacci and Maniezzo [9] proposed an exact method to solve the CARP that consists of transforming it into a CVRP (see Section 9.2.2) and then solving the resulting CVRP with a slight adaptation of the exact method implemented by Lysgaard, Letchford, and Eglese [62]. Recall that this transformation produces a graph with $2|E_R|+1$ nodes, where each required edge $\{i,j\}$ generates two nodes, s_{ij} and s_{ji}, and the edge $\{s_{ij},s_{ji}\}$ to which a cost $-M$ is assigned. The motivation of this cost is to assure that one route of the CVRP solution will traverse edge $\{s_{ij},s_{ji}\}$. The method of [62] is a branch-and-cut algorithm based on a formulation of the CVRP that uses one variable for each edge and this variable is equal to one if the corresponding edge is used by the solution. Then, Baldacci and Maniezzo [9] exploit this fact and avoid using the cost $-M$ for edges $\{s_{ij},s_{ji}\}$ by fixing the corresponding variables to one in the CVRP formulation. Besides this trick, they apply the branch-and-cut implemented by Lysgaard, Letchford, and Eglese [62] with few changes. They use the package of routines developed by Lysgaard [61] to separate capacity inequalities, homogeneous multistar inequalities, generalized large multistar inequalities, framed capacity inequalities, strengthened comb inequalities, and comb inequalities. The branching strategy adopted is the CPLEX strong branching strategy, and the node selection rule is the best first rule that selects the node with the smallest lower bound to be processed next.

The authors tested their method with the instance sets *val* and *egl*, and an initial upper bound equal to the cost of the best heuristic solution was used for each instance. Twenty-eight out of the 34 *val* instances were solved to optimality within two hours. This means six optimums more than the cutting plane method of Belenguer and Benavent [17] that is described in 9.5.3. Nevertheless, the lower bounds obtained at the root node are slightly worse on average than those obtained by [17]. The second set of instances contains 15 out of the 24 *egl* instances, and the branch-and-cut was able to prove the optimality of the corresponding heuristic solutions for five instances. No new better solution was found for any instance.

9.5.2 ▪ Branch-and-cut for the two-index formulation

Belenguer and Benavent [15] developed a branch-and-cut algorithm for the CARP based on the two-index formulation (9.1)–(9.6) including also the disaggregate version of capacity constraints (9.12) and odd-cut constraints (9.13) from the one-index formulation. Any valid inequality from the one-index formulation can be converted into a valid inequality in the two-index formulation by using equations (9.10). In [15], separation procedures were applied to identify violated inequalities (9.12) and (9.13) as well as connectivity constraints (9.3), cocircuit constraints (9.7), constraints (9.8), and minimal cover constraints. The algorithm was tested on 24 of the 34 *val* instances (those instances for which the number of vehicles is at most five) and 16 of them were solved to optimality. Nevertheless, it was observed that the instances optimally solved were precisely those for which the lower bound at the root node was equal to the optimum. The tightness of this lower bound was mainly due to constraints (9.12) and (9.13). The constraints of the two-index formulation are useful only to cut off fractional solutions, but fail to increase the objective in the linear relaxation; for details see Benavent, Corberán, and Sanchis [22].

The two-index formulation has the important drawback of allowing symmetric solutions: The routes of different vehicles can be permuted, leading to different solutions which are in fact identical. This fact causes serious problems in the branch-and-cut algorithm, as identical solution might be found in $|K|!$ nodes of the search tree. The number of branch-and-bound nodes can thus become prohibitively large.

Welz [74] also implemented a branch-and-bound algorithm based on his two-index formulation that has directed variables. In his implementation, he uses a cutting-plane algorithm in the root node in order to identify violated connectivity and directed odd-cut constraints. The largest instance solved with this method has a fleet of two vehicles on a graph with 27 nodes and 82 edges.

9.5.3 ▪ Cutting-plane methods for the one-index formulation

Belenguer and Benavent [17] developed a cutting-plane algorithm to solve the linear relaxation of the one-index formulation to obtain a lower bound for the CARP. They implemented heuristic algorithms to separate the capacity constraints (9.12), odd-cut constraints (9.13), and disjoint-path inequalities (9.15). Odd-cut constraints are also separated with an exact procedure based on the method described in Padberg and Rao [67].

The heuristics used to separate capacity and odd-cut constraints work by generating promising sets S of nodes, with different methods, and then checking (9.12) and (9.13) for possible violation. Let $G(y)$ be the graph induced by the edges $e \in E$ with $y_e > 0$ in the current LP solution. The methods to generate those promising subsets of nodes are (1) starting with $S = \{d\}$, generate a sequence of new node subsets S, where each node subset is equal to the preceding one extended by the adjacent nodes in G, until S contains all nodes; (2) compute the node subsets of the connected components of $G(y)$; and (3) the *fractional capacity inequality* is a weaker version of the capacity constraint (9.12) where $K(S)$ is substituted by $D(S)/Q$; violated fractional capacity inequalities can be separated in polynomial time by solving a max-flow/min-cut problem on a transformed graph. This procedure is applied repeatedly, increasing each time the original demands by different percentages.

The heuristic for separating the disjoint-path inequalities (9.15) uses the family of node subsets generated by the preceding procedures. From this family, different sequences of node subsets $S_0 \subset S_1 \subset \cdots \subset S_t \subseteq V \setminus \{d\}$ are built. The edge set E' is usually defined as $E' = \{e \in E(V \setminus S_0) \setminus \bigcup_{i=1}^{t} \delta(S_i) : y_e = 0\}$. Then the corresponding disjoint-path inequality is first checked for violation, and, if it is violated, it is checked for validity.

The lower bound obtained with this algorithm outperformed any other lower bounding procedure for the CARP at that time. Furthermore, the solution provided is usually integer. Nevertheless, as it was pointed out in Section 9.4.2, this integer solution may not correspond to a CARP solution. Thus, the optimality of the computed lower bound can be stated only if a heuristic CARP solution is known with the same cost.

Comparing this lower bound with the best upper bound known, it was optimal for 22 of the 34 *val* instances with an average gap of 0.41%. For the *egl* instances, the gap between the lower bound and the best known upper bound was 2.40% on average. The cutting-plane algorithm was also tested on the 23 *gdb* instances, and the lower bound was optimal in 20 cases, with an average gap of 0.14%. The effectiveness of this method comes probably from the fact that the inequalities force the LP solution to use deadheading edges, which are the ones that cause the lower bound to increase. Furthermore, the method is fast because it works directly on the original graph, usually a sparse one.

Ahr [2] developed an exact separation routine for capacity constraints (9.12) by formulating the separation problem as a mixed-integer program and solving it several times, once for each possible number of vehicles. His algorithm was able to detect violated capacity constraints that were not found by the heuristic separation routines of Belenguer and Benavent [17]. In fact, for nine *egl* instances, the lower bound obtained using the exact separation routines outperformed the lower bound obtained by [17] (that includes the separation of the disjoint-path inequalities). Martinelli, Poggi de Aragão, and Subramanian [65] developed a new formulation of the separation problem of capacity constraints that is inspired by Ahr's formulation, but is capable of identifying violated capacity cuts for any number of vehicles at once. Martinelli, Poggi de Aragão, and Subramanian [65] report about a 36% saving on average in the computing time needed to apply the complete cutting-plane algorithm on the *egl* instances when using their exact separation formulation for capacity constraints instead of Ahr's formulation. Nevertheless, they remark that the use of any of these exact separation procedures is only practical for reasonably sized instances. They also propose a *Dual-Ascent* (DA) heuristic that works on the dual of the linear relaxation of the one-index formulation of the CARP, where each variable, say π_S, is associated to a subset $S \subseteq V \setminus \{d\}$. At each iteration, DA generates a large set of edge cutsets, selects one, say S^*, assigns $\pi^*_S = \min\{c_e : e \in \delta(S^*)\}$, and modifies the cost of edges $e \in \delta(S^*)$ by subtracting π^*_S. At each iteration, edges with zero cost are shrunk in order to guarantee that no zero cost edges appear in the edge cutsets of future iterations. Then, DA performs at most $|V|-1$ iterations. Four different strategies were used to generate edge cutsets at each iteration: (1) simple cuts containing only one node, (2) complete cuts containing all nodes except one, (3) random cuts, and (4) cuts generated by the edges of a minimum spanning tree. The lower bounds obtained by the DA method are not competitive, but Martinelli, Poggi de Aragão, and Subramanian [65] use the set of edge cutsets generated during the application of the method as an initial set of capacity and odd-cut constraints in the cutting plane algorithm for the one-index formulation. With this strategy, they reduce the number of times that the exact separation procedures are used, thus allowing one to apply the method to the set of large CARP instances *egl*-g, for which no lower bound was known before.

9.6 ▪ Column generation and branch-and-price

The very first column-generation approach tailored to the CARP was implemented and tested by Gómez-Cabrero, Belenguer, and Benavent [47] in 2005. Later in 2009, another column-generation-based algorithm for the CARP by Letchford and Oukil [57], still not a complete exact method, was published. In this section, we present in detail three recent but very different branch-and price algorithms for the CARP. Although the three approaches are based on the same basic methodology (for an introduction to integer column generation, see Lübbecke and Desrosiers [60] and Desaulniers, Desrosiers, and Solomon [36]), they vary significantly as they (re-)formulate the CARP on different graphs and apply different relaxations, pricing and cutting techniques, and schemes to finally end up with feasible integer solutions. The three implementations were developed independently from each other and published in the years 2012 and 2013.

Recently, Bartolini, Cordeau, and Laporte [13] proposed another branch-and-price algorithm for an extension of the CARP, the so-called *CARP with Deadheading Demand* (CARPDD). This algorithm is in fact a hybrid of the latter two approaches presented in the following Sections 9.6.2 and 9.6.3. It can also be used to solve the CARP, resulting in some slight improvements compared to Bartolini, Cordeau, and Laporte [12] and Bode and Irnich [24]. However, we see the main contribution of the work of Bartolini, Cordeau, and Laporte [13] in the presentation and solution of the CARPDD, and therefore address it in Section 9.7.3 on CARP variants.

9.6.1 ▪ Branch-and-price for CVRP

In principle, the transformations into CVRP presented in Section 9.2 permit the solution of the CARP by any exact or heuristic CVRP algorithm. Methods that are based on SP or SC formulations hereby produce relatively large models with $2|E_R|$ or $3|E_R|$ covering constraints. Longo, Poggi de Aragão, and Uchoa [59] devised an adapted CVRP branch-and-price-and-cut algorithm that solves significantly smaller SP problems with only $|E_R|$ covering constraints. Their approach is based on the transformation of the CARP into a CVRP with $2|E_R|+1$ nodes, where each required edge $e=\{i,j\}\in E_R$ induces two separate nodes s_{ij} and s_{ji}. The new edges $\{s_{ij},s_{ji}\}$ are fixed to one (see Section 9.2). The column-generation formulation combines something that is equivalent to the SP model with the symmetric and undirected two-index model of the CVRP (see, e.g., Toth and Vigo [72, p. 14]). For the transformed CARP instances, this two-index model states that every customer node s_{ij} is incident to two edges (one is the fixed edge $\{s_{ij},s_{ji}\}$), that $2|K|$ edges are incident to the depot 0, and that tours must be feasible. Feasibility is guaranteed by *Generalized Subtour-Elimination Constraints* (GSEC) in the form of rounded capacity cuts. Because of fixing the edges $\{s_{ij},s_{ji}\}$ to one for all $\{i,j\}\in E_R$, the node-degree constraints can be reduced to $|E_R|$ partitioning constraints, one for each required edge $e=\{i,j\}\in E_R$, stating that one of the new nodes, say s_{ij}, has to be connected to another node ($\neq s_{ji}$) exactly once.

For this subsection, let Ω be the set of those feasible tours r that visit the nodes s_{ij} and s_{ji} always consecutively, either from s_{ij} to s_{ji} or vice versa. The cost of tour $r\in\Omega$ is c_r. Moreover, we define $\bar{x}_{\{s_{ij},s_{kl}\},r}$ as the number of times that the route $r\in\Omega$ traverses the edge $\{s_{ij},s_{kl}\}$. For edges $\{s_{ij},s_{ji}\}$ induced by a required edge $e=\{i,j\}\in E_R$, we simply write $\bar{x}_{er}$. Then, the extended SP and column-generation formulation is (equivalent to) (9.16)–(9.19). Longo, Poggi de Aragão, and Uchoa [59] model slightly differently with the node-degree constraints (=2); they also use $\geq |K|$ in (9.19). Defining the customer set as

$$N=\bigcup_{\{i,j\}=e\in E_R}\{s_{ij},s_{ji}\},$$

the linear relaxation of the model can be strengthened by the following sets of constraints:

$$\sum_{r\in\Omega}\sum_{e\in\delta(S)}\bar{x}_{er}\lambda_r\geq 2k(S)\quad\forall\, S\subseteq N, \tag{9.21}$$

$$\sum_{r\in\Omega}\bar{x}_{\{s_{ij},s_{kl}\},r}\lambda_r\leq 1\quad\forall\, s_{ij},s_{kl}\in N \text{ with } \{i,j\}\neq\{k,l\}. \tag{9.22}$$

The transformed GSEC (9.21) ensure that any subset S of customers is visited by $k(S)$ tours, where $k(S)$ is (a lower bound on) the minimum number of tours needed to serve customers S (in Longo, Poggi de Aragão, and Uchoa [59] $k(S)$ is $\lceil\frac{1}{Q}\sum_{s\in S}q_s\rceil$). The constraints (9.22) state that no edge is traversed more than once. Additional valid inequalities for the CVRP such as those presented in Lysgaard, Letchford, and Eglese [62] can be formulated using the coefficients $\bar{x}_{\{s_{ij},s_{kl}\},r}$ and can be dynamically added in order to strengthen the linear relaxation of the model.

The branch-and-price-and-cut algorithm of Fukasawa et al. [42] for the CVRP is then applied to solve the column-generation formulation. In fact, a relaxation was solved where instead of elementary routes, nonelementary q-routes (Christofides, Mingozzi, and Toth [32]) without 3-cycles are allowed in the set Ω. In this way, one can exploit the tradeoff between hardness of the pricing problem and the strength of the resulting column-generation formulation. For a more detailed discussion, see Irnich and Villeneuve [51].

In their computational tests, Longo, Poggi de Aragão, and Uchoa [59] considered the *kshs*, *gdb*, *val*, and *egl* benchmark sets. All instances from the first two sets were solved to proved optimality. For the *val* benchmark set, the lower bound obtained from the linear relaxation closed the optimality gap for 21 of the 34 instances. For five more instances, optimal solutions were computed in the branch-and-bound tree. Finally, two of the *egl* instances were solved to optimality in the branch-and-bound root node.

9.6.2 ▪ Column generation for GVRP

The approach of Bartolini, Cordeau, and Laporte [12] is based on an extended SP formulation of the asymmetric GVRP. Recall from Section 9.2 that a GVRP results from the transformation where each required edge $e \in E_R$ is represented by a cluster of two nodes, one for each direction in which e can be serviced. More precisely, let Ω be the set of all feasible GVRP routes, $\bar{x}_{er}$ be the number of times that route $r \in \Omega$ visits the nodes from the cluster $\overline{V}_e = \{i_e, j_e\}$ for $e \in E_R$, and c_r be the tour costs. Then, the initial formulation in [12] is again (9.16)–(9.18) together with (9.19) where $= |K|$ is replaced by $\geq K_{min} = \lceil \frac{1}{Q} q(E_R) \rceil$.

Overall, Bartolini, Cordeau, and Laporte [12] provide a series of four lower *Bounding Procedures* (BPs), BP_1, BP_2, BP_3, and BP_4, that all solve increasingly stronger relaxations of extended/strengthened SP models. The resulting lower bounds fulfill $LB_1 \leq LB_2 \leq LB_3 \leq LB_4$, typically with strict inequality, and the last bound LB_4 is often very close to the optimum opt. Since each BP is limited to a linear relaxation of the (extended) SP model, the bounding procedures generally do not solve the integer problem. If, however, a valid upper bound UB for the integer model is known, the lower bounding procedures can be complemented to provide an exact CARP algorithm. The final remaining gap $UB - LB_4$ is closed by solving the last extended SP formulation explicitly. This requires on the one hand that all relevant tours with a reduced cost not exceeding the gap $UB - LB_4$ are enumerated. They form the set Ω then. On the other hand, an exact solver for the corresponding extended SP problem is needed, which is the standard IP-solver CPLEX in Bartolini, Cordeau, and Laporte [12]. Both components might fail, i.e., it can happen that the number of tours is too large to be generated, or that the IP-solver is unable to compute an optimal SP solution. In the first case, the algorithm just provides the lower bound LB_4, while in the latter case, the IP-solver might provide an improved integer solution and therewith a better upper bound.

Set-partitioning relaxations Relaxations of SP can be obtained by allowing non-elementary tours in Ω, while strengthened formulations are linear relaxations of SP that incorporate valid inequalities. As both the number of tours and inequalities is huge, variables and constraints are dynamically added in a cut-and-column generation fashion. The reason for using a series of BPs is that solving the extended SP formulation directly and even its linear relaxation is practically intractable. Instead, one can start with a weaker relaxation that can be solved relatively fast, but typically provides just a weaker lower bound. Besides this lower bound LB_i for the ith BP, every BP also provides a dual solution, which is helpful in order to generate all relevant or the best n_{max} (for a given bound n_{max}) tours, in the sense that these tours have minimum reduced cost. Note that a tour cannot be part of an optimal solution if its reduced cost exceeds the integrality gap $opt - LB_i$. Because opt is generally unknown, one can replace the integrality gap by the gap $UB - LB_i$. If an incumbent bounding procedure BP_i fails to enumerate all relevant tours with a reduced cost not exceeding the integrality gap $opt - LB_i$, the set of the already generated tours can initialize the next BP_{i+1} (at least if the two bounding procedures use identical tour relaxations). Details on the four different BP can be found in Bartolini, Cordeau, and Laporte [12] and will also be given later in this subsection.

The fact that the GVRP results from a CARP is exploited by Bartolini, Cordeau, and Laporte [12] in two ways: First, one can a priori compute shortest paths P_{ij} between any two nodes $i, j \in V$ in the original CARP graph. As the CARP is symmetric, $P_{ij} = P_{ji}$ holds without loss of generality. Moreover, one can arrange that any CARP tour, which services two required edges e, e' consecutively, deadheads from the end of e, say, node i, to the start of e', say j, through the path P_{ij}. For a given sequence of serviced edges (including direction of service), a feasible CARP tour therefore has a corresponding unique GVRP tour, and vice versa. Bartolini, Cordeau, and Laporte [12] point out that even under these assumptions there exists a huge number of equivalent and so redundant CARP tours. In order to overcome this redundancy, they define *forward-service (backward-service) routes* as partial tours starting at the depot, where service is performed as early (late) as possible, i.e., if an edge is serviced and deadheaded, then service must always precede (succeed) any deadheading. This predefinition is beneficial when enumerating GVRP tours as discussed later on.

Second, it is possible to count both the number $\bar{y}_{er}$ of times a tour r deadheads through an edge $e \in E$ and the number $\bar{p}_{ij,r}$ of times it deadheads through the paths P_{ij} and P_{ji} (more precisely, the sum of traversals through both paths). Associated variables (of a compact formulation) are

$$y_e = \sum_{r \in \Omega} \bar{y}_{er} \lambda_r \quad (e \in E) \qquad \text{and} \qquad p_{ij} = \sum_{r \in \Omega} \bar{p}_{ij,r} \lambda_r \quad (i, j \in V, i < j).$$

Bartolini, Cordeau, and Laporte [12] prove that the odd-cut constraints (9.13) can be reformulated in the p_{ij} variables and constitute lifted versions of these valid inequalities. They dominate the ones formulated in the original deadheading variables y_e (leading to cuts of the form (9.20)) and are useful to further strengthen the SP formulation. The same applies to the capacity constraints (9.12), which are reformulated by Bartolini, Cordeau, and Laporte [12] in terms of aggregate variables indicating whether or not two required edges e and e' are served consecutively.

Subset-Row (SR) inequalities (Jepsen et al. [52]) are added as a third class of valid inequalities to the SP formulation. In contrast to odd-cut and capacity constraints, these cuts cannot be written in the variables of a compact formulation, but refer directly to tour variables λ_r. In Bartolini, Cordeau, and Laporte [12] only the subclass of SR inequalities for triplets $T = \{e_1, e_2, e_3\} \subset E_R$ of required edges is considered: Let Ω_T be the subset of tours that cover at least two of three edges of the triplet, i.e., $\Omega_T = \{r \in \Omega : \bar{x}_{e_1 r} + \bar{x}_{e_2 r} + \bar{x}_{e_3 r} \geq 2\}$. Obviously, there cannot be more than one tour from a triplet set Ω_T in any solution as expressed by the SR inequalities

$$\sum_{r \in \Omega_T} \lambda_r \leq 1 \quad \forall\, T \subset E_R, |T| = 3. \tag{9.23}$$

Summarizing, the extended SP formulation used for cut-and-column generation in Bartolini, Cordeau, and Laporte [12] consists of (9.16)–(9.18) together with lifted odd-cut and capacity constraints, and SR inequalities (9.23).

Lower bounding procedures Bartolini, Cordeau, and Laporte [12] apply four lower bounding procedures BP_1, BP_2, BP_3, and BP_4 that are sketched in the following:

BP_1 Model: Model M_1 is the linear relaxation of SP (9.16)–(9.19) without additional valid inequalities, where nonelementary 3-loop free CARP tours (a shortest possible cycle over required edges is $(e_1, e_2, e_3, e_4 = e_1)$) form the set Ω. The dual of formulation M_1 is (heuristically) solved using a tailored column-generation procedure based on the dual-ascent method.

Init.: The dual-ascent method is a specialized subgradient algorithm that produces penalties (λ_e, λ_0) to the constraints (9.17) and (9.19).

Output: Besides the lower bound $LB_1 \leq z(M_1)$, the result is a dual feasible solution. Bartolini, Cordeau, and Laporte [12] show how the penalties (λ_e, λ_0) can be transformed into a feasible dual solution $D^1 = (\pi_e^1, \mu^1)$ to the dual of M_1.

BP_2 Input: Bound LB_1 and dual solution $D^1 = (\pi_e^1, \mu^1)$.

Model: Model M_2 differs from M_1 as it adds the lifted versions of the odd-cut (9.13) and capacity inequalities (9.12), formulated in the tour variables, similar to (9.20). M_2 is solved using cut-and-column generation, where the simplex algorithm is used for reoptimization.

Init.: The column-generation master program is initialized with the following set Ω of tours: For each required edge $e \in E_R$ and both of its endpoints v, a minimum reduced cost w.r.t. D^1 (open) CARP q-route with last edge e and ending at node v is computed. Routes are then combined into feasible CARP tours. Furthermore, odd-cut constraints for all singleton sets $S = \{i\}, |\delta_R(S)|$ odd, are used as initial valid inequalities.

Output: The bounding procedure produces the bound $LB_2 = z(M_2)$ and a dual solution $D^2 = (\pi_e^2, \mu^2, \alpha_s^2)$, where α_s^2 are optimal dual prices to the resulting odd-cut and capacity inequalities. Finally, a subset of valid inequalities will be reused in the next bounding procedures.

BP_3 Input: Bound LB_2 and dual solutions $D^1 = (\pi_e^1, \mu^1)$ and $D^2 = (\pi_e^2, \mu^2, \alpha_s^2)$.

Model: The Model M_3 in the third bounding procedure is based on elementary tours Ω. As for M_2, lifted odd-cut and capacity constraints extend the SP formulation. Again, the simplex algorithm is used to reoptimize the master program after adding new tours and cuts.

Init.: Those odd-cut and capacity constraints found by LB_2 are used to initialize M_3. Moreover, a sophisticated tour generation procedure, called GenRoutes (briefly described in the next paragraph), is applied in order to generate the initial tour set Ω for the cut-and-column generation master program. GenRoutes computes all relevant or a limited number (using an a priori bound) of routes with a reduced cost w.r.t. D^2 not exceeding the gap $UB - LB_2$. Fathoming rules in GenRoutes utilize both dual solutions D^1 and D^2.

Output: The result of BP_3 is the bound $LB_3 = z(M_3)$, a dual solution $D^3 = (\pi_e^3, \mu^3, \alpha_s^3)$, and the subset of identified valid inequalities.

BP_4 Input: Bound LB_3 and dual solutions $D^1 = (\pi_e^1, \mu^1)$, $D^2 = (\pi_e^2, \mu^2, \alpha_s^2)$, and $D^3 = (\pi_e^3, \mu^3, \alpha_s^3)$.

Model: The Model M_4 additionally incorporates SR inequalities (9.23), while routes are required to be elementary.

Init.: Valid inequalities from M_3 are used to initialize M_4. Similarly, GenRoutes generates the initial tour set Ω based on the dual solution D^3 not exceeding the gap $UB - LB_3$. Fathoming in GenRoute uses dual solutions D^1, D^2, and D^3.

Output: The result of BP_4 is the bound $LB_4 = z(M_4)$, a dual solution $D^4 = (\pi_e^4, \mu^4, \alpha_s^4, \beta_T^4)$ (where (β_T) are duals to the SR inequalities (9.23)), and the subset of identified valid inequalities.

Tour generation procedure GenRoutes The bounding procedures BP_3 and BP_4 and also the final SP solver need to be initialized with a subset of tours Ω with small reduced cost. Given a dual solution D and parameters $\eta \geq 0$ and $\Delta \in \mathbb{Z}_+$, the task of the procedure GenRoutes is to generate all relevant, properly different GVRP tours with a reduced cost w.r.t. D not exceeding the bound η. As the number of these tours might be huge, the parameter Δ controls that GenRoutes produces not more that Δ tours (favoring those with small reduced cost) whenever the number of GVRP tours is larger than Δ.

GenRoutes has two phases. The idea is that in phase one, open GVRP forward and backward routes are generated first that are combined into tours in phase two. More precisely, phase one is a dynamic programming (labeling) procedure that produces partial paths P (= open GVRP routes) that start at the depot 0, visit single nodes from clusters $\overline{V}_e = \{i_e, j_e\}$, and end at one of these cluster nodes $v = v(P)$. In order to generate only relevant and properly different routes ending at a node v, different dominance rules and fathoming techniques are applied: At first, for each partial path P, it must be possible to complete it to a tour with a reduced cost not exceeding η, which is tested by computing q-route and ng-path bounds for completions (see Christofides, Mingozzi, and Toth [32], Baldacci, Mingozzi, and Roberti [10]). Moreover, only a single least reduced cost path must be kept among all paths performing the same set of services. In order to *not* generate identical or symmetric counterparts of the same final tour, partial paths are restricted to have not more that $\lceil Q/2 \rceil$ load (similar to bidirectional shortest-path computations; see Righini and Salani [69]). Additionally, the open GVRP routes must correspond to forward-service (backward-service) routes, i.e., when traversing an edge multiple times, deadheading must always follow (precede) the corresponding service. The second phase then combines, whenever feasible, the generated forward-service routes with backward-service routes into GVRP tours having reduced cost not larger than η. Supplementary dominance and fathoming rules are applied in both phases. These are crucial for the effectiveness of the overall approach, but are too intricate to be summarized in this survey (many more details are given in Bartolini, Cordeau, and Laporte [12]).

Overall, the algorithm by Bartolini, Cordeau, and Laporte [12] is one of the most effective methods for computing optimal CARP solutions. Their computational study reports results for the *kshs*, *gdb*, *val*, *egl*, and *bmcv* instances. All *kshs* and *gdb* instances were solved on average within less than one second. The hierarchy of lower bounding procedures left an average small optimality gap measured by $100 \cdot LB/UB$, namely, 99.83% for *val*, 99.71% for *egl*, and 99.73% for *bmcv*. Moreover, optimal solutions were computed for 29 out of 34 *val*, 10 out of 24 *egl*, and 66 out of 100 *bmcv* instances. For several more instances, optimality of known heuristic solutions were proved with the lower bounds.

9.6.3 ▪ Column generation and branch-and-price for the CARP

A complementary approach for solving the CARP is based on the fact that real-world street networks modeled by $G = (V,E)$ are very sparse. A node has typically a degree that does not exceed four. The fundamental idea of exploiting the sparsity of G was first coined by Letchford and Oukil [57] and later fully implemented in a cut-first branch-and-price-second approach by Bode and Irnich [24], which will be discussed in the following. Independently, Martinelli et al. [64] followed a seemingly very similar approach, for which we will outline the major differences in each respective paragraph. Before presenting details on cutting, pricing, and branching, we give a general overview of the major differences compared to the cut-and-column-generation approach of Bartolini, Cordeau, and Laporte [12] presented in the last subsection.

First, instead of using four different relaxations, only one linear relaxation is used. In Bode and Irnich [24], it is basically the formulation (9.16)–(9.20) where Ω is the set of all nonelementary CARP tours that do not contain 2-loops. A 2-loop means that the tour services the same required edge $e \in E_R$ twice and no other service is performed between the two. (Similarly, a k-loop has no more than $k-2$ other services performed in between.) For tours defined on the original graph, a required edge $e = \{i,j\} \in E_R$ can be traversed in the same or in opposite directions and any kind of deadheading between the two endpoints i and/or j of e can occur. It is even more important that the relaxation is not only used for bounding, but throughout the entire branch-and-bound process. While the exact algorithm by Bartolini, Cordeau, and Laporte [12] requires an upper bound UB in order to enumerate all relevant tours to finally compute an integer solution using a MIP solver, no upper bound is needed for the branch-and-price in Bode and Irnich [24].

Second, the pricing of CARP tours can be performed efficiently on the original sparse graph. For the SC formulation (9.16)–(9.18) (without inequalities (9.19) and cuts (9.20)), Letchford and Oukil [57] developed an efficient pricing algorithm with worst-case complexity $\mathcal{O}(Q(|E|+|V|\log|V|))$. Details on variants of the pricing problem and efficient solution approached will be discussed later.

Third, finding integer solutions to the CARP using a linear relaxation of (9.16)–(9.20) is not straightforward: On the one hand, the column-generation master program is an aggregate formulation, i.e., it has no variables indexed by vehicles. Opposed to many node-routing formulations, fractional tour variables λ of the SP formulation for the CARP can induce integral aggregate edge-flow variables such as the y_e variables of the one-index formulation. In such a situation, standard branching rules based on bounding edge-flow variables are in vain, and additional branching rules have to be employed in order to deliver feasible integer solutions. On the other hand, as pointed out in Bode and Irnich [24], a convex combination of fractional tour variables λ can sometimes be reinterpreted and lead to a feasible integer solution. Both complications have been solved using branching based on follower and nonfollower constraints saying that two required edges are either serviced consecutively or with at least one other service between them.

In the following we present the details of the cut-first branch-and-price-second approach of Bode and Irnich [24]:

Cutting The first phase of the approach requires the solution of the linear relaxation of the one-index formulation (9.11)–(9.14). The subset of those valid inequalities that are finally active are then reformulated in the λ variables of the column-generation master program and used for the initialization of the second phase. Violated odd-cut inequalities and capacity cuts as well as disjoint-path inequalities are separated, and those binding in the very last iteration of the cutting-plane procedure are considered active. Note that

Martinelli et al. [64] have no separation procedure for disjoint-path inequalities, but they conduct cut generation also after solving each node of the branch-and-price procedure (presented below).

For the remaining of this subsection, we assume that $\mathcal{S}$ is the set of active inequalities, i.e., the set $\mathcal{S}$ is kept fixed, and no additional cuts are generated dynamically. There is nothing special about the cutting-plane procedure in Bode and Irnich [24]; however, details such as the building blocks of the separation procedures, the overall ordering of separation procedures, and their parameters can be found in the electronic companion of [24].

Relaxations and pricing Bode and Irnich [24] derive the extended SP formulation (9.16)–(9.20) from a Dantzig–Wolfe decomposition of the two-index formulation (9.1)–(9.6) and a subsequent aggregation. In this initial formulation, CARP tours are elementary, i.e., no tour can service any edge more than once. Let $\pi = (\pi_e)_{e \in E_R}$ be dual prices associated with the partitioning constraints (9.17), $\beta = (\beta_s)_{s \in \mathcal{S}}$ to the cuts (9.20), and μ to the constraint (9.19). The task of the pricing problem is the generation of one or several feasible CARP tours with negative reduced cost or to prove that no such tour exists. The reduced cost of a tour $r \in \Omega$ can easily be expressed with variables and coefficients defined over the original graph, using the correspondences

$$x_e = \sum_{r \in \Omega} \bar{x}_{er} \lambda_r \quad (\forall\, e \in E_R) \quad \text{and} \quad y_e = \sum_{r \in \Omega} \bar{y}_{er} \lambda_r \quad (\forall\, e \in E), \tag{9.24}$$

and edge reduced costs

$$\bar{c}_e^{serv} = c_e^{serv} - \pi_e \quad (\forall\, e \in E_R) \quad \text{and} \quad \bar{c}_e = c_e - \sum_{s \in \mathcal{S}} d_{es} \beta_s \quad (\forall\, e \in E). \tag{9.25}$$

The *Pricing Problem* (PP) is then

$$z_{PP} = \min \sum_{e \in E_R} \bar{c}_e^{serv} x_e + \sum_{e \in E} \bar{c}_e y_e - \mu \qquad \text{s.t.} \qquad (9.3)\text{–}(9.6),$$

where the vehicle index k is omitted (with $K = \{1\}$).

This problem is in fact *not* a routing problem, i.e., a problem where a path has to be determined, as can be seen using a small example: The solution $x_{23} = x_{24} = 1$, $y_{d3} = 2$, $y_{34} = 1$ represents a CARP tour, feasible if $q_{23} + q_{24} \le Q$. There are, however, two paths $(d - 3 = 2 = 4 - 3 - d)$ and $(d - 3 - 4 = 2 = 3 - d)$ that realize the solution (as before, "=" means service and "−" deadheading). In general, there might exist a huge number of paths leading to the same CARP tour, as already discussed in the preceding Section 9.6.2. In order to avoid paths as (partial) solutions, one can solve the pricing problem directly using MIP as reported in Letchford and Oukil [57]. These and other experiments with branch-and-cut procedures for solving pricing problems, e.g., Drexl and Irnich [37], indicate that MIP-based approaches are relatively slow. Often, the pricing problem instances tend to require more computation time when approaching optimal dual solutions. Therefore, MIP-based algorithms are currently not competitive with dynamic programming approaches that can solve relaxations of pricing problem efficiently, but are based on constructing partial paths/routes.

Such an efficient dynamic program for solving the nonelementary version of the pricing problem has been developed by Letchford and Oukil [57]. It is a dynamic programming labeling procedure in which labels L represent partial paths $p(L)$ that start at the depot d and service and deadhead some edges before they arrive at their incumbent node $n(L)$.

A labeling algorithm iteratively performs two kinds of procedures (see Irnich and Desaulniers [50]): In the *path extension step*, an unprocessed label L is selected and extensions of this label along edges $e \in \delta(n(L))$ are created. The extension step produces one or two new labels L' for the path $p(L') = (p(L), j)$ with endpoint $j = n(L')$ for each edge. A new label L' results from either a deadheading through edge $e = \{n(L), j\}$ or a service if e is a required edge. Thus, L' has the accumulated reduced cost $\overline{c}(L') = \overline{c}(L) + \overline{c}_e^{serv}$ or $\overline{c}(L') = \overline{c}(L) + \overline{c}_e$, and the accumulated demand $q(L') = q(L) + q_e$ or $q(L') = q(L)$, respectively, according to (9.25). The second component is the *dominance procedure* that eliminates labels that finally lead to tours not having smaller reduced cost and load. The algorithm can be summarized as follows:

Efficient Pricing Algorithm, Letchford and Oukil [57]:

For $q = 0, 1, 2, \ldots, Q$

  `// Dijkstra-like extension; Use reduced costs` $\overline{c}_e \geq 0$

  Extend labels L with $q(L) = q$ along edges $e \in \delta(n(L))$

    (Among labels with $q(L) = q$, keep only one least-cost label at each node)

  `// Service extension; Use reduced cost` $\overline{c}_e^{serv}$

  Extend labels L with $q(L) = q$ along required edges $e \in \delta_R(n(L))$

    (Among labels with identical load, keep only one least-cost label at each node)

Apply dominance algorithm for filtering out Pareto-optimal labels at depot node d

---

The Dijkstra-like extension step produces partial paths with one or several deadheading edges at the end. In the iteration with load q, it initializes labels, exactly one label for each node, with the minimum reduced cost of a path with load q ending at that node. These labels are selected and extended in a Dijkstra-like fashion, i.e., smallest (reduced) cost first. In these extension steps, all newly generated labels L' have the same load $q(L') = q$ and have to be reconsidered for path-extension steps in the same iteration (for load q). The dominance step is implicitly performed by keeping only one least-cost label with load q (which requires only constant time $\mathcal{O}(1)$ if labels are stored in vectors indexed by load q), and the final filtering step, which can be implemented in linear time $\mathcal{O}(Q)$. Since the Dijkstra algorithm can be implemented running in $\mathcal{O}(|E| + |V| \log |V|)$ time using Fibonacci heaps (see, e.g., Ahuja, Magnanti, and Orlin [3]), the overall worst-case bound for Algorithm 9.1 is $\mathcal{O}(Q(|E| + |V| \log |V|))$.

The reduced costs $\overline{c}_e$ of the y_e variables modeling deadheading can become negative when valid inequalities (9.20) are present (see equations (9.25) and consider large duals β_s). As the Dijkstra algorithm requires nonnegative reduced deadheading costs ($\overline{c}_e$), the presence of active valid inequalities (9.20) seems to render efficient pricing impossible. Bode and Irnich [24] resolved this problem by introducing one additional unbounded variable z_e for each $e \in E$ into the column-generation formulation. The variable $z_e \geq 0$ models a deadheading cycle (e, e) (going back and forth through edge $e \in E$). It contributes to the extended SP model with a cost of $2c_e$ and coefficients $2\sum_{e \in E} d_{es}$ in the inequalities (9.20). The z_e variables relax the primal model, whereas the associated dual model is more constrained. In fact, the variable z_e corresponds to the constraint $\sum_{s \in \mathcal{S}} d_{es} \beta_s \leq c_e$ formulated in the dual variables β_s. Such inequalities that are fulfilled by optimal dual solutions are known as dual-optimal inequalities (Ben Amor, Desrosiers and Valério de Carvalho [20]). Their direct consequence is nonnegative reduced costs $\overline{c}_e = c_e - \sum_{s \in \mathcal{S}} d_{es} \beta_s \geq 0$ and the stabilization of the dual variables β_s in the column-generation process. It remains to be mentioned that z_e variables correspond to extreme rays in the domain of pricing problem (the ray for $e = \{i, j\} \in E$ is given by $y_e = 2$ and $p_i = p_j = 1$). The respective tours have redundant deadheading cycles and were denoted as *extended k-routes* in Belenguer and Benavent [16]; they are possible in the pricing problem here, the one-index formulation, and also the two-index formulation.

In two recent studies, Bode and Irnich [26, 25] show that one should further exploit the tradeoff between the strength of the (relaxed) SP formulation and the difficulty of solving the corresponding pricing problems. Instead of 2-loop free tours, k-loop free tours (for $k \geq 3$) (Bode and Irnich [26]) and tours resulting from a ng-path relaxation (Baldacci, Mingozzi, and Roberti [10]) are allowed. Both k-loop elimination and ng-paths for the CARP are based on considering (service) tasks associated to a subset of edges/arcs instead of all edges, arcs, or nodes, as conceptually introduced and discussed in Desaulniers et al. [35], Irnich and Desaulniers [50], and Irnich and Villeneuve [51].

Martinelli et al. [64] use a slightly different pricing approach based on dynamic programming with a worst-case complexity of $\mathcal{O}(Q|E_R|^2)$, where the sparsity of the network is not fully exploited. However, 2-loop elimination (see below) is used to improve the quality of the generated CARP tours.

Integrality If the solution of the linear relaxation of (9.16)–(9.20) contains fractional tour variables λ_r, additional effort is needed for computing an integer CARP solution. As the number of CARP tours is fixed to $|K|$ in Bode and Irnich [24], the first two levels of the proposed branching scheme directly refer to edge flows x_e and y_e, which can be retrieved from the column-generation model using (9.24):

First, if the node degree of any node $v \in V$, given by

$$d_v := \sum_{e\in\delta_R(v)} x_e + \sum_{e\in\delta(v)} y_e = |\delta_R(v)| + y(\delta(v)),$$

is *not* an even integer, i.e., $d_v \in (2k, 2k+2)$ for an integer $k \in \mathbb{Z}_+$, two new problems with the additional constraints $y(\delta(v)) \leq 2k - |\delta_R(v)|$ and $y(\delta(v)) \geq 2k + 2 - |\delta_R(v)|$ are created. Priority is given to small node degrees d_v close to the middle $2k+1$ of the interval.

Second, if all node degrees are even integers, but any edge flow y_e for an edge $e \in E$ is fractional, i.e., $y_e \in (k, k+1)$ for an integer $k \in \mathbb{Z}_+$, then two new problems with constraints $y_e \leq k$ and $y_e \geq k+1$ are constructed. Here, priority is given to flows with a high fractional part computed as $\min\{k+1-y_e, y_e-k\}$.

If there is no node with non-even node degree and no edge with fractional flow, the tour variables λ_r and herewith the column-generation master problem may still be fractional. (A similar behavior can, for example, be observed in the split-delivery VRP(TW) Desaulniers [34], where, likewise, customers can be visited more than once.) In this respect, the exploitation of sparsity, with tours defined directly on the original graph, creates additional complications when computing feasible integer solutions, but has led to much more efficient pricing procedures (as discussed in the last paragraph).

A first complication arises for the relaxation with 2-loop free tours. Even fractional tour variables can impose a unique and feasible sequence in which required edges are served. Let $f_{ee'r}$ be the number of times that two required edges $e, e' \in E_R$ are served consecutively, i.e., with a possible deadheading connecting the two edges, but no other service between them. For example, if the tour r services edges in the sequence $(e_1, e_2, e_1, e_3, e_2, e_1)$, then $f_{e_1e_2r} = 3$ and $f_{e_1e_3r} = f_{e_2e_3r} = 1$. It is proved in Bode and Irnich [24] that

$$f_{ee'} := \sum_{r\in\Omega} f_{ee'r}\lambda_r \in \{0,1\} \quad \forall e, e' \in E_R \tag{9.26}$$

is a necessary and sufficient condition for an integer feasible CARP solution. The reason is that integer values $f_{ee'}$ imply unique sequences of services (except for reversal). Given one sequence, a least-cost tour can easily be reconstructed using a simple DP. This DP decides the orientation in which each serviced edge is traversed, taking into account that possible intermediate deadheadings must use shortest paths.

The second complication is the more serious situation that integral edges flows are implied by fractional tour variables λ. In this case, branching on follower and nonfollower decisions (9.26) allows the continuation of the branch-and-price procedure, and finally guarantees an integer CARP solution. Therefore, the third level of the branching scheme is branching on follower and nonfollower decisions, meaning that for $f_{ee'} = 1$ the two required edges are followers because they are served consecutively, while for $f_{ee'} = 0$ they are nonfollowers.

The most important advantage of this branching scheme is that it can be implemented so that the basic structure of the column-generation master as well as the pricing problem remains unchanged: For branching on node degrees and edge flows, the associated inequalities in the y variables are all of the form (9.20) and, thus, just modify the cost (9.25) for deadheading. For branching on followers and nonfollowers, the modification requires a longer explanation. We start by describing the two simpler cases of having just one active follower or nonfollower constraint. Thereafter, we briefly discuss the case of several active constraints.

In the case that there is only one follower constraint, say, that edges $e = \{i,j\}, e' = \{i',j'\} \in E_R$ must be served consecutively, both edges are removed from the graph $G = (V,E)$. Four new edges are instead inserted in order to model the paths $(i,j,\ldots,i',j')$, $(i,j,\ldots,j',i')$, $(j,i,\ldots,i',j')$, and $(j,i,\ldots,j',i')$, where "…" stands for a least-cost deadheading along a shortest path w.r.t. reduced costs $(\overline{c}_e)$. In order to prevent 2-loops w.r.t. the original edges e and e', the four new edges all represent the same task and task-2-loops have to be excluded in the pricing (Irnich and Desaulniers [50] and Irnich and Villeneuve [51]).

A single nonfollower constraint for two edges $e,e' \in E_R$ can be handled by leaving both edges in the graph $G=(V,E)$. Again, the same task is assigned to both edges so that a task-2-loop free tour never serves e and e' consecutively.

The most intricate situation is that several follower and nonfollower constraints are active at the same time. Now, the first step is to identify clusters of edges $F \subset E_R$ that are directly or indirectly linked by follower and nonfollower constraints. A cluster F with only one active constraint, i.e., $F = \{e,e'\}$ can be handled as seen above. Larger clusters with $|F| > 2$ require that in a preparative step, all compatible sequences of required edges of the cluster F are enumerated. For example, the follower constraints $f_{e_1e_2} = f_{e_3e_4} = 1$ and the nonfollower constraints $f_{e_2,e_3} = f_{e_2,e_4} = 0$ give rise to the compatible sequences (e_1,e_2), (e_3,e_4), (e_2,e_1,e_3,e_4), and (e_2,e_1,e_4,e_3), but no other sequence except for reversal. In the modified graph $G=(V,E)$, all edges F are removed, and for each computed sequence four new edges are inserted as described for single follower constraints. The cost of a new edge is the cost of a least-cost sequence that can again be computed with a simple DP. As the number of compatible sequences can increase exponentially with the cluster size $|F|$, the selection of a concrete pair e and e' in follower and nonfollower branching should favor small clusters.

The two recent studies by Bode and Irnich [26, 25] make use of the stronger k-loop free and ng-path relaxations. Both relaxations are stronger than the 2-loop free tour relaxation, and they refer to the required edges as tasks (see above). Since branching requires exactly 2-loop free tours with respect to the tasks introduced for branching, the above relaxations have to be combined with 2-loop elimination. While for ng-path relaxations the 2-loop free constraints can be integrated in a straightforward manner, a combined

$(k,2)$-loop free relaxation results in the other case. The paper by Bode and Irnich [26] introduces an efficient labeling algorithm for the solution of the $(k,2)$-loop free SP and compares the resulting branch-and-price algorithms.

The results of the computational study presented in Bode and Irnich [24] show that an approach working on the original network is competitive with that by Bartolini, Cordeau, and Laporte [12]. All *kshs* and *gdb* instances were solved within three seconds, on average in less than one second. Moreover, optimal solutions were computed for 32 out of 34 *val* instances, and the lower bounds for the two remaining instances closed the optimality gap. For the *egl* instances, 5 out of 24 instances were solved to optimality with branch-and-price. In the remaining cases, some best known lower bounds were improved, also compared to the good bounds provided by Bartolini, Cordeau, and Laporte [12]. The benchmark set *bmcv* was not included in the experiments.

The distinctive difference between the approach presented by Bode and Irnich [24] and that of Martinelli et al. [64] is that the latter does not provide a *complete* branching scheme. In fact, in [64] branching is only performed on the edge flows y_e so that route variable λ_r may remain fractional. Therefore, integer solutions are solutions for the one-index formulation, and therefore might be infeasible for the CARP. As a consequence, the entire approach of Martinelli et al. [64] generally outputs a lower bound only. The computational study presented optimal lower bounds for 3 of the 24 *egl* instances. No other benchmark sets were considered.

The very recent computational results presented in [13, 26, 25] show that improvements of the exact algorithms presented in this chapter are possible using different underlying models, valid inequalities, SP relaxations, and branching schemes. Depending on the type of CARP instance, sometimes a GVRP-based modeling approach and sometimes a sparse original network based approach turns out to best suited. It is also very instance specific whether the use of a stronger relaxation, which generally requires more time to solve a single SP, is beneficial at the end.

9.7 ▪ Variants and extensions

Up to now, we have studied methods that were proposed to solve the basic CARP defined on an undirected graph and without additional constraints and options. In the real-world applications, for which the CARP is the reference model, many additional characteristics and constraints have to be taken into account, thus giving rise to different problems that can be referred to as CARP variants. First of all, some CARP applications need to consider a directed or mixed graph. As far as we know, no exact approach has been proposed for the directed case. A further generalization consists of considering not only required arcs and edges but also required nodes; this leads to the mixed capacitated general routing problem whose study has recently begun. In the next two subsections, we will focus on the mixed and general cases. Other variants are briefly described afterwards.

9.7.1 ▪ Mixed CARP

The *Mixed Capacitated Arc Routing Problem* (MCARP) is a variant of the CARP which is defined on a mixed graph $G=(V,E\cup A)$ with a set V of nodes, an edge set E, and an arc set A. Elements of the union $E\cup A$ are called links.

Let us introduce some additional notation. Given a subset of nodes $S\subseteq V$, we denote by $\delta^+(S)$ the set of arcs leaving S, by $\delta^-(S)$ the set of arcs entering S, and by $\delta(S)$ the set of edges between S and $V\setminus S$. The cutset induced by S is defined as $\delta^L(S)=\delta^+(S)\cup\delta^-(S)\cup\delta(S)$. Let $L(S)$ be the set of links having both end nodes in S. A subscript R

in any of these sets indicates the corresponding subset of required edges/arcs/links; for instance, $\delta_R^+(S)$ denotes the set of required arcs leaving S. The unbalance of the cutset $\delta^L(S)$ is defined as $b(S) = |\delta_R^-(S)| - |\delta_R^+(S)| - |\delta_R(S)|$.

Two methods have been proposed to solve the MCARP exactly. On the one hand, Belenguer et al. [19] adapt the one-index formulation of the undirected CARP to the mixed case and embed it in a cutting-plane algorithm. On the other hand, Gouveia, Mourão, and Pinto [48] propose a compact two-index formulation for the MCARP based on flow variables and solve the corresponding aggregate model.

Mixed CARP approach by Belenguer et al. Belenguer et al. [19] adapt the one-index formulation of the undirected CARP to the mixed case by defining y_u as the total number of times the link $u \in E \cup A$ is deadheaded by the vehicles. The mathematical model includes the so called balanced-set constraints, proposed by Nobert and Picard [66] for the mixed Chinese postman problem. These constraints assure that the balanced-set conditions, which are necessary conditions for a mixed graph to be Eulerian, are satisfied:

$$y(\delta^+(S)) - y(\delta^-(S)) + y(\delta(S)) \geq b(S) \quad \forall\, \emptyset \neq S \subseteq V. \tag{9.27}$$

Belenguer et al. [19] showed that any valid inequality for the CARP can be translated into a valid inequality for the MCARP. Therefore, the separation routines used in Belenguer and Benavent [17], and briefly explained in Section 9.4.2, can be also used here. In addition, an exact separation procedure for constraints (9.27) was implemented (see [19] for details). All the identification algorithms were combined in a cutting-plane scheme, using the following strategy. The heuristics described in Section 9.4.2 that generate node sets $S \subseteq V \setminus \{d\}$ are called first; these node sets are used to check the violation of capacity, odd-cut, and also balanced-set constraints. In case that these heuristics do not find any violated constraint, the exact identification procedures for balanced-set and odd-cut constraints are called. Finally, when all the previous procedures fail, the separation procedure for disjoint-path inequalities is invoked.

The method was tested in two sets of benchmark instances. The first set, the so-called *mval* files, contains 34 instances with between 24 and 50 nodes and between 43 and 138 links, which are all required. In these instances, the number of required edges is always greater than the number of required arcs. The second set, the so-called *lpr* files, includes 15 instances with between 28 and 401 nodes and 50 to 1056 links, and for two-thirds of them, the number of required arcs is greater than the number of required edges. The lower bound obtained showed an average gap from the best upper bound known of 0.51% and 0.33%, respectively. Furthermore, the optimality of the corresponding solutions of 23 of the 34 *mval* instances and of 6 of the 15 *lpr* instances was proved.

Mixed CARP approach by Gouveia, Mourão, and Pinto Gouveia, Mourão and Pinto [48] propose using flow variables to get a compact formulation for the MCARP. They consider that the depot is also a dumpsite and a cost λ is associated to the garbage dump. They work on a directed graph $\hat{G} = (N, \hat{A})$ obtained from the original mixed graph by replacing each edge from E by two antiparallel arcs, i.e., $\hat{A} = A \cup \{(i,j),(j,i) : \{i,j\} \in E\}$. For simplicity, we use the notation $\hat{\delta}^+(S)$ for the set of arcs leaving S in $\hat{G}$, and similarly $\hat{\delta}^-(S)$.

Gouveia, Mourão, and Pinto [48] assume that the depot cannot be used as an intermediate node (see constraints (9.32) below). It can be easily shown that this assumption does not imply a loss of generality.

Their formulation uses decision variables $x_{ij}^k \in \{0,1\}$ to indicate whether or not vehicle k for $k \in K$ serves arc $(i,j) \in \hat{A}_R$, and decision variables y_{ij}^k to count the number of times vehicle k deadheads through arc $(i,j) \in \hat{A}$. Furthermore, they use the variables f_{ij}^k as the flow trough arc $(i,j) \in \hat{A}$ that is related with the total demand that remains to be served in tour k.

The model Gouveia, Mourão, and Pinto [48] propose for the MCARP is the following:

$$\text{(9.28)} \quad \min \sum_{k\in K}\left[\sum_{(i,j)\in\hat{A}_R} c_{ij}^{serv} x_{ij}^k + \sum_{(i,j)\in\hat{A}} c_{ij} y_{ij}^k + \lambda\Big(x^k(\hat{\delta}_R^-(d)) + y^k(\hat{\delta}^-(d))\Big)\right]$$

$$\text{(9.29)} \quad \text{s.t.} \quad \sum_{k\in K} x_{ij}^k = 1 \quad \forall\,(i,j)\in A_R,$$

$$\text{(9.30)} \quad \sum_{k\in K}(x_{ij}^k + x_{ji}^k) = 1 \quad \forall\,\{i,j\}\in E_R,$$

$$\text{(9.31)} \quad x^k(\hat{\delta}_R^+(i)) + y^k(\hat{\delta}^+(i)) = x^k(\hat{\delta}_R^-(i)) + y^k(\hat{\delta}^-(i)) \quad \forall\, i\in V \text{ and } k\in K,$$

$$\text{(9.32)} \quad x^k(\hat{\delta}_R^+(d)) + y^k(\hat{\delta}^+(d)) \le 1 \quad \forall\, k\in K,$$

$$\text{(9.33)} \quad f^k(\hat{\delta}^-(i)) - f^k(\hat{\delta}^+(i)) = \sum_{(j,i)\in\hat{\delta}_R^-(i)} q_{ji} x_{ji}^k \quad \forall\, i\in V\setminus\{d\} \text{ and } k\in K,$$

$$\text{(9.34)} \quad f^k(\hat{\delta}^+(d)) = \sum_{(i,j)\in\hat{A}_R} q_{ij} x_{ij}^k \quad \forall\, k\in K,$$

$$\text{(9.35)} \quad f^k(\hat{\delta}^-(d)) = \sum_{(i,j)\in\hat{\delta}_R^-(d)} q_{id} x_{id}^k \quad \forall\, k\in K,$$

$$\text{(9.36)} \quad f_{ij}^k \le Q\big(x_{ij}^k + y_{ij}^k\big) \quad \forall (i,j)\in\hat{A},\, k\in K,$$

$$\text{(9.37)} \quad x^k \in \{0,1\}^{|\hat{A}_R|},\, f^k \in \mathbb{Z}_+^{|\hat{A}|},\, y^k \in \mathbb{Z}_+^{|\hat{A}|} \quad \forall\, k\in K.$$

The objective function (9.28) includes the service and deadheading costs as well as costs of the dumping. Constraints (9.29) and (9.30) guarantee the service of each required arc and edge, respectively. Constraints (9.31) impose the continuity of tours at each node, and (9.32) imply that each vehicle departs from the depot (and returns) at most once so that the dumping cost is adequately charged in the objective function. Constraints (9.33)–(9.35) are flow-conservation constraints that together with the linking constraints (9.36) force the connectivity of the tours and assure that the capacity constraints are satisfied. Note that (9.33) are generalized flow-conservation constraints guaranteeing that, if arc (j,i) is serviced by vehicle k, then q_{ji} units of flow are absorbed by node i.

The following sets of valid inequalities can strengthen the linear relaxation of model (9.28)–(9.37):

$$\text{(9.38)} \quad \sum_{k\in K}\Big(x^k(\hat{\delta}_R^+(d)) + y^k(\hat{\delta}^+(d))\Big) \ge \left\lceil \frac{D_T}{Q} \right\rceil,$$

$$\text{(9.39)} \quad f_{ij}^k \ge q_{ij} x_{ij}^k \quad \forall (i,j)\in\hat{A}_R,\, k\in K,$$

$$\text{(9.40)} \quad f_{ij}^k \ge y_{ij}^k - 1 \quad \forall (i,j)\in\hat{A}\setminus\hat{A}_R,\, k\in K,$$

$$\text{(9.41)} \quad x^k(\hat{\delta}_R^+(d)) + y^k(\hat{\delta}^+(d)) \ge x^{k+1}(\hat{\delta}_R^+(d)) + y^{k+1}(\hat{\delta}^+(d)), \quad k = 1,\dots,|K|-1.$$

Constraint (9.38) establishes the minimum number of nonempty tours in any feasible solution. Constraints (9.39) and (9.40) are obvious. Finally, constraints (9.41) assure that, if p nonempty tours are used in the solution, it will be precisely those with indices from 1 to p.

The number of variables is equal to $|K|(2|\hat{A}| + |\hat{A}_R|)$, and the number of constraints is $|K|(2|V| + 2|\hat{A}| + 3) + |A_R| + |E_R| + 1$. These numbers are too large for solving the integer programm directly. For this reason, the authors proposed an aggregate one-index formulation using variables $x_{ij} = \sum_{k\in K} x_{ij}^k$, $y_{ij} = \sum_{k\in K} y_{ij}^k$, and $f_{ij} = \sum_{k\in K} f_{ij}^k$, and adding each family of constraints (9.31)–(9.36), for all vehicles. This formulation is also reinforced by the aggregate version of constraints (9.38)–(9.40). It has the same drawback as the preceding one-index formulation for the MCARP: It is a relaxation as it may contain solutions that are not MCARP solutions.

Let F_1 be the valid formulation (9.28)–(9.37) and F_2 the aggregate model. The authors prove that the optimum values of the linear relaxations of F_1 and F_2 are identical. A similar result is also proved for the strengthened formulations that are denoted by F_1R and F_2R, respectively. Gouveia, Mourão, and Pinto [48] report computational results obtained by F_1R and F_2R, and their linear relaxations, when applied to the same benchmark sets analyzed by Belenguer et al. [19]. They report the best lower bound obtained with each model when solved with the IP solver CPLEX with a time limit of one hour. The best results, in terms of both computation time and lower bounds, were obtained with formulation F_2R, which shows similar results to those reported by Belenguer et al. [19]: They proved the optimality of 23 of the 34 *mval* instances and 6 of the 15 *lpr* instances. The average gaps obtained with the best upper bounds known were 0.69% and 0.34%, respectively.

9.7.2 ▪ Mixed capacitated general routing problem

Recently, several papers have appeared that deal with a more general problem. In the Mixed Capacitated General Routing Problem (MCGRP) the demands can be located in the links (arcs or edges), but also on a subset of nodes that are called required nodes. This problem generalizes both the CARP and the CVRP and models certain situations occurring in garbage collection or postal delivery where, while most of the demand can be located along links, some demand points (like hospitals, for instance) can be better modeled as demand nodes. This problem is sometimes also referred to as the Capacitated General Routing Problem on mixed graphs (CGRP-m) or the Node, Edge, and Arc Routing Problem (NEARP). In this subsection the required edges, arcs, and nodes will be jointly called required entities. Four exact methods to solve the MCGRP have been appeared recently, up to our knowledge: Bosco et al. [28], Gaze [43], Bode et al. [27], and Bach [8].

Bosco et al. [28] solve the MCGRP with a branch-and-cut algorithm based on a two-index formulation. Their formulation follows the lines of the two-index formulation shown in Subsection 9.4.1 although it is adapted to the mixed case using directed traversal and service variables for the arcs and edges (both possible directions of service and traversal are considered for the edges as in the MCARP formulation (9.28)–(9.37)). Moreover, it includes service variables for required nodes. The formulation comprises capacity and connectivity constraints that are similar to those shown in Subsection 9.4.1, and also symmetry constraints similar to (9.31). In order to strengthen the linear relaxation, the authors also used aggregate constraints that were adapted from the one-index formulation of Subsection 9.4.2: aggregate capacity and odd-cut constraints. The performance of the branch-and-cut algorithm is evaluated by carrying out computational experiments on 12 datasets derived from classical datasets for the undirected and mixed CARP. The algorithm of Bosco et al. optimally solves 154 of the 264 instances. The authors point out that it is less effective when the number of vehicles increases, so their computational results are restricted to instances with at most seven vehicles.

Gaze [43] has also studied the Mixed Capacitated General Routing Problem (MCGRP). A formulation of the MCGRP is proposed that is very similar to that of Bosco et al. [28]. Nevertheless the solution method proposed for the MCGRP is based on a SP formulation that is solved with a branch-and-price procedure. The pricing problem is formulated as a compact ILP formulation (with a number of variables and constraints that is polynomial in the size of the instance) where the capacity and connectivity constraints are imposed using commodity-flow variables similar to those used by Gouveia, Mourão, and Pinto [48]. On the other hand, branching is carried out by adding *pairing* and *non-pairing* constraints: One node is created where two required entities must be in the same route, and another one where these two entities cannot be in the same route. These constraints are easily imposed as additional constraints in the pricing problem, and the authors prove that they are enough to guarantee integrality. Computational results show that the method can solve medium-size instances and is more efficient when the number of vehicles increases (or the number of required entities serviced per vehicle decreases).

Recently, Bach [8] proposed a branch-and-cut-and-price algorithm for the MCGRP based on a combined model inspired by the modeling technique proposed by Fukasawa et al. [42] for the CVRP. The model combines a generalization of the two-index formulation of Subsection 9.4.1 with a set partitioning model. Both types of variables, the ones from the compact and SP formulation, are coupled with additional coupling constraints. The pricing problem is solved as a shortest path problem with resource constraints using a labeling algorithm on the original graph (see paragraph on Relaxations and Pricing in Section 9.6.3). Valid inequalities are added to the master problem as cuts without changing the structure of the pricing problem. The valid constraints used are adapted from the one-index formulation of Subsection 9.4.2: capacity and odd-cut constraints. In order to ensure nonnegative reduced costs for deadheading, Bach relaxes the coupling equalities into inequalities (see [8, Sect. 2.4.4]). The branching strategy is based on *follower* and *nonfollower* rules as in Bode and Irnich [24]. The algorithm is tested on 215 instances, and 122 are solved to optimality. For all instances also tested by Bosco et al. [28], his algorithm provides better lower bounds.

Bode et al. [27] present a mathematical model for the MCGRP based on a two-index formulation that is similar to the one proposed by Bosco et al. [28], although they use undirected edge variables and include balance constraints that are a disaggregate version of constraints (9.27). The authors propose a two-phase branch-and-cut algorithm that uses first an aggregate formulation, similar to the one proposed by Belenguer et al. [19], to develop an effective lower bounding procedure. This procedure also provides strong valid inequalities for the two-index model. Two kinds of branching are used: the standard branching on fractional variables and also branching on the vertex degree whenever it is odd. Their algorithm improves the results of Bosco et al. [28], and it is also applied to instances with a number of vehicles greater than seven, although it is not able to optimally solve any of them.

9.7.3 ▪ Other variants

Other CARP variants are fully described in Chapter 10 of this book. We summarize in Table 9.1 variants for which branch-and-price methods have been proposed, whereas Table 9.2 shows those variants that have been solved exactly by other techniques. The first four columns are identical in both tables and display the reference, a short description of the variant, the underlying graph, and the formulation used.

The last column in Table 9.1 shows the method for solving the column-generation subproblem. In general, the subproblem has to be solved at each iteration to obtain routes with negative reduced cost. The only exception is the paper by Johnson and Wøhlk [53] in which all feasible routes are computed at the beginning and, in each iteration, a minimum reduced cost route to be added is determined by direct inspection of the precomputed routes. In Bartolini, Cordeau, and Laporte [13], the subproblem is solved by transforming the *CARP with Deadheading Demand* (CARPDD) into a generalization of the GVRP defined on a directed multigraph, and searching for GVRP q-routes without loops of two consecutive clusters, which correspond to 3-loop free CARPDD q-routes.

The last column in Table 9.2 lists the methods used to solve the proposed formulations. In two of them, the problem is solved by a classical cutting-plane algorithm. The two papers related with refill points apply a different approach based also on cutting-plane techniques, but solving an integer program instead of a linear relaxation. Finally, Eskandarzadeh, Tavakkoli-Moghaddam, and Azaron [41] propose a Lagrangean relaxation that is solved with methods derived from monotropic programming.

9.8 ▪ Conclusions and outlook

This chapter has shown that over the years several groups of researchers developed a variety of different exact solution approaches for the CARP. Some methods, such as those using combinatorial lower bounds (Section 9.3) or the one-index formulation (Section 9.4.2), basically provide lower bounds on optimal solutions. They have to be complemented with good (meta)heuristics for computing solutions and upper bounds that in some cases close the optimality gap and so provide a proof of optimality. The other methods presented here are all variants of branch-and-bound. They are complete exact methods, meaning that they guarantee the computation of an optimal solution at termination, if the underlying branching scheme is complete. The recently most successful ones are either based on cutting-plane methods and the two-index formulation (Section 9.5), or on cut-and-column-generation techniques and an extensive formulation (Section 9.6).

The one-index formulation solved with cutting-plane algorithms alone is incomplete, but provides a very fast way to compute generally excellent lower bounds. It is questionable if it can be modified into a complete formulation. However, even if incomplete, it would be interesting to further strengthen it by identifying new classes of valid inequalities. The respective lower bounds can then be used alone or for tentative bounding purposes in more complex exact solution approaches.

We have seen that the two-index formulation (Section 9.4.1) solved by branch-and-cut gives a complete method. Conceptually, it is simple to adapt to all kinds of extensions. Unfortunately however, the two-index formulation suffers from the inherent symmetry. Hence, new ideas on symmetry-breaking constraints might help to make branch-and-cut approaches attractive, in particular if instances with only very few vehicles need to be optimized.

The existing approaches that use column-generation techniques (Section 9.6) rely on different extensive formulations and, more importantly, are either based on (explicit or implicit) transformations of the CARP into a node routing problem or work on the CARP graph directly exploiting sparsity. At the moment, there is no clear winner and we can just point out the trade-offs between these methods: First, the direct transformation of the CARP into a CVRP can make use of all kinds of cut-and-column-generation procedures for the CVRP, and is therefore perfectly suited to provide very strong lower bounds.

Bartolini, Cordeau, and Laporte [12] pointed out that, by lifting some valid inequalities of the CARP in the node routing–based extensive formulation, the resulting linear relaxation can produce improved bounds compared to column-generation methods working directly on the original graph. Second, the latter methods such as the algorithm by Bode and Irnich [24] can best exploit the sparsity of typical CARP instances and, thus, they have the potential of being solved fastest. However, while branching schemes are long established and uncritical for VRP branch-and-price procedures, the direct CARP graph-based methods require more intricate branching schemes to make them complete exact methods.

Finally, the discussed trade-offs for the CARP will have to be analyzed also for all kinds of extensions of the CARP as sketched in Section 9.7. We expect that in these cases the column-generation subproblems become harder to solve depending on the kind of extension. Effective pricing heuristics will therefore become a key success factor.

Certainly, the research on rich VRP, i.e., unifying approaches for modeling and solving VRP with many types of standard and application-specific constraints, will have an impact when tackling challenging old and new CARP variants.

Acknowledgments

José Manuel Belenguer wishes to thank the Generalitat Valenciana (project GVPROMETEO2013-049) and the Ministerio de Ciencia e Innovación (project DPI2011-24977) for their support. Enrique Benavent wishes to thank the Ministerio de Economía y Competitividad of Spain (project MTM2012-36163-C06-02) and the Generalitat Valenciana (project GVPROMETEO2013-049) for their support. Project MTM2012-36163-C06-02 is funded in cooperation with the European Community Fund FEDER.

Table 9.1. *Branch-and-price approaches for CARP variants.*

Reference	CARP variant	Graph	Formulation	Subproblem
Johnson and Wøhlk [53]	time windows	undirected	set partitioning	exhaustive
Afsar [1]	flexible time windows	undirected	set partitioning	nonelementary shortest path subproblem on a complete graph
Archetti et al. [6]	profits and max time route	undirected	set partitioning	elementary shortest path subproblem on the original graph
Bartolini, Cordeau, and Laporte [13]	deadheading demands	undirected	set covering	GVRP q-routes without loops of two consecutive clusters
Christiansen, Lysgaard, and Wøhlk [30]	stochastic demands	undirected	set covering	nonelementary shorthest path subproblem without 1-cycles

Table 9.2. *Other approaches for CARP variants.*

Reference	CARP variant	Graph	Formulation	Solution method
Amaya, Langevin, and Trépanier [4]	refill points	directed	two-index	IP cutting plane
Amaya, Langevin, and Trépanier [5]	refill points and multiple loads	directed	two-index	IP cutting plane
Ghiani, Improta, and Laporte [44]	intermediate facilities	undirected	one-index	cutting plane
Belenguer et al. [18]	split delivery	undirected	one-index	cutting plane
Eskandarzadeh, Tavakkoli-Moghaddam, and Azaron [41]	split delivery	undirected	one-index with flow variables	Lagrangian relaxation

Bibliography

[1] H.M. AFSAR, *A branch-and-price algorithm for capacitated arc routing problem with flexible time windows*, Electronic Notes in Discrete Mathematics, 36 (2010), pp. 319–326.

[2] D. AHR, *Contributions to Multiple Postmen Problems*, Ph.D. dissertation, Ruprecht-Karls-Universität Heidelberg, Heidelberg, Germany, 2004.

[3] R.K. AHUJA, T.L. MAGNANTI, AND J.B. ORLIN, *Network Flows: Theory, Algorithms, and Applications*, Prentice–Hall, Englewood Cliffs, NJ, 1993.

[4] C.-A. AMAYA, A. LANGEVIN, AND M. TRÉPANIER, *The capacitated arc routing problem with refill points*, Operations Research Letters, 35 (2007), pp. 45–53.

[5] ——, *A heuristic method for the capacitated arc routing problem with refill points and multiple loads*, The Journal of the Operational Research Society, 61 (2010), pp. 1095–1103.

[6] C. ARCHETTI, D. FEILLET, A. HERTZ, AND M.G. SPERANZA, *The undirected capacitated arc routing problem with profits*, Computers & Operations Research, 37 (2010), pp. 1860–1869.

[7] A.A. ASSAD AND B.L. GOLDEN, *Arc routing methods and applications*, in Network Routing, M.O. Ball, T.L. Magnanti, C.L. Monma, and G.L. Nemhauser, eds., Handbooks in Operations Research and Management Science 8, Elsevier Science Publishers B.V., North–Holland, Amsterdam, 1995, pp. 375–483.

[8] L. BACH, *Routing and Scheduling Problems: Optimization using Exact and Heuristic Methods*, Ph.D. dissertation, School of Business and Social Sciences, Aarhus University, Aarhus, Denmark, 2014.

[9] R. BALDACCI AND M. MANIEZZO, *Exact methods based on node-routing formulations for undirected arc-routing problems*, Networks, 47 (2006), pp. 52–60.

[10] R. BALDACCI, A. MINGOZZI, AND R. ROBERTI, *New route relaxation and pricing strategies for the vehicle routing problem*, Operations Research, 59 (2011), pp. 1269–1283.

[11] F. BARAHONA AND M. GRÖTSCHEL, *On the cycle polytope of a binary matroid*, Journal of Combinatorial Theory, Series B, 40 (1986), pp. 40–62.

[12] E. BARTOLINI, J.-F. CORDEAU, AND G. LAPORTE, *Improved lower bounds and exact algorithm for the capacitated arc routing problem*, Mathematical Programming, 137 (2013), pp. 409–452.

[13] ——, *An exact algorithm for the capacitated arc routing problem with deadheading demand*, Operations Research, 61 (2013), pp. 315–327.

[14] J.M. BELENGUER, *El Poliedro del Problema de Rutas por Arcos con Capacidades (The Capacitated arc routing problem polyhedron)*, Ph.D. dissertation, University of Valencia, Valencia, Spain, 1990.

[15] J.M. BELENGUER AND E. BENAVENT, *Branch and cut for the CARP*, in Workshop on Algorithmic Approaches to Large and Complex Combinatorial Optimization Problems, Giens, France, 1994.

[16] ——, *The capacitated arc routing problem: Valid inequalities and facets*, Computational Optimization and Applications, 10 (1998), pp. 165–187.

[17] ——, *A cutting plane algorithm for the capacitated arc routing problem*, Computers & Operations Research, 30 (2003), pp. 705–728.

[18] J.M. BELENGUER, E. BENAVENT, N. LABADI, C. PRINS, AND M. REGHIOUI, *Split-delivery capacitated arc-routing problem: Lower bound and metaheuristic*, Transportation Science, 44 (2010), pp. 206–220.

[19] J.M. BELENGUER, E. BENAVENT, P. LACOMME, AND C. PRINS, *Lower and upper bounds for the mixed capacitated arc routing problem*, Computers & Operations Research, 33 (2006), pp. 3363–3383.

[20] H. BEN AMOR, J. DESROSIERS, AND J.M. VALÉRIO DE CARVALHO, *Dual-optimal inequalities for stabilized column generation*, Operations Research, 54 (2006), pp. 454–463.

[21] E. BENAVENT, V. CAMPOS, A. CORBERÁN, AND E. MOTA, *The capacitated arc routing problem: Lower bounds*, Networks, 22 (1992), pp. 669–669.

[22] E. BENAVENT, A. CORBERÁN, AND J.M. SANCHIS, *Linear programming based methods for solving arc routing problems*, in Arc Routing: Theory, Solutions and Applications, M. Dror, ed., Kluwer, Boston, 2000, pp. 231–275.

[23] P. BEULLENS, L. MUYLDERMANS, D. CATTRYSSE, AND D. VAN OUDHEUSDEN, *A guided local search heuristic for the capacitated arc routing problem*, European Journal of Operational Research, 147 (2003), pp. 629–643.

[24] C. BODE AND S. IRNICH, *Cut-first branch-and-price-second for the capacitated arc-routing problem*, Operations Research, 60 (2012), pp. 1167–1182.

[25] ——, *In-depth analysis of pricing problem relaxations for the capacitated arc-routing problem*, Transportation Science, (2014). In Press.

[26] ——, *The shortest-path problem with resource constraints with $(k,2)$-loop elimination and its application to the capacitated arc-routing problem*, European Journal of Operational Research, 238 (2014), pp. 415–426.

[27] C. BODE, S. IRNICH, D. LAGANÀ, AND F. VOCATURO, *Two-phase branch-and-cut for the mixed capacitated general routing problem*, Technical Report LM-2014-02, Chair of Logistics Management, Gutenberg School of Management and Economics, Johannes Gutenberg University Mainz, Mainz, Germany, 2014.

[28] A. BOSCO, D. LAGANÀ, R. MUSMANNO, AND F. VOCATURO, *Modeling and solving the mixed capacitated general routing problem*, Optimization Letters, 7 (2013), pp. 1451–1469.

[29] J. BRANDÃO AND R.W. EGLESE, *A deterministic tabu search for the capacitated arc routing problem*, Computers & Operations Research, 35 (2008), pp. 1112–1126.

[30] C.H. CHRISTIANSEN, J. LYSGAARD, AND S. WØHLK, *A branch-and-price algorithm for the capacitated arc routing problem with stochastic demands*, Operations Research Letters, 37 (2009), pp. 392–398.

[31] N. CHRISTOFIDES, *The optimum traversal of a graph*, Omega, 1 (1973), pp. 719–732.

[32] N. CHRISTOFIDES, A. MINGOZZI, AND P. TOTH, *State-space relaxation procedures for the computation of bounds to routing problems*, Networks, 11 (1981), pp. 145–164.

[33] A. CORBERÁN AND C. PRINS, *Recent results on arc routing problems: An annotated bibliography*, Networks, 56 (2010), pp. 50–69.

[34] G. DESAULNIERS, *Branch-and-price-and-cut for the split-delivery vehicle routing problem with time windows*, Operations Research, 58 (2010), pp. 179–192.

[35] G. DESAULNIERS, J. DESROSIERS, I. IOACHIM, M.M. SOLOMON, F. SOUMIS, AND D. VILLENEUVE, *A unified framework for deterministic time constrained vehicle routing and crew scheduling problems*, in Fleet Management and Logistics, T.G. Crainic and G. Laporte, eds., Kluwer Academic Publishers, Dordrecht, The Netherlands, 1998, pp. 57–93.

[36] G. DESAULNIERS, J. DESROSIERS, AND M.M. SOLOMON, eds., *Column Generation*, Springer, New York, 2005.

[37] M. DREXL AND S. IRNICH, *Solving elementary shortest-path problems as mixed-integer programs*, OR Spectrum, 36 (2014), pp. 281–296.

[38] M. DROR, ED., *Arc Routing: Theory, Solutions, and Applications*, Kluwer, Boston, 2000.

[39] R.W. EGLESE AND A.N. LETCHFORD, *Polyhedral theory for arc routing problems*, in Arc Routing: Theory, Solutions, and Applications, M. Dror, ed., Kluwer, Boston, 2000, pp. 199–230.

[40] H.A. EISELT, M. GENDREAU, AND G. LAPORTE, *Arc routing problems, part II: The rural postman problem*, Operations Research, 43 (1995), pp. 399–414.

[41] S. ESKANDARZADEH, R. TAVAKKOLI-MOGHADDAM, AND A. AZARON, *An extension of the relaxation algorithm for solving a special case of capacitated arc routing problems*, Journal of Combinatorial Optimization, 17 (2009), pp. 214–234.

[42] R. FUKASAWA, H. LONGO, J. LYSGAARD, M. POGGI DE ARAGÃO, M. REIS, E. UCHOA, AND R.F. WERNECK, *Robust branch-and-cut-and-price for the capacitated vehicle routing problem*, Mathematical Programming Series A, 106 (2006), pp. 491–511.

[43] K. GAZE, *Exact Optimization Methods for the Mixed Capacitated General Routing Problem*, Master thesis, Norwegian University of Science and Technology, Trondheim, Norway, 2013.

[44] G. GHIANI, G. IMPROTA, AND G. LAPORTE, *The capacitated arc routing problem with intermediate facilities*, Networks, 37 (2001), pp. 134–143.

[45] B.L. GOLDEN, J.S. DEARMON, AND E.K. BAKER, *Computational experiments with algorithms for a class of routing problems*, Operations Research, 10 (1983), pp. 47–59.

[46] B.L. GOLDEN AND R.T. WONG, *Capacitated arc routing problems*, Networks, 11 (1981), pp. 305–315.

[47] D. Gómez-Cabrero, J.M. Belenguer, and E. Benavent, *Cutting plane and column generation for the capacitated arc routing problem*, in Proceedings of the ORP3 2005 (Operational Research Peripathetic Postgraduate Programme), Valencia, Spain, 2005, pp. 259–266.

[48] L. Gouveia, M.C. Mourão, and L.S. Pinto, *Lower bounds for the mixed capacitated arc routing problem*, Computers & Operations Research, 37 (2010), pp. 692–699.

[49] R. Hirabayashi, Y. Saruwatari, and N. Nishida, *Tour construction algorithm for the capacitated arc routing problem*, Asia-Pacific Journal of Operational Research, 9 (1992), pp. 155–175.

[50] S. Irnich and G. Desaulniers, *Shortest path problems with resource constraints*, in Column Generation, G. Desaulniers, J. Desrosiers, and M.M. Solomon, eds., Springer, New York, 2005, pp. 33–65.

[51] S. Irnich and D. Villeneuve, *The shortest path problem with resource constraints and k-cycle elimination for $k \geq 3$*, INFORMS Journal on Computing, 18 (2006), pp. 391–406.

[52] M. Jepsen, B. Petersen, S. Spoorendonk, and D. Pisinger, *Subset-row inequalities applied to the vehicle-routing problem with time windows*, Operations Research, 56 (2008), pp. 497–511.

[53] E.L. Johnson and S. Wøhlk, *Solving the capacitated arc routing problem with time windows using column generation*, Coral working papers L-2008-09, Aarhus School of Business, Department of Business Studies, University of Aarhus, Aarhus, Denmark, 2009. Available at http://pure.au.dk/portal/files/3875/L_2008_09.PDF.

[54] M. Kiuchi, Y. Shinano, R. Hirabayashi, and Y. Saruwatari, *An exact algorithm for the capacitated arc routing problem using parallel branch and bound method*. In Abstracts of the 1995 Spring National Conference of the Operational Research Society of Japan, 1995 (In Japanese).

[55] J.K. Lenstra and A.H.G. Rinnooy Kan, *On general routing problems*, Networks, 6 (1976), pp. 273–280.

[56] A.N. Letchford, *Polyhedral Results for Some Constrained Arc-Routing Problems*, Ph.D. dissertation, Department of Management Science, Lancaster University, Lancaster, UK, 1997.

[57] A.N. Letchford and A. Oukil, *Exploiting sparsity in pricing routines for the capacitated arc routing problem*, Computers & Operations Research, 36 (2009), pp. 2320–2327.

[58] L.Y.O. Li and R.W. Eglese, *An interactive algorithm for vehicle routing for winter-gritting*, The Journal of the Operational Research Society, 47 (1996), pp. 217–228.

[59] H. Longo, M. Poggi de Aragão, and E. Uchoa, *Solving capacitated arc routing problems using a transformation to the CVRP*, Computers & Operations Research, 33 (2006), pp. 1823–1837.

[60] M.E. LÜBBECKE AND J. DESROSIERS, *Selected topics in column generation*, Operations Research, 53 (2005), pp. 1007–1023.

[61] J. LYSGAARD. *A CVRPSEP package*. Available at http://www.hha.dk/lys/CVRPSEP.htm, 2004.

[62] J. LYSGAARD, A.N. LETCHFORD, AND R.W. EGLESE, *A new branch-and-cut algorithm for the capacitated vehicle routing problem*, Mathematical Programming, 100 (2004), pp. 423–445.

[63] S. MARTELLO AND P. TOTH, *Knapsack Problems: Algorithms and Computer Implementations*, Wiley Intersience Series in Discrete Mathematics and Optimization, Wiley, New York, 1990.

[64] R. MARTINELLI, D. PECIN, M. POGGI DE ARAGÃO, AND H. LONGO, *A branch-cut-and-price algorithm for the capacitated arc routing problem*, in Experimental Algorithms, P. Pardalos and S. Rebennack, eds., Lecture Notes in Computer Science 6630, Springer, Berlin, Heidelberg, 2011, pp. 315–326.

[65] R. MARTINELLI, M. POGGI DE ARAGÃO, AND A. SUBRAMANIAN, *Improved bounds for large scale capacitated arc routing problem*, Computers & Operations Research, 40 (2013), pp. 2145–2160.

[66] Y. NOBERT AND J.-C. PICARD, *An optimal algorithm for the mixed Chinese postman problem*, Networks, 27 (1996), pp. 95–108.

[67] M.W. PADBERG AND M.R. RAO, *Odd minimum cut-sets and b-matchings*, Mathematics of Operations Research, 7 (1982), pp. 67–80.

[68] W.L. PEARN, A.A. ASSAD, AND B.L. GOLDEN, *Transforming arc routing into node routing problems*, Computers & Operations Research, 14 (1987), pp. 285–288.

[69] G. RIGHINI AND M. SALANI, *Symmetry helps: Bounded bi-directional dynamic programming for the elementary shortest path problem with resource constraints*, Discrete Optimization, 3 (2006), pp. 255–273.

[70] Y. SARUWATARI, R. HIRABAYASHI, AND N. NISHIDA, *Node duplication lower bounds for the capacitated arc routing problems*, Journal of the Operational Research Society of Japan, 35 (1992), pp. 119–133.

[71] ———, *Subtour elimination algorithm for the capacitated arc routing problem*, in Operations Research Proceedings 1990: Papers of the 19th Annual Meeting of DGOR (Vienna, 1992), W. K. Bühler, G. Feichtinger, R. F. Hartl, F. J. Radermacher, and P. Stahly, eds., Springer-Verlag, New York, 1992, pp. 334–341.

[72] P. TOTH AND D. VIGO, EDS., *The Vehicle Routing Problem*, SIAM Monographs on Discrete Mathematics and Applications, SIAM, Philadelphia, 2001.

[73] P. TOTH AND D. VIGO, eds., *Vehicle Routing: Problems, Methods, and Applications*, MOS-SIAM Series on Optimization 18, SIAM, Philadelphia, 2014.

[74] S.A. WELZ, *Optimal solutions for the capacitated arc routing problem using integer programming*, Ph.D. dissertation, Department of Quantitative Analysis and Operations Management, University of Cincinnati, Cincinnati, OH, 1994.

[75] S. WØHLK, *A decade of capacitated arc routing*, in The Vehicle Routing Problem: Latest Advances and New Challenges, B.L. Golden, S. Raghavan, and E.A. Wasil, eds., Operations Research/Computer Science Interfaces Series 43, Springer, Boston, 2008, pp. 29–48.

Chapter 10

Variants of the Capacitated Arc Routing Problem

Luc Muyldermans
Gu Pang

10.1 ▪ Introduction

The *Undirected Capacitated Arc Routing Problem* (UCARP or CARP), first introduced by Golden and Wong [46], considers an undirected network or graph where every edge has a nonnegative length or cost and some of the edges (also called the required edges) have a positive demand for service. A fleet of identical vehicles, each vehicle with limited capacity, is based at the depot. The aim in the UCARP is to find a set of vehicle tours of total minimum length, each tour starting and ending at the depot, such that every required edge is serviced in exactly one tour, and that the total load on each route does not exceed the vehicle capacity. Heuristic solution procedures, lower bounds, and exact methods for CARP were covered in Chapters 7, 8, and 9 of this book.

Over the past few decades several variants and extensions of the standard UCARP have been proposed in the literature. More than 30 publications just dealing with CARP variants were discussed in the recent bibliography on arc routing problems by Corberán and Prins [26]. Most of these articles were printed after the publication of the well-known book *Arc Routing: Theory, Solutions, and Applications*, edited by Dror [30] in 2000. Wøhlk [102] also provides an overview of new developments in the area of capacitated arc routing till 2007. We were able to identify more than 45 other and new articles related to CARP variants since the publication of the overview by Corberán and Prins [26]. The amount of publications in recent years shows that the study of CARP variants is a thriving and rich area of research, to which many researchers contribute. It also gives an indication of the importance of this domain for practical applications, as, essentially, these CARP variants are studied in order to better model more complex, more realistic, or new real-life applications.

This chapter aims to provide an overview of recently developed CARP variants. Its focus is on both the articles discussed in Corberán and Prins [26] and on the newly published material. We review the solution procedures and techniques proposed, highlight key algorithmic features, and briefly discuss results from computational experiments. Whenever possible, we will also report on new managerial insights obtained from experimentation or applying these CARP variants to real-life case studies. We have intentionally left out all mathematical formulations and notations in this chapter, simply because it is impossible to present all these details in limited space.

Table 10.1 provides an overview of the different CARP variants discussed in this chapter. A variant can originate by modifying one or several of the key assumptions in the standard UCARP definition, including network characteristics (directed and mixed graphs), vehicle characteristics (multi-compartment, heterogeneous fleet), tour and facility characteristics (open tour, multiple depots, intermediate facilities, mobile depots), demand characteristics (demand location, demand splitting, time windows, time-dependent service costs, periodic demand, stochastic demand, deadheading demand), and objective function (minimize the number of vehicles, minimize fixed and variable routing cost, minimize makespan or the longest tour length, multiobjective routing, profit maximization). The different CARP variants, their acronyms, and the corresponding section in this chapter discussing the variant are indicated in Table 10.1. In Section 10.7 we report our conclusions and point to some areas for further research.

We note that a number of CARP variant studies differ from the UCARP in several features, which makes our classification of articles according to Table 10.1 a bit more cumbersome. In that case, we discuss the relevant study in the section that captures (in our opinion) the most distinguishing feature from the UCARP.

Table 10.1. *CARP variants.*

Characteristic	Option	Acronym	Section
Network	Directed	DCARP	10.2.1
	Mixed graph	MCARP	10.2.2
Vehicle	Multi-compartment	MCCARP	10.3.1
	Heterogeneous fleet	CARPVSD	10.3.2
Tour / Facility	Open tour	OCARP	10.4.1
	Multiple depots	MDCARP	10.4.2
	Intermediate facilities	CARPIF	10.4.3
	Mobile depots	CARPMD	10.4.4
Demand	Demand location	CGRP	10.5.1
	Demand splitting	SDCARP	10.5.2
	Time windows	CARPTW	10.5.3
	Time-dependent service costs	CARPTDSC	10.5.4
	Periodic demand	PCARP	10.5.5
	Stochastic demand	CARPSD	10.5.6
	Deadheading demand	CARPDD	10.5.7
Objective function	Multiobjective	MOCARP	10.6.1
	Profits	CARPP	10.6.2

Many CARP variant algorithms are tested on modified UCARP instances. In order to avoid too much repetition, we provide below the key to well-known UCARP instances. These test sets can be obtained from http://www.uv.es/~belengue/carp.html.

gdb: 23 instances with 7 to 27 nodes and 11 to 55 edges, all required (Golden et al. [45])

val: 34 instances with 24 to 50 nodes and 34 to 97 edges, all required (Benavent et al. [13])

egl: 24 instances with 77 to 140 nodes, 98 to 190 edges, and 51 to 190 required edges (Li and Eglese [66])

kshs: 6 instances with 6 to 9 nodes, 15 edges, all required (Kiuchi et al. [58])

10.2 ▪ CARP variants: Network characteristics

10.2.1 ▪ The directed Capacitated Arc Routing Problem

The *Directed Capacitated Arc Routing Problem* (DCARP) differs from the UCARP in that the network links are directed arcs (instead of undirected edges) so that direction of travel is fixed. Compared with undirected networks, there is usually less flexibility (or fewer paths) to travel between any two destinations in directed networks (unless every undirected edge is replaced by two arcs with opposite direction). There are relatively few studies in the literature that deal with the DCARP.

Mourão and Almeida [73] model a household waste collection problem in Lisbon as a DCARP with additional side constraints. One-way streets are modelled by single arcs, and two-way streets are represented by two arcs with opposite direction (assuming collection on two-way streets requires two passes). Household waste collection is planned by quarter, with vehicles departing from a central garage located outside the quarter and first picking up the collection crew at the crew base. The first collection tour starts at the crew base and collects waste along the streets until its capacity is reached and then unloads at a recycling facility. Subsequent tours start and end at the recycling facility, while in the final tour the crew is first dropped off before unloading the vehicle at the recycling site and returning to the garage. The authors present two lower bounding procedures based on the transportation model and a route-first, cluster-second heuristic to generate solutions. Tests on instances with up to 50 nodes and 97 required arcs show that the heuristic produces solutions with an average gap of around 12% and that running times were negligible.

Maniezzo and Roffilli [67] also model a solid waste collection problem in urban areas as DCARP. Arcs are used to model one-way streets and forbidden turns, and like in the previous study, two-way streets are represented by two arcs with opposite direction. The authors [67] transform the DCARP into an *Asymmetric Capacitated Vehicle Routing Problem* (ACVRP) and solve the ACVRP by a modified iterated *Variable Neighborhood Search* (VNS). Three different diversification or shaking strategies are explored in VNS: random *Multi Start* (MS), *Genetic Algorithm* (GA), and *Data Perturbation* (DP). Initial solutions were constructed by a two-phase algorithm (Christofides et al. [21]). The three VNS variants show comparable performance on the small DCARP instances based on the *gdb* test set, and DP performed best on larger *val*-based instances. VNS with DP is then tested on five large real-life instances with up to 12,388 nodes and 19,045 arcs. The computational results show that VNS with DP improve the initial solutions by 12% to 32%, with running times around 9,000 to 17,000 seconds.

10.2.2 ▪ The mixed Capacitated Arc Routing Problem

There are more studies that deal with the CARP on mixed graphs, most of these inspired by operations to be carried out on road networks such as household refuse collection, mail collection or delivery, snow plowing, street sweeping, and winter gritting. A mixed network contains both arcs and edges and is therefore a much more realistic representation of a real street network, whereas the pure UCARP or DCARP may be too simplistic. In a mixed graph, arcs are introduced to model one-way streets, forbidden turns, windy edges (travel cost dependent on direction of travel), or two-way streets that cannot be serviced in one pass; edges are used to model two-way streets where service can be carried out in one pass.

Lacomme et al. [62] introduced the *Extended Capacitated Arc Routing Problem* (ECARP) in which they consider several additional constraints: a mixed graph with two types of links (edges and arcs) and parallel links, two distinct costs per link (deadheading and collecting), prohibited turns and turn penalties (penalise left turns), and a maximum trip length. In order to deal with these features, the authors present several modifications to the *Memetic Algorithm* (MA) of Lacomme et al. [61] and three well-known UCARP heuristics, which are later tested in Belenguer et al. [12].

Belenguer et al. [12] present a cutting-plane algorithm to obtain a lower bound for the MCARP. The formulation is based on the supersparse formulation by Belenguer and Benavent [10] for the UCARP yet generalised for mixed graphs.

Gouveia et al. [47] present two MCARP formulations (the nonvalid aggregate model F2 and valid disaggregate F1 model) from which they derive two lower bounds (F2R and F1R). The main differences between the aggregate model in Gouveia et al. [47] and the model proposed by Belenguer and Benavent [10] are the network type (mixed vs. undirected) and the size of the models (compact vs. an exponential number of constraints). Belenguer and Benavent [10] use an exponentially sized set of constraints to force connectivity, whereas Gouveia et al. [47] formulate the MCARP by a compact model in which capacity and connectivity constraints are enforced by additional flow variables. The authors [47] test their procedures on two sets of MCARP instances from Belenguer et al. [12]: the 34 *mval* instances (containing 24–50 nodes and 43–138 links) which are derived from the *val* instances and 15 *lpr* instances, containing large, sparse planar networks with up to 401 nodes and 1056 links. With respect to the results for the 34 *mval* instances, Gouveia et al. [47] find 23 optimal values. Belenguer et al. [12] also find 23 optima. For the remaining 11 *mval* instances, the BBLP bounds produced in Belenguer et al. [12] are better than the F2R bounds of Gouveia et al. [47] in four cases. For the larger *lpr* instances, it appears that the F2R bounds are better when the number of required arcs is larger than the number of required edges, while the BBLP bounds are better when the number of required edges increases. On these two sets of instances, the FR2 lower bounds deviate no more than 5% and 2% from the upper bounds, respectively (upper bounds were obtained by an MA (Lacomme et al. [62] and Belenguer et al. [12]). Both F2R and BBLP provide optimal values for the smaller instances with up to 53 nodes and 169 links.

Mourão and Amado [74] generalise the procedure for DCARP studied in Mourão and Almeida [73] to mixed graphs for the application of refuse collection in Lisbon. First, the *Service Direction* algorithm transforms a mixed graph into an Eulerian directed graph. The procedure starts by directing the required edges while maintaining or improving the node balance. The directed graph is then made Eulerian by adding minimum cost deadheading paths between unbalanced nodes through the solution of a transportation problem. Next a three-phase heuristic is applied to construct a feasible solution. The first phase decomposes the Eulerian graph into small circuits, which may have to be linked to the depot and dump site for feasibility. The second phase tries to aggregate small trips together through the solution of a matching problem in a trips multigraph (nodes correspond to the small trips; edges are defined between trips that can be merged; edge costs are the savings in routing distance). In the final phase, trip adjustments and improvements are made. The procedure was tested on 37 randomly generated DCARP instances (with up to 401 nodes and 1215 links) and on 15 MCARP instances of Belenguer et al. [12]. The heuristic generates good upper bounds. On the first 26 DCARP instances (with dump site outside the collection area) the average gap is about 18%, while on the remaining 11

DCARP instances (with dump site inside the collection area) the average gap is about 1.90%. Mourão and Amado [74] explain the larger gaps on the first set of DCARP instances by the lower quality of their lower bound. On the 15 MCARP instances the gap varies between 0.28% and 5.47%. On these instances the procedure performed similarly to the *Modified Augment-Merge*, *Modified Path-Scanning*, *Modified Ulusoy's Tour Partitioning*, and *Modified Merge* algorithms, which produce average gaps in the range of 2–4% (Belenguer et al. [12]).

Lacomme et al. [62] and Belenguer et al. [12] generalize three well-known UCARP heuristics: *Path-Scanning* (PS) (Golden et al. [45]), *Augment-Merge* (AM) (Golden and Wong [46]), and *Ulusoy's Tour Partitioning* procedure (Ulusoy [97]) for the MCARP. Modifying these procedures to work on mixed graphs is not trivial. The Merge phase in AM-based arc routing heuristics is similar to the Clarke and Wright savings procedure (Clarke and Wright [25]) for node routing. In an undirected arc routing context, however, there are four ways of merging two edges (or trips) together: Every edge (trip) can be inverted or not. In a mixed graph, the cost of a trip may change with inversion, and in case the trips contain required arcs, the inversion is not possible. Path Scanning heuristics for arc routing are inspired by the *Nearest Neighbor* heuristic for the *Traveling Salesman Problem* (TSP). Belenguer et al. [12] test their modified procedures on the *mval* and *lpr* MCARP instances and conclude that Merge-based heuristics often do better than PS-based procedures and that the Augment phase in AM can have a bad impact on performance. The average gaps on the *mval* instances are around 25.6%–21.1% for PS variants, around 9.9%–17% for (A)-M variants, and 14.3%–21.3% for Ulusoy variants. On the *lpr* instances the average gaps are 3.3%–3.8% for PS, 1.6%–2.5% for (A)-M variants, and 2.5%–3.1% for Ulusoy variants. Belenguer et al. [12] also upgrade the *Memetic Algorithm* (MA) for CARP by Lacomme et al. [61], [62]. As in Merge-based algorithms, moves such as 2-opt on two trip segments become much more intricate when the cost of the inverted segment is different. Computational tests reveal that MA produces high quality solutions: The average gaps between the MA solution and the lower bound on the 34 *mval* and 15 *lpr* instances are 0.51% and 0.33%, respectively, and 23 *mval* and 6 *lpr* instances are solved to optimality.

Bautista et al. [9] also model an urban waste collection problem in a town within the metropolitan area of Barcelona, Spain, as an MCARP. The road map of the town results in a mixed graph with 777 vertices, 504 edges, and 715 arcs. The MCARP is transformed into a node routing problem with additional constraints for forbidden turns. The problem is then solved by two ant colony heuristics based on nearest neighbor, nearest insertion, and local search. Computational results show a reduction in the total length of 35% (Ant1 nearest neighbor) and 37% (Ant2 nearest insertion) compared with the total length of the current urban waste collection routes.

Xing et al. [104] also consider an ECARP which accounts for deadheading costs, service costs, penalty costs for forbidden turns, fixed vehicle cost, and service time constraints. The authors propose a *Hybrid Ant Colony Optimization Algorithm* (HACOA). They test their procedure on instances derived from the *gdb*, *val*, *egl*, and *kshh* test bed, yet do not explain what modifications were made.

10.3 ▪ CARP variants: Vehicle characteristics

10.3.1 ▪ The multi-compartment Capacitated Arc Routing Problem

In the *Multi-Compartment Capacitated Arc Routing Problem* (MCCARP) the required edges have a demand for different commodities and these commodities must be kept segregated during transportation. Multi-compartment vehicles (each compartment with limited capacity and dedicated to a specific commodity) are available at the depot to organize transportation. The objective in the MCCARP is to find a set of minimum cost routes, with all routes starting and ending at the depot, and such that all edge demands are serviced from the designated compartments without violating the compartment capacities. A well-known MCCARP application is the collection of source-separated waste streams from households by partitioned trucks. When householders generate different waste streams (e.g., recyclables and general waste), waste collection firms have the opportunity to deploy different vehicles. One option is to collect each waste stream separately (separate collection) using unpartitioned trucks. Another option is to co-collect these different waste streams using vehicles with multiple compartments (with one compartment dedicated to each type of waste).

Muyldermans and Pang [81] present a local search algorithm that exploits well-known moves (2-opt, reinsert, relocate, exchange, and cross), adapted to work well in an arc routing context. Speed-up tricks such as marking and neighbor lists are used, and the procedure is embedded within the *Guided Local Search* metaheuristic in order to reach high quality solutions. Muyldermans and Pang [81] assume that the different commodities associated with a required edge can be serviced (collected) by different vehicles. This complicates the solution representation and tour length calculations significantly, depending on whether or not all the commodities on a required edge are collected in the same pass by the same vehicle. Muyldermans and Pang [81] test three sets of instances: 24 *egl* UCARP instances, 24 MCCARP instances with two commodities derived from the *egl* instances, and a new set of randomly generated MCCARP instances. The average gap on the single and dual commodity *egl* instances is 1.23% and 1.32%, respectively. The authors carried out further experiments on the third set to gain more managerial insight into the potential benefits from deploying multi-compartment vehicles. The results reveal that co-collection is in many cases better than separate collection and that the improvement in routing distance (compared with separate collection) increases when the number of commodities is higher, when the vehicle capacity increases, when the items are less bulky, when more edges have a demand for all commodities, when the required edge density is lower, and when the depot is more centrally located in the road network. The benefits from co-collection reduce under opposite conditions.

10.3.2 ▪ The Capacitated Arc Routing Problem with vehicle/site dependencies

Another variant of the UCARP is the *Capacitated Arc Routing Problem with Vehicle/Site Dependencies* (CARPVSD), where a heterogeneous vehicle fleet consists of at least two vehicle classes (a vehicle class is a set of vehicles with identical characteristics) and at least one arc in the network has a vehicle/site dependency. A vehicle/site dependency on an arc is a constraint that prohibits a vehicle of a certain class from servicing or traversing the arc due to some limitation. The vehicles from different classes can differ in capacity, daily capital cost, length of workday, crew size, or other characteristics. The vehicles that can service an arc, travel along an arc without servicing, and neither travel along nor service an arc are determined by the vehicle/site dependency on the arc. The CARPVSD with no vehicle/site dependencies is known as the *Mixed Fleet Capacitated Arc Routing Problem* (Dror [30]). One application of the CARPVSD is the problem of routing residential waste collection vehicles where vehicle/site dependencies occur due to factors such as some streets being too narrow for certain vehicles, or bridges or overpasses not being able to support the weight of some vehicles (Sniezek [87]; Sniezek et al. [89]). Another application of the CARPVSD is the scheduling of field service or repair services, where the staff is broken down by skill level and some arcs can only be serviced by employees with certain skill levels (Sniezek and Bodin [88]).

Sniezek et al. [89] study the CARPVSD for the application of waste collection in Philadelphia, US. The problem is defined on a directed graph with several vehicle classes. Each vehicle class is characterised by a capacity and the number of vehicles available. For each required arc, the largest vehicle type that can traverse the arc and the largest type that can service it are given. The objective is to design a set of tours with minimum cost and to assign one vehicle to each tour, so that the total demand serviced by the tour does not exceed the vehicle capacity and all edges traversed and serviced are compatible with the vehicle. Sniezek et al. [89] also consider additional compactness and workload balance constraints. They solve the CARPVSD by applying a complex multiphase *Vehicle Decomposition Algorithm* (VDA), which decomposes the CARPVSD into several smaller single vehicle class CARPs. Each of these smaller CARPs is solved by the partitioning procedures contained in RouteSmart (a vehicle routing software). The solutions to the smaller problems are then integrated to form a final CARPVSD solution. Once the partitions are formed, travel-path-generation techniques in RouteSmart are used to find an approximate minimum deadhead time travel path for the arcs and edges in each partition. Despite the success of using VDA in Philadelphia, the development of VDA is an ongoing effort and several issues regarding the improvement of VDA are explored in Sniezek [87].

In a subsequent study, Sniezek and Bodin [88] solve the CARPVSD by a solution procedure based on two *Mixed Integer Program* (MIPs): the *Mathematical Programming Procedure* (MPP)—a MIP formulation of the CARPVSD, and the *Initial Fleet Mix* (IFM)—a generator of an initial fleet mix for the CARPVSD. Sniezek and Bodin [88] find that just minimizing total distance or deadheading does not produce good CARPVSD solutions which can be implemented in practice. Therefore, the authors define a multicriteria function that evaluates solutions based on the daily fixed costs (capital and crew salary), variable labor costs including overtime, and variable routing cost (based on total distance and number of trips to the dump site). Through experimentation, Sniezek and Bodin [88] show that the MPP was not able to solve problems having more than 500 arcs, 10 vehicles, and 4 vehicle classes within reasonable time. Therefore, the authors use the MPP as a between-route route improvement procedure. The overall solution approach works sequentially: (1) the IFM determines a good fleet mix, (2) the VDA (Sniezek et al. [89]) determines partitions and travel paths for the fleet mix, and (3) variants of the MPP (where some routes remain fixed and some are extracted) are solved to improve the solution.

Ghiani et al. [41] report on a real-life solid waste collection case study in Southern Italy. The problem is modelled on a mixed network and involves streets divided into three classes (each class to be serviced within a specified deadline), three types of trucks with different capacities, and two types of bins. The smallest trucks cannot handle the largest bins, and the largest trucks cannot traverse narrow streets. The authors develop a cluster-first, route-second heuristic. In the first stage, required arcs and edges are allocated to vehicles so as to satisfy deadlines and bin-to-vehicle compatibility constraints. In the second stage, vehicle tours are built by inserting arcs and edges according to their service class and linking the tours with the dump site. The procedure is tested on a real-life road network with 273 nodes, 292 edges, and 84 arcs. Compared with the solution generated by the municipality, the heuristic reduced the travel distance by 10.6%, the working time by 13.5%, and total cost by 7.8% and eliminated overtime and improved workload balance among the vehicles. The improvements were mainly due to a better assignment of containers to vehicles: The allocation of streets with higher demand to the two larger trucks reduced the number of trips to the dump site made by the smaller vehicles.

10.4 ▪ CARP variants: Tour and facility characteristics

10.4.1 ▪ The open Capacitated Arc Routing Problem

The *Open Capacitated Arc Routing Problem* (OCARP) differs from the UCARP in that routes in OCARP are not constrained to form closed circuits (i.e., start and end at the same node, usually the depot) but they are open (or can start and/or end at any node in the network). Two applications of the OCARP are the *Meter Reader Routing Problem* (MRRP) (Stern and Dror [91]) and the *Cutting Path Determination Problem* (CPDP) (Moreira et al. [72]). The MRRP does not consider a depot because the employees (the meter readers) travel from the office to the address of their first meter reading, and after completing their routes, they return home. In the CPDP, the trajectories of a set of blowtorches must be determined for a cut pattern on a quadrilateral steel plate in order to produce a predefined set of polygonal pieces in minimum time. A piece is produced when its shape is fully traversed by one or more blowtorches. The blowtorches have a limited amount of energy to spend and must not traverse the interior of any shape, but they may dislocate above the plate level, reflecting additional elevating and lowering manoeuvres times.

Stern and Dror [91] present a two-phase heuristic which starts by constructing an Eulerian walk covering the required edges, and then partitioning the walk into smaller work segments such that each work segment does not exceed the work shift time limit.

Usberti et al. [98] propose an integer linear programming model to derive lower bounds, and develop a reactive path-scanning heuristic for the OCARP (with no prespecified start or end point for the routes). Some special features of the proposed heuristic include a cost-demand edge-selection rule, self-tuned according to the instance hardness to achieve feasible solutions; an ellipse rule, which shortens the routes in favor of solution cost; and a reactive parameter, responsible for regulating how often the ellipse rule is applied. Usberti et al. [98] selected five sets of UCARP instances 23 *gdb*, 34 *val*, 24 *egl*, *32 A*, and *24 B* (Wøhlk [100]) with three different minimum numbers of vehicles, and thus derive 411 OCARP instances (69 *ogdb*, 102 *oval*, 72 *oegl*, 96 *oA*, and 72 *oB*). Their heuristic outperforms the path-scanning heuristic with random selection of tied edges (Belenguer et al. [12]) and the path-scanning heuristic with ellipse rule (Santos et al. [86]). Out of 411 instances, 133 were solved optimally. Usberti et al. [98] also show that OCARP can be reduced polynomially into UCARP and, vice versa, that UCARP can be reduced polynomially into OCARP.

Fung et al. [40] recently presented a *Memetic Algorithm* (MA) to solve a directed OCARP with a depot from where the tours must start. The authors transform the problem into an equivalent node routing problem and propose an IP model, which they use to compute tight lower bounds. Initial solutions are based on savings, nearest neighbor, and insertion heuristics, and the *Split* algorithm ([97] and [62]) is used to evaluate chromosomes in the MA. The authors experiment with randomly generated instances and 20 modified *gdb* and *val* instances. For this last test set, the authors report an average gap of 1.9% (the gap between lower bound and their best solution).

10.4.2 ▪ The multi-depot Capacitated Arc Routing Problem

In the *Multi-Depot Capacitated Arc Routing Problem* (MDCARP) vehicles can start and end their routes at several depots. It is usually assumed that vehicles must return to the depot from which they departed. The MDCARP arises in large-scale applications, including waste disposal, snow removal and winter gritting, and street sweeping.

One approach to dealing with several depots is to apply a clustering or districting procedure (see Muyldermans [78] for a review), which, essentially, allocates edges or streets to depots, and then, in a second stage, calculates routes for each district separately using CARP heuristics. Districting, however, typically belongs to a more tactical decision level and, apart from supporting good vehicle routing, good districts may aim for other objectives such as workload balance, compactness, no overlap, and contiguity.

Muyldermans et al. [80] discuss the problem of district design for the application of winter gritting and propose a heuristic procedure. First, the road network is partitioned (at minimum cost) into small cycles. The motivation for using cycles (instead of individual edges) is due to their potential to support good or efficient arc routing. Cycles are then allocated to depots in two phases: Phase 1 expands the district with the least workload by the largest-weight adjacent cycle; phase 2 allocates remaining cycles through a multicriteria approach, taking into account a measure for compactness, an estimate for the number of vehicles, and an indication for the imbalance in workload. The eligible cycles in both phases are based on the radial distances for reaching the edges on the cycle from each of the facilities. On a small, real-life sample network, the districts obtained by this procedure were shown (for a curative gritting intervention) to incur about 14% less deadheading compared with the actual district borders.

In follow up work, Muyldermans et al. [79] explore three districting heuristics, either using cycles or individual edges in the aggregation phase. The cycles (or edges) are assigned to the nearest facility in a rather greedy manner, or through the solution of an integer linear programming model (minimizing a lower bound on the number of vehicles to be used). The procedures are tested on large graphs constructed from the road network in Flanders (Belgium), and the quality of the district partitions was evaluated by solving CARPs [14] in the districts. Muyldermans et al. [79] recommend different districting policies depending on the size of the vehicle capacity or the relative importance of local and radial routing distance.

More recently, Mourão et al. [75] study the *Sectoring Arc Routing Problem* (SARP), where the aim is to partition the service territory into a number of sectors so that each sector can be covered by a set of vehicle trips. The authors propose two two-phase heuristics. In the first phase, the sectors are determined; in the second phase, vehicle routes are obtained by solving a MCARP. The two variants differ by the sector partition strategy, using either small circuits or individual network links, respectively. A third heuristic builds sectors and trips simultaneously. Sectors are initialized by selecting seed locations as far as possible away from each other. In order to ensure workload balance, the sector with the least work load is expanded next. The three heuristics are tested on three groups with five MCARP instances each, resembling waste collection operations in modern towns, old historical town centers, and low-traffic suburban areas. The authors evaluate the sector partitions based on routing distance, imbalance in workload, sector and partition diameter, and dispersion metrics to measure compactness.

Amberg et al. [3] were among the first to study the CARP with multiple centers directly. The problem was defined on an undirected network with M depots and a given number of vehicles stationed in each depot. The authors convert the problem into a special capacitated minimum-cost spanning tree problem. Each depot is represented as a root node, and from each root node there are as many branches as vehicles in the depot. In each subtree emanating from a root node, the sum of node weights is limited by the vehicle capacity and the sum of the edge weights by the maximum length allowed for a route. This special capacitated minimum cost spanning tree problem is solved by a constructive heuristic and two metaheuristics based on a node exchange improvement procedure (Simulated Annealing and Tabu Search). A feasible UCARP route is then deduced from each subtree. The solution methods are applied to two real-world networks.

Wøhlk [100] considers the undirected MDCARP when the vehicles have various fixed costs from a theoretical point of view and proposes a mathematical model with two lower bounds for the routing cost and for the fixed cost. Zhu et al. [108] propose a hybrid genetic algorithm for the MDCARP assuming that vehicles may start and end at different depots. The solution procedure is based on the MA by Lacomme et al. [62], yet computational experiments are limited and not very convincing.

More recently, Kansou and Yassine [56] explored two approaches for MDCARP. The first approach is based on Ant Colony Optimization combined with an insertion heuristic, called *Hybrid Ant Colony Algorithm* (HACA). An initial MDCARP solution is obtained by allocating edges to the nearest depot and applying an insertion heuristic for each depot separately. The HACA procedure constructs a giant tour for each depot and applies a tour splitting procedure similar to Lacomme et al. [62] to obtain capacity feasible routes. Several local search moves, including an inter-depot move (moving edges between giant tours), are explored to improve the solution. The second approach is a multi-depot MA based on a special crossover. Solutions are represented as sequences of subchromosomes, each corresponding to a giant tour for a depot. The crossover operator is the classical linear order crossover in which the child inherits the sizes of the subchromosomes (number of edges allocated to each depot) of the first parent. A local search, which includes relocating one or two edges to a different position in the giant tour, exchanging two edges and 2-opt moves, is used to improve the solution. The authors experiment with both procedures using the *gdb* instances in which they add a depot on the first and last node of each instance, and conclude that the MA performs better.

Xing et al. [103] propose an *Evolutionary Algorithm* (EA) for the MDCARP which may also deal with routing costs, service costs, turn penalties, and maximum trip lengths and maximum service time durations. Several classical UCARP heuristics are extended (including Path-Scanning, Augment-Merge, and Ulusoy's heuristic) and integrated in the EA framework. The authors test their procedures on 107 instances with up to 140 nodes and 380 arcs; 87 instances were derived from well-known UCARP problems (*gdb*, *kshs*, *val*, and *egl*). However, most instances have only one depot and other features such as penalty costs and service time limits are varied. The results show that the EA performs better than the simple heuristics. The short communication by Xu et al. [105] is based on the work presented in Xing et al. [103].

Doulabi and Seifi [29] recently proposed lower and upper bounds for the *(Multi-Depot Location Capacitated Arc Routing Problem* (MDLCARP), which involves locating a number of depots and designing CARP routes so as to service the network links at minimal cost. The authors present integer programming models (based on the models by Gouveia et al. [47]) both for the single and multiple depot problem. Relaxations of these formulations provide lower bounds on the optimal distribution cost. Doulabi and Seifi [29] also present an insertion algorithm and location-allocation heuristic, which are incorporated in a *Simulated Annealing* (SA) framework. The algorithms are tested on problems derived from Mixed CARP instances (*mval* and *lpr* [12]) and for up to three depots. Average gaps range between 4 and 6% for *mval*-based instances and between 10 and 15% for *lpr*-based instances. The authors also show that the potential cost saving from incorporating location decisions in CARP can be very significant.

10.4.3 ▪ The Capacitated Arc Routing Problem with intermediate facilities

Ghiani et al. [43] introduce the *Capacitated Arc Routing Problem with Intermediate Facilities* (CARPIF), where intermediate facilities (IFs) are available to empty or replenish vehicles along their routes. An IF can be a box or silo of salt or grit in the context of winter gritting, or a landfill site or incinerator plant in the context of waste collection. The vehicle routes in the CARPIF are composed of different segments: The first starts at the depot and ends at an IF; optional intermediate segments link two IFs; and the last segment links an IF to the depot. The total demand serviced on each segment is limited by the vehicle capacity. It may be possible that no demand is serviced on the last segment, for example, in the case that vehicles cannot be emptied at the main depot in the application of waste collection.

Ghiani et al. [43] deal with a single vehicle CARPIF and propose two lower bounds: The first is based on the *Rural Postman Problem* (RPP); the second one is based on a relaxation of an integer linear programming formulation for the CARPIF which uses only deadheading variables. Both lower bounds are calculated in a cutting plane fashion. The authors also develop two heuristics: one based on solving the RPP and cutting the giant tour into appropriate sections while connecting to intermediate facilities; the second based on solving the UCARP in a modified network, and transforming the solution into a feasible CARPIF solution. The UCARP is solved by the Tabu Search procedure (CARPET) of Hertz et al. [52]. Ghiani et al. [43] test their procedures on CARPIF instances derived from the *gdb* and *val* test sets with, respectively, one and two IFs added, and on a set of randomly generated instances as in Hertz et al. [52] with up to 263 required edges and 3 IFs. Their results indicate that the second lower bound is superior to the first, and that in most cases, the upper bounds obtained by the CARP-based solution approach are better than their RPP-inspired algorithm. The authors report average gaps of 1.25%, 2.78%, and 0.81% for the three test sets.

In a follow-up study, Ghiani et al. [42] also investigate the *Capacitated Arc Routing Problem with Intermediate Facilities and Tour Length Restrictions* (CLARPIF), which involves multiple vehicles and is a restricted version of CARPIF in the sense that the length of any route cannot exceed a predetermined bound. The authors develop a construction procedure based on partitioning a RRP tour and two Tabu Search approaches TS1 and TS2. TS1 includes moves which transfer required edges to other route segments, IF insertion and ejection, and route splitting and merging moves. TS2 is a modification of CARPET. The procedures are tested on the *gdb* and *val* CARPIF instances in which tour length restrictions are added. For the *gdb* CLARPIF problems (with one IF) they also compute a lower bound based on a relaxation of an ILP formulation (De Rosa et al. [27]). Their results indicate that the CARPET-based solution procedure was best. On the *gdb* test set the average gap was 14.22%.

In a related work, De Rosa et al. [27] study the *Arc Routing and Scheduling Problem with Transshipment* (ARPT), inspired by the application of municipal waste collection. The collection vehicles must unload their waste at an IF, where it is compacted and loaded onto bigger vehicles that travel between the IF and a landfill site. When no truck is available at the IF, the collection vehicles queue. The number of vehicles of each type is limited. The goal is to design a set of routes for the collection vehicles and a set of truck schedules of minimum total cost, subject to a maximum duration for each collection vehicle route and each truck schedule. The authors develop a Tabu Search heuristic and lower bound procedure based on an integer programming formulation and branch-and-cut. The procedures are tested on four sets of instances derived from the *gdb* and *val* test sets. The average gaps are 3.0%, 6.7%, 2.9%, and 6.6% respectively.

Polacek et al. [84] propose *Variable Neighborhood Search* (VNS) to tackle the UCARP, CARPIF, and CLARPIF. Initial solutions are giant tours, which are converted into feasible solutions using Ulusoy's tour partitioning procedure. Different neighborhood structures are defined by the number of consecutive edges exchanged between two solutions using the cross-exchange operator. Local search is also employed to further improve the solutions. The authors compare their VNS with the CARPET-inspired TS heuristics for CARPIF (Ghiani et al. [43]) and CLARPIF (Ghiani et al. [42]) and report an average improvement in solution quality of 1.67% (*gdb* instances) and 13.60% (*val* instances), respectively.

Ghiani et al. [44] also developed a sophisticated *Ant Colony Optimization* (ACO) procedure for the CLARPIF. The authors test the procedure on the instances introduced by Ghiani et al. [42]. In almost all cases ACO performs better than the TS implementation [42]. On the *gdb* instances the ACO produces solutions with an average gap of 11.76%, compared with 14.22% for TS [42]. However, on the *val* instances, which are also investigated by Polacek et al. [84], the last authors obtain better results with their VNS procedure.

10.4.4 ▪ The Capacitated Arc Routing Problem with mobile depots

Del Pia and Filipi [28] consider a real waste collection problem encountered in Due Carrare, a town located in Northern Italy. Two types of vehicles (compactors and satellites) are involved. The largest trucks (the compactors) cannot access the town center with narrow streets, whereas each satellite vehicle can serve any street but its capacity is very limited. Because the depot is far away from the town, appointments have to be organized between satellite vehicles and compactors. The compactors can be seen as mobile depots; hence, the problem is called the *CARP with Mobile Depots* (CARPMD). The authors develop a solution based on the *Variable Neighborhood Descent* algorithm for the UCARP by Hertz and Mittaz [53]. The routes of satellite vehicles may violate vehicle capacity restrictions but feasibility is restored by a rendezvous procedure: Each infeasible satellite route is corrected by the least-cost insertion of one rendezvous node visited by a compactor. Applied to the actual network with 351 nodes and 422 edges, the metaheuristic reduces the total deadheading time by 30%.

Amaya et al. [2] consider another problem with rendezvous, called the *Capacitated Arc Routing Problem with Refill Points* (CARPRP). An application of this problem involves road maintenance in which road markings have to be painted or repainted every year, and two types of vehicles are used. The first type is a fleet of servicing vehicles with a finite capacity which mark the roads. The second type is a refilling vehicle which meets and replenishes the servicing vehicles so that the latter can continue their road marking service when they run out of paint. The authors model the CARPRP on a mixed graph because the central line can be painted when traversing a road in any direction while painting lane separators requires the vehicles to follow the same direction as the road lane. Amaya et al. [2] propose an integer linear programming model for the CARPRP, which they solve through cutting planes. The procedure is tested on randomly generated DCARP instances with up to 70 nodes and 595 arcs. Most instances are solved to optimality, and the maximum gap is less than 4%. MCARP instances based on the *val* instances (replacing some edges by arcs) were much harder to solve, with gaps reaching 10%.

10.5 ▪ CARP variants: Demand characteristics

10.5.1 ▪ The capacitated general routing problem

In the *Capacitated General Routing Problem* (CGRP), both edges and nodes have a demand for service. The aim is to construct tours of total minimum length, starting and ending at the depot and such that all required edges and nodes are serviced by vehicles without violating the vehicle capacity constraint. The CGRP can be seen as a combination of the UCARP and the CVRP.

Pandi and Muralidharan [83] present a heuristic procedure for the CGRP on a mixed graph with additional constraints on the tour duration (maximum travel and service time). The authors test a route-first, partition-second approach. In the first phase, they construct a giant tour covering all required edges and nodes. This tour is then partitioned into routes taking time and capacity considerations into account. The procedure was tested on a set of instances generated for a residential waste collection problem where waste is to be collected from the curb-sides (arcs and edges) and convenience stores (nodes).

Prins and Bouchenoua [85] solve a CGRP on a mixed graph, called the *Node, Edge, and Arc Routing Problem* (NEARP). The NEARP allows a mixed network with required nodes, edges, and arcs and is inspired by practical waste collection applications in which household waste is located along street segments and waste generated by businesses (e.g., hospitals, schools, supermarkets) is located at some locations in the road network. Prins and Bouchenoua [85] propose a *Memetic Algorithm* (MA) and apply a common data structure for the required links and nodes (i.e., without conversion into a pure arc routing or node routing problem). Nearest neighbor, merge, and tour splitting heuristics provide the initial population for the MA. A tour splitting heuristic is also used in the MA for chromosome evaluation. Computational tests on standard CVRP and UCARP instances indicate that the MA is competitive with the best metaheuristics published for these particular cases. A set of 23 new NEARP instances involving up to 150 nodes, 311 links, and 212 tasks was generated and investigated. Experimental results show that the MA was able to significantly improve the solutions obtained by their nearest neighbor, merge, and tour splitting heuristics followed by local search.

Santos et al. [86] solve a waste collection problem in Coimbra, Portugal, as a constrained CARP due to various network-specific conditions/constraints. These constraints include one-way streets, prohibited turns, demand at nodes and along the arcs, multiple-vehicle/ multi-stop, and maximum time constraints. They incorporate a modified version of the Path-Scanning heuristic and a geographical information system in a *Spatial Decision Support System* (SDSS). The SDSS can be used to provide system analysts with data and maps regarding the total system performance and information for the waste collection crew regarding their individual routes. This is a step forward to enable waste collection planners to analyze the benefits and costs of possible changes to system parameters including vehicle capacity and working shift duration constraints.

Bräysy et al. [18] study a CGRP on a mixed graph considering turn penalties and forbidden turns. The authors transform the problem into an *Asymmetric Capacitated Vehicle Routing Problem* (ACVRP) and present a heuristic algorithm based on the MA for the CVRP proposed by Nagata and Bräysy [82]. Bräysy et al. [18] apply the MA to three sets of CGRP instances with up to 700 arcs, 160 (required) edges, 160 vertices, 225 required arcs, and 28 required vertices, which are transformed into ACVRP instances with up to 623 vertices. The MA was able to find feasible solutions for all 336 instances including the large-size instances. The computation times varied from a few seconds to more than one hour depending on the problem size.

Bosco et al. [16] present an integer programming formulation for the *Mixed Capacitated General Routing Problem* (MCGRP). The authors extend valid inequalities for CARP to MCGRP and incorporate these in a branch-and-cut procedure. The procedure was tested on MCGRP instances derived from the *gdb* and *mval* test set. The algorithm reaches an optimal solution in 154 of the 264 instances. For the instances that were not solved optimally, the average percentage gap was 2.56%. Further experiments were carried out by considering some small NEARP from [85], of which two instances were solved to optimality.

Bach et al. [7] recently proposed a new lower bound for the NEARP, based on the *Multiple Cuts Node Duplication Lower Bound* (MCNDLB) for the CARP [101]. The authors also present new NEARP instances derived from the *gdb*, *val*, and *egl* test set, and larger instances with up to 347 required nodes and 486 required edges based on real-life data for a newspaper delivery application. The authors report an average gap of around 25%, and believe that the main reason for the rather large gap is due to the quality of the upper bounds, for which they rely on the *Spider* VRP solver [51].

10.5.2 ▪ The split delivery Capacitated Arc Routing Problem

In the UCARP, the goal is to determine a set of trips serving all the required edges such that each required edge is serviced by one vehicle. In the *Split Delivery Capacitated Arc Routing Problem* (SDCARP) this constraint is relaxed. That is, each edge may be serviced by different trips or vehicles. This extension of the UCARP can be found in urban waste collection (or in salt-spreading operations) where the vehicle capacity can be exhausted in the middle of a street segment (or between two exits on a motorway). In these situations, servicing an edge with a single vehicle is sometimes not realistic. The works dealing with split demands in a node routing context, studied in Dror and Trudeau [32] and [33], show that if the average demands are large compared with the vehicle capacity (e.g., larger than 10% of the vehicle capacity), allowing split demands may cause significant cost savings both in the total distance and the number of vehicles required.

Mullaseril et al. [77] propose an adaptation of the SDVRP heuristic of Dror and Trudeau [33] for the SDCARP with time windows. Their method is used to solve a livestock feed distribution problem. Guéguen [50] transforms the SDCARP with time windows into a node routing formulation, which is solved by column generation. Labadi et al. [60] investigate SDCARP, in which each required edge must be completely traversed for each partial service. In other words, U-turns on an edge are prohibited. Labadi et al. [60] propose a linear model derived from the sparse formulation for the UCARP (Belenguer and Benavent [10]), an insertion heuristic, and a Memetic Algorithm. Computational tests show that split deliveries can bring important savings if the average demand per customer is large enough (30%) compared with vehicle capacity.

Eskandarzadeh et al. [36] propose an exact method for SDCARP developed in the context of monotropic programming theory, generalizing the relaxation algorithm by Tseng and Bertsekas [96]. However, computational experiments are inconclusive so far. Belenguer et al. [11] propose an integer programming model for the SDCARP inspired by the supersparse formulation for the UCARP (Belenguer and Benavent [10]). A lower bound, computed via a cutting-plane algorithm, is compared with an upper bound obtained through an evolutionary local search with a multistart procedure and a variable neighborhood descent. The computational tests involve 126 instances obtained by modifying demands and allowing partial services in the *gdb*, *val*, and *egl* instances. On average, the new metaheuristic outperforms the MA developed by Labadi et al. [60] and reaches small average deviations to the lower bounds (below 1.5%). Out of 126, 39 instances are solved to optimality. Belenguer et al. [11] also compare SDCARP and UCARP solution values and conclude that demand splitting can result in significant distance savings. The magnitude of these savings depends on the bin-packing characteristics within the UCARP, yet network configuration issues may play a role as well.

10.5.3 ▪ The Capacitated Arc Routing Problem with time windows

The problem of servicing a set of demand edges using a fleet of capacitated vehicles, with the restriction that the service of each required edge must occur within a prespecified time slot, is known as the *CARP with Time Windows* (CARPTW). The CARPTW arises in practical applications such as street sweeping (Bodin and Kursh [15]), where each street has to be swept within a specified time window. The problem of routing winter gritting vehicles (e.g., Eglese and Li [35] and Eglese [34]), where some streets must be serviced within two hours and others within four hours, can also be considered as an application of the CARPTW.

Mullaseril [76] and Dror and Leung [31] study a split-delivery directed CARPTW inspired by a cattle feed distribution problem, and propose heuristic approaches and a transformation into a node routing equivalent, VRPTW. Guéguen [50] proposes an integer linear programming formulation for the undirected CARPTW and another transformation into a VRPTW, but without computational results. Wøhlk [100] presents two modelling approaches for the undirected CARPTW: an integer linear programming model based on constructing a node duplicated network and a transformation into an equivalent VRPTW. In addition, several heuristics are developed, among which is the *Preferable Neighbor Heuristic* (PNH). In PNH, promising vehicle tours are constructed first and then a *Set Covering Problem* is solved to select the best tours. Wøhlk [100] also presents a column generation approach to obtain tight lower bounds. More recently, Labadi et al. [59] developed a *Greedy Randomized Adaptive Search Procedure* (GRASP) with *Path Relinking* (PR) for CARPTW. On a set of 24 CARPTW instances proposed by Wøhlk [100] with up to 60 nodes, 90 edges, and 81 required edges, PNH and GRASPPR both find 17 optimal solutions. The average gaps are 1.2% and 0.71%, respectively.

Johnson and Wøhlk [55] propose two column generation methods to solve CARPTW optimally and the Preferable Follow-on heuristic to obtain near optimal solutions to the problem when the time windows are wide or the instances too large. The optimal approaches involve generating feasible tours first and then solving a large Set Partitioning Problem. The latter is solved by a two-phase or iterative method. A set of 20 *egl* instances (with 51 to 190 required edges) with varying time window sizes is tested. The computational results show that the iterative method is superior to the two-phase method both in terms of computation time and ability to solve hard instances. The Preferable Follow-on heuristic generates solutions which deviate on average no more than 2.4% from optimality.

In both Wøhlk [100], Labadi et al. [59], and Johnson and Wøhlk [55] the time windows are hard although vehicles can wait at the depot before beginning the service. Afsar [1] introduces the *CARP with flexible or soft Time Windows* (CARPfTW) on undirected graphs. The cost of servicing an edge is modelled as a piecewise linear function over time. This makes the subproblem—a nonelementary capacitated shortest path problem—more difficult to solve. A branch-and-price algorithm is proposed to solve CARPfTW optimally.

10.5.4 ▪ The Capacitated Arc Routing Problem with time-dependent service costs

Tagmouti, Gendreau, and Potvin [94] study the *CARP with Time-Dependent Service Costs* (CARPTDSC) and underline that their problem is similar to the CARP with flexible time windows when the penalty cost of violating the time windows is linear. This extension is motivated by winter gritting applications where the timing of each intervention is crucial.

That is, if the intervention is too early or too late, the cost in time and de-icing material increases. Tagmouti et al. [94] transform the CARPTDSC into an equivalent node routing problem and solve the problem exactly with a column generation procedure. The authors test their procedure on modified VRPTW benchmark problems (Solomon [90]). All instances with 25 customers and some instances with 35 and 40 customers were solved to optimality.

More recently, Tagmouti et al. [92] proposed a *Variable Neighborhood Descent* (VND) heuristic for solving the CARPTDSC. The algorithm explores different neighborhoods based on reinsertion, cross, and block moves and also applies a modified version of the *Shorten* procedure for CARP proposed by Hertz et al. [52]. The authors report results on modified *gdb* and *egl* instances and two types of service cost functions. They also compare their results with the adaptive multistart Local Search algorithm of Ibaraki et al. [54], where the arc routing problem is solved after being transformed into an equivalent node routing problem. For the first type of service costs, the results on *gdb* instances show that when the VND heuristic was applied without explicitly considering the number of vehicles, the algorithm of Ibaraki et al. [54] generates fewer vehicles on five instances. For the same number of vehicles, VND provides an average reduction of 3.8% on the total cost. The penalty cost for visiting required arcs outside their optimal time interval accounts for 7.7% and 11.3% of the total cost for VND and the algorithm of Ibaraki et al. [54], respectively. For the first type of service costs, the results on the *egl* instances show that VND produces a solution with the same number of vehicles as the algorithm by Ibaraki et al. [54], but with a smaller total routing cost. For the second type of service costs, the computational results reported on *gdb* instances indicate that for the same number of vehicles, an average reduction of 1.6% in total cost is obtained when comparing VND with Ibaraki et al. [54]. The penalty cost for visiting required arcs outside their optimal time interval accounts for 6.1% and 7.3% of the total cost for VND and the algorithm of Ibaraki et al. [54].

In subsequent work, Tagmouti et al. [93] study the dynamic CARP with *Time-Dependent Service Costs*. The problem is inspired by the application of winter gritting, where some information (e.g., weather conditions or road conditions) might not be available when the vehicles start their tours. When new information (e.g., weather report updates) is unveiled during the execution of the routes, the problem becomes dynamic. The authors adapt the VND heuristic to solve the dynamic variant of the CARPTDSC. The starting solution is computed with the VND using service time cost functions based on some initial forecast. When vehicles execute their routes, regular weather report updates lead to modifications to the service time interval associated with each required arc. To address the dynamic nature of the problem, the VND reoptimizes the routes each time a new weather report becomes available, ignoring the street segments that were serviced before and using the updated service time costs for unserviced required edges.

Vansteenwegen et al. [99] discuss an interesting application of the CARP with soft time windows, involving a mapping vehicle which has to traverse all streets in a region and take pictures of streets and road signs within a minimum number of days. The soft time windows in this problem arise from the fact that taking pictures in the direction of the sun should be avoided as much as possible. Vansteenwegen et al. [99] convert this CARP with soft TW into a VRP with soft TW. Then, a two step metaheuristic, consisting of a local search phase to minimize the number of days and an iterated local search phase to minimize the time window violations, is designed to solve the problem. A linear program calculates the ideal starting time of the vehicle in each street. The approach is tested on the Solomon [90] instances with 100 clients and on real-life problems involving up to 893 landmarks, 1289 edges, and 91 arcs. The results for the test set of Solomon [90] indicate that the time window violations are reduced by 32% compared with the initial violations, for an average computation time of 7 seconds. For the real-life problems, the computation time is acceptable and the time window violations are reduced by 40%.

10.5.5 ▪ The periodic Capacitated Arc Routing Problem

In the *Periodic Capacitated Arc Routing Problem* (PCARP) each required edge has a given service frequency over a prespecified planning horizon. For instance, household waste has to be collected every day in the center of a town but only twice a week in the residential suburbs where waste can be stored for a few days in larger containers. The goal in the PCARP is to determine which edges are serviced each day, taking into account the edge frequencies, and to solve the CARP for each day so as to minimize the total cost of the trips over the planning horizon. The PCARP is therefore a natural extension of the CARP when operations have to be planned over a longer time horizon.

Lacomme et al. [63] present a classification of PCARP problems. Class A problems have period-independent demands and costs, and mainly model preventive problems like winter gritting, inspection of power lines, or the treatment of rails with herbicides. In class B problems, demand is generated by each edge in each period, giving period-dependent demands and costs. In subclass B1, the horizon is acyclic, for instance, when mowing vegetation that grows along the roads: The vegetation stops growing in winter and restarts in spring. Problems in subclass B2 are characterized by a cyclic planning horizon; for instance, waste collection is often planned in a one-week cycle. In all class B problems, the amount serviced by a vehicle depends on the time spent since the last visit. In practice, minimal and maximum spacing between two successive services must be respected. A common policy involves specifying a set of allowable day combinations for each edge, for instance, Monday, Thursday, or Tuesday, Friday for a street with frequency 2 in waste collection. Minimum and maximum spacing can be converted into day combinations and vice versa, but the day combination system is simpler, because the demands for each day in a combination can be precomputed and the spacing constraints are implicitly satisfied by each combination. Monroy et al. [71] suggest a further subdivision of class A problems regarding whether services are regular or irregular within the time horizon, and discuss the application of the road network monitoring operations as an example of period-independent demand and cost and irregular services (e.g., major roads may require more visits during the weekend due to heavier traffic and less during the working week).

Chu et al. [22] propose an integer linear programming model and preliminary lower bounds based on a transformation of PCARP into a special case of UCARP. Lower bounding techniques originally designed for UCARP (e.g., the *Hierarchical Relaxation Lower Bound* by Amberg and Voss [4]) are applied to the transformed problem. The authors also develop a *Scatter Search* heuristic and apply this to PCARP instances derived from the 23 *gdb* instances. The average deviation from the lower bound is about 9.6%. In subsequent work, Chu et al. [23] propose an alternative integer linear programming model based on the day combination system, and three constructive heuristics: a decreasing frequency nearest neighbor method, a best insertion heuristic, and a two-phase algorithm. In the first two approaches, starting from empty trips in each day, each iteration selects one edge or task and one of its day combinations according to a greedy criterion, and inserts one occurrence of the task in a trip, for each day of the combination. The two-phase method computes one cluster of tasks for each day, satisfying the day combinations, and then solves the resulting UCARP in each period, using the MA of Lacomme et al. [62].

The heuristics were evaluated by adding a multiperiod horizon and frequencies to the *gdb* UCARP instances.

Lacomme et al. [63] describe an MA in which each chromosome is composed of several giant tours (one per day) and satisfies frequencies and day combinations. A special crossover operator generates offspring solutions with the same properties. The offspring solution is evaluated optimally, subject to the sequence, by applying for each day the UCARP splitting procedure of Lacomme et al. [62]. To save time, local search is applied for each day separately. The crossover operation deals with the tactical or planning level, which defines the set of edges to be serviced in each period, while the local search focuses on the operational level or the detailed composition of the trips in each day.

In follow-up work, Chu et al. [24] study a more ambitious PCARP in which the main objective is to minimize the fleet size cost whilst the total cost of the trips is considered as a secondary objective. The encoding scheme is the same as in the previous algorithm (Lacomme et al. [62]), and it is still possible to design a polynomial time splitting procedure that converts a chromosome into a feasible PCARP solution, subject to the sequence imposed by this chromosome. Chu et al. [24] propose a Scatter Search procedure for this problem.

Marzolf et al. [68] study the PCARP with irregular service duties, inspired by the application of monitoring of road networks. A single vehicle must inspect all the road segments of a network over a two-week horizon comprising 21 working shifts per week, subject to frequencies. For instance, highways must be monitored once every weekend and once every five-day week. National roads must be covered once every seven days. The length of a route is limited by the duration of a shift. During a shift, the vehicle may have to leave the planned route to answer emergency calls and then come back to resume its route, which it may not be able to complete during its shift. Marzolf et al. [68] describe a decision support system for vehicles equipped with global positioning systems, including planning heuristics and replanning procedures at the end of each shift or day.

The PCARP with irregular intervals in the context of street surveillance is also studied in Monroy et al. [71]. The roads to be visited are divided into classes according to the surveillance requirements. The objective is to design a set of routes satisfying the passage frequencies for each class of roads in each period and not exceeding the vehicle capacity. Monroy et al. [71] propose an integer programming formulation for small-sized problems and a cluster-first, route-second heuristic for larger instances. The heuristic was tested on five sets of instances. The authors report a worst gap of 24%, and running times are smaller than 32 minutes for all tested instances, but mention that the bound against which solutions are evaluated is rather weak. Also, 82% of cases have a gap lower than 15%, and 85% of cases have a running time smaller than one minute. For the smaller instances, the heuristic is able to find four optimal solutions from eight instances.

Finally, Mei et al. [69] proposed a new MA for the PCARP which adopts an explicit solution representation and a new crossover operator. The authors consider minimizing the number of vehicles as the primary objective and minimizing service and routing cost as secondary objective. They note that the first objective is rather insensitive to many existing local search moves and propose a *Route-Merging* (RM) procedure to tackle this issue. They compare the MA with RM (MARM) with other metaheuristic approaches (SS by Chu et al. [24] and MA by Lacomme et al. [63]) on two PCARP benchmark sets (periodic *gdb* and periodic *val* test sets) and a real-world data set. The results show that MARM generally obtained better solutions than the other two procedures, in less computational time.

10.5.6 ▪ The Capacitated Arc Routing Problem with stochastic demand

For some applications, the demand on each edge may be better described by a random variable. Christiansen et al. [20] give the examples of stochastic snow heights (due to drifting) on roads and stochastic amounts of waste from households and commercial or industrial waste generators. Stochastic edge demands make the CARPSD different from the deterministic CARP in two aspects. First, at some point along a route, the actual accumulated demand might exceed the vehicle capacity. This event is generally denoted as a failure, and a recourse action must be taken in order to complete the originally planned routes (e.g., a back and forth trip from the point of failure to the depot). Second, the objective in the CARPSD is to find a set of routes that minimizes the expected cost, that is, the cost of the planned solution plus the expected cost of the recourse action.

Fleury et al. [37] and [38] investigate the robustness of deterministic CARP solutions when demands are randomized. The authors developed expressions for the expected number of trip interruptions and the expected total cost, which they incorporated in the objective function guiding their *Memetic Algorithm*. Their tests show that the planned solution of the deterministic MA undergoes on the field an increase in cost of 8%, with 70% of trips interrupted. The planned solution of the MA with the stochastic objective is 3.5% more expensive, yet very robust. The cost variation on the field is only 0.01%, with a probability of trip interruption of 0.5%. Hence, the new MA saves 4.5% on the actual operating costs.

Fleury et al. [39] study the CARPSD with normally distributed edge demands (truncated to avoid negative demands and demands that exceed the vehicle capacity). It is assumed that a route can fail only once, and if a failure occurs on a route it will always be immediately before serving the last edge. The solution method was based on the MA by Lacomme et al. [62].

Christiansen et al. [20] consider the CARPSD in which the demands on required edges are described by Poisson distributions. Failure may occur anywhere on an edge, and this complicates the calculation of the recourse cost in two ways: accounting in the expected cost for the serviced fraction of the edge where failure occurs, and deciding via which edge endpoint to return to the depot. Christiansen et al. [20] formulate the CARPSD as a set-partitioning problem and develop a branch-and-price algorithm in which the pricing is carried out by dynamic programming. They test the algorithm on several, modified UCARP instances (*gdb*, *kshs*, *bccm*, *egl* instances with less than 80 vertices, and *A* instances from Wøhlk [100]). The largest instance solved (gap less than 1%) has 40 vertices and 69 required edges. Their pricing procedure is most successful on instances with tight capacity constraints, which is in line with their experience with branch-and-price algorithms for the node routing counterpart of this problem (Christiansen and Lysgaard [19]).

Laporte et al. [65] study the CARPSD in the context of waste collection. Routes can fail only once, and failures can occur anywhere on a required edge. The authors derive closed-form expressions for the expected cost of recourse and develop an *Adaptive Large-scale Neighborhood Search* (ALNS) heuristic for the problem. The ALNS is a local search approach where simple heuristics compete to modify the current solution. In each iteration, an algorithm is chosen to destroy the current solution, and another one is selected to repair it. The new solution is accepted in an annealing based fashion. A modified path scanning procedure was used to compute starting solutions. The authors test their procedure on two sets of instances derived from the *gdb* instances, assuming that edge demands follow Poisson distributions. In the first test, the dump site is collocated with the depot; in the second set, the dump site is moved to another vertex. For the first set of instances,

Laporte et al. [65] compare the solution value by the ALNS heuristic with the optimal cost of the corresponding deterministic UCARP instance plus the expected recourse cost associated with this optimal solution. For all the instances, the ALNS algorithm provides better solutions and the average improvement over the deterministic solutions is 2.29%. The authors use the second test set to investigate the sensitivity of solutions with respect to changes in the vehicle capacity. The experiments show that the cost of a recourse increases when the capacity is higher, but that the total solution cost decreases.

10.5.7 ▪ The Capacitated Arc Routing Problem with deadheading demand

The *Capacitated Arc Routing Problem with Deadheading Demand* (CARPDD) was recently introduced by Kirlik and Sipahioglu [57] and extends the classical CARP by introducing an additional capacity consumption incurred by a vehicle deadheading an edge. The main difference from the standard CARP is that, in addition to the service demand associated with required edges, there is a travel demand associated with every edge in the graph, which is consumed whenever it is deadheaded.

Bartolini et al. [8] show how the CARPDD can be used to model a variety of time-, distance-, or energy-constrained arc routing problems. Practical applications of the CARPDD include automated bridge inspection, road maintenance, mail delivery, snow plowing, meter reading, and robotics. Complete coverage path planning [57] (e.g., in the context of patrolling and search operations) with a fleet of robots moving (and consuming energy) so that all points or aisles of the search area are covered is an example of the last category.

Kirlik and Sipahioglu [57] consider the CARPDD where the travel demands are consumed every time an edge is traversed (and not only when it is deadheaded as in Bartolini et al. [8]). The authors propose an integer programming formulation which is used to solve small instances to optimality. They also develop a heuristic based on a modification of Ulusoy's splitting algorithm for the CARP [97] and two post-optimization procedures. The algorithms are tested on CARPDD instances derived from the *gdb*, *val*, and *egl* test sets. The post-optimization procedures improve the intial Ulusoy-based solutions by 7.33%, 18.59%, and 63.22% for the three test sets. For the smaller *gdb*-based instances the authors report an average gap of 17.76% (using lower bounds obtained through a relaxation of their IP model).

Bartolini et al. [8] propose an exact algorithm for the CARPDD. The authors introduce CARPDD-specific capacity constraints for which they developed a heuristic separation procedure. CARPDD nonelementary routes are generated through dynamic programming, and used in a cut-and-column based lower bounding procedure. The latter is integrated in a branch-and-price algorithm to solve the CARPDD exactly. The authors perform extensive computational experiments on CARPDD (derived from the *gdb*, *val*, *egl*, and *bmcv* [14] test sets) and on standard CARP instances. Out of 227 CARPDD instances, 139 are solved to optimality including all *gdb*-based instances and all but two *val*-based instances from Kirlik and Sipahioglu [57]. The procedure also works very well for standard CARP problems, solving an additional 22 *bmcv* instances for the first time.

10.6 ▪ CARP variants: Objectives

10.6.1 ▪ The multi-objective Capacitated Arc Routing Problem

The standard objective function in many routing problems is to minimize the total distance travelled. In many real-life situations, however, the total distance travelled is not the only consideration and other cost components or objectives, such as service and labor costs, fixed vehicle costs, minimizing the number of vehicles, overtime and penalty costs, equalizing workload, minimizing the length or duration of the longest route, etc. may be more or equally important. Several studies discussed in the previous sections combine different cost components in a single objective function. In this section, we review *CARP* variants that deal with alternative, less common objective functions and that explicitly deal *with Multiple Objectives* (MOCARP). In the latter, one typically ends up with a set of Pareto (near) optimal solutions.

Lacomme et al. [62] present a modified tour splitting procedure, which they incorporate in their MA, to deal with makespan minimization subject to a limited fleet size. The authors report an average gap of about 6% on the *gdb* instances, but mention that their simple lower bound is likely to be rather weak.

Amponsah and Salhi [5] consider a bi-objective CARP for waste collection in developing countries with hot weather and propose a Path-Scanning inspired heuristic, minimizing a convex combination of two objective functions. The first objective is the sum of the demand and service cost ratios for the required edges, nominator and denominator exponentially weighted by factors α and $(1-\alpha)$, respectively, with $\alpha \in [0,1]$. The rationale is to minimize the environmental impact and trade off the quantity to be collected and the service cost. Cheap edges are processed first if α is close to 0, whereas priority is given to largest demands for α close to 1. The second objective, which intends to minimize the smell, is the sum of the collection quantity multiplied by the time lag between the vehicle's departure time and the edge's service starting time.

Lacomme et al. [64] consider a bi-objective CARP where the first objective involves minimizing the total duration or length of the trips while the second objective minimizes the maximum trip duration or makespan. The aim of the second criterion is to better balance the trips and create a fair workload distribution among drivers. The authors propose a *Multi-Objective Genetic Algorithm* (MOGA), which presents the decision maker with a set of nondominated solutions. The procedure employs a nondominated sorting algorithm which computes successive population generations and partitions the solutions in nondominated fronts. The GA is based on the MA for UCARP by Lacomme et al. [62], yet the authors adapt their local search and evaluate several ways of integrating it in the MOGA. For instance, the local search must not disturb the overall search mechanism of the MOGA by concentrating on the solutions into small clusters. Experiments show that, although designed for a bi-objective problem, the best MOGA version is able to retrieve most best-known solutions for the single objective UCARP, on the *gdb*, *val*, and *egl* instances.

Grandinetti et al. [48] and [49] consider three objectives: minimizing the total transportation cost, minimizing the longest route (makespan), and minimizing the number of vehicles used to service all the required edges (i.e., the total number of routes). They formulate the MOCARP as an integer linear programming model using route variables. A set of different feasible routes was generated using Path-Scanning approaches (Santos et al. [86]) and the Tabu Search procedure for UCARP by Brandão and Eglese [17]. The MOCARP is solved by the ϵ-constraint method, which transforms MOCARP in a set of single objective CARPs with the other objective expressed as constraints. Experiments show that the procedure was able to reproduce the Pareto front solutions found by Lacomme et al. [64] for quite a large number of benchmark instances, and improve the Pareto front for a few instances.

Mei et al. [70] solve a bi-objective CARP with total distance and makespan minimization through *Decomposition-based Memetic Algorithm with Extended Neighborhood Search* (D-MAENS). The MOCARP is decomposed into N single objective CARPs in which the objective is a linear combination of the different objectives considered, yet each with different weights. That is, some of the N single objective CARPs put more weight on distance minimization, whereas others focus on makespan minimization. The procedure maintains a subpopulation for each of the N single objective CARPs and exploits the MEANS for UCARP (Tang et al. [95]). Mei et al. [70] experiment with three well-known UCARP benchmark sets (*gdb*, *val*, and *egl* instances) and report that their D-MAENS procedure performs better than the MOGA approach by Lacomme et al. [64].

10.6.2 ▪ The Capacitated Arc Routing Problem with profits

Arc routing problems with profits are very novel extensions in the routing literature and will be discussed in greater detail in Chapter 12. We are aware of two recent studies that deal with the *Capacitated Arc Routing Problem with Profits* (CARPP). In the CARPP, introduced by Archetti et al. [6], some edges (called the profitable edges) have a profit and demand for service. Vehicles have limited capacity, and a limitation on the total tour length or travel time is also imposed. The aim is to construct a set of routes, starting and ending at the depot, such that the profit is maximized and route length and vehicle capacity constraints are satisfied. The vehicles do not have to service all demand edges, and profit is only collected if the edge is serviced. Applications of the CARPP involve service providers that have to propose bids to potential customers. The service provider must be able to assess the impact of adding new clients in their existing routes subject to capacity and/or driver time restrictions, and is interested in identifying the most profitable customers (Archetti et al. [6]).

Archetti et al. [6] propose a branch-and-price procedure for the CARPP, and three metaheuristics: The first one is a *Variable Neighborhood Search* (VNS) procedure; the other two algorithms use *Tabu Search* (TS). The TS implementations differ in how they handle capacity and maximum duration constraints. The first TS scheme explores feasible routes only, whereas the second broadens the solution space by allowing infeasibilities, which are dealt with via infeasibility penalization terms in the objective function. The authors test their algorithms on modified *val* instances. The exact approach can solve instances with up to 97 profitable edges in a few minutes in the case that the tours are rather short. In the case that the routes are longer, the instances are much harder. The VNS appears to be the best metaheuristic and generates solutions with an average deviation from the optimal or best-known solution less than 1%.

Zachariadis and Kiranoudis [107] consider the CARPP with a hierarchical objective function in which the first goal is to maximize the collected profits and the second goal is to minimize the total travel time. The authors propose a local search metaheuristic which explores two neighborhoods: inserting or removing a profitable link in/from the solution and exchanging a profitable arc currently in the solution with another (unrouted) profitable arc. The overall search is coordinated through the "promises" concept, which is based on the aspiration criteria of Tabu Search and originally introduced for the *Vehicle Routing Problem with Backhauls* (Zachariadis and Kiranoudis [106]). The procedure employs a best admissible move strategy and associates with every move a solution score or threshold value. A move is considered admissible only if it leads to a solution value of higher quality than its corresponding threshold value. The authors test their solution approach on the instances introduced by Archetti et al. [6]. The results reveal that they were able to match most best-known solutions and improve several of them.

10.7 ▪ Conclusions and outlook

In this chapter we have reviewed variants of the well-known *Capacitated Arc Routing Problem*. More than 80 articles were discussed, and almost all of them were printed after the publication of the book *Arc Routing: Theory, Solutions and Applications*, edited by Dror [30] in 2000. The extent and variety of publications on CARP variants in recent years show clear evidence of the importance of this domain, both from a research and an applications point of view. Indeed, CARP variants are studied in order to better model and understand more complex, more realistic, or new real-life arc routing applications. Among the variants discussed in this chapter are the CARP on mixed graphs; the CARP with a heterogeneous fleet and the multi-compartment CARP; CARPs with multiple, intermediate, and mobile depots; CARPs with split demands, time windows, time-dependent service costs, periodic, stochastic, and deadheading demand; and the CARP with profits and multiple objectives. For many of these variants, the number of publications to date is fairly limited, so there are definitely lots of opportunities to further explore these areas by developing new solution approaches or extending and improving existing models and solution techniques.

Regarding the solution techniques, it should come as no surprise that most research has concentrated on developing effective metaheuristics approaches. Indeed, the CARP itself is an NP-hard problem, and adding more complexity to the model will, in most cases, not make it easier to solve. Several metaheuristic tools have been proposed by various authors. Memetic Algorithms (or Genetic Algorithms hybridized with local search) in combination with Tour Splitting procedures (Ulusoy [97], Lacomme et al. [61] and [62]) have been explored by a number of researchers and have been shown to perform well on several CARP variants. One of the strengths of this approach is that the Tour Splitting procedures are quite flexible and can incorporate a number of extra constraints when the sequence of the links is fixed. Other popular metaheuristic frameworks are Variable Neighborhood Search and Tabu Search. Regarding the exact approaches, the most popular are column generation techniques (e.g., Set-Covering and Set-Partitioning Problems with route variables) and cutting planes. The latter usually aim to provide a good lower bound. Another trendy solution strategy is to convert the arc routing problem into an equivalent node routing problem and solve it through a better, faster, or existing node routing algorithm. This approach seems particularly popular when dealing with problems with asymmetric cost structure and/or time windows. At present, however, it is not clear that these node routing transformations are the best strategy to all problems. Indeed, the transformation into a node routing problem may disguise some of the structural properties of the underlying network (number of connected components formed by required edges, node degree, etc.) which may be easier exploited by tackling the problem directly through an arc routing algorithm. More research is needed into the benefits of both modelling strategies.

It is good to see that several studies show the benefits of their solution approach by investigating and comparing their solutions with real-life cases. An interesting area of research would be to compare the academic solution procedures on real-life problems with off-the-shelf commercial routing software. A number of articles also perform interesting sensitivity studies to show how the routing efficiency or solution is impacted by changing some of the problem parameters. These new insights are of particular importance to managers responsible for designing and operating distribution systems. We would like to encourage researchers to perform more work in this area, and try to uncover the key trade-offs between different cost components or parameters for the distribution system under study. For some CARP variants (e.g., CARP with multiple depots), there is also a need for large and more realistic instances.

Finally, while there are many opportunities for more research in the CARP variants discussed in this chapter, there are several other aspects (such as environmental considerations, CO_2-emissions, and driver familiarity) which have not found their way as yet into CARP literature. We look forward to seeing these developments in the next decade.

Bibliography

[1] H.M. AFSAR, *A branch-and-price algorithm for capacitated arc routing problem with flexible time windows*, Electronic Notes in Discrete Mathematics, 36 (2010), pp. 319–326.

[2] A. AMAYA, A. LANGEVIN, AND M. TRÉPANIER, *The capacitated arc routing problem with refill points*, Operations Research Letters, 35 (2007), pp. 45–53.

[3] A. AMBERG, W. DOMSCHKE, AND S. VOSS, *Multiple centre capacitated arc routing problems: A tabu search algorithm using capacitated trees*, European Journal of Operational Research, 124 (2000), pp. 360–376.

[4] A. AMBERG AND S. VOSS, *A hierarchical relaxation lower bound for the capacitated arc routing problem*, in Proceedings of the 35th Annual Hawaii International Conference on System Sciences, R.H. Sprague, ed., IEEE, Washington, DC, 2002, pp. 1–10.

[5] S.K. AMPONSAH AND S. SALHI, *The investigation of a class of capacitated arc routing problems: The collection of garbage in developing countries*, Waste Management, 24 (2004), pp. 711–712.

[6] C. ARCHETTI, D. FEILLET, A. HERTZ, AND M.G. SPERANZA, *The undirected capacitated arc routing problem with profits*, Computers & Operations Research, 37 (2010), pp. 1860–1869.

[7] L. BACH, G. HASLE, AND S. WØHLK, *A lower bound for the node, edge, and arc routing problem*, Computers & Operations Research, 40 (2013), pp. 943–952.

[8] E. BARTOLINI, J.-F. CORDEAU, AND G. LAPORTE, *An exact algorithm for the capacited arc routing problem with deadheading demand*, Operations Research, 61 (2013), pp. 315–327.

[9] J. BAUTISTA, E. FERNÁNDEZ, AND J. PEREIRA, *Solving an urban waste collection problem using ant heuristics*, Computers & Operations Research, 35 (2008), pp. 3020–3033.

[10] J.M. BELENGUER AND E. BENAVENT, *A cutting plane algorithm for the capacitated arc routing problem*, Computers & Operations Research, 30 (2003), pp. 705–728.

[11] J.M. BELENGUER, E. BENAVENT, N. LABADI, C. PRINS, AND M. REGHIOUI, *Split-delivery capacitated arc-routing problem: Lower bound and metaheuristic*, Transportation Science, 44 (2010), pp. 206–220.

[12] J.M. BELENGUER, E. BENAVENT, P. LACOMME, AND C. PRINS, *Lower and upper bounds for the mixed capacitated arc routing problem*, Computers & Operations Research, 33 (2006), pp. 3363–3383.

[13] E. BENAVENT, V. CAMPOS, Á. CORBERÁN, AND E. MOTA, *The capacitated Chinese postman problem: Lower bounds*, Networks, 22 (1992), pp. 669–690.

[14] P. BEULLENS, L. MUYLDERMANS, D. CATTRYSSE, AND D. VAN OUDHEUSDEN, *A guided local search heuristic for the capacitated arc routing problem*, European Journal of Operational Research, 147 (2003), pp. 629–643.

[15] L.D. BODIN AND S.J. KURSH, *A computer-assisted system for the routing and scheduling of street sweepers*, Operations Research, 26 (1987), pp. 525–537.

[16] A. BOSCO, D. LAGANÀ, R. MUSMANNO, AND F. VOCATURO, *Modelling and solving the mixed capacited general routing problem*, Optimization Letters, 7 (2013), pp. 1451–1469.

[17] J. BRANDÃO AND R. EGLESE, *A deterministic tabu search algorithm for the capacitated arc routing problem*, Computers & Operations Research, 35 (2008), pp. 1112–1126.

[18] O. BRÄYSY, E. MARTÍNEZ, Y. NAGATA, AND D. SOLER, *The mixed capacitated general routing problem with turn penalties*, Expert Systems with Applications, 38 (2011), pp. 12954–12966.

[19] C.H. CHRISTIANSEN AND J. LYSGAARD, *A branch-and-price algorithm for the capacitated vehicle routing problem with stochastic demands*, Operations Research Letters, 35 (2007), pp. 773–781.

[20] C.H. CHRISTIANSEN, J. LYSGAARD, AND S. WØHLK, *A branch-and-price algorithm for the capacitated vehicle routing problem with stochastic demands*, Operations Research Letters, 37 (2009), pp. 392–398.

[21] N. CHRISTOFIDES, A. MINGOZZI, AND P. TOTH, *The vehicle routing problem*, in Combinatorial Optimisation, N. Christofides, A. Mingozzi, P. Toth, and C. Sandi, eds., Wiley, New York, 1979, pp. 315–338.

[22] F. CHU, N. LABADI, AND C. PRINS, *The periodic capacitated arc routing problem: Linear programming model, metaheuristic and lower bounds*, Journal of Systems Science and Systems Engineering, 13 (2004), pp. 423–435.

[23] ——, *Heuristics for the periodic capacitated arc routing problem*, Journal of Intelligent Manufacturing, 16 (2005), pp. 243–251.

[24] ——, *A scatter search for the periodic capacitated arc routing problem*, European Journal of Operational Research, 169 (2006), pp. 586–605.

[25] G. CLARKE AND J.W. WRIGHT, *Scheduling of vehicles from a central depot to a number of delivery points*, Operations Research, 12 (1964), pp. 568–581.

[26] Á. CORBERÁN AND C. PRINS, *Recent results on arc routing problems: An annotated bibliography*, Networks, 56 (2010), pp. 50–69.

[27] B. DE ROSA, G. GHIANI, AND R. MUSMANNO, *The arc routing and scheduling problem with transhipment*, Transportation Science, 36 (2002), pp. 301–313.

[28] A. DEL PIA AND C. FILIPI, *A variable neighbourhood descent algorithm for a real waste collection problem with mobile depots*, International Transactions in Operational Research, 13 (2006), pp. 125–141.

[29] S.H.H. DOULABI AND A. SEIFI, *Lower and upper bounds for location-arc routing problems with vehicle capacity constraints*, European Journal of Operational Research, 224 (2013), pp. 189–208.

[30] M. DROR, *Arc Routing: Theory, Solutions and Applications*, Kluwer Academic Publishers, Boston, MA, 2000.

[31] M. DROR AND J.M.Y. LEUNG, *Combinatorial optimization in a cattle yard: Feed distribution, vehicle scheduling, lot sizing and dynamic pen assignment*, in Industrial Applications of Combinatorial Optimization, G. Yu, ed., Kluwer Academic Publishers, Boston, MA, 1998, pp. 142–171.

[32] M. DROR AND P. TRUDEAU, *Savings by split delivery routing*, Transportation Science, 23 (1989), pp. 141–145.

[33] ——, *Split delivery routing*, Naval Research Logistic Quarterly, 37 (1990), pp. 383–402.

[34] R.W. EGLESE, *Routing winter gritting vehicles*, Discrete Applied Mathematics, 48 (1994), pp. 231–244.

[35] R.W. EGLESE AND L.Y.O. LI, *Efficient routing for winter gritting*, The Journal of the Operational Research Society, 43 (1992), pp. 1031–1034.

[36] S. ESKANDARZADEH, R. TAVAKKOLI-MOGHADDAM, AND A. AZARON, *An extension of the relaxation algorithm for solving a special case of capacitated arc routing problems*, Journal of Combinatorial Optimization, 17 (2009), pp. 214–234.

[37] G. FLEURY, P. LACOMME, AND C. PRINS, *Evolutionary algorithms for stochastic arc routing*, in Applications of Evolutionary Computing, G.R. Raidl, F. Rothlauf, G.D. Smith, G. Squillero, Cagnoni, S., J. Branke, D.W. Corne, R. Drechsler, Y. Jin, and C.G Johnson, eds., Lecture Notes in Computer Science 3005, Springer-Verlag, Berlin, 2004, pp. 501–512.

[38] ——, *Stochastic Capacitated Arc Routing Problem*, Research report LIMOS, Université Blaise Pascal, Clermont-Ferrand, France, 2005.

[39] G. FLEURY, P. LACOMME, C. PRINS, AND W. RAMDANE-CHÉRIF, *Improving robustness of solutions to arc routing problems*, The Journal of the Operational Research Society, 56 (2005), pp. 526–538.

[40] R.Y.K. FUNG, R. LIU, AND Z. JIANG, *A memetic algorithm for the open capacited arc routing problem*, Transportation Research E, 50 (2013), pp. 53–67.

[41] G. GHIANI, F. GUERRIERO, G. IMPROTA, AND R. MUSMANNO, *Waste collection in southern Italy: Solution of a real-life arc routing problem*, International Transactions in Operational Research, 12 (2005), pp. 135–144.

[42] G. GHIANI, F. GUERRIERO, AND G. LAPORTE, *Tabu search heuristics for the arc routing problem with intermediate facilities under capacity and length restrictions*, Journal of Mathematical Modelling and Algorithms, 3 (2004), pp. 209–223.

[43] G. GHIANI, G. IMPROTA, AND G. LAPORTE, *The capacitated arc routing problem with intermediate facilities*, Networks, 37 (2001), pp. 134–143.

[44] G. GHIANI, D. LAGANÀ, G. LAPORTE, AND F. MARI, *Ant colony optimization for the arc routing problem with intermediate facilities under capacity and length restrictions*, Journal of Heuristics, 16 (2010), pp. 211–233.

[45] B.L. GOLDEN, J.S. DEARMON, AND E.K. BAKER, *Computational experiments with algorithms for a class of routing problems*, Computers & Operations Research, 10 (1983), pp. 47–59.

[46] B.L. GOLDEN AND R.T. WONG, *Capacitated arc routing problems*, Networks, 11 (1981), pp. 305–315.

[47] L. GOUVEIA, M.C. MOURÃO, AND L.S. PINTO, *Lower bounds for the mixed capacitated arc routing problem*, Computers & Operations Research, 37 (2010), pp. 692–699.

[48] L. GRANDINETTI, F. GUERRIERO, D. LAGANÀ, AND O. PISACANE, *An approximate ϵ-constraint method for the multi-objective undirected capacitated arc routing problem*, in Experimental Algorithms, P. Festa, ed., Lecture Notes in Computer Science 6049, Springer-Verlag, Berlin, 2010, pp. 214–225.

[49] ——, *An optimization-based heuristic for the multi-objective undirected capacitated arc routing problem*, Computers & Operations Research, 39 (2012), pp. 2300–2309.

[50] C. GUÉGUEN, *Exact Solution Methods for Vehicle Routing Problems*, Ph.D. dissertation, Central School of Paris, Paris, France, 1999.

[51] G. HASLE, O. KLOSTER, M SMEDSRUD, AND K. GAZE, *Experiments on the Node, Edge, and Arc Routing Problem*, Technical report SINTEF, Trondheim, Norway, 2012.

[52] A. HERTZ, G. LAPORTE, AND M. MITTAZ, *A tabu search heuristic for the capacitated arc routing problem*, Operations Research, 48 (2000), pp. 129–135.

[53] A. HERTZ AND M. MITTAZ, *A variable neighborhood descent algorithm for the undirected capacitated arc routing problem*, Transportation Science, 35 (2001), pp. 425–434.

[54] T. IBARAKI, S. IMAHORI, M. KUDO, T. MASUDA, T. UNO, AND M. YAGIURA, *Effective local search algorithms for routing and scheduling problems with general time window constraints*, Transportation Science, 39 (2005), pp. 206–232.

[55] E.L. JOHNSON AND S. WØHLK, *Solving the capacitated arc routing problem with time windows using column generation*, Research report CORAL, University of Aarhus, Aarhus School of Business, Aarhus, Denmark, 2009.

[56] A. KANSOU AND A. YASSINE, *New upper bounds for the multi-depot capacitated arc routing problem*, International Journal of Metaheuristics, 1 (2010), pp. 81–95.

[57] G. KIRLIK AND A. SIPAHIOGLU, *Capacitated arc routing problem with deadheading demands*, Computers & Operations Research, 39 (2012), pp. 2380–2394.

[58] M. KIUCHI, Y. SHINANO, R. HIRABAYASHI, AND Y. SARUWATARI, *An exact algorithm for the capacitated arc routing problem using parallel branch and bound method*, in Abstracts of the Spring National Conference of the Operational Research Society of Japan, 1995, pp. 28–29.

[59] N. LABADI, C. PRINS, AND M. REGHIOUI, *Grasp with path relinking for the capacitated arc routing problem with time windows*, in Advances in Computational Intelligence in Transport, Logistics, and Supply Chain Management, A. Fink and F. Rothlauf, eds., Studies in Computational Intelligence 144, Springer-Verlag, Berlin, 2008, pp. 111–135.

[60] ——, *An evolutionary algorithm with distance measure for the split delivery capacitated arc routing problem*, in Recent Advances in Combinatorial Optimization, C. Cotta and J. Van Hemert, eds., Studies in Computational Intelligence 153, Springer-Verlag, Berlin, 2008, pp. 275–294.

[61] P. LACOMME, C. PRINS, AND W. RAMDANE-CHÉRIF, *A genetic algorithm for the capacitated arc routing problem and its extensions*, in Applications of evolutionary computing, E.J.W. Boers, ed., Lecture Notes In Computer Science 2037, Springer-Verlag, Berlin, 2001, pp. 473–483.

[62] ——, *Competitive memetic algorithms for arc routing problems*, Annals of Operations Research, 131 (2004), pp. 159–185.

[63] ——, *Evolutionary algorithms for periodic arc routing problems*, European Journal of Operational Research, 165 (2005), pp. 535–553.

[64] P. LACOMME, C. PRINS, AND M. SEVAUX, *A genetic algorithm for a bi-objective capacitated arc routing problem*, Computers & Operations Research, 33 (2006), pp. 3473–3493.

[65] G. LAPORTE, R. MUSMANNO, AND F. VOCATURO, *An adaptive large neighbourhood search heuristic for the capacitated arc routing problem with stochastic demands*, Transportation Science, 44 (2010), pp. 125–135.

[66] L.Y.O. LI AND R.W. EGLESE, *An iterative algorithm for vehicle routeing for winter-gritting*, Journal of the Operational Research Society, 47 (1996), pp. 217–228.

[67] V. MANIEZZO AND M. ROFFILLI, *Algorithms for large directed capacitated arc routing problem instances*, in Recent Advances in Evolutionary Computation for Combinatorial Optimization, C. Cotta and J. van Hemert, eds., Studies in Computational Intelligence 153, Springer-Verlag, Berlin, 2008, pp. 259–274.

[68] F. MARZOLF, M. TRÉPANIER, AND A. LANGEVIN, *Road network monitoring: Algorithms and a case study*, Computers & Operations Research, 33 (2006), pp. 3494–3507.

[69] Y. MEI, K. TANG, AND X. YAO, *A memetic algorithm for periodic capacitated arc routing problem*, IEEE Transactions on Systems, Man, and Cybernetics—Part B, 41 (2011), pp. 1654–1667.

[70] ——, *Decomposition-based memetic algorithm for multi-objective capacitated arc routing problem*, IEEE Transactions on Evolutionary Computation, 15 (2011), pp. 151–165.

[71] I.M. MONROY, C.A. AMAYA, AND A. LANGEVIN, *The periodic capacitated arc routing problem with irregular services*, Discrete Applied Mathematics, 161 (2013), pp. 691–701.

[72] L.M. Moreira, J.F. Oliveira, A.M. Gomes, and J.S. Ferreira, *Heuristics for a dynamic rural postman problem*, Computers & Operations Research, 34 (2007), pp. 3281–3294.

[73] M.C. Mourão and T. Almeida, *Lower-bounding and heuristic methods for a refuse collection vehicle routing problem*, European Journal of Operational Research, 121 (2000), pp. 420–434.

[74] M.C. Mourão and L. Amado, *Heuristic method for a mixed capacitated arc routing problem: A refuse collection application*, European Journal of Operational Research, 160 (2005), pp. 139–153.

[75] M.C. Mourão, A.C. Nunes, and C. Prins, *Heuristic methods for the sectoring arc routing problem*, European Journal of Operational Research, 197 (2009), pp. 856–868.

[76] P.A. Mullaseril, *Capacitated Rural Postman Problem with Time Windows and Split Delivery*, Ph.D. dissertation, University of Arizona, Tucson, AZ, 1996.

[77] P. Mullaseril, M. Dror, and J. Leung, *Split-delivery routing heuristics in livestock feed distribution*, The Journal of the Operational Research Society, 48 (1997), pp. 107–116.

[78] L. Muyldermans, *Routing, Districting and Location for Arc Traversal Problems*, Ph.D. dissertation, Catholic University of Leuven, Leuven, Belgium, 2003.

[79] L. Muyldermans, D. Cattrysse, and D. Van Oudheusden, *District design for arc-routing applications*, The Journal of the Operational Research Society, 54 (2003), pp. 1209–1221.

[80] L. Muyldermans, D. Cattrysse, D. Van Oudheusden, and T. Lotan, *Districting for salt spreading operations*, European Journal of Operational Research, 139 (2002), pp. 521–532.

[81] L. Muyldermans and G. Pang, *A guided local search procedure for the multi-compartment capacitated arc routing problem*, Computers & Operations Research, 37 (2010), pp. 1662–1673.

[82] Y. Nagata and O. Bräysy, *Edge assembly based memetic algorithm for the capacitated vehicle routing problem*, Networks, 54 (2009), pp. 205–215.

[83] R. Pandi and B. Muralidharan, *A capacitated general routing problem on mixed graphs*, Computers & Operations Research, 22 (1995), pp. 465–478.

[84] M. Polacek, K.F. Doerner, R.F. Hartl, and V. Maniezzo, *A variable neighbourhood search for the capacitated arc routing problem with intermediate facilities*, Journal of Heuristics, 14 (2008), pp. 405–423.

[85] C. Prins and S. Bouchenoua, *A memetic algorithm solving the vrp, the carp and general routing problems with nodes, edges and arcs*, in Recent Advances in Memetic Algorithms, W. Hart, N. Krasnogor, and J. Smith, eds., Studies in Fuzziness and Soft Computing 166, Springer-Verlag, Berlin, 2005, pp. 65–85.

[86] L. Santos, J. Coutinho-Rodrigues, and J.R. Current, *An improved heuristic for the capacitated arc routing problem*, Computers & Operations Research, 36 (2009), pp. 2632–2637.

[87] J. SNIEZEK, *The Capacitated Arc Routing Problem with Vehicle/Site Dependencies: An Application of Arc Routing and Partitioning*, Ph.D. dissertation, University of Maryland, College Park, MD, 2001.

[88] J. SNIEZEK AND L. BODIN, *Using mixed integer programming for solving the capacitated arc routing problem with vehicle/site dependencies with an application to the routing of residential sanitation collection vehicles*, Annals of Operations Research, 144 (2006), pp. 33–58.

[89] J. SNIEZEK, L. BODIN, L. LEVY, AND M. BALL, *Chapter 11: Capacitated arc routing problem with vehicle-site dependencies: The Philadelphia experience*, in The Vehicle Routing Problem, P. Toth and D. Vigo, eds., SIAM Monograph on Discrete Mathematics and Applications 9, 2001, pp. 287–308.

[90] M.M. SOLOMON, *Algorithms for the vehicle routing and scheduling problems with time window constraints*, Operations Research, 35 (1987), pp. 254–265.

[91] H.I. STERN AND M. DROR, *Routing electric meter readers*, Computers & Operations Research, 6 (1979), pp. 209–223.

[92] M. TAGMOUTI, M. GENDREAU, AND J.Y. POTVIN, *A variable neighbourhood descent heuristic for arc routing problems with time-dependent service costs*, Computers and Industrial Engineering, 59 (2010), pp. 954–963.

[93] ——, *A dynamic capacitated arc routing problem with time-dependent service costs*, Transportation Research C, 19 (2011), pp. 20–28.

[94] M. TAGMOUTI, M. GENDREAU, AND J.-Y. POTVIN, *Arc routing problems with time-dependent service costs*, European Journal of Operational Research, 181 (2007), pp. 30–39.

[95] K. TANG, Y. MEI, AND X. YAO, *Memetic algorithm with extended neighborhood search for capacitated arc routing problems*, IEEE Transactions on Evolutionay Computation, 13 (2009), pp. 1151–1166.

[96] P. TSENG AND D.P. BERTSEKAS, *Relaxation methods for linear programs*, Mathematics of Operations Research, 12 (1987), pp. 569–596.

[97] G. ULUSOY, *The fleet size and mix problem for capacitated arc routing*, European Journal of Operational Research, 22 (1985), pp. 329–337.

[98] F.L. USBERTI, P.M. FRANÇA, AND A.L.M. FRANÇA, *The open capacitated arc routing problem*, Computers & Operations Research, 38 (2011), pp. 1543–1555.

[99] P. VANSTEENWEGEN, W. SOUFFRIAU, AND K. SÖRENSEN, *Solving the mobile mapping van problem: A hybrid meta-heuristic for capacitated arc routing with soft time windows*, Computers & Operations Research, 37 (2010), pp. 1870–1876.

[100] S. WØHLK, *Contributions to Arc Routing*, Ph.D. dissertation, University of Southern Denmark, Odense, Denmark, 2005.

[101] ——, *New lower bound for the capacited arc routing problem*, Computers & Operations Research, 33 (2006), pp. 3458–3472.

[102] ——, *A decade of capacitated arc routing*, in The Vehicle Routing Problem: Latest Advances and New Challenges, B. Golden, S. Raghavan, and E. Wasil, eds., Springer, New York, 2008, pp. 29–48.

[103] L. XING, P. ROHLFSHAGEN, Y. CHEN, AND X. YAO, *An evolutionary approach to the multi-depot capacitated arc routing problem*, IEEE Transactions on Evolutionay Computation, 14 (2010), pp. 356–374.

[104] L.N. XING, P. ROHLFSHAGEN, Y.W. CHEN, AND X. YAO, *A hybrid ant colony optimization algorithm for the extended capacitated arc routing problem*, IEEE Transactions on Systems, Man, and Cybernetics—Part B, 41 (2011), pp. 1110–1123.

[105] H. XU, C.H. ZHANG, Y.A. TAN, AND J. LU, *An improved evolutionary approach to the extended capacitated arc routing problem*, Expert Systems with Applications, 38 (2011), pp. 4637–4641.

[106] E.E. ZACHARIADIS AND C.T. KIRANOUDIS, *An innovative meta-heuristic solution approach for the vehicle routing problem with backhauls*, Technical Report, National Technical University of Athens, Athens, Greece, 2009.

[107] ——, *Local search for the undirected capacitated arc routing problem with profits*, European Journal of Operational Research, 210 (2011), pp. 358–367.

[108] Z. ZHU, X. LI, Y. YANG, X. DENG, M. XIA, Z. XIE, AND J. LIU, *A hybrid genetic algorithm for the multiple depot capacitated arc routing problem*, in Proceedings of the IEEE International Conference on Automation and Logistics, IEEE, Washington, DC, 2007, pp. 2253–2258.

Chapter 11

Arc Routing Problems with Min-Max Objectives

Enrique Benavent
Ángel Corberán
Isaac Plana
José María Sanchis

11.1 ▪ Introduction

Given a road network, the *Chinese Postman Problem* (CPP) consists of finding a shortest closed walk traversing all the roads in the network at least once. Since real-life applications of the CPP include road maintenance, garbage collection, mail delivery, etc., large road networks for which only one vehicle is not enough have to be considered. In these cases, the workload has to be distributed among $K \geq 2$ vehicles, and then a set of K routes, each starting and ending at the depot, must be found. If the objective is to minimize the total distance traveled by the K vehicles, we have the *K vehicles Chinese Postman Problem* (K-CPP) which, like the CPP itself, is polynomially solvable when the network is fully undirected or fully directed. For the undirected case, this was first observed by Assad, Pearn, and Golden [8], and for the directed case by Pearn [48]. Zhang [52] also presented polynomial algorithms for these problems.

Since the objective in the K-CPP is to minimize the total distance traveled by the K vehicles, it tends to produce "unfair" solutions in which the routes are unbalanced. There are, however, situations where an equity criterion to obtain routes of approximately the same length is more appropriate. In the *Min-Max K-Chinese Postman Problem* (MM K-CPP), the goal is to minimize the length of the longest of the K routes. Thereby, the routes tend to be more balanced, resulting in a "fairer" solution. Furthermore, with this kind of min-max objective function, if the travel times are proportional to the travel distance, the last customer serviced is serviced as early as possible.

The min-max objective was first introduced for several node and arc routing problems by Frederickson, Hecht, and Kim [34]. In the same paper, the authors showed that these problems are NP-hard. Later, in Applegate et al. [6], a min-max node routing problem in the context of newspaper delivery was considered. The problem was proposed by Aarts,

Hurkens, and Lenstra [1] as part of the Whizzkids '96 mathematics challenge. The newspaper routing instance specified in the competition consisted of a set of 120 customers to be serviced by four vehicles. The goal was to ensure that all the customers were serviced as soon as possible. The instance needed 10 days of computing time to be solved by a sophisticated branch-and-cut algorithm implemented on a distributed network of 188 processors. This gives an idea of the difficulty that branch-and-cut algorithms have with solving problems with a min-max objective.

Regarding arc routing problems with min-max objective functions, in addition to the undirected MM K-CPP that will be presented in Section 11.2, the undirected and windy versions of the *Min-Max K-Rural Postman Problem* (MM K-RPP) have been studied. These are more general problems than the MM K-CPP, in which the edges are divided into two sets: required edges that must be serviced and nonrequired edges that can be used for deadheading. The results known for these problems are presented in Sections 11.3 and 11.4.

11.2 ▪ The min-max K-CPP

The MM K-CPP can be defined as follows. Given an undirected graph $G = (V,E)$, a cost $c_e \geq 0$ for each edge $e \in E$, a distinguished depot vertex, and a fixed number $K \geq 2$ of vehicles, find K tours, i.e., closed walks starting and ending at the depot, jointly traversing each edge $e \in E$ at least once, in such a way that the length of the longest of the K tours is minimal.

The heuristic and exact algorithms proposed for this problem are described in what follows.

11.2.1 ▪ Heuristics and metaheuristics

Frederickson, Hecht, and Kim [34] proposed a heuristic algorithm which first solves the CPP on graph G and computes a single tour R traversing all the edges $e \in E$. The tour R is then partitioned by selecting $K-1$ splitting vertices on R that determine K tour segments of R of approximately the same length. Finally, the K tours are completed by connecting these tour segments to the depot with shortest paths. It was shown in [34] that this is a $(2-\frac{1}{K})$-approximation algorithm.

More sophisticated heuristic algorithms were developed by Ahr and Reinelt [3] and are described in the following:

The augment-merge algorithm: This algorithm is based on the augment-merge algorithm for the CARP given in [14]. It starts with a closed walk C_e for each edge $e = (i,j) \in E$, formed with the edges of a shortest path between the depot and i, the edge e itself, and the edges of a shortest path between vertex j and the depot. Then two closed walks are successively merged until K tours remain, trying to keep the tour lengths low and balanced. Note that for some edges e a tour C_e is explicitly created while for other edges it is not, since they are contained in an already existing tour. The first kind of edges are termed critical edges and the second noncritical edges.

The cluster algorithm: In the first step, the edge set E is divided into K clusters, and in the second step, a tour for each cluster is built. The cluster step first performs a K-clustering of the critical edges into edge sets $F_1,\dots,F_K$. Then each cluster F_i is supplemented by edges in shortest paths connecting it with the depot. The routing step solves the CPP for each subgraph induced by F_i.

Two improvement procedures: These are applied to the solutions obtained by the previous algorithms as a post-optimization step, trying to reduce the length of the longest tour in a greedy way. The first improvement procedure, *Eliminate Redundant Edges*, eliminates edges traversed at least twice in the longest tour if these edges are already covered by other tours. The second improvement procedure, *Two Edge Exchange*, tries to shorten the longest tour by removing edges from it and integrating them into a shorter tour.

Later, Ahr and Reinelt [4] presented a tabu search algorithm for the MM K-CPP. It is based on a set of basic procedures that modify and improve single tours. Furthermore, different neighborhoods are considered. Initial solutions are taken from the heuristics described above. The tabu search algorithm was run on a variety of test instances from the literature as well as on random instances. The authors were able to improve the results obtained by the heuristics in Ahr and Reinelt [3] for almost all instances. The improvement was, in many cases, about 10%, and for some instances as much as 20–30%, yielding many near-optimal or even optimal solutions. Furthermore, the behavior of the different neighborhoods was compared and the effect of the improvement procedures on the solution quality and computing time was investigated.

11.2.2 ▪ A branch-and-cut algorithm

Ahr [2] adapted his branch-and-cut algorithm for the *Capacitated Arc Routing Problem* (CARP) to solve the MM K-CPP. This algorithm is based on the sparse formulation given in Belenguer and Benavent [12] and uses two sets of variables: binary variables y_e^k for each edge $e \in E$ and each vehicle $k=1,\dots,K$, taking value one when vehicle k services edge e, and integer variables x_e^k for each edge $e \in E$ and each vehicle $k=1,\dots,K$, which indicate the number of times edge e is traversed by vehicle k without servicing it.

To present the IP formulation proposed in Ahr [2], we introduce the following notation: Vertex $1 \in V$ represents the depot. For a subset of vertices $S \subseteq V$, $\delta(S)$ denotes the set of edges with exactly one endpoint in S, while $E(S)$ denotes the set of edges with both endpoints in S. For a subset of edges $F \subseteq E$, $x^k(F) = \sum_{e\in F} x_e^k$ and $y^k(F) = \sum_{e\in F} y_e^k$. The IP formulation of Ahr [2] is

$$\begin{aligned}
&(11.1) && \text{(MM K-CPP)} \quad \min \ z && \\
&(11.2) && \text{s.t.} \quad \sum_{e\in E} c_e(x_e^k + y_e^k) \leq z && \forall\, k=1,\dots,K,\\
&(11.3) && \sum_{k=1}^{K} y_e^k = 1 && \forall e \in E,\\
&(11.4) && x^k(\delta(S)) + y^k(\delta(S)) \equiv 0 \pmod 2 && \forall S \subset V\setminus\{1\} \text{ and } \forall k=1,\dots,K,\\
&(11.5) && x^k(\delta(S)) + y^k(\delta(S)) \geq 2y_e^k && \forall S \subset V\setminus\{1\},\ \forall e \in E(S),\ \forall k,\\
&(11.6) && x_e^k \geq 0 \text{ and integer} && \forall e \in E,\ \forall k=1,\dots,K,\\
&(11.7) && y_e^k \in \{0,1\} && \forall e \in E,\ \forall k=1,\dots,K,\\
&(11.8) && z \geq 0. &&
\end{aligned}$$

The min-max objective function is modeled by introducing an additional variable z to be minimized (11.1) and K inequalities (11.2) which ensure that the length of the K tours is at most z. Equations (11.3) force each edge to be serviced, and connectivity inequalities (11.5) ensure that each single tour is connected and connected to the depot. Note that constraints (11.4), which guarantee the parity of each vertex in the solution, are not linear. Although there is a way of expressing them as linear constraints, it involves the use of additional integer variables. Instead of doing this, the following two families of inequalities are used: The tour parity inequalities

$$y^k(\delta(S)\setminus F)+x^k(\delta(S))\geq y^k(F)-|F|+1 \tag{11.9}$$

for all $S\subseteq V\setminus\{1\}$ and $F\subseteq \delta(S)$ with $|F|$ odd, and for all k, and the aggregate parity inequalities

$$\sum_{k=1}^{K} x^k(\delta(S))\geq 1 \tag{11.10}$$

for all $S\subseteq V\setminus\{1\}$ with $|\delta(S)|$ odd.

Ahr [2] noted that there might be no need for two variable types, x-variables and y-variables, so that only one type of variable indicating the number of times an edge is traversed or serviced could be enough. However, with only one such type of variable the tour parity inequalities (11.9) cannot be formulated, since binary variables are required. The tour parity inequalities can be shown to be valid as follows. If all edges in F are serviced by vehicle k (that is, $y^k(F)=|F|$), then, given that $|F|$ is odd, vehicle k must cross $\delta(S)$ at least once more, so $y^k(\delta(S)\setminus F)+x^k(\delta(S))$ should be at least 1. On the other hand, if $y^k(F)\leq |F|-1$, the inequality is trivial.

Let us suppose that, given a subset $S\subseteq V$, we can compute a lower bound $L(S)$ for the number of vehicles needed to traverse all the edges in $E(S)$ and to visit the depot in an *optimal* K-vehicles solution. That is, any feasible solution for which all the edges in $E(S)$ are traversed by less than $L(S)$ vehicles would be nonoptimal. Then, in each optimal solution the cutset $\delta(S)$ has to be crossed at least $2L(S)$ times and the so-called aggregate L-tour constraints

$$\sum_{k=1}^{K} x^k(\delta(S))+\sum_{k=1}^{K} y^k(\delta(S))\geq 2L(S) \quad \forall\, S\subseteq V\setminus\{1\} \tag{11.11}$$

can be used in the branch-and-cut algorithm. In Ahr [2] an iterative algorithm for the computation of $L(S)$ is described. Note that these constraints might cut off feasible solutions but no optimal solutions. For $L(S)=1$, these constraints are dominated by the connectivity inequalities (11.5).

The branch-and-cut algorithm described in Ahr [2] incorporates separation algorithms for inequalities (11.5), (11.9), and (11.10) that were adapted from the corresponding procedures designed for the CARP. In addition, some heuristic separation algorithms for the aggregate L-tour constraints (11.11) were implemented. Computational results proved the L-tour constraints to be very effective. The algorithm also incorporates the shortest path tour lower bound (SPTLB). This lower bound is based on the observation that in an optimal solution the length of the longest tour must be at least as long as a shortest tour traversing an edge farthest away from the depot.

The algorithm was tested on a set of 74 instances both from the literature and randomly generated, with up to 200 vertices, 392 edges, and $K=10$ vehicles, obtaining a total of 571 instances. It was capable of solving 54% of all 571 instances (within two hours CPU time) exactly. It turned out that instances with medium-sized K (usually $K=4, 5, 6, 7$) were especially hard to solve. This is due to the fact that the lower bounds obtained by the LP relaxation are only good for K small and the SPTLB is only good for K large. For medium values of K, both bounds are rather bad.

11.3 ▪ The min-max K-RPP

The MM K-RPP is a more general problem than the MM K-CPP, since not all the edges in the graph have to be serviced. The problem can be defined as follows. Given an undirected graph $G = (V, E)$, a cost $c_e \geq 0$ for each edge $e \in E$, a subset $E_R \subseteq E$ of required edges, a distinguished depot vertex, and a fixed number $K \geq 2$ of vehicles, find K tours each starting and ending at the depot and jointly traversing each edge $e \in E_R$ at least once, in such a way that the length of the longest of the K tours is minimal.

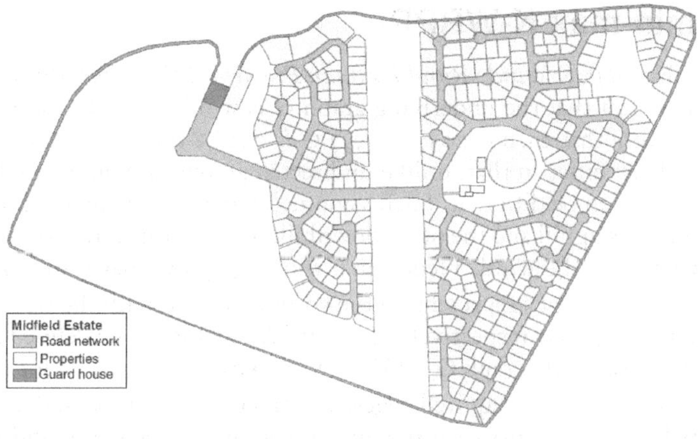

Figure 11.1. *Midfield estate in Gauteng, South Africa [51].*

A heuristic algorithm for the MM K-RPP was proposed by Willemse and Joubert [51] in the context of the design of routes for security guard patrols in South Africa. According to the authors, secure estates in that country have become a popular choice of residence, mostly because of the estates' ability to provide 24/7 security for their inhabitants. The security system's main objective is that criminals perceive a higher risk of detection and detention. This approach requires highly visual security initiatives such as the patrolling of the estate's inner roads and pathway network by security guards. Designing patrol routes for security guards is a complex task because *"security guards cannot follow the same patrolling route every day as this predictable routing information could be used by unlawful parties to side-step the security guards. Moreover, all roads and paths have to be patrolled evenly as information regarding lesser patrolled roads and paths can be exploited. Patrolling of the roads and paths should also be evenly distributed among the guards to avoid discontentment and possible work overload."*

The authors modeled this problem as an MM K-RPP and proposed a tabu search algorithm which was tested on a real problem instance from an estate in Gauteng, South Africa (see Figure 11.1, taken from Willemse and Joubert [51]). The algorithm is an extension to the MM K-RPP of the algorithm by Ahr and Reinelt [4] for the MM K-CPP (see Section 11.2.1). It uses the encoding scheme of Lacomme, Prins, and Ramdane-Chérif [41] for the CARP in which only required edges are explicitly modeled. It is assumed that a shortest path between consecutive required edges is always followed. Therefore, the removal and insertion procedures are only applied to required edges. Moreover, this algorithm generates multiple solutions for the same problem instance, which is important in this application to enable the security manager to randomly choose which patrol routes to implement during a shift while taking into consideration guard availability. The algorithm works in three phases. During the first phase it generates a specified number of different initial solutions using a constructive heuristic, and these solutions are improved in the second phase. In the third phase, a tabu search algorithm is applied.

The results produced by this algorithm on the instance depicted in Figure 11.1 showed a significant improvement on the existing routes and schedules implemented at the estate. In particular, the difference between the most and least patrolled edges for the new generated routes is 10 traversals, whereas it was 50 for the old ones. The algorithm was also tested on instances generated from eight CARP instances proposed in Li and Eglese [43], with up to 140 vertices and 190 required edges.

11.4 ▪ The min-max K-WRPP

The *Min-Max K-Windy Rural Postman Problem* (MM K-WRPP) is similar to the MM K-RPP, except that the underlying graph G is windy. Windy and undirected graphs share the feature that each edge $e = (i, j) \in E$ can be traversed either from i to j or from j to i. They differ in that, unlike undirected graphs, in windy graphs the costs of the two traversals can be different. Such asymmetry in the costs of an edge may occur when one direction is uphill and the other downhill, or when one direction is with the wind and the other against the wind. Hence, in a windy graph we have two costs $c_{ij}, c_{ji} \geq 0$ for each edge $e = (i, j) \in E$, representing the costs associated with the traversal of $e \in E$ from i to j and from j to i, respectively. The MM K-WRPP can be defined as follows.

Given a windy graph $G = (V, E)$ with two costs $c_{ij}, c_{ji} \geq 0$ for each edge $e = (i, j) \in E$, a subset $E_R \subseteq E$ of required edges, a distinct depot vertex, and a fixed number $K \geq 2$ of vehicles, find K tours each starting and ending at the depot and jointly traversing each edge $e \in E_R$ at least once, in such a way that the length of the longest of the K tours is minimal.

Note that an arc routing problem on a windy graph includes the corresponding problem on an undirected ($c_{ij} = c_{ji}$ for each edge $(i, j) \in E$), directed ($c_{ji} = +\infty$ for each arc $(i, j) \in A$), or mixed graph as special cases.

Several papers have been devoted to the MM K-WRPP. The metaheuristic algorithm implemented in Benavent, Corberán, and Sanchis [19] is described in the next section. The two exact procedures, a branch-and-cut algorithm proposed in Benavent et al. [17, 18] and a branch-price-and-cut algorithm developed in Benavent et al. [15], are detailed in Sections 11.4.2 and 11.4.3, respectively.

11.4.1 ▪ A heuristic

In this section we briefly describe the metaheuristic for the MM K-WRPP proposed in Benavent, Corberán, and Sanchis [19], whose main components are a *Multi-Start* (MS) algorithm, a *Variable Neighborhood Descent* (VND), and an *Iterated Local Search* (ILS) procedure. It is based on the multi-start iterated local search algorithm proposed in Belenguer et al. [13] for the split delivery capacitated arc routing problem and provides high quality solutions that have also been used in the exact procedures described in Sections 11.4.2 and 11.4.3 to prune the enumeration tree.

Overall the algorithm works as follows. The MS algorithm of Benavent et al. [16] produces, at each iteration, a single route traversing all the required edges (a feasible solution for the *Windy Rural Postman Problem* (WRPP) on graph G). The route is then split into K small routes, one for each vehicle, to obtain an MM K-WRPP solution. Then, two different VNDs and a simplification procedure are applied to this solution. The ILS procedure is applied to this improved solution. At any iteration of the ILS, the MM K-WRPP solution is randomly perturbed and is then improved using procedures similar to the VNDs and simplification procedures mentioned above. The best solution found during the ILS is used as the starting solution for the next iteration. The best solution found in the whole MS procedure is the output of the algorithm. All these procedures are described in more detail in what follows.

In the first step of the algorithm, a fixed number of WRPP solutions are generated with the MS algorithm for the WRPP described in Benavent et al. [16]. Each WRPP solution (a single route) is encoded as a sequence of the required edges $(s_1, \ldots, s_m)$, $m = |E_R|$, each one traversed in a given direction. It is assumed that the route follows a shortest path from the end vertex of any required edge in the sequence to the initial vertex of the next one. A variable neighborhood descent procedure (called VND1), consisting of several local search methods, is applied to each WRPP solution. The first method, called *2-interchange*, is a steepest-descent algorithm in which a simple move consisting of interchanging the positions of two required edges in the sequence is performed at each iteration. The second procedure is also a steepest-descent algorithm in which each move is an *Or-interchange* (Or [46]). In this move, a section consisting of at most L consecutive required edges in the sequence is inserted elsewhere between two consecutive required edges, the first one of which must be among the M edges closest to the first edge of the section. This insertion is performed in such a way that the direction of traversal remains unchanged. After some testing, the values chosen for parameters L and M were 4 and 11, respectively. The third improvement procedure finds an optimal direction in which each required edge has to be traversed for a given order of the required edges. It is called the *reversal procedure* and was proposed in Lacomme, Prins, and Ramdane-Chérif [41]. VND1 uses these improvement procedures in the following order: the Or-interchange, the reversal procedure, and the 2-interchange.

The single route obtained is split into K routes, applying the procedure described in [41]. The algorithm builds a directed graph H with $m + 1$ vertices indexed from 0 to m. An arc $(i - 1, j)$ in H corresponds to a route starting at the depot, traversing the required edges in the subsequence $(s_i, \ldots, s_j)$ in this order, and coming back to the depot. The arc weight is given by the route cost. Then, a path between vertices 0 and m, which minimizes the cost of the largest arc weight and contains at most K arcs, is computed to obtain an MM K-WRPP solution.

Thereafter, two different VNDs and a simplification procedure are applied to this solution. The first one is VND1, described above, and is applied to each one of the K routes in the solution. The simplification procedure is applied next. It works on an expanded representation of the solution: The shortest paths that are traversed between two consecutive required edges are included in the solution, which is then represented as a set of K closed walks. The closed walks are then simplified by removing multiple traversals of some edges and changing the service of some required edges from the longest route to other routes. The second VND (called VND2) uses three inter-route moves called *change2to0*, *change1to0*, and *change1to1*, in that order. The first two consist of moving two and one consecutive required edges, respectively, from the longest route to another route. In the third move, one required edge of the longest route and another of another route are interchanged.

The resulting improved MM K-WRPP solution is used as the starting point of the ILS procedure. At each iteration of the ILS, the MM K-WRPP solution is randomly modified using a perturbation procedure that works as follows. The K routes are first merged to form a single giant tour. Then the single tour is perturbed by interchanging the required edges at two randomly selected positions, before it is finally split again into K routes. The resulting MM K-WRPP solution is then improved by using procedures similar to those described above. The best solution found during the ILS is used as the starting point for the next iteration. The best solution found in the whole MS procedure is the output of the algorithm.

This metaheuristic has been applied on a set of 144 WRPP instances presented in Benavent et al. [14] with up to 50 vertices, 184 edges, and 78 required edges, grouped into 24 subsets of six instances each. Each instance has been tested for two, three, four, five, and six vehicles.

The average percentage gaps between the cost of the solutions provided by the algorithm and the optimum values, if known, or the lower bounds obtained with the branch-and-cut algorithm described in Benavent et al. [17], are very small. In particular, the gaps for the instances with two or three vehicles are less than 0.40% on average. Instances with four, five, and six vehicles present average gaps that are about 1.0%, 1.6%, and 1.46%, respectively. Furthermore, the number of solutions obtained by the heuristic proved to be optimal by the exact algorithms was 112, 116, 118, 120, and 115 for two, three, four, five, and six vehicles, respectively. Regarding CPU times, the maximum average time is achieved on the 2-vehicle instances and is 56.2 seconds on an Intel Core 2 CPU 2.40GHz and 2GB RAM machine. The authors point out that, probably due to the longer computing times needed by the improvement algorithms when applied to longer routes, instances with more vehicles are less time consuming.

11.4.2 ▪ A branch-and-cut algorithm

A branch-and-cut algorithm for the MM K-WRPP was proposed in Benavent et al. [17] and improved in [18]. It is based on a formulation using variables representing the traversal of the graph edges. In what follows, this formulation is presented, the associated polyhedron is described, and the branch-and-cut algorithm based on this polyhedral description is summarized.

11.4.2.1 ▪ Formulation

In Benavent et al. [17] the following formulation is proposed. For each edge $e=(i,j)\in E$ and for each vehicle $k=1,\dots,K$, two variables x_{ij}^k, x_{ji}^k are defined, representing the number of times edge e is traversed by vehicle k from i to j and from j to i, respectively. In addition, for each required edge $e=(i,j)\in E_R$ and for each vehicle $k=1,\dots,K$, there is a binary variable y_{ij}^k taking value 1 if edge $e=(i,j)$ is serviced by vehicle k and 0 otherwise. The min-max objective function is modeled by introducing an additional variable z representing the length of the longest route.

As in other arc routing problems, one can assume that each vertex in V is incident to at least one required edge. This is not a restriction, since there is a simple way of transforming an instance into an equivalent one which satisfies this assumption (see, e.g., Christofides et al. [21]). To present the IP formulation for the MM K-WRPP we need some notation. Let (V,E_R) be the graph induced by the required edges. In general, (V,E_R) is not connected. The sets of vertices $V_1,\dots,V_p$ of the p connected components of (V,E_R) are called R-sets. Given two subsets of vertices $S_1,S_2\subseteq V$, $(S_1:S_2)$ denotes the edge set with one endpoint in S_1 and the other in S_2. Again, given $S\subseteq V$, $\delta(S)=(S:V\setminus S)$ and $E(S)=(S:S)$. The previous sets restricted to the required edges are denoted by $\delta_R(S)$, $E_R(S)$ and $(S_1 : S_2)_R$. Finally, for any subset of edges $F \subseteq E$, $x^k(F)$ denotes $\sum_{(i,j)\in F}(x^k_{ij} + x^k_{ji})$. The MM K-WRPP formulation is

$$\begin{aligned}
&(11.12)\quad \text{(MM K-WRPP)} \quad \min \quad z \\
&(11.13)\quad \text{s.t.} \quad \sum_{(i,j)\in E} \left(c_{ij}x^k_{ij} + c_{ji}x^k_{ji}\right) \le z, && k=1,\dots,K, \\
&(11.14)\quad \sum_{k=1}^{K} y^k_e = 1 && \forall e\in E_R, \\
&(11.15)\quad x^k_{ij} + x^k_{ji} \ge y^k_e && \forall e=(i,j)\in E_R \text{ and } k=1,\dots,K, \\
&(11.16)\quad \sum_{(i,j)\in\delta(i)} (x^k_{ij} - x^k_{ji}) = 0 && \forall i\in V \text{ and } k=1,\dots,K, \\
&(11.17)\quad x^k(\delta(S)) \ge 2y^k_e && \forall S\subset V\setminus\{1\},\ \forall e\in E_R(S),\ k=1,\dots,K, \\
&(11.18)\quad x^k_{ij}, x^k_{ji} \ge 0 \text{ and integer} && \forall (i,j)\in E,\ k=1,\dots,K, \\
&(11.19)\quad y^k_e \in \{0,1\} \text{ and integer} && \forall e\in E,\ k=1,\dots,K.
\end{aligned}$$

Inequalities (11.13) imply that the length of the longest route is minimized. Equations (11.14) ensure that each required edge is serviced by exactly one vehicle. Traversing inequalities (11.15) force a vehicle to traverse the edges it services. Flow conservation constraints (11.16) force each vehicle route to be symmetric, while connectivity constraints (11.17) ensure that each vehicle route is connected and connected to the depot.

11.4.2.2 ▪ The MM K-WRPP polyhedron

A vector $(x^1,y^1,\dots,x^K,y^K)$ with $(2|E|+|E_R|)K$ components satisfying (11.14)–(11.19) will be called an MM K-WRPP solution on G. Note that the above formulation allows solutions in which a vehicle tour x^k is formed by several disconnected subtours, one of them connecting all the edges it services to the depot and the others traversing edges not serviced by this vehicle.

In Benavent et al. [17] it is shown that the convex hull of all the MM K-WRPP solutions, KWRPP(G), is a polyhedron of dimension $K\big(2|E|-|V|+1\big)+(K-1)|E_R|$. This polyhedron is highly symmetric: If $(x^1,y^1,x^2,y^2,\dots,x^K,y^K)$ is a vector in KWRPP(G), then any permutation, e.g., $(x^2,y^2,x^1,y^1,\dots,x^K,y^K)$, is also a vector in KWRPP(G).

The following inequalities are facet-inducing for KWRPP(G) under mild conditions:

- Trivial inequalities, $x^k_{ij} \ge 0$, $x^k_{ji} \ge 0$ $\forall (i,j) \in E$ and $\forall k$ and $y^k_e \ge 0$ $\forall e \in E_R$ and $k=1,\dots,K$.
- Traversing inequalities (11.15).
- Connectivity inequalities (11.17).

In order to strengthen the linear formulation, other large families of valid inequalities have been described and proved to induce facets of KWRPP(G) in Benavent et al. [17, 18]. Some of these families are derived from already known inequalities for the (single vehicle) WRPP polyhedron.

Let us start by describing the aggregate inequalities. Let $F(x)\ge c_0$ be any valid inequality for the WRPP defined on graph G. Given an MM K-WRPP solution $(x^1,y^1,\dots,x^K,y^K)$, the vector $x=\sum_{k=1}^{K} x^k$ is a WRPP tour on G and therefore it satisfies $F(x)\ge c_0$. This means that the inequality $\sum_{k=1}^{K} F(x^k) \geq c_0$ is valid for KWRPP(G) and is called an aggregate inequality. Hence, the R-odd cut, K-C, honeycomb, and path-bridge inequalities of the WRPP (Corberán, Plana, and Sanchis [23]) result in valid aggregate inequalities for KWRPP(G).

Parity inequalities

The aggregate R-odd cut (or aggregate parity) inequalities are

$$\sum_{k=1}^{K} x^k(\delta(S)) \geq |\delta_R(S)|+1 \quad \forall S \subset V \text{ such that } |\delta_R(S)| \text{ is odd.} \tag{11.20}$$

Moreover, versions of these families corresponding either to a single vehicle, called disaggregate inequalities, or a subset of P vehicles, called P-aggregate inequalities, are presented in Benavent et al. [17]. To introduce these inequalities, let us consider the aggregate parity inequalities (11.20). They imply that each MM K-WRPP solution crosses a given cutset $\delta(S)$ with an odd number $|\delta_R(S)|$ of required edges at least $|\delta_R(S)|+1$ times. Note that there can be MM K-WRPP solutions in which a particular vehicle, say, vehicle k, does not cross $\delta(S)$. Hence, inequality $x^k(\delta(S)) \geq |\delta_R(S)|+1$ is not valid for KWRPP(G). Nevertheless, it can be seen that the inequality

$$x^k(\delta(S)) \geq 2y^k(\delta_R(S)) - |\delta_R(S)|+1 \tag{11.21}$$

is indeed valid for KWRPP(G). Note that for the MM K-WRPP solutions in which vehicle k services all the required edges in $\delta_R(S)$, the right-hand side of the inequality takes the value $|\delta_R(S)|+1$ and the inequality implies that vehicle k crosses the cutset $\delta(S)$ at least $|\delta_R(S)|+1$ times. This example illustrates the idea behind a disaggregate inequality obtained from a known inequality $F(x) \geq c_0$ for the WRPP: While preserving the left-hand side, $F(x^k)$, define a right-hand side using some y^k-variables such that, when all these variables take the value 1, the right-hand side takes the value c_0.

Moreover, as the edges in E_R are not "required" to be serviced by a given vehicle k, inequalities (11.21) can be generalized in the following way. Let $\delta(S)$ be a cutset, and let $F \subset \delta_R(S)$ be a subset of required edges with $|F|$ odd. For each vehicle k, the disaggregate parity inequality is

$$x^k(\delta(S)) \geq 2y^k(F) - |F|+1. \tag{11.22}$$

Inequalities (11.22) are also known as cocircuit inequalities in Barahona and Grötschel [10] and have also been used in other arc routing problems (Belenguer and Benavent [12], Ghiani and Laporte [35]). They can be generalized by considering a subset of vehicles instead of only one vehicle. For each subset $\Omega \subset \{1,\ldots,K\}$ of $P = |\Omega| > 1$ vehicles, the P-aggregate parity inequality is

$$\sum_{k\in\Omega} x^k(\delta(S)) \geq 2\sum_{k\in\Omega} y^k(F) - |F|+1. \tag{11.23}$$

In Benavent et al. [17] it is shown that the aggregate parity inequalities (11.20), the disaggregate parity inequalities (11.22), and the P-aggregate parity inequalities (11.23) are facet-inducing for KWRPP(G) under mild conditions.

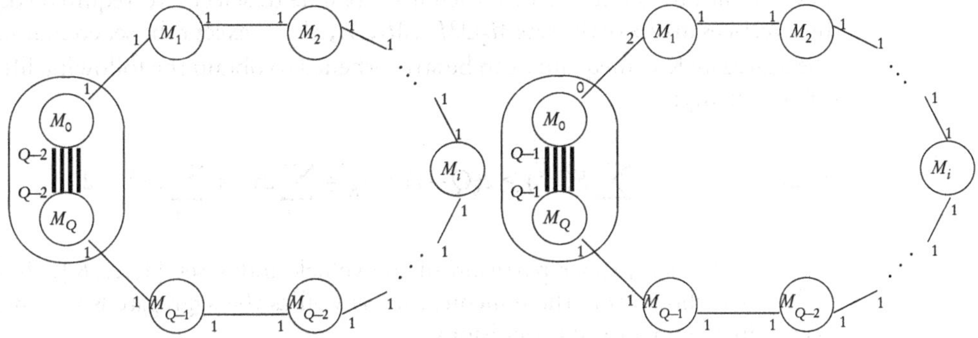

Figure 11.2. *K-C and K-C_{02} inequalities.*

K-C and K-C_{02} inequalities

K-C (K-component) inequalities were introduced for the undirected RPP in Corberán and Sanchis [24] (see Chapter 5) and generalized for the WRPP in Corberán, Plana, and Sanchis [22] (see Chapter 6). K-C_{02} inequalities for the WRPP were presented in [22]. The name of these families of inequalities is motivated by the number of clusters into which V is partitioned, which is usually denoted by K. To avoid confusion with the number of vehicles, the symbol Q will be used instead.

K-C and K-C_{02} inequalities for the WRPP (see Figure 11.2) are defined by a partition of V into $Q+1$ sets, $\{M_0, M_1, \ldots, M_Q\}$, with $Q \geq 3$ for the K-C and $Q \geq 2$ for the K-C_{02}, satisfying the following condition:

(11.24) *The only required edges with their two endpoints in different sets M_i are the required edges in $(M_0 : M_Q)$, which are a positive and even number.*

These required edges are depicted in bold lines in Figure 11.2. The coefficients a_{ij} of the variables in the inequality are also shown in Figure 11.2, where each number attached to vertex i and edge (i,j) represents the coefficient a_{ij} of variable x_{ij} associated with the traversal from i to j. The coefficient of a variable associated with an edge not shown in Figure 11.2 between sets M_i and M_j can be calculated as the length of a shortest path from M_i to M_j using the given coefficients. The coefficients corresponding to the variables associated with edges in sets $E(M_i)$ are zero. The K-C and K-C_{02} inequalities for the WRPP are

$$F(x) = \sum_{(i,j)\in E} (a_{ij}x_{ij} + a_{ji}x_{ji}) \geq 2(Q-1) + \alpha_R, \tag{11.25}$$

where $\alpha_R = (Q-2)|(M_0 : M_Q)_R|$ for the K-C and $\alpha_R = (Q-1)|(M_0 : M_Q)_R|$ for the K-C_{02} inequalities. Then, the corresponding aggregate K-C and K-C_{02} inequalities that are valid for the MM K-WRPP are

$$\sum_{k=1}^{K} F(x^k) \geq 2(Q-1) + \alpha_R. \tag{11.26}$$

It can be seen (Benavent et al. [18]) that the aggregate K-C inequalities are facet-inducing only when the depot $1 \in M_0 \cup M_Q$ and $|E_R(M_i)| = 1$ for each $i = 1, 2, \ldots, Q-1$. If any of these conditions is not satisfied, then it is possible to select two required edges e, f with their vertices in one of the sets $M_0 \cup M_Q, M_1, \dots, M_{Q-1}$, except the set containing the depot. The aggregate K-C inequality can be strengthened to obtain the following lifted aggregate K-C inequality:

$$\sum_{k=1}^{K} F(x^k) \geq 2(Q-1) + \alpha_R + \sum_{k\in I_1} 2y_e^k + \sum_{k\in I_2} 2y_f^k - 2, \tag{11.27}$$

where $\{I_1, I_2\}$ is a proper partition of the vehicle index set $\{1,\dots,K\}$. In Benavent et al. [18] it is shown that these inequalities as well as the aggregate K-C_{02} inequalities are facet-inducing under mild conditions.

Now we describe the disaggregate versions. As has already been mentioned for parity inequalities (11.22), all the edges in E_R are not "required" to be serviced by a given vehicle k. Hence, for the disaggregate (and P-aggregate) K-C and K-C_{02} inequalities, the partition $\{M_0, M_1, \dots, M_Q\}$ of V does not need to satisfy condition (11.24). This means that edges joining different sets M_i could be required. In particular, all the edges in E could be required. Therefore, unlike aggregate inequalities, the disaggregate and P-aggregate K-C and K-C_{02} inequalities we describe next are also valid for the *Min-Max K-Windy Postman Problem* (MM K-WPP).

Let us first assume that $1 \in M_0 \cup M_Q$. Let $W \subseteq (M_0 : M_Q)_R$ be an edge set such that $|W|$ is positive and even, and let $H = \{e_1, \dots, e_{Q-1}\} \subset E_R$ be another edge set such that each $e_i \in E(M_i)$. It is shown in Benavent et al. [18] that, for each vehicle k, the following disaggregate K-C and K-C_{02} inequalities are valid and facet-inducing for KWRPP(G):

$$F(x^k) \geq 2y^k(H) + (Q-2)\big(2y^k(W) - |W|\big), \tag{11.28}$$

$$F'(x^k) \geq 2y^k(H) + (Q-1)\big(2y^k(W) - |W|\big), \tag{11.29}$$

where $F(x)$ and $F'(x)$ are the left-hand sides of the standard K-C and K-C_{02} inequalities corresponding to the partition $\{M_0, M_1, \dots, M_Q\}$, respectively.

If the depot $1 \in M_i$, $i \neq 0, Q$, then the following disaggregate K-C and K-C_{02} inequalities are valid and facet-inducing for KWRPP(G) (see Benavent et al. [18]):

$$F(x^k) \geq 2 + 2y^k(H) + (Q-2)\big(2y^k(W) - |W|\big), \tag{11.30}$$

$$F'(x^k) \geq 2 + 2y^k(H) + (Q-1)\big(2y^k(W) - |W|\big), \tag{11.31}$$

where $H = \{e_1, \dots, e_{Q-1}\} \setminus \{e_i\}$.

It is shown in Benavent et al. [18] that, for each subset $\Omega \subset \{1,\dots,K\}$ of $P = |\Omega| > 1$ vehicles, the following P-aggregate K-C inequalities are valid and facet-inducing if $1 \in M_0 \cup M_Q$:

$$\sum_{k\in\Omega} F(x^k) \geq \sum_{k\in\Omega} 2y^k(H) + (Q-2)\left(\sum_{k\in\Omega} 2y^k(W) - |W|\right). \tag{11.32}$$

For the case $1 \notin M_0 \cup M_Q$, a similar inequality can be found in Benavent et al. [18] as well as the corresponding versions for the P-aggregate K-C_{02} inequalities.

Honeycomb and Honeycomb$_{02}$ inequalities

Honeycomb (HC) inequalities were introduced for the undirected *Rural Postman Problem* (RPP) in Corberán and Sanchis [25] (see Chapter 5) and presented for the WRPP in Corberán, Plana, and Sanchis [23] (see Chapter 6). In the K-C inequalities, an R-set (or a cluster of R-sets) is divided into two parts, M_0 and M_Q. In the HC inequalities for the WRPP described next, an R-set (or a cluster of R-sets) is divided into $L \geq 2$ parts. They are defined (see Figure 11.3) by

- a partition $\{W_1, \ldots, W_L, M_1, \ldots, M_{Q-1}\}$ of V, with $L \geq 2$, $Q \geq 3$, such that $(W_1 \cup \ldots \cup W_L)$, $M_1, \ldots, M_{Q-1}$ are clusters of one or more R-sets, and the graph induced by the required edges on the vertex set $\{W_1, \ldots, W_L\}$ is even and connected, and
- a tree T spanning the sets $W_1, \ldots, W_L, M_1, \ldots, M_{Q-1}$ such that the degree in T of every vertex set W_i is one, the degree of vertex sets M_j is at least two, and the number of edges of the unique path in the tree connecting any pair of sets $A, B \in \{W_1, \ldots, W_L, M_1, \ldots, M_{Q-1}\}$, which will be denoted by $d(A,B)$, is at least three.

The HC inequality is

$$F(x) = \sum_{(i,j)\in E} (a_{ij}x_{ij} + a_{ji}x_{ji}) \geq 2(Q-1) + \sum_{(i,j)\in \mathcal{E}_R} a_{ij}, \tag{11.33}$$

where the coefficient a_{ij} of each variable x_{ij} in the HC inequality is

- $a_{ij} = d(M_p, M_q)$ if $i \in M_p$ and $j \in M_q$, $i \neq j$,
- $a_{ij} = d(M_p, W_q)$ if $i \in M_p$, $j \in W_q$ or $i \in W_q$, $j \in M_p$,
- $a_{ij} = d(W_p, W_q) - 2$ if $i \in W_p$ and $j \in W_q$, $i \neq j$,

and $\mathcal{E}_R$ is the set of required edges joining two distinct sets W_i, W_j. Note that, for the edges $(i,j) \in \mathcal{E}_{\mathcal{R}}$, $a_{ij} = a_{ji}$ holds, but only one coefficient is added in the right-hand side of inequality (11.33). Figure 11.3 represents a honeycomb inequality. The bold lines represent the required edges, and the thin lines represent the edges in the spanning tree T. The right-hand side for this example is $2(8-1)+(1+4+1+6)$.

Honeycomb$_{02}$ (HC$_{02}$) inequalities are associated with a similar partition but with different coefficients. Details can be found in Corberán, Plana, and Sanchis [23].

Let $F(x) \geq 2(Q-1) + \sum_{(i,j)\in\mathcal{E}_R} a_{ij}$ be an HC inequality as above. The corresponding aggregate HC inequality is therefore valid for the MM K-WRPP

$$\sum_{k=1}^{K} F(x^k) \geq 2(Q-1) + \sum_{(i,j)\in\mathcal{E}_R} a_{ij}. \tag{11.34}$$

Let us consider a set $W \subset \mathcal{E}_R$ such that the graph induced by edges in W on the vertex set $\{W_1, \ldots, W_L\}$ is even and connected. Assume that $1 \in \bigcup W_i$, and let $H = \{e_1, \ldots, e_{Q-1}\} \subset E_R$ be a subset of edges such that each $e_i \in E(M_i)$. For each subset $\Omega \subset \{1, \ldots, K\}$ of $P = |\Omega|$ vehicles, $1 < P < K$, the following P-aggregate HC inequality is valid for KWRPP(G) (see Benavent et al. [18]):

$$\sum_{k\in\Omega} F(x^k) \geq \sum_{k\in\Omega} 2y^k(H) + \sum_{e=(i,j)\in W} a_{ij} \left(\sum_{k\in\Omega} 2y_e^k - 1 \right). \tag{11.35}$$

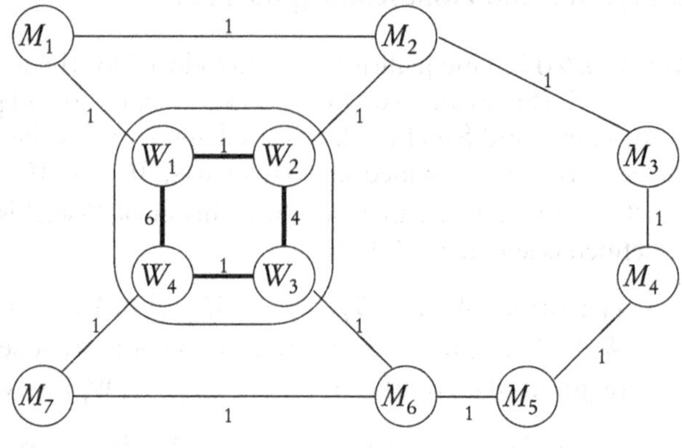

Figure 11.3. *A Honeycomb inequality.*

In the case $\Omega = \{k\}$, we have the disaggregate HC inequalities for each vehicle k. If the depot is located in a set M_i, there are similar versions of the above inequalities. As far as the HC_{02} inequalities are concerned, the corresponding aggregate, disaggregate, and P-aggregate versions can be found in Benavent et al. [18]. Finally, as for the K-C inequalities, it can be seen that the disaggregate and P-aggregate HC and HC_{02} inequalities also apply to the MM K-WPP.

Path-Bridge inequalities

Path-Bridge (PB) inequalities were introduced by Letchford [42] for the undirected RPP (see Chapter 5) and are based on the path inequalities for the graphical TSP (Cornuéjols, Fonlupt, and Naddef [26]). They were extended to the WRPP in Corberán, Plana, and Sanchis [22] (see Chapter 6). In what follows, we introduce the aggregate, disaggregate, and P-aggregate PB inequalities for the MM K-WRPP, which have not been published before.

Let $p \geq 1$ and $b \geq 0$ be integers such that $p+b \geq 3$ is an odd number, and let $n_1, \ldots, n_p$, be integer numbers greater than or equal to 2. Consider a partition $\{A, Z, \{M_q^s\}_{q=1,\ldots,n_s}^{s=1,\ldots,p}\}$ of V satisfying the condition that each R-set is contained either in $A \cup Z$ or in a set M_q^s and $(A : Z)$ contains b required edges. For the sake of simplicity, we identify A with M_0^s and Z with $M_{n_s+1}^s$ for all s. A PB inequality for the WRPP is defined by p *paths* from A to Z, each of them with $n_s + 2$ vertex sets, and by b required edges joining the sets A and Z, which form the *bridge*. This is shown in Figure 11.4, where the number near each edge joining sets M_q^s, M_r^t represents the coefficient a_{ij} of all the variables in the PB inequality corresponding to edges $(i,j) \in (M_q^s : M_r^t)$. These coefficients are $a_{ij} = c(M_q^s, M_r^t)$, where

- $c(A, Z) = 1$,
- $c(V_q^s, V_r^s) = \frac{|q-r|}{n_s-1}$ for all $q, r \in \{0, 1, \ldots, n_{s+1}\}$, $s \in \{1, \ldots, p\}$,
- $c(V_q^s, V_r^t) = \frac{1}{n_s-1} + \frac{1}{n_t-1} + \left|\frac{q-1}{n_s-1} - \frac{r-1}{n_t-1}\right|$ for all $s \neq t \in \{1, \ldots, p\}, q \in \{1, \ldots, n_s\}$, $r \in \{1, \ldots, n_t\}$,

and $a_{ij} = 0$ otherwise. The PB inequality for the WRPP is

$$F(x) = \sum_{(i,j)\in E} \left(a_{ij}\, x_{ij} + a_{ji}\, x_{ji}\right) \geq 1 + \sum_{s=1}^{p} \frac{n_s + 1}{n_s - 1} + b. \tag{11.36}$$

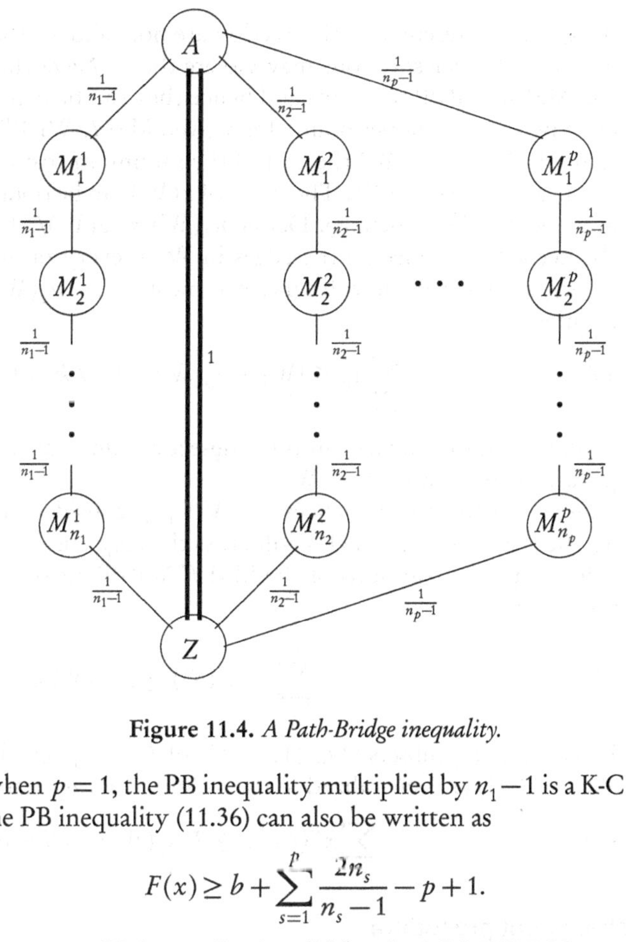

Figure 11.4. *A Path-Bridge inequality.*

Note that when $p = 1$, the PB inequality multiplied by $n_1 - 1$ is a K-C inequality with $Q = n_1 + 1$. The PB inequality (11.36) can also be written as

$$F(x) \geq b + \sum_{s=1}^{p} \frac{2n_s}{n_s - 1} - p + 1. \tag{11.37}$$

The following aggregate PB inequality is valid for the MM K-WRPP:

$$\sum_{k=1}^{K} F(x^k) \geq b + \sum_{s=1}^{p} \frac{2n_s}{n_s - 1} - p + 1. \tag{11.38}$$

Now we describe the disaggregate and P-aggregate versions. Again, no condition on the distribution of the required edges in the partition $\{A, Z, \{M_q^s\}_{q=1,\ldots,n_s}^{s=1,\ldots,p}\}$ of V is needed beyond the existence of sets W and H^s defined below, and the disaggregate and P-aggregate PB inequalities we present next also apply to the MM K-WPP. Let us assume first that $1 \in A \cup Z$. Let $W \subseteq (A : Z)_R$ be such that $p + |W| \geq 3$ is an odd number, and, for each $s = 1, \ldots, p$, let $H^s = \{e_1^s, \ldots, e_{n_s}^s\} \subseteq E_R$ be such that $e_q^s \in E_R(M_q^s)$. For each subset $\Omega \subset \{1, \ldots, K\}$ of $P = |\Omega|$ vehicles, the following P-aggregate PB inequality is valid for KWRPP(G):

$$\sum_{k\in\Omega} F(x^k) \geq \left(2\sum_{k\in\Omega} y^k(W) - |W|\right) + \sum_{s=1}^{p} \frac{2}{n_s - 1} \sum_{k\in\Omega} y^k(H_s) - p + 1. \tag{11.39}$$

In the particular case $\Omega = \{k\}$, we have the disaggregate PB inequality for vehicle k. If the depot is not located in $A \cup Z$ but in another set M_q^s, similar versions of the above inequalities can be obtained.

Max-length constraints

The inequalities presented in this section are not valid for the MM K-WRPP polyhedron because some *feasible* solutions may violate them. Nevertheless, they are satisfied by all *optimal* MM K-WRPP solutions and hence they can be used in an exact procedure.

Let $\overline{z}$ be a known upper bound for a given MM K-WRPP instance. Let W be a subset of required edges and $z(W)$ the cost of the minimum cost tour starting at the depot and traversing all the edges in W. The value of $z(W)$ can be obtained by solving a WRPP (see Benavent et al. [17] for details). Define $n_v(W) = \lceil z(W)/\overline{z} \rceil$; then at least $n_v(W)$ vehicles will be needed to service all the edges in W in every optimal MM K-WRPP solution. Therefore, a single vehicle will service at most $|W| - n_v(W) + 1$ of these edges, and the inequalities

$$\sum_{e \in W} y_e^k \leq |W| - n_v(W) + 1 \quad \forall\, k = 1,\dots,K \tag{11.40}$$

are satisfied by any optimal solution. These inequalities are called disaggregate max-length inequalities associated with set W.

Now, given any set of vertices $S \subseteq V \setminus \{1\}$, consider the set of edges $W_S = E_R(S) \cup \delta_R(S)$. At least $n_v(W_S)$ vehicles will cross the edge cutset $\delta(S)$ in any optimal solution. Therefore, optimal solutions of the MM K-WRPP satisfy the following aggregate max-length inequality:

$$\sum_{k=1}^{K} x^k(\delta(S)) \geq 2\, n_v(W_S). \tag{11.41}$$

By considering subsets $\Omega \subset \{1,\dots,K\}$ of $P = |\Omega|$ vehicles instead of all of them, inequalities (11.41) are generalized to the following P-aggregate max-length inequalities:

$$\sum_{k \in \Omega} x^k(\delta(S)) \geq 2\big(n_v(W_S) - (K - P)\big). \tag{11.42}$$

11.4.2.3 ▪ Branch-and-cut procedure

Now we describe the branch-and-cut procedure applied in Benavent et al. [17, 18]. The initial linear program contains constraints (11.13) to (11.16), some additional constraints that avoid symmetric solutions, and those aggregate max-length constraints (11.41) associated with the R-sets V_i for which $n_v(E_R(V_i)) > 1$.

At each iteration of the cutting-plane algorithm, separation procedures for the inequalities described before were applied in the following way. Given an LP solution $(\overline{x}^1, \overline{y}^1, \dots, \overline{x}^K, \overline{y}^K) \in \mathbb{R}^{(2|E|+|E_R|)K}$ and a subset $\Omega \subseteq \{1,\dots,K\}$ of P vehicles, vectors $\overline{x}^\Omega = \sum_{k\in\Omega} \overline{x}^k$ and $\overline{y}^\Omega = \sum_{k\in\Omega} \overline{y}^k$ are defined. Then the support graph $\overline{G}^\Omega = (\overline{V}^\Omega, \overline{E}^\Omega, \overline{x}^\Omega, \overline{y}^\Omega)$ is the associated weighted graph, where $\overline{V}^\Omega, \overline{E}^\Omega$ are the sets of vertices and edges of the subgraph of G induced by the edges $e \in E$ with $\overline{x}_e^\Omega = \overline{x}_{ij}^\Omega + \overline{x}_{ji}^\Omega > 0$, plus the depot, if necessary.

In the case $P = K$, the aggregate graph $\overline{G}^\Omega$ and the set of required edges E_R are the input for the separation procedures to find connectivity, R-odd cut, KC, KC_{02}, HC, and HC_{02} inequalities for the *Windy General Routing Problem* (WGRP) violated by $\overline{x}^\Omega$.

The graphs $\overline{G}^{\Omega}$ corresponding to the cases $P=1$ and $1<P<K$ are the input for the separation of the disaggregate and P-aggregate inequalities, respectively. These algorithms are an adaptation of the separation procedures for the corresponding inequalities for the WGRP in which the edges considered required are those $e \in E_R$ with $\overline{y}_e^{\Omega} > 1-\epsilon$, where ϵ is a given parameter.

Connectivity inequalities (11.17) can be exactly separated in polynomial time with the following procedure. Given a vehicle k, for each edge $e \in E_R$ with $\overline{y}_e^k > 0$, the minimum cut in graph $\overline{G}^k$ separating edge e from the depot is computed. If the weight of this cut is less than $2\overline{y}_e^k$, the corresponding inequality (11.17) is violated. Disaggregate and P-aggregate parity inequalities (11.22) can also be exactly separated with a polynomial algorithm similar to the one proposed in Padberg and Rao [47] based on the computation of a minimum weight odd cut on an auxiliary graph. Furthermore, heuristic separation algorithms for the max-length inequalities (11.40), (11.41), and (11.42) have also been implemented in Benavent et al. [17].

When no new violated inequalities are found at a node of the search tree, authors branch using the Strong Branching strategy [5] implemented in Cplex. The branch-and-cut procedure uses upper bounds obtained by the metaheuristic algorithm presented in Benavent, Corberán, and Sanchis [19].

This branch-and-cut algorithm was tested on the same 144 MM K-WRPP instances used for the metaheuristic algorithm presented in Benavent, Corberán, and Sanchis [19] and described in Section 11.4.1. Each instance was tested for two, three, four, five, and six vehicles. The algorithm is able to solve 144, 130, 124, 113, and 101 instances to optimality, respectively, within a time limit of one hour. These computational results are depicted in Table 11.1 in Section 11.4.3.5, where they are compared with the results obtained by the branch-price-and-cut algorithm (Benavent et al. [15]) described in the next section.

11.4.3 ▪ A branch-price-and-cut algorithm

In the previous section a branch-and-cut algorithm providing good results, mainly when the number of vehicles is small, has been described. Although branch-and-cut is among the leading methodologies for solving complex arc routing problems, its performance is worse when the number of vehicles increases. In this section we describe the branch-price-and-cut method proposed in Benavent et al. [15], which obtains better results on MM K-WRPP instances with a larger number of vehicles.

11.4.3.1 ▪ Formulation

In this section, with each edge joining vertices i and j we associate two arcs (i,j) and (j,i) of costs (lengths) c_{ij} and c_{ji}, respectively.

A route is a path in G starting and ending at the depot. Its length is given by the sum of the lengths of the arcs it traverses. A solution of the MM K-WRPP consists of K routes such that each required edge $e \in E_R$ is included in at least one of them. If a solution of the MM K-WRPP of cost U is known and costs c_{ij} are integer, then the set of routes can be restricted to those whose length is less than or equal to $U-1$.

The MM K-WRPP is formulated in Benavent et al. [15] using the concept of a *route/service pattern* (RSP). An RSP is a sequence of serviced required edges and shortest paths starting and ending at vertex 1. For instance, the RSP $(1,2)-(4,3)-1$ services two required edges traversed from 1 to 2 and from 4 to 3, respectively, and uses the shortest paths from 2 to 4 and from 3 to 1. The solutions of the MM K-WRPP are then sets of RSPs.

Using the RSPs, the MM K-WRPP can be formulated as an extended set-partitioning problem. Let P be the set of RSPs and, for each $p \in P$, let c_p be the length of the route associated with p. For each $e \in E_R$ and each $p \in P$, let b_p^e be a binary parameter equal to 1 if RSP p services edge e, and 0 otherwise. For each $p \in P$ and each vehicle $k \in \{1,\dots,K\}$, define a binary variable θ_p^k that takes value 1 if p is chosen for vehicle k in the solution, and 0 otherwise. Finally, let z_k be the length of the route assigned to vehicle $k \in \{1,\dots,K\}$. Without loss of generality (and to reduce symmetry), it is assumed that route lengths are nonincreasing for vehicles $1,\dots,K$. Hence, the first vehicle is assigned to a longest route and vehicle k to a shortest route. The MM K-WRPP can be formulated as the following integer program:

$$\text{(11.43)} \qquad \text{(MM K-WRPP col)} \qquad \min \quad z_1$$

$$\text{(11.44)} \qquad \text{s.t.} \qquad \sum_{k=1}^{K}\sum_{p\in P} b_p^e \theta_p^k = 1 \qquad \forall e \in E_R,$$

$$\text{(11.45)} \qquad \sum_{p\in P} \theta_p^k = 1 \qquad \forall k \in \{1,\dots,K\},$$

$$\text{(11.46)} \qquad z_k - \sum_{p\in P} c_p \theta_p^k = 0 \qquad \forall k \in \{1,\dots,K\},$$

$$\text{(11.47)} \qquad z_k - z_{k-1} \le 0 \qquad \forall k \in \{2,\dots,K\},$$

$$\text{(11.48)} \qquad \theta_p^k \in \{0,1\} \qquad \forall k \in \{1,\dots,K\},\ \forall p \in P.$$

The objective function minimizes the length of a longest route. Equations (11.44) force each required edge to be covered by one vehicle, and (11.45) ensure that one route is assigned to each vehicle. Equations (11.46) determine the length of each vehicle route, and constraints (11.47) ensure that routes are ordered in nonincreasing length. They serve as symmetry-breaking constraints.

11.4.3.2 ▪ Solution method

Formulation (11.43) to (11.48) contains a huge number of RSP variables that, in practice, cannot be enumerated a priori. However, it can be solved by an exact branch-and-price method (Barnhart et al. [11], Desaulniers, Desrosiers, and Solomon [28], Lübbecke and Desrosiers [45]), that is, a column generation algorithm embedded into a branch-and-bound algorithm.

Column generation (Dantzig and Wolfe [27], Gilmore and Gomory [36]) is an iterative process for solving a linear programming problem such as the linear relaxation of (11.43)–(11.48), which is called the *master problem* (MP). At each iteration, the MP restricted to a (small) subset of its RSP variables, called the *restricted master problem* (RMP), is solved to obtain an optimal primal solution and its corresponding dual solution. The dual solution is then used to solve several *subproblems* (also called pricing problems) in order to find RSP variables with negative reduced costs among those not considered by the RMP. If no such variable exists, then the current RMP solution is also optimal for the MP. Otherwise, variables (columns) with negative reduced costs are added to the RMP before starting a new iteration.

For the MM K-WRPP, the subproblem for each vehicle $k \in \{1,\dots,K\}$ can be solved as a constrained elementary shortest path problem defined on a directed graph $H^k =$

(N^k, A^k). Vertex set N^k contains a source vertex s, a sink vertex t, and a pair of vertices n_e^I and n_e^J for each required edge e representing the traversal of e in each direction. Arc set A^k contains the arcs described now:

- For each edge $e \in E_R$, two arcs (s, n_e^I) and (s, n_e^J) representing a shortest path in graph G between the depot and each end-vertex of e (no edges are serviced along this path). Two arcs (n_e^I, t) and (n_e^J, t) representing a shortest path that starts with edge e between each end-vertex of e and the depot (only edge e is serviced along this path).

- For each pair of edges $e, f \in E_R$, four arcs (n_e^I, n_f^I), (n_e^J, n_f^J), (n_e^I, n_f^J), and (n_e^J, n_f^I) representing a shortest path between the corresponding end-vertices of e and f. Each path starts with edge e (only edge e is serviced along this path).

With each arc $(u, v) \in A^k$, we associate a length $\bar{d}_{uv}$ equal to the length of the corresponding required edge plus that of a shortest path, and a reduced cost $\bar{c}_{uv}^k$ that is given by

$$\bar{c}_{uv}^k = \begin{cases} \bar{d}_{uv}\sigma_k - \mu_k & \text{if } u = s, \\ \bar{d}_{uv}\sigma_k - \pi_e & \text{if } u = n_e^I \text{ or } u = n_e^J \text{ for an edge } e \in E_R, \end{cases}$$

where π_e, μ_k, and σ_k are the dual variables associated with constraints (11.44), (11.45), and (11.46), respectively. The subproblem consists of finding a minimum reduced-cost path from s to t in H^k such that its length is less than U and it services each required edge at most once. This problem is a constrained elementary shortest path problem, which is NP-hard in the strong sense (Dror [31]). When the number of required edges in E_R is relatively large, the subproblems can be very difficult to solve. To overcome this difficulty, a relaxation of the subproblems called the *ng-path relaxation* (Baldacci, Mingozzi, and Roberti [9]) is used, which allows the generation of paths with certain cycles (this means that a given required edge can be serviced several times on these paths). The resulting subproblems can be solved exactly using an exact *labeling algorithm* (see, e.g., Irnich and Desaulniers [39]). In such an algorithm, partial paths from the source to any other vertex are represented by labels. Starting from a label with all components equal to 0 at the source vertex, the labeling process extends the labels forward in the network by using label extension functions. The labeling procedure is accelerated with two techniques. Bidirectional labeling (Righini and Salani [49]) extends labels forward from the source vertex and backward from the sink vertex, simultaneously. Both extension processes stop when crossing a midpoint that is defined for the path length as $0.5\,U$. The second acceleration strategy, called decremental search space (Boland, Dethridge, and Dumitrescu [20], Righini and Salani [50]), starts by solving the nonelementary version of the constrained shortest path problem in which any kind of cycles are allowed. If the shortest source-to-sink path contains no cycle (with respect to the required edges), then it is optimal. Otherwise, the optimal path contains at least one cycle starting with a required edge and, to avoid this cycle, the labels of the edges in the cycle are recomputed. The process is repeated until an optimal path without cycles is found.

Furthermore, in Benavent et al. [15] two fast pricing heuristics are used: a multi-start tabu search developed in Desaulniers, Lessard, and Hadjar [30] for the *Vehicle Routing Problem with Time Windows* (VRPTW) and a labeling algorithm with a relaxed dominance rule that allows many more labels to be discarded, potentially excluding a label yielding the RSP with the least reduced cost (see [30] for additional details). The exact and heuristic algorithms are used as follows. At each column generation iteration, the tabu search heuristic is invoked first. If it finds negative reduced cost columns, these columns are added to the RMP, which is reoptimized. Otherwise, the heuristic labeling algorithm is called. If it does not succeed in generating negative reduced cost columns, then the exact labeling algorithm is executed. As reported in Desaulniers, Lessard, and Hadjar [30], the computationally expensive exact labeling algorithm is rarely used with such a strategy.

11.4.3.3 ▪ Valid inequalities

The branch-price-and-cut algorithm proposed in Benavent et al. [15] uses four families of valid inequalities to tighten the linear relaxation.

Lower bound inequality

In addition to an upper bound U on the optimal value, which implies that $z_k \leq U-1$ for all $k \in \{1,\ldots,K\}$, a lower bound L is also used only for vehicle 1, that is, $z_1 \geq L$. Note that the lower bound L is not used in the labeling algorithm that solves the subproblem because imposing a lower and an upper bound simultaneously is computationally expensive. During the solution process, the upper bound U is updated every time a new best incumbent integer solution is found. The lower bound L is updated at each node of the search tree according to the optimal value of the corresponding MP, z_{MP}. Specifically, L is set to $\lceil z_{MP} \rceil$ for the descendant nodes.

Aggregate R-odd cut inequalities

Aggregate R-odd cut inequalities (11.20) ensure that each feasible integer solution traverses the edges in a cutset $\delta(S)$ with $|\delta_R(S)|$ odd at least $|\delta_R(S)|+1$ times. This is equivalent to deadheading at least one edge in $\delta(S)$. Using the RSP variables, it can be written as follows:

$$\sum_{k=1}^{K}\sum_{p\in P} q_{p,S}^{\text{OC}}\,\theta_p^k \geq 1 \qquad \forall S\subseteq V \text{ such that } |\delta_R(S)| \text{ is odd}, \tag{11.49}$$

where $q_{p,S}^{\text{OC}} = \sum_{(i,j)\in p} q_{ij,S}^{\text{OC}}$ and $q_{ij,S}^{\text{OC}}$ are equal to 1 if a shortest path between i and j traverses at least one edge in $\delta(S)$ in deadhead, and 0 otherwise. The notation $(i,j)\in p$ indicates that the RSP p contains a shortest path deadheaded between vertices i and j in G.

Although these inequalities can be separated exactly in polynomial time (see Padberg and Rao [47]), in Benavent et al. [15] they are separated by enumerating all odd cutsets $\delta(S)$ with $|S| \leq 5$. The most violated cuts are added to the MP, and their dual variables are taken into account in the column generation subproblems.

Clique inequalities based on three aggregate arcs

The clique inequalities proposed in Benavent et al. [15] are a special case of the valid inequalities introduced in Hadjar, Marcotte, and Soumis [37] for the multi-depot vehicle scheduling problem, and also of the 2-cut or 1-path inequalities presented in Löbel [44]. They are as follows. Given two required edges $e_i, e_j \in E_R$ and a vehicle k, let us denote by a_{ij}^k the set of arcs from a vertex representing e_i ($n_{e_i}^I$ or $n_{e_i}^J$) to a vertex representing e_j ($n_{e_j}^I$ or $n_{e_j}^J$) in graph H^k. Now, consider three required edges $e_1, e_2, e_3 \in E_R$ and a partition of the set of vehicles into two subsets Ω_{12} and Ω_{23}. Furthermore, let $\Omega_{13} = \{1,\ldots,K\}$. It can be seen that all arcs in $\left(\bigcup_{k\in\Omega_{12}} a_{12}^k\right)\bigcup\left(\bigcup_{k\in\Omega_{13}} a_{13}^k\right)\bigcup\left(\bigcup_{k\in\Omega_{23}} a_{23}^k\right)$ are in conflict pairwise, that is, two arcs cannot be part of a feasible integer solution. Hence, they induce a clique in a conflict graph (see Hoffman and Padberg [38]). Let W be the set of all those cliques and denote by Q_w the set of arcs inducing clique $w \in W$. The inequality for clique $w \in W$ stipulates that the total flow on the arcs in Q_w must not exceed one. In terms of the RSP variables, this family of inequalities can be expressed as follows:

$$\sum_{k=1}^{K}\sum_{p\in P} q_{wp}^{\mathrm{CQ}}\,\theta_p^k \le 1 \quad \forall w \in W, \tag{11.50}$$

where q_{wp}^{CQ} is equal to the number of arcs in Q_w traversed by the RSP p. These inequalities are separated by enumeration. All violated cuts are added to the MP, and the dual variables obtained are considered in the column generation subproblems (see Benavent et al. [15] for details).

Subset-row inequalities based on three required edges

Subset-row inequalities were introduced in Jepsen et al. [40] for the VRPTW, and are a special case of Chvátal-Gomory inequalities of rank 1. For computational reasons, in Benavent et al. [15] only the inequalities defined by three required edges are considered. Given a subset F of three required edges and assuming that all routes are elementary, such a constraint stipulates that at most one route that services at least two of the edges in F can be part of a feasible integer solution. When cycles are allowed in routes, these inequalities are expressed as follows:

$$\sum_{k=1}^{K}\sum_{p\in P}\left\lfloor \frac{q_{p,F}^{\mathrm{SR}}}{2}\right\rfloor \theta_p^k \le 1 \qquad \forall F \subseteq E_R \text{ such that } |F| = 3, \tag{11.51}$$

where $q_{p,F}^{\mathrm{SR}}$ is equal to the number of times that the route p services a required edge in subset F. Enumeration is used to separate the subset-row inequalities (11.51). As opposed to the two previous types of inequalities, the dual variables associated with these inequalities cannot be transferred directly to the arcs of the networks, and the labeling algorithm has to be modified to handle these dual variables (see Jepsen et al. [40] or Desaulniers, Desrosiers, and Spoorendonk [29] for details).

The four families of valid inequalities described above are used at each node of the branch-and-bound procedure as follows. Once the bounds $\lceil L, U\rceil$ have been adjusted, the separation algorithms for each family of inequalities (11.49)–(11.51) are invoked in this order. When violated inequalities are identified for a family, the separation routines for the subsequent families are not called. Note that the family order favors the generation of aggregate R-odd cut and clique inequalities over that of subset-row inequalities, which are more complex to handle in the pricing process.

11.4.3.4 ▪ Branching decisions

Three types of branching decisions are taken. Their description relies on the following notation: y_e is equal to the total flow in deadhead on edge $e \in E$ (required or not), and x_{uv}^k is equal to the flow on arc $(u,v) \in A^k$ for a specific vehicle k.

The first type consists of branching on the variable y_e whose fractional part is closest to 0.5. The second type of decision forbids or requires the assignment of a required edge

$e \in E_R$ to a (specific) subset Ω_e of vehicles. The third type consists of branching on the value x^k_{uv} which is closer to 0.5 for $u \neq s$ (the source vertex), where (u,v) is an arc in A^k for a specific vehicle k.

11.4.3.5 • Computational comparison

The *branch-price-and-cut* (BP) method was tested on the same 144 MM K-WRPP instances used for the *branch-and-cut* (BC) algorithm. The number of instances solved to optimality within one hour by both BP and BC methods for two, three, four, five, and six vehicles are given in Table 11.1. It can be seen that the performance of the BP and BC procedures is very good and that the two methods are complementary. The BP procedure stands out for instances with $K \geq 5$, where it outperforms the BC algorithm. The authors explain this behavior by noting that, in general, BP provides better lower bounds than BC at the root node of the search tree. However, computing these lower bounds can be highly time-consuming for small K because the upper bounds U increase as K decreases, yielding subproblems that are harder to solve. The quality of the above results benefits from the use of the tight upper bound provided by the metaheuristic described in Benavent, Corberán, and Sanchis [19].

Table 11.1. *Number of optimally solved instances (out of 144).*

Algorithm	2 veh.	3 veh.	4 veh.	5 veh.	6 veh.
branch-and-cut	144	130	124	113	101
branch-and-price	113	119	120	124	121

11.5 • Related problems

In Arkin, Hassin, and Levin [7], approximation algorithms for the so-called *Min-Max Postmen Cover* and *Min-Max Rural Postmen Cover* problems are presented. Given a complete graph $G = (V,E)$ with positive lengths associated with its edges, the Min-Max Postmen Cover Problem consists of finding K paths $P_1,\dots,P_K$ satisfying $E \subseteq P_1 \cup \dots \cup P_K$ and such that the length of the longest path is minimum. The authors propose a 3-approximation algorithm for this problem. A 7-approximation algorithm is also described for the Min-Max Rural Postmen Cover Problem, which is the more general case where only the edges in a subset of required edges $E_R \subseteq E$ have to be covered by the K paths.

A memetic algorithm for an extended version of the CARP, *Extended CARP*, is proposed in Lacomme, Prins, and Ramdane-Chérif [41]. Recall that the CARP consists of determining a set of routes of minimum total cost such that each route starts and ends at the depot, each required edge is serviced by one single route, and the total demand processed by a route does not exceed the vehicle capacity (see Chapter 7 for more details). The authors present powerful memetic algorithms for CARP extensions considering parallel arcs, turn penalties, a maximum route length, or a limited fleet. Moreover, besides the usual objective of minimizing the total cost, the proposed algorithms are able to minimize the number of vehicles used or handle a min-max objective.

Another arc routing problem with a min-max objective related to the design of snow plow routes is described in Dussault, Golden, and Wasil [33]. Previously, in Dussault et al. [32], the same authors presented a variant of the WPP called the *Plowing with Precedence Problem* (PPP, see Chapter 4). A relaxation of the PPP which ignores the restriction that if a street is completely unplowed, a plow must clear the street before deadheading, is also presented in Dussault et al. [32]. This relaxation is called the *Downhill Plow Problem* (DPP) in Dussault, Golden, and Wasil [33]. In this latter paper, the *Min-Max Downhill Plow Problem with K Plows* (MM K-DPP) is presented. This problem is similar to the MM K-WRPP in that it has a min-max objective function for multiple plows. However, the MM K-DPP incorporates four different costs for traversing and servicing an edge, and the plow may have the option of deadheading any street, whether it has been plowed or not. It can be defined as follows. Let $G = (V,E)$ be an undirected graph with a distinguished vertex as the depot, where each edge $(i,j) \in E$ has four costs: c^+_{ij} and c^+_{ji} are the costs of plowing from i to j and from j to i, respectively, while c^-_{ij} and c^-_{ji} are the costs of deadheading from i to j and from j to i. Given a number $K > 1$ of vehicles, the MM K-DPP consists of finding K routes, each beginning and ending at the depot, jointly servicing each edge twice (for plowing each side of the street), and satisfying the condition that the length of the longest route is minimum. The MM K-DPP is solved in Dussault, Golden, and Wasil [33] with a procedure based on the local search algorithm proposed in Dussault et al. [32], extended to consider multiple vehicles. It starts with a single cycle that is split into K separate routes to obtain an initial solution for the MM K-DPP. A local search procedure then tries to balance the individual routes to get equal, or near equal, route lengths.

Acknowledgments

The authors would like to thank Stefan Irnich for his careful reading of the manunscript. His many comments and suggestions have improved the readability and contents of the chapter. The authors also wish to thank the Spanish Ministerio of Economía y Competitividad (project MTM2012-36163-C06-02) and the Generalitat Valenciana (project GVPROMETEO2013-049) for their support.

Bibliography

[1] E.H.L. AARTS, C.A.J. HURKENS, AND J.K. LENSTRA, *Whizzkids: Two exercises in computational discrete optimization*, in ICIAM 99: Proceedings of the Fourth International Congress on Industrial and Applied Mathematics, Edinburgh, M. Ball and J.C.R. Hunt, eds., Oxford University Press, Oxford, UK, 2000, pp. 141–152.

[2] D. AHR, *Contributions to Multiple Postmen Problems*, Ph.D. dissertation, University of Heidelberg, Heidelberg, Germany, 2004.

[3] D. AHR AND G. REINELT, *New heuristics and lower bounds for the min-max k-Chinese postman problem*, in Algorithms—ESA 2002, R. Möhring and R. Raman, eds., Lecture Notes in Computer Science 2461, Springer, Berlin, 2002, pp. 64–74.

[4] ——, *A tabu search algorithm for the min-max k-Chinese postman problem*, Computers & Operations Research, 33 (2006), pp. 3403–3422.

[5] D. APPLEGATE, R.E. BIXBY, V. CHVÁTAL, AND W. COOK, *Finding cuts in the tsp*, Technical report DIMACS 95–05, Rutgers University, Piscataway, NJ, 1995.

[6] D. APPLEGATE, W. COOK, S. DASH, AND A. ROHE, *Solution of a min-max vehicle routing problem*, INFORMS Journal on Computing, 14 (2002), pp. 132–143.

[7] E.M. ARKIN, R. HASSIN, AND A. LEVIN, *Approximations for minimum and min-max vehicle routing problems*, Journal of Algorithms, 59 (2006), pp. 1–18.

[8] A.A. ASSAD, W.L. PEARN, AND B.L. GOLDEN, *The capacitated Chinese postman problem: Lower bounds and solvable cases*, American Journal of Mathematics and Management Science, 7 (1987), pp. 63–88.

[9] R. BALDACCI, A. MINGOZZI, AND R. ROBERTI, *New route relaxation and pricing strategies for the vehicle routing problem*, Operations Research, 59 (2011), pp. 1269–1283.

[10] F. BARAHONA AND M. GRÖTSCHEL, *On the cycle polytope of a binary matroid*, Journal of Combinatorial Theory, 40 (1986), pp. 40–62.

[11] C. BARNHART, E.L. JOHNSON, G.L. NEMHAUSER, M.W.P. SAVELSBERGH, AND P.H. VANCE, *Branch-and-price: Column generation for solving huge integer programs*, Operations Research, 46 (1998), pp. 316–329.

[12] J.M. BELENGUER AND E. BENAVENT, *The capacitated arc routing problem: Valid inequalities and facets*, Computational Optimization and Applications, 10 (1998), pp. 165–187.

[13] J.M. BELENGUER, E. BENAVENT, N. LABADI, C. PRINS, AND M. REGHIOUI, *Lower and upper bounds for the mixed capacitated arc routing problem*, Transportation Science, 44 (2010), pp. 206–220.

[14] E. BENAVENT, A. CARROTTA, Á. CORBERÁN, J.M. SANCHIS, AND D. VIGO, *Lower bounds and heuristics for the windy rural postman problem*, European Journal of Operational Research, 176 (2007), pp. 855–869.

[15] E. BENAVENT, Á. CORBERÁN, G. DESAULNIERS, F. LESSARD, I. PLANA, AND J.M. SANCHIS, *A branch-price-and-cut method for the min-max k-vehicle windy rural postman problem*, Networks, 63 (2014), pp. 34–45.

[16] E. BENAVENT, Á. CORBERÁN, E. PIÑANA, I. PLANA, AND J.M. SANCHIS, *New heuristic algorithms for the windy rural postman problem*, Computers & Operations Research, 32 (2005), pp. 3111–3128.

[17] E. BENAVENT, Á. CORBERÁN, I. PLANA, AND J.M. SANCHIS, *Min-max k-vehicles windy rural postman problem*, Networks, 54 (2009), pp. 216–226.

[18] ——, *New facets and an enhanced branch-and-cut for the min–max k-vehicles windy rural postman problem*, Networks, 58 (2011), pp. 255–272.

[19] E. BENAVENT, Á. CORBERÁN, AND J.M. SANCHIS, *A metaheuristic for the min-max windy rural postman problem with k vehicles*, Computational Management Science, 7 (2010), pp. 269–287.

[20] N. BOLAND, J. DETHRIDGE, AND I. DUMITRESCU, *Accelerated label setting algorithms for the elementary resource constrained shortest path problem*, Operations Research Letters, 34 (2006), pp. 58–68.

[21] N. CHRISTOFIDES, E. BENAVENT, V. CAMPOS, Á. CORBERÁN, AND E. MOTA, *An optimal method for the mixed postman problem*, in System Modelling and Optimization, P. Thoft-Christensen, ed., Lecture Notes in Control and Information Sciences 59, Springer, Berlin, 1984, pp. 641–649.

[22] Á. Corberán, I. Plana, and J.M. Sanchis, *A branch & cut algorithm for the windy general routing problem and special cases*, Networks, 49 (2007), pp. 245–257.

[23] ——, *The windy general routing polyhedron: A global view of many known arc routing polyhedra*, SIAM Journal on Discrete Mathematics, 22 (2008), pp. 606–628.

[24] Á. Corberán and J.M. Sanchis, *A polyhedral approach to the rural postman problem*, European Journal of Operational Research, 79 (1994), pp. 95–114.

[25] ——, *The general routing problem polyhedron: Facets from the RPP and GTSP polyhedra*, European Journal of Operational Research, 108 (1998), pp. 538–550.

[26] G. Cornuéjols, J. Fonlupt, and D. Naddef, *The traveling salesman problem on a graph and some related integer polyhedra*, Mathematical Programming, 33 (1985), pp. 1–27.

[27] G.B. Dantzig and P. Wolfe, *Decomposition principle for linear programs*, Operations Research, 8 (1960), pp. 101–111.

[28] G. Desaulniers, J. Desrosiers, and M.M. Solomon, *Column generation*, Springer, New York, 2005.

[29] G. Desaulniers, J. Desrosiers, and S. Spoorendonk, *Cutting planes for branch-and-price algorithms*, Networks, 58 (2011), pp. 301–310.

[30] G. Desaulniers, F. Lessard, and A. Hadjar, *Tabu search, partial elementarity, and generalized k-path inequalities for the vehicle routing problem with time windows*, Transportation Science, 42 (2008), pp. 387–404.

[31] M. Dror, *Note on the complexity of the shortest path models for column generation in the VRPTW*, Operations Research, 42 (1994), pp. 977–978.

[32] B. Dussault, B. Golden, C. Gröer, and E. Wasil, *Plowing with precedence: A variant of the windy postman problem*, Computers & Operations Research, 40 (2013), pp. 1047–1059.

[33] B. Dussault, B. Golden, and E. Wasil, *The downhill plow problem with multiple plows*, The Journal of the Operational Research Society, 65 (2014), pp. 1465–1474.

[34] G.N. Frederickson, M.S. Hecht, and C.E. Kim, *Approximation algorithms for some routing problems*, SIAM Journal on Computing, 7 (1978), pp. 178–193.

[35] G. Ghiani and G. Laporte, *A branch-and-cut algorithm for the undirected rural postman problem*, Mathematical Programming, 87 (2000), pp. 467–481.

[36] P.C. Gilmore and R.E. Gomory, *A linear programming approach to the cutting-stock problem*, Operations Research, 9 (1961), pp. 849–859.

[37] A. Hadjar, O. Marcotte, and F. Soumis, *A branch-and-cut algorithm for the multiple depot vehicle scheduling problem*, Operations Research, 54 (2006), pp. 130–149.

[38] K.L. Hoffman and M. Padberg, *Solving airline crew scheduling problems by branch-and-cut*, Management Science, 39 (1993), pp. 657–682.

[39] S. IRNICH AND G. DESAULNIERS, *Shortest path problems with resource constraints*, in Column Generation, G. Desaulniers, J. Desrosiers, and M.M. Solomon, eds., Springer, Berlin, 2005, pp. 33–65.

[40] M. JEPSEN, B. PETERSEN, S. SPOORENDONK, AND D. PISINGER, *Subset-row inequalities applied to the vehicle-routing problem with time windows*, Operations Research, 56 (2008), pp. 497–511.

[41] P. LACOMME, C. PRINS, AND W. RAMDANE-CHÉRIF, *Competitive memetic algorithms for arc routing problems*, Annals of Operations Research, 131 (2004), pp. 159–185.

[42] A.N. LETCHFORD, *New inequalities for the general routing problem*, European Journal of Operational Research, 96 (1997), pp. 317–322.

[43] L.Y.O. LI AND R.W. EGLESE, *An interactive algorithm for vehicle routing for winter-gritting*, The Journal of the Operational Research Society, 47 (1996), pp. 1031–1034.

[44] A. LÖBEL, *Optimal Vehicle Scheduling in Public Transit*, Ph.D. thesis, Technische Universität Berlin, Berlin, Germany, 1997.

[45] M. LÜBBECKE AND J. DESROSIERS, *Selected topics in column generation*, Operations Research, 53 (2005), pp. 1007–1023.

[46] I. OR, *Traveling Salesman-Type Combinatorial Problems and their Relation to the Logistics of Regional Blood Banking*, Ph.D. thesis, Northwestern University, Evanston, IL, 1976.

[47] M.W. PADBERG AND M.R. RAO, *Odd minimum cut-sets and b-matchings*, Mathematics of Operations Research, 7 (1982), pp. 67–80.

[48] W.L. PEARN, *Solvable cases of the k-person Chinese postman problem*, Operations Research Letters, 16 (1994), pp. 241–244.

[49] G. RIGHINI AND M. SALANI, *Symmetry helps: Bounded bidirectional dynamic programming for the elementary shortest path problem with resource constraints*, Discrete Optimization, 3 (2006), pp. 255–273.

[50] ——, *New dynamic programming algorithms for the resource constrained elementary shortest path problem*, Networks, 51 (2008), pp. 155–209.

[51] E.J. WILLEMSE AND J.W. JOUBERT, *Applying min–max k postmen problems to the routing of security guards*, The Journal of the Operational Research Society, 63 (2012), pp. 245–260.

[52] L. ZHANG, *Polynomial algorithms for the k-Chinese postman problem*, in Proceedings of the IFIP 12th World Computer Congress. Volume 1: Algorithms, Software, Architecture, J. Van Leeuwen, ed., Elsevier, Amsterdam, 1992, pp. 430–435.

Chapter 12
Arc Routing Problems with Profits

Claudia Archetti
M. Grazia Speranza

12.1 ▪ Introduction

The class of routing problems with profits contains a large variety of problems which share the characteristic that, contrary to classical routing problems, the customers to serve have to be chosen from a set of potential customers. A profit is associated with each potential customer. The objective function may be the collected profit, the traveling cost, or a combination of both. The research activity has been mainly focused on node routing problems with profits, i.e., problems where customers are identified with specific locations which can be represented as vertices of a graph. The word "node" is usually not specified because having customers on nodes is the most common situation. We will specify it in this chapter as we will mention both classes of problems, with customers on nodes and on arcs. A first survey on single vehicle node routing problems with profits can be found in Feillet, Dejax, and Gendreau [21]. These problems are called *Traveling Salesman Problems* (TSPs) with profits. In this paper, the authors distinguish three classes of problems on the basis of the objective function. In the *Orienteering Problem* (OP) the objective is to maximize the collected profit with the constraint that the traveling time of the route does not exceed a given threshold. The *Prize-Collecting Traveling Salesman Problem* (PCTSP) is the problem of finding the route of minimum cost that collects a profit higher than a given lower bound. Finally, the *Profitable Tour Problem* (PTP) is the problem of finding the route that maximizes the difference between the total collected profit and the traveling cost. More recently, in Vansteenwegen, Souffriau, and Van Oudheusden [31] the authors focused on the OP and on its multiple vehicle version, i.e., the *Team Orienteering Problem* (TOP). A recent survey on node routing problems with profits, both with a single vehicle and with multiple vehicles, can be found in Archetti, Speranza, and Vigo [9].

From the previously cited surveys it can be observed that the interest in node routing problems with profits is growing. We observe a similar situation for the arc routing problems with profits. However, few papers have been published and several problems remain to be studied. This reflects the same phenomenon which occurs in classical routing problems: The literature on node routing problems is much broader than that on arc routing problems. The interest in arc routing problems with profits is motivated by a variety of applications that concern situations where demand is associated with links (as in all arc routing problems), and, whereas some links may be required to be traversed, others may be traversed only if beneficial. Arc routing problems have peculiarities that make them deeply different from their node routing counterparts (see Corberán and Prins [15] for a recent survey on arc routing problems). This implies that models and solution approaches cannot be directly inherited and need to be specifically studied.

In this chapter we present a survey of the literature on arc routing problems with profits. In Section 12.2 we introduce the notation that is common to all problems. In Section 12.3 we focus on problems with a single vehicle and we then move, in Section 12.4, to the multiple vehicle case. We present the different models analyzed in each class as well as their applications and solution approaches. A summary of the state of the art is made in Section 12.5, and proposals on the names for the problems are advanced. Conclusions are drawn in Section 12.6.

12.2 ▪ Problem representation and notation

For the ease of presentation, we will use the same notation throughout the chapter even if there is no uniformity in the literature. We have chosen a notation that is consistent with the one used for the node routing problems with profits (see Archetti, Speranza, and Vigo [9]).

A connected directed graph $G=(V,A)$ is given where $V=\{0,\dots,n\}$ is the set of vertices and A is the set of arcs. Vertex 0 is the starting point, and vertex n is the ending point of each route. Two nonnegative values may be associated with each arc $(i,j)\in A$, a traveling cost c_{ij} and a traveling time t_{ij}. In most of the problems only one of these two values is relevant. When the traveling time t_{ij} is relevant, a time limit $T_{\max}$ is set on the total traveling time of a route. A subset of arcs $A'\subseteq A$ is defined that corresponds to the set of profitable arcs. A nonnegative profit p_{ij}^k is associated with the kth traversal of arc $(i,j)\in A'$, for $k=1,\dots,n_{ij}$. Thus, while cost c_{ij} is paid whenever an arc $(i,j)\in A$ is traversed, profit p_{ij}^k depends on the number of traversals of arc $(i,j)\in A'$ and is gained only a maximum of n_{ij} times.

In the case of an undirected graph, we will use the notation $G=(V,E)$ for the graph and an edge is identified with $e=(i,j)\in E$ instead of $(i,j)\in A$.

As a general notation, for the case of an undirected graph, given two sets $S,S'\subseteq V$, $(S:S')$ denotes the set of edges with one endpoint in S and the other in S'. Moreover, we define $\delta(S)=\{e=(i,j)\in E:(i\in S,j\notin S)\text{ or }(i\notin S,j\in S)\}$ and $E(S)=\{e=(i,j)\in E:i\in S,j\in S\}$. We will use, for simplicity, $\delta(i)$ to denote the set $\delta(S)$ when S contains the vertex i only. Similarly, for a directed graph we have $A(S)=\{(i,j)\in A:i\in S,j\in S\}$, $\delta^+(S)=\{(i,j)\in A:(i\in S,j\notin S)\}$, and $\delta^-(S)=\{(i,j)\in A:(i\notin S,j\in S)\}$.

12.3 ▪ Single vehicle arc routing problems with profits

As in the case of node routing problems with profits, the first papers that appeared on arc routing problems with profits consider the case of one vehicle only. Three problems were studied, called by the authors the Maximum Benefit Chinese Postman Problem, the Prize-Collecting Rural Postman Problem, and the Arc Orienteering Problem.

12.3.1 ▪ The maximum benefit Chinese Postman Problem

The *Maximum Benefit Chinese Postman Problem* (MBCPP) was introduced in Malandraki and Daskin [25] where the directed version of the problem is studied. Different formulations of the problem were proposed in the literature. We refer here to the formulation given in Corberán et al. [14] where the undirected version of the problem is studied.

The MBCPP is defined on a connected undirected graph $G = (V, E)$. Vertex 0 is the depot and coincides with the ending vertex n. Thus, the vehicle route has to start and end at the depot. Set E' coincides with E, meaning that all edges are profitable. The cost c_e of each edge $e \in E$ depends on the number of times the edge is traversed. More specifically, a traversal cost c_e^i is associated with the ith traversal of edge e, $i = 1, \ldots, n_e$, and a deadheading cost c_e^d with the ith traversal of edge e when $i > n_e$. The deadheading cost is the cost of traversing the edge without servicing it. The profit associated with the ith traversal of edge e is denoted as p_e^i. The net benefit of the ith traversal of edge e is $\bar{p}_e^i = p_e^i - c_e^i$ for $i \leq n_e$ and $\bar{p}_e^i = -c_e^d$ for $i > n_e$. The MBCPP is the problem of finding a route starting and ending at the depot with maximum total net benefit. The problem is NP-hard as the *Rural Postman Problem* (RPP) is a special case. In the RPP a set R of required edges is defined and the problem is to find the minimum cost route that traverses each edge in R at least once. The RPP is obtained from the MBCPP by setting, for each $e \in R$, $p_e^1 = M$ and $p_e^k = 0$ for $k = 2, \ldots, n_e$, where M is a large positive value, and $c_e^i = c_e^d = c_e$ for each $e \in E$ and $i = 1, \ldots, n_e$.

Applications of the MBCPP arise, for example, in the planning of the routes of street cleaners and of garbage collection, and in the design of street snow-plowing and snow-salting routes. Typically, these problems are represented as arc routing problems, either with a single vehicle or multiple vehicles, where required edges represent the streets that have to be served. Modeling the problem as an MBCPP allows the possibility of not serving secondary streets (with a low benefit) and serving multiple times primary streets (with a high benefit). Also, by considering multiple profits related to each edge, it is possible to model the situation where the profit depends on the number of times that the edge is traversed. For example, in street snow-plowing, the first traversal is related to a higher profit as it is the one that usually removes the larger portion of the snow, while the subsequent traversals are less effective. On the other hand, the first traversal of the street is more costly as the higher level of snow slows down the vehicles.

In the following we present two mathematical formulations of the problem and discuss the solution approaches proposed in the literature.

12.3.1.1 ▪ Problem formulation

The formulation proposed in Corberán et al. [14] is based on the following theorem.

Theorem 12.1. *Solving the MBCPP on a graph with $n_e + 1$ net benefits associated with each edge e is equivalent to solving it with only two net benefits, b_e^{odd} and b_e^{even}, defined as*

$$b_e^{odd} = \max\left\{\sum_{l=1}^{k} \bar{p}_e^l : k \text{ is odd and } k \leq n_e + 1\right\}, \tag{12.1}$$

$$b_e^{even} = \max\left\{\sum_{l=1}^{k} \bar{p}_e^l : k \text{ is even and } k \leq n_e + 1\right\} - b_e^{odd}. \tag{12.2}$$

When $n_e = 0$, $b_e^{odd} = b_e^{even} = -c_e^d$.

In the problem formulation b_e^{odd} will be associated with the first traversal of edge e. If e is traversed an odd number of times, then b_e^{odd} is the total net benefit corresponding to the number of traversals that maximize the sum of the net benefits. If e is traversed an even number of times, then b_e^{even} has the same meaning. In this case, however, b_e^{even} is associated with the second traversal of edge e. Thus, the total net benefit is decreased by the net benefit gained from the first traversal (b_e^{odd}).

As a consequence of Theorem 12.1, the MBCPP can be formulated using two binary variables associated with each edge e. Variable x_e takes value 1 if edge e is traversed at least once, 0 otherwise. Variable y_e takes value 1 if edge e is traversed twice.

The MBCPP can be modeled as

$$(12.3) \qquad \max \sum_{e \in E} (b_e^{odd} x_e + b_e^{even} y_e),$$

$$(12.4) \qquad \sum_{e \in \delta(i)} (x_e + y_e) \equiv 0 \pmod 2, \qquad i \in V,$$

$$(12.5) \qquad \sum_{e \in \delta(S)} (x_e + y_e) \geq 2x_f, \qquad S \subset V \setminus \{0\}, f \in E(S),$$

$$(12.6) \qquad x_e \geq y_e, \qquad e \in E,$$

$$(12.7) \qquad x_e, y_e \in \{0,1\}, \qquad e \in E.$$

The objective function (12.3) aims at maximizing the total net benefit. Constraints (12.4) ensure that each vertex is visited an even number of times, while connectivity is guaranteed by (12.5). Finally, (12.6) establish that each edge can be traversed the second time only if it has been traversed the first time, and (12.7) are variable definitions.

In Corberán et al. [14] a polyhedral study of formulation (12.3)–(12.7) was presented and the authors proved that most of the inequalities are facet-inducing (sometimes under special conditions on graph G). Also, note that constraints (12.4) are not linear. In [14] the authors proved that they can be substituted by the following linear inequalities, called parity inequalities:

$$(12.8) \qquad \sum_{e \in \delta(S) \setminus F} (x_e - y_e) \geq \sum_{e \in F} (x_e - y_e) - |F| + 1, \qquad S \subset V, F \subset \delta(S) \text{ with } |F| \text{ odd.}$$

Two classes of valid inequalities were also proposed to strengthen the formulation, namely, the K-C inequalities and the p-connectivity inequalities.

The K-C inequalities are derived from the ones previously proposed for the RPP in Corberán and Sanchis [16]. We refer the reader to Chapter 6 of this book for details on K-C inequalities.

The p-connectivity inequalities are instead derived specifically for the MBCPP and are defined as follows. Let $\{S_0, \dots, S_p\}$ be a partition of V. Assume that $0 \in S_d$, $d \in \{0, \dots, p\}$, and consider one edge $e_j \in E(S_j)$ for each $j \in \{0, \dots, p\} \setminus \{d\}$. The associated p-connectivity inequality is defined as

$$(12.9) \qquad \sum_{e \in \delta(S_0)} (x_e + y_e) + 2 \sum_{1 \leq r < t \leq p} \sum_{e \in (S_r : S_t)} x_e \geq 2 \sum_{i=0, i \neq d}^{p} x_{e_i}.$$

A p-connectivity inequality enforces the connectivity of sets S_j, $j \in \{0, \dots, p\} \setminus \{d\}$, with the depot. In fact, it detects fractional solutions which satisfy inequalities (12.5) but for which sets S_j, $j \in \{0, \dots, p\} \setminus \{d\}$, are not properly connected to the depot.

Both classes of inequalities are proven to be facet-inducing.

We reported the formulation proposed in Corberán et al. [14] as it is the most recent one and was proved to be more efficient than the ones previously proposed in the literature. For the sake of completeness, we report also the formulation proposed by Malandraki and Daskin [25], who were the first authors who studied the MBCPP. They considered a directed graph and the following further assumptions: $p_{ij}^k < p_{ij}^{k-1}, k = 2, \ldots, n_{ij}$, and $c_{ij}^k = c_{ij}^t, k = 1, \ldots, n_{ij}$ and $c_{ij}^t > c_{ij}^d$, where c_{ij}^t is the cost of traversing (i, j) when a profit is collected. Thus, the benefit is decreasing with the number of traversals and the traversal cost related to the service of an edge is constant and bigger than the deadheading cost. They proposed a minimum cost flow formulation with subtour elimination constraints which makes use of the following decision variables:

- x_{ij}^k is a binary variable which takes value 1 if arc (i, j) is traversed for the kth time, 0 otherwise;
- y_{ij} is an integer variable representing the number of times arc (i, j) is deadheaded.

The formulation is the following:

$$\text{(12.10)} \quad \min \sum_{(i,j)\in A} \sum_{k=1}^{n_{ij}} (c_{ij}^t - p_{ij}^k) x_{ij}^k - \sum_{(i,j)\in A} c_{ij}^d y_{ij},$$

$$\text{(12.11)} \quad \sum_{(i,j)\in\delta^+(i)} \sum_{k=1}^{n_{ij}} x_{ij}^k + \sum_{(i,j)\in\delta^+(i)} y_{ij} - \sum_{(j,i)\in\delta^-(i)} \sum_{k=1}^{n_{ji}} x_{ji}^k - \sum_{(j,i)\in\delta^-(i)} y_{ji} = 0, \qquad i \in V,$$

$$\text{(12.12)} \quad y_{0n} \geq 1,$$

$$\text{(12.13)} \quad y_{ij} \geq 0 \text{ and integer}, \qquad (i,j) \in A,$$

$$\text{(12.14)} \quad x_{ij}^k \in \{0,1\}, \qquad (i,j) \in A, k = 1, \ldots, n_{ij},$$

plus subtour elimination constraints. The objective function aims at minimizing the difference between the traversal cost and the benefit. Constraints (12.11) are flow conservation constraints while (12.12) imposes to visit the depot.

12.3.1.2 ▪ Solution approaches

As already mentioned, the directed version of the MBCPP was introduced in Malandraki and Daskin [25]. The proposed solution procedure is a branch-and-bound algorithm that generates the subtour elimination constraints when needed while retaining the minimum cost flow formulation. The approach is used to solve an instance with 25 vertices.

Heuristic and approximation algorithms were proposed in Pearn and Wang [27] and Pearn and Chiu [26]. In both papers, as in Malandraki and Daskin [25], it was assumed that the benefit is decreasing with the number of traversals. Moreover, the traversal cost related to the service of an edge is constant and is greater than or equal to the deadheading cost. In [27] an undirected graph was considered and a heuristic was presented which first finds a minimum spanning tree on an expanded graph obtained by replacing each edge of the original graph with a set of at most n_e edges with positive net benefit given by the difference between the benefit and the traversal cost. In particular, each edge of the expanded graph has an associated benefit which is collected only at the first traversal of the edge, and this edge is created only if the benefit is positive. Then, a minimum cost matching on odd-degree vertices is solved to obtain a route. Finally, negative net benefit cycles are removed. The algorithm was tested on an instance with 15 vertices and 26 edges. In [26] the problem was studied for the case of a directed graph. Three heuristic algorithms were presented: a branch-and-scan algorithm, a connection algorithm, and a directed tree expansion algorithm. The branch-and-scan algorithm is essentially based on the exact algorithm presented in [25] where the branch-and-bound tree is searched through a depth-first strategy, and stopping rules, like maximum run time or feasible solution found, are introduced. The connection algorithm consists of two phases. In the first phase the graph is expanded by splitting each arc into several arcs, each representing a single traversal. Then, a minimum cost flow problem is solved, without subtour elimination constraints. In the second phase, subtours which are possibly created in the first phase are eliminated by linking all connected components of the solution. Different linking strategies are considered. Finally, the directed tree expansion algorithm also consists in two phases. The first phase is the same as in the connection algorithm. In the second phase, a shortest spanning arborescence is solved to link all the components. Improvement procedures, like one-opt, two-opt, component-exchange, and component-drop, are applied to improve the solution obtained by the previous algorithm. Tests are performed on instances with up to 30 vertices and 783 arcs. The results show that the branch-and-scan algorithm outperforms the others for small- or moderate-size instances, while the connection algorithm is the best for large instances.

A second exact approach for the solution of the MBCPP was more recently proposed in Corberán et al. [14]. The problem formulation and assumptions are the one described in Section 12.3.1.1. The approach is a branch-and-cut algorithm. Polynomial time exact algorithms are implemented to separate connectivity inequalities (12.5) and parity inequalities (12.8). For the K-C inequalities and the p-connectivity inequalities, instead, as the separation problem is NP-hard, heuristic separation algorithms are implemented. Computational tests were made on instances derived from benchmark instances for the RPP. The algorithm is able to solve instances with up to 1000 vertices and 3000 edges in one hour of computing time.

12.3.2 ▪ The prize-collecting Rural Postman Problem

The *Prize-Collecting Rural Postman Problem* (PCRPP) was introduced in Aráoz, Fernández, and Zoltan [4] with the name Privatized Rural Postman Problem. It is a special case of the MBCPP where the profit of an edge can be collected at most once. Thus, $n_e = 1$ for each $e \in E$ and $p_e^1 = p_e$. The objective of the PCRPP is to find a route starting and ending at the depot that maximizes the total net benefit.

As for the MBCPP, we now present a mathematical formulation and discuss a solution approach based on this formulation. Then, we will describe some problem variants.

12.3.2.1 ▪ Problem formulation and solution

In Aráoz, Fernández, and Zoltan [4] different formulations for the PCRPP were proposed. We present here the one that was subsequently strengthened and used in Aráoz, Fernández, and Meza [2] to solve the problem. The formulation is based on the property, shown in [4], that there always exists an optimal solution to the PCRPP where each edge is traversed at most twice.

As in the case of the MBCPP, two binary variables are associated with each edge e. Variable x_e takes value 1 if edge e is traversed at least once, 0 otherwise. Variable y_e takes value 1 if edge e is traversed twice. Moreover, let $\varphi_e = p_e - c_e$. The PCRPP can be formulated as

$$\max \sum_{e\in E} \varphi_e x_e - \sum_{e\in E} c_e y_e, \tag{12.15}$$

$$\sum_{e\in\delta(v)\setminus F} (x_e + y_e) \geq \sum_{e\in F} (x_e + y_e) - |F| + 1, \quad v \in V, F \subset \delta(v) \text{ with } |F| \text{ odd}, \tag{12.16}$$

$$\sum_{e\in\delta(S)} (x_e + y_e) \geq 2x_f, \quad S \subset V \setminus \{0\}, f \in E(S), \tag{12.17}$$

$$x_e \geq y_e, \quad e \in E, \tag{12.18}$$

$$x_e, y_e \in \{0,1\}, \quad e \in E. \tag{12.19}$$

The objective function (12.15) aims at maximizing the total net benefit. Constraints (12.16) and (12.17) are parity and connectivity constraints, respectively. Finally, (12.18) establish that each edge can be traversed the second time only if it has been traversed the first time, and (12.19) are variable definitions.

In Aráoz, Fernández, and Meza [2] formulation (12.15)–(12.19) is strengthened with inequalities derived from dominance rules which are based on the partition of the profitable edges into different sets on the basis of their net benefit φ_e. A solution approach was proposed which consists of two phases. In both phases, parity and connectivity inequalities are relaxed and inserted only once violated through exact polynomial time separation algorithms. In the first phase, integrality constraints are also relaxed while they are reintroduced in the second phase. Heuristic algorithms were also proposed to obtain lower bounds. Tests were made on two sets of instances derived from benchmark instances for the RPP, with 118 instances in each set. All instances were solved to optimality within one hour of computing time for each phase, and the solution of the first phase suffices to solve 94 instances of the first set and 117 of the second set. More recently, Schaeffer, Ríos-Mercado, and Fernández [28] studied the PCRPP on a Windy graph (WPCRPP) and proposed an ant colony heuristic algorithm. Tests were made on benchmark instances for the RPP and the General Routing Problem (GRP), adapted to the WPCRPP. Results were compared with a lower bound and show that the heuristic gives good solutions in a short computing time.

12.3.2.2 • Variants

A variant of the PCRPP was studied in Aráoz, Fernández, and Meza [3] where customers are clustered in groups and the profit is associated with each group instead of with each customer. If a customer of a group is served, all the customers of the same group must be served. The problem is called the *Clustered Prize-Collecting Arc Routing Problem* (CPCARP). The authors analyzed the problem properties, proposed a mathematical formulation, and developed a branch-and-cut algorithm. Tests were made by adapting the instances proposed in Aráoz, Fernández, and Meza [2] for the PCRPP. Results show that 75% of the instances are solved at the root node while the others are solved with a small additional computational effort, and only one instance requires more than one hour of computing time. The same problem was studied in Corberán et al. [13] on a windy graph. The authors proposed a mathematical formulation and presented polyhedral results including facet-defining and valid inequalities. The problem was solved through a cutting-plane algorithm which is able to solve instances with up to 196 vertices and 316 edges. More recently, Aráoz, Fernández, and Franquesa [1] proposed a heuristic algorithm for the CPCARP which combines a GRASP heuristic with different construction procedures and path relinking. Computational tests were made on the benchmark instances proposed in [3] and show that the heuristic gives high quality solutions in rather short computing time.

In Guastaroba, Mansini, and Speranza [23] and Archetti, Guastaroba and Speranza [8] a variant of the directed version of the PCRPP was studied where the objective is the minimization of the sum of the total traveling cost and the penalties for the unserved arcs. This objective is equivalent to the maximization of the difference between the total collected profit and the total traveling cost. In [23] the problem faced by shippers that have to select the lanes to serve was presented as a motivating application. A shipper has to decide which transportation requests to undertake with its own fleet of vehicles and which ones to outsource. Two formulations were proposed and compared on a set of randomly generated instances. Both formulations use binary variables to represent the traversal of each arc. In [8] a formulation was proposed which uses integer variables to model the traversal of each arc. Moreover, the authors proposed a matheuristic to solve the problem which combines a tabu search heuristic with the optimal solution of an integer linear programming model for the intensification phase. Computational tests were made on instances generated from the PCRPP instances proposed in Aráoz, Fernández, and Meza [2] and from benchmark instances for the Capacitated Chinese Postman Problem studied in Benavent et al. [10]. Computational results show that the average error with respect to the optimal solution, when available, or to a lower bound, is below 1% and the maximum error is 6.10%. Recently, Colombi and Mansini [12] studied the same problem considered in [8]. They proposed a matheuristic that uses the optimal solution of a problem relaxation to identify different subsets of service arcs on which to formulate and solve a sequence of RPPs. Moreover, they developed a branch-and-cut algorithm which introduces violated subtour elimination constraints in a "lazy way," i.e., only when an integer solution is found. Computational results show that the matheuristic produces high quality solutions whereas the exact algorithm solves all instances proposed in [8] in one hour of computing time.

Black, Eglese, and Wøhlk [11] studied a variant of the directed version of the PCRPP with time-dependent costs, i.e., the cost of traversing each arc depends on the time at which the traversal is made. The authors presented a mathematical formulation of the problem and three heuristic algorithms: a variable neighborhood search, a variable neighborhood descent algorithm, and a tabu search. Tests were made on instances generated from the road network of two areas: the northwest and the southeast of England. These instances have up to 50 vertices, 2500 arcs, and 350 profitable arcs. Computational results show that the variable neighborhood search outperforms the other heuristic approaches.

12.3.3 ▪ The arc orienteering problem

The *Arc Orienteering Problem* (AOP) was introduced in Souffriau et al. [29] and is defined on a directed graph $G=(V,A)$. Set A' coincides with A, i.e., all arcs are profitable. For each arc $(i,j)\in A$, $n_{ij}=1$ and $p_{ij}^1=p_{ij}$. A value t_{ij} is associated with each arc $(i,j)\in A$, and a maximum value T_{max} is set on the total traveling time of the route. In the application presented in [29] the values t_{ij} and T_{max} are interpreted as costs. Moreover, the starting and ending nodes coincide.

The AOP consists in finding a route from vertex 0 to vertex n with maximum profit and total traveling time not greater than T_{max}.

12.3.3.1 ▪ Problem formulation

In Souffriau et al. [29] a problem formulation was proposed which makes use of the following decision variables:

- x_{ij} is a binary variable which takes value 1 if arc (i, j) is traversed, and 0 otherwise;
- u_i is a variable indicating the position of vertex i along the route.

We now present the formulation of the AOP proposed in Souffriau et al. [29]. We note that, although this is not specified in [29], this formulation is based on the assumption that the graph G is complete. In fact, if this is not the case, an arc may be traversed more than once in the optimal solution. The formulation is the following:

$$\max \sum_{i=0}^{n}\sum_{j=0}^{n} p_{ij}x_{ij}, \tag{12.20}$$

$$\sum_{j=1}^{n} x_{0j} = \sum_{i=0}^{n-1} x_{in} = 1, \tag{12.21}$$

$$\sum_{j=0}^{n} x_{kj} = \sum_{i=0}^{n} x_{ik} \leq 1, \qquad k = 1, \ldots, n-1, \tag{12.22}$$

$$\sum_{i=0}^{n}\sum_{j=0}^{n} t_{ij}x_{ij} \leq T_{max}, \tag{12.23}$$

$$1 \leq u_i \leq n, \qquad i = 1, \ldots, n, \tag{12.24}$$

$$u_i - u_j + 1 \leq n(1 - x_{ij}), \qquad i, j = 1, \ldots, n, i \neq j, \tag{12.25}$$

$$x_{i,j} \in \{0, 1\}, \qquad i, j = 0, \ldots, n. \tag{12.26}$$

The objective function (12.20) aims at maximizing the total collected profit. Constraints (12.21) impose that the route starts in vertex 0 and ends in vertex n. Inequalities (12.22) establish that each vertex can be visited at most once. Constraint (12.23) limits the total traveling time of the route to be not greater than T_{max}. Inequalities (12.24) and (12.25) eliminate subtours. Finally, (12.26) arc variable definitions.

12.3.3.2 ▪ Applications and solution approaches

Applications of the AOP were studied in Souffriau et al. [29] and Deitch and Ladany [19]. In the following we describe those applications and the solution approaches proposed to solve them.

In Souffriau et al. [29] the AOP was used to model the problem of planning cycle trips in the province of East Flanders. The East Flanders network is a concatenation of five regions and is composed of 989 nodes and 2963 arcs, for a total of 3585 km. An online version of this network is available which facilitates creating personalized routes. Furthermore, an automated route planning tool was developed and incorporated into the software. This tool is the focus of the analysis carried out in [29]. The problem arising from the application is a relaxation of the AOP as multiple visits to each vertex are allowed, contrary to what is stated by constraints (12.22). The proposed solution approach is based on a *Greedy Randomized Adaptive Search Procedure* (GRASP) where the following two phases are repeated until a stopping criterion is met:

- randomly generate a feasible route;

- insert profitable arcs until no arc can be feasibly inserted.

The insertion phase is performed on the basis of a list of profitable arcs ordered by their greediness value. This value is updated at each iteration. Tests were made on instances generated from the East Flanders network. To validate the approach, the solution value is compared with the optimal solution obtained from formulation (12.20)–(12.26). The stopping criterion for the GRASP algorithm is set to 1 s of computing time, as the algorithm should be used in online applications and, consequently, must provide solutions in a very short time. Computational results show that the algorithm gives an average error with respect to the optimal solution below 1% for all classes of instances except one where it is 3.83%.

In Deitch and Ladany [19] the *one-period Bus Touring Problem* (BTP) was analyzed, which is the problem of finding a route for a bus that has to visit tourist sites and scenic routes, each associated with a nonnegative attractiveness value, in such a way that the total attractiveness of the route is maximized while side constraints, on maximum touring time or cost, are satisfied. The BTP can be seen as a combination of the AOP and the OP, as in the BTP profits are associated with both vertices and arcs of the corresponding graph. Moreover, a visiting time is required each time a profitable vertex of the graph is visited. To solve the BTP, in [19] the problem is first reduced to the OP by means of a transformation of the original graph. Then, two algorithms proposed in the literature for the solution of the OP were compared: The stochastic algorithm proposed in Tsiligirides [30] and the 1-step improvement algorithm proposed in Deitch and Ladany [18]. Computational tests were conducted on a set of 11 instances for the BTP. Results show that the 1-step improvement algorithm outperforms the stochastic algorithm, giving better solutions on 7 instances and the same solution on the remaining 4 instances in a much shorter amount of time. However, as mentioned in [19], the transformation of the BTP into an OP does not always guarantee a successful retransformation from a given OP solution to the corresponding BTP solution.

12.4 ▪ Multiple vehicle arc routing problems with profits

Multiple vehicle problems arise as generalizations of the single vehicle cases, with possible additional constraints. Three multiple vehicle arc routing problems were considered in the literature: the Profitable Arc Tour Problem, the Undirected Capacitated Arc Routing Problem with Profits, and the Team Orienteering Arc Routing Problem.

12.4.1 ▪ The profitable arc tour problem

The *Profitable Arc Tour Problem* (PATP) was introduced in Feillet, Dejax, and Gendreau [22] and is defined on a complete directed graph $G=(V,A)$. Here, two values c_{ij} and t_{ij} are associated with each arc $(i,j)\in A$ because cost and time are both relevant. For each profitable arc $(i,j)\in A'$ the profit is constant, i.e., $p_{ij}^k = p_{ij}$, $k=1,\dots,n_{ij}$. Starting and ending points are not defined. Thus, each route may start from and end at any vertex of the graph. The objective of the PATP is to find a set of routes maximizing the difference between the total collected profit and the traveling cost such that the traveling time of each route does not exceed T_{max}. No limit on the number of vehicles used, and thus on the number of routes constructed, is imposed.

12.4.1.1 ▪ Problem formulation

A route-based mathematical formulation for the PATP, suitable for a branch-and-price approach, was presented in Feillet, Dejax, and Gendreau [22]. Let Ω be the set of routes having a total traveling time not greater than T_{max} in G and such that each route $r_k \in \Omega$ corresponds to an elementary cycle in G. For each route $r_k \in \Omega$, let p_k and c_k be the corresponding profit and cost, respectively. Moreover, for each $(i,j) \in A'$ and each $r_k \in \Omega$, let $a_{ij}^k = 1$ if r_k collects a profit on arc (i,j), 0 otherwise. Note that, since r_k corresponds to an elementary cycle, the profit of arc (i,j) for which $a_{ij}^k = 1$ is collected only once. Finally, let x_k be an integer variable indicating the number of times route r_k is used in the optimal solution. The PATP can be formulated as

$$\max \sum_{r_k \in \Omega} (p_k - c_k) x_k, \tag{12.27}$$

$$\sum_{r_k \in \Omega} a_{ij}^k x_k \leq n_{ij}, \qquad (i,j) \in A', \tag{12.28}$$

$$x_k \in \mathbb{N}, \qquad r_k \in \Omega. \tag{12.29}$$

The objective function in (12.27) is the difference between the collected profit and the traveling cost. Constraints (12.28) enforce that the profit of arc $(i,j) \in A'$ is collected n_{ij} times at most. Finally, (12.29) are variable definitions.

As mentioned in Feillet, Dejax, and Gendreau [22], the applications of the PATP are related to the domain of the tactical freight transportation-planning problem in the car industry as well as to telecommunication network optimization.

12.4.1.2 ▪ Solution approaches

The only exact approach proposed for the solution of the PATP was presented in Feillet, Dejax, and Gendreau [22]. In this paper it was assumed that $t_{ij} = c_{ij}$, $(i,j) \in A$. The approach is a branch-and-price algorithm based on formulation (12.27)–(12.29). The pricing problem turns out to be an elementary shortest path problem with resource constraints and is solved through a dynamic programming algorithm. Exact and heuristic solution algorithms were proposed for the solution of the pricing problem. Moreover, a specific branching rule, called flow splitting, is proposed to define the flow on an arc when branching. Computational tests were carried out on randomly generated instances. The branch-and-price algorithm was compared with an exact approach based on a node routing formulation of the PATP. Computational results prove that the branch-and-price algorithm for the PATP outperforms the algorithm based on the node routing formulation.

Heuristics for the PATP were proposed in Euchi and Chabchoub [20]. They are hybrid algorithms which combine an *Adaptive Memory Procedure* (AMP) with *Tabu Search* (TS) or *Variable Neighborhood Search* (VNS). A scheme of the adaptive memory procedure is the following:

1. Initialize the memory M.
2. While a stopping criterion is not met do:
   (a) construct a new solution s combining components of M;
   (b) apply a local search procedure to s (TS or VNS). Let s^* be the improved solution;

(c) update M using components of s^*.

The initialization phase consists in applying the nearest neighbor method to generate a set of cycles. For the selection of the next arc to be inserted into a cycle, an advantage measure av_{ij} is associated with each profitable arc in A' which is defined as

$$av_{ij} = \begin{cases} \frac{(p_{ij}-c_{ij})n_{ij}}{t_{ij}} & \text{if } n_{ij} > 0, \\ \frac{1}{t_{ij}} & \text{otherwise.} \end{cases}$$

The TS and VNS are defined on a neighborhood created by classical swapping and insertion moves. Improvement procedures like 2-opt are also implemented.

Computational tests were made on the same set of instances used in Feillet, Dejax, and Gendreau [22] and on a new set of instances. The results show that the hybrid algorithm which uses VNS outperforms the one using TS and gives a percentage error with respect to the best known solution which is always lower than 1%.

12.4.2 ▪ The undirected Capacitated Arc Routing Problem with profits

In the *Undirected Capacitated Arc Routing Problem with Profits* (UCARPP) we have an undirected graph $G = (V, E)$, where vertex 0 is the depot which corresponds to the starting and ending point of each route. For each $e \in E'$, $n_e = 1$ and $p_e^1 = p_e$. A traveling time t_e is associated with each edge $e \in E$. Moreover, a nonnegative demand d_e is associated with each profitable edge $e \in E'$. A set of m vehicles is available. Each vehicle has a capacity Q, and the traveling time of the corresponding route cannot be greater than T_{max}. If the profit of a profitable edge is collected by a vehicle, the same vehicle has to serve the corresponding demand. The objective of the UCARPP is to find a set of at most m routes maximizing the total collected profit and such that

- each route starts and ends at the depot;
- the capacity and maximum traveling time constraints are satisfied.

12.4.2.1 ▪ Problem formulation

The UCARPP was introduced in Archetti et al. [7] where a route-based mathematical formulation was proposed. The notation is similar to the one used in the formulation of the PATP. Let Ω be the set of feasible routes, and let p_k be the profit associated with route $r_k \in \Omega$. Moreover, let $a_{ek} = 1$ if edge $e \in E'$ is served by route $r_k \in \Omega$, 0 otherwise. A binary decision variable x_k is defined for each route $r_k \in \Omega$, where $x_k = 1$ if route r_k is used in the optimal solution, 0 otherwise. The UCARPP can be formulated as

(12.30) $$\max \sum_{r_k \in \Omega} p_k x_k,$$

(12.31) $$\sum_{r_k \in \Omega} a_{ek} x_k \leq 1, \qquad e \in E',$$

(12.32) $$\sum_{r_k \in \Omega} x_k \leq m,$$

(12.33) $$x_k \in \{0,1\}, \qquad r_k \in \Omega.$$

The objective function (12.30) aims at maximizing the total collected profit. Constraints (12.31) impose that each profitable edge is served at most once while (12.32) limits to m the maximum number of routes to be chosen. Finally, (12.33) are variable definitions.

Applications of the UCARPP are described in Archetti et al. [7] in the domain of truckload transportation services where carriers can exploit fleet spare capacity by looking for requests of service posted on the Internet.

12.4.2.2 ▪ Solution approaches

The UCARPP was studied in Archetti et al. [7] and in Zachariadis and Kiranoudis [32]. An exact solution approach and several heuristic algorithms were proposed in [7] while a local search procedure was presented in [32].

The exact approach proposed in [7] is a branch-and-price algorithm which is an adaptation of the one proposed in Feillet, Dejax, and Gendreau [22] for the PATP. The algorithm is based on formulation (12.30)–(12.33). As before, the pricing problem is an elementary shortest path problem with resource constraints and is solved through dynamic programming. In the same paper, the authors proposed three heuristic algorithms for the solution of the UCARPP: two TS and a VNS heuristics. The main idea of all algorithms is to keep all profitable edges organized in routes. Obviously, the number of routes needed to serve all profitable edges in E' is typically much larger than m. Thus, at the end only the m most profitable routes will form the final solution. All the algorithms make use of the same main procedures: the construction of the initial solution, the internal tabu search phase, and the jump phase. The construction of the initial solution is made through a simple greedy algorithm that inserts profitable edges into the current route while it is feasible; otherwise, it creates a new route. The internal tabu search phase is based on two moves:

- 1-move, which consists in moving a profitable edge from its current route to a different route;
- swap-move, which consists in swapping two profitable edges served in two different routes.

The jump phase destroys the current solution and builds a completely different one, moving the search to a different part of the solution space. The difference between the VNS and TS is that the VNS heuristic performs a large number of jumps with a short internal tabu search phase while the TS heuristic performs a small number of jumps with a long internal tabu search. The difference between the two proposed TS heuristics is that one explores only feasible solutions while infeasibility is allowed in the second. A computational study was made on instances adapted from benchmark instances for the arc routing problem. The results show that the branch-and-price algorithm can solve instances with up to 97 profitable edges when the time and capacity constraints are binding in one hour of computing time. The VNS turns out to be the best heuristic and gives an average error of 1% with respect to the best known solution.

In Zachariadis and Kiranoudis [32] a different objective function was considered which is defined as

$$\min -M \sum_{r_k \in \Omega} p_k x_k + \sum_{r_k \in \Omega} t_k x_k, \tag{12.34}$$

where t_k is the total traveling time of route k. The constraints are the same as those defined for the UCARPP. Objective function (12.34) is a hierarchical function that aims at maximizing the total collected profit first and minimizing the traveling time second.

Note that a solution which is optimal with respect to (12.34) is optimal also for (12.30) but not vice versa. The introduction of this new objective function was made in order to discriminate between different solutions collecting the same amount of total profit. A metaheuristic algorithm, called *Moving Promise Algorithm* (MPA), was proposed to solve the problem. The approach tackles the UCARPP within a node routing setting and to do so the original graph G is transformed in such a way that each profitable edge corresponds to two vertices in the new graph. An initial solution is constructed through a simple constructive algorithm. Then a local search phase is started where the operators are similar to the 1-move and swap-move defined above. The cardinality of the neighborhood is reduced in order to reduce the computing time. Diversification components were introduced to explore different parts of the solution space. Computational tests were made on the set of instances proposed in Archetti et al. [7]. The results show that the errors generated are comparable to the ones of the heuristics proposed in [7] and that the computing time of the algorithm is lower.

More recently, Cura [17] proposed an artificial bee colony algorithm for the solution of the UCARPP. The algorithm is a metaheuristic scheme which was recently introduced in the literature and used to solve combinatorial optimization problems. The idea is to replicate the behavior of bees looking for food sources, in a similar way as in ant colony algorithms. The algorithm was proved to be competitive with respect to the previous algorithms proposed in the literature.

12.4.3 ▪ The team orienteering arc routing problem

The *Team Orienteering Arc Routing Problem* (TOARP) is the extension to the multiple vehicle case of the AOP. It is defined on a directed graph $G=(V,A)$, where vertex 0 is the depot, which is the starting and ending point of each vehicle route. In addition to the set of profitable arcs A', another subset of arcs A_R is defined which corresponds to the set of required arcs, i.e., arcs that must be traversed. A traveling time t_{ij} is associated with each arc $(i,j) \in A$. For each $(i,j) \in A'$, $n_{ij}=1$ and $p_{ij}^1 = p_{ij}$. A fleet of m vehicles is available, and a maximum traveling time T_{max} is imposed on each vehicle route. The objective of the TOARP is to find at most m routes that maximize the total collected profit and such that

- each route starts and ends at the depot;
- all arcs in A_R are served;
- the traveling time of each route does not exceed T_{max}.

Note that, contrary to what happens in the AOP, in the TOARP the routes start and end at the same vertex, the depot. Also, in the TOARP there is a subset of arcs, A_R, which must be traversed while this is not the case in the AOP. However, the TOARP can be transformed into the same setting of the AOP as follows. First, we define a vertex $n+1$ which corresponds to the depot and require that each route starts from vertex 0 and ends in vertex $n+1$. Second, we assign a very high profit to each arc $(i,j) \in A_R$ and we move all the arcs from A_R to A'.

12.4.3.1 ▪ Problem formulation

The TOARP was introduced in Archetti et al. [6] where a mathematical formulation is proposed. The decision variables are

- x_{ij}^k, an integer variable representing the number of times that vehicle k traverses arc $(i,j) \in A$;
- y_{ij}^k, a binary variable which takes value 1 if vehicle k serves arc $(i,j) \in A_R \cup A'$, 0 otherwise.

The TOARP can be modeled as

$$\max \sum_{k=1}^{m} \sum_{(i,j) \in A'} p_{ij} y_{ij}^k, \tag{12.35}$$

$$\sum_{j \in V \setminus \{i\}} x_{ij}^k = \sum_{j \in V \setminus \{i\}} x_{ji}^k, \quad i \in V, k = 1, \ldots, m, \tag{12.36}$$

$$\sum_{i \in V \setminus \{S\}, j \in S} x_{ij}^k \geq y_{tz}^k, \quad S \subset V \setminus \{0\}, (t,z) \in A_R(S) \cup A'(S), k = 1, \ldots, m, \tag{12.37}$$

$$x_{ij}^k \geq y_{ij}^k, \quad (i,j) \in A_R \cup A', k = 1, \ldots, m, \tag{12.38}$$

$$\sum_{k=1}^{m} y_{ij}^k = 1, \quad (i,j) \in A_R, \tag{12.39}$$

$$\sum_{k=1}^{m} y_{ij}^k \leq 1, \quad (i,j) \in A', \tag{12.40}$$

$$\sum_{(i,j) \in A} t_{ij} x_{ij}^k \leq T_{max}, \quad k = 1, \ldots, m, \tag{12.41}$$

$$x_{ij}^k \in \mathbb{N}, \quad (i,j) \subset A, k = 1, \ldots, m, \tag{12.42}$$

$$y_{ij}^k \in \{0,1\}, \quad (i,j) \in A_R \cup A', k = 1, \ldots, m. \tag{12.43}$$

The objective function (12.35) aims at maximizing the total collected profit. Constraints (12.36) are symmetry constraints while (12.37) ensure connectivity. Inequalities (12.38) force each vehicle to traverse the arcs that it serves. Constraints (12.39) ensure that required arcs are served while (12.40) establish that each profitable arc can be served at most once. The traveling time constraint is imposed through (12.41). Finally, (12.42)–(12.43) are variable definitions.

12.4.3.2 ▪ Solution approaches

The TOARP is the most recent arc routing problem with profits studied in the literature. An exact and a heuristic solution approach were proposed in Archetti et al. [6] and Archetti et al. [5], respectively. We describe them in this section.

In [6] a deep polyhedral study of formulation (12.35)–(12.43) was performed. In order to define the dimension of the polyhedron and to find facet-inducing inequalities, constraints (12.41) were relaxed. The authors proved that, with this relaxation, the inequalities

$$x_{ij}^k \geq 0, \quad (i,j) \in A \setminus (A_R \cup A'), k = 1,\ldots,m, \tag{12.44}$$

$$y_{ij}^k \geq 0, \quad (i,j) \in A_R \cup A', k = 1,\ldots,m, \tag{12.45}$$

$$x_{ij}^k \geq y_{ij}^k, \quad (i,j) \in A_R \cup A', k = 1,\ldots,m, \tag{12.46}$$

$$\sum_{k=1}^{m} y_{ij}^k \leq 1, \quad (i,j) \in A', \tag{12.47}$$

$$\sum_{i \in V \setminus \{S\}, j \in S} x_{ij}^k \geq y_{tz}^k, \quad S \subset V \setminus \{0\}, (t,z) \in A_R(S) \cup A'(S), k = 1,\ldots,m, \tag{12.48}$$

are facet-inducing for the TOARP. They also proved that the K-C inequalities and path-bridge inequalities, previously introduced for the RPP in Corberán and Sanchis [16] and for the general routing problem in Letchford [24], respectively, can be adapted to the TOARP. Then, new classes of valid inequalities were introduced which are related to the constraint on the maximum traveling time of each route.

A branch-and-cut approach was developed which is based on formulation (12.35)–(12.43) with the addition of the valid inequalities mentioned above. An initial lower bound, which helps to speed up the solution process, was provided and corresponds to the solution obtained through the algorithm proposed in Archetti et al. [5]. Computational tests were made on a large set of instances generated from benchmark instances for the RPP. The results show that the algorithm is able to solve instances with up to 100 vertices, 800 arcs, and four vehicles within one hour of computing time. Instances where the set A_R is large are simpler to solve.

In Archetti et al. [5] a matheuristic approach was proposed to solve the TOARP. The approach is based on the same scheme as the one of the VNS heuristic presented in Archetti et al. [7] for the solution of the UCARPP. Improvement procedures consisting in the exact solution of different MILP models are introduced to obtain better solutions. Computational tests were made on the same set of instances proposed in Archetti et al. [6]. The results show that the algorithm gives an average percentage error with respect to the optimal solution which is lower than 1%.

12.5 ▪ Summary

To clarify the state of the art on arc routing problems with profits and the several not-yet-studied problems, we now give a classified summary of the main ones surveyed in this chapter. The classification is shown in Table 12.1, where the name of each problem is reported. Then, the following columns refer to the objective function, multiple visits to each arc/edge, and constraints. An "X" indicates that the problem on the corresponding row has the characteristic on the corresponding column.

Table 12.1: Summary of arc routing problems with profits.

Problem	Objective function			Multiple	Constraints	
	Max profit	Max (profit–cost)	Min cost	visits	Traveling time	Capacity
MBCPP		X		X		
PCRPP		X				
AOP	X				X	
PATP		X		X	X	
UCARPP	X				X	X
TOARP	X				X	

In the node routing literature the problem where the traveling cost is minimized subject to a minimum profit to be collected has been studied as the prize-collecting TSP. From Table 12.1 one can notice that no paper deals with the minimization of the traveling cost.

The names proposed by the authors for the problems surveyed in this chapter often come from the inspiring applications and are disconnected from their node routing counterparts. Sometimes the same authors have used different names for the same problem. This is due to the fact that the literature in this area is not well established. As the literature on node routing problems with profits is wider and more established, in Table 12.2 we propose new names for the problems studied in this survey which reflect the terminology used in node routing with profits (rightmost column in Table 12.2).

Table 12.2: New terminology for arc routing problems with profits.

Names used in the literature	Names proposed
MBCPP	Profitable Rural Postman Problem with Multiple Visits
PCRPP	Profitable Rural Postman Problem
AOP	Arc Orienteering Problem
PATP	Profitable Arc Routing Problem
UCARPP	Undirected Capacitated Team Orienteering Arc Routing Problem
TOARP	Team Orienteering Arc Routing Problem

We hope that this can contribute to a better understanding of the problems studied and will encourage the study of the many that still remain to be tackled.

12.6 ▪ Conclusions

In this chapter we have reviewed the literature on arc routing problems with profits. The number of papers dealing with this class of problems is very limited. However, we believe that arc routing problems with profits will attract substantial attention from the research community in the near future, as it has already been the case for the node routing problems with profits. In fact, the potential applications of this class of problems cover a wide range of problems. Many problems have not been addressed at all, especially the ones where only the travel cost appears in the objective function while the profit is treated as a constraint. Finally, concerning exact solution approaches, all algorithms proposed till now in the literature are based on the branch-and-cut scheme. Since branch-and-price is the leading methodology for the exact solution of routing problems, a promising line of research could be to apply this methodology also to arc routing problems with profits.

Bibliography

[1] J. ARÁOZ, E. FERNÁNDEZ, AND C. FRANQUESA, *GRASP and path relinking for the clustered prize-collecting arc routing problem*, Journal of Heuristics, 19 (2013), pp. 343–371.

[2] J. ARÁOZ, E. FERNÁNDEZ, AND O. MEZA, *Solving the prize-collecting rural postman problem*, European Journal of Operational Research, 196 (2009), pp. 886–896.

[3] ——, *The clustered prize-collecting arc routing problem*, Transportation Science, 43 (2009), pp. 287–300.

[4] J. ARÁOZ, E. FERNÁNDEZ, AND C. ZOLTAN, *Privatized rural postman problem*, Computers & Operations Research, 33 (2006), pp. 3432–3449.

[5] C. ARCHETTI, Á. CORBERÁN, I. PLANA, J.M. SANCHIS, AND M.G. SPERANZA, *A matheuristic for the team orienteering arc routing problem*, Tech. Report WPDEM 2013/9, Department of Economics and Management, University of Brescia, Brescia, Italy, 2013.

[6] ——, *The team orienteering arc routing problem*, Transportation Science, 48 (2014), pp. 442–457.

[7] C. ARCHETTI, D. FEILLET, A. HERTZ, AND M.G. SPERANZA, *The undirected capacitated arc routing problem with profits*, Computers & Operations Research, 37 (2010), pp. 1860–1869.

[8] C. ARCHETTI, G. GUASTAROBA, AND M.G. SPERANZA, *An* ILP-*refined tabu search for the directed profitable rural postman problem*, Discrete Applied Mathematics, 163 (2014), pp. 3–16.

[9] C. ARCHETTI, M.G. SPERANZA, AND D. VIGO, *Vehicle routing problems with profits*, Tech. Report WPDEM 2013/3, Department of Economics and Management, University of Brescia, Brescia, Italy, 2013.

[10] E. BENAVENT, V. CAMPOS, Á. CORBERÁN, AND E. MOTA, *The capacitated Chinese postman problem: Lower bounds*, Networks, 22 (1992), pp. 669–690.

[11] D. BLACK, R. EGLESE, AND S. WØHLK, *The time-dependent prize-collecting arc routing problem*, Computers & Operations Research, 40 (2013), pp. 526–535.

[12] M. COLOMBI AND R. MANSINI, *New results for the directed profitable rural postman problem*, European Journal of Operational Research, 238 (2014), pp. 760–773.

[13] Á. CORBERÁN, E. FERNÁNDEZ, C. FRANQUESA, AND J.M. SANCHIS, *The windy clustered prize-collecting arc-routing problem*, Transportation Science, 45 (2011), pp. 317–334.

[14] Á. CORBERÁN, I. PLANA, A.M. RODRÍGUEZ-CHÍA, AND J.M. SANCHIS, *A branch-and-cut algorithm for the maximum benefit Chinese postman problem*, Mathematical Programming A, 141 (2013), pp. 21–48.

[15] Á. CORBERÁN AND C. PRINS, *Recent results on arc routing problems: An annotated bibliography*, Networks, 56 (2010), pp. 50–69.

[16] Á. CORBERÁN AND J.M. SANCHIS, *A polyhedral approach to the rural postman problem*, European Journal of Operational Research, 79 (1994), pp. 95–114.

[17] T. CURA, *An artificial bee colony approach for the undirected capacitated arc routing problem with profits*, International Journal of Operational Research, 17 (2013), pp. 483–508.

[18] R. DEITCH AND S.P. LADANY, *A heuristic improvement process algorithm for the touring problem*, SCIMA, 23 (1994), pp. 61–73.

[19] ——, *The one-period bus touring problem: Solved by an effective heuristic for the orienteering tour problem and improvement algorithm*, European Journal of Operational Research, 127 (2000), pp. 69–77.

[20] J. EUCHI AND H. CHABCHOUB, *Hybrid metaheuristics for the profitable arc tour problem*, The Journal of the Operational Research Society, 62 (2011), pp. 2013–2022.

[21] D. FEILLET, P. DEJAX, AND M. GENDREAU, *Traveling salesman problems with profits*, Transportation Science, 39 (2005), pp. 188–205.

[22] ——, *The profitable arc tour problem: Solution with a branch-and-price algorithm*, Transportation Science, 39 (2005), pp. 539–552.

[23] G. GUASTAROBA, R. MANSINI, AND M.G. SPERANZA, *Modeling the pre-auction stage: The truckload case*, in Innovations in Distribution Logistics, Lecture Notes in Economics and Mathematical Systems Series 619, L. Bertazzi, M.G. Speranza, and J. Van Nunen, eds., Springer-Verlag, Berlin, 2009, pp. 219–233.

[24] A.N. LETCHFORD, *New inequalities for the general routing problem*, European Journal of Operational Research, 96 (1997), pp. 317–322.

[25] C. MALANDRAKI AND M.S. DASKIN, *The maximum benefit Chinese postman problem and the maximum benefit traveling salesman problem*, European Journal of Operational Research, 65 (1993), pp. 218–234.

[26] W.L. PEARN AND W.C. CHIU, *Approximate solutions for the maximum benefit Chinese postman problem*, International Journal of Systems Science, 36 (2005), pp. 815–822.

[27] W.L. PEARN AND K.H. WANG, *On the maximum benefit Chinese postman problem*, OMEGA, 31 (2003), pp. 269–273.

[28] S.E. SCHAEFFER, R.Z. RÍOS-MERCADO, AND E. FERNÁNDEZ, *The windy prize-collecting rural postman problem: An ant-colony based heuristic*, Journal of Heuristics. To appear.

[29] W. SOUFFRIAU, P. VANSTEENWEGEN, G. VANDEN BERGHE, AND D. VAN OUDHEUSDEN, *The planning of cycle trips in the province of East Flanders*, OMEGA, 39 (2011), pp. 209–213.

[30] T. TSILIGIRIDES, *Heuristic methods applied to orienteering*, The Journal of the Operational Research Society, 35 (1984), pp. 797–809.

[31] P. VANSTEENWEGEN, W. SOUFFRIAU, AND D. VAN OUDHEUSDEN, *The orienteering problem: A survey*, European Journal of Operational Research, 209 (2011), pp. 1–10.

[32] E.E. ZACHARIADIS AND C.T. KIRANOUDIS, *Local search for the undirected capacitated arc routing problem with profits*, European Journal of Operational Research, 210 (2011), pp. 358–367.

Part III

Applications

Chapter 13

Route Optimization for Meter Reading and Salt Spreading

Richard Eglese
Bruce Golden
Edward Wasil

13.1 ▪ Introduction

In this chapter, we focus on the application of arc routing models to problems in meter reading and salt spreading for winter maintenance of roads.

On a daily basis, utility companies send out employees to read the electric, gas, and water meters of thousands of customers. The goal is to develop efficient and balanced meter-reading routes. We start by describing an early modeling effort of the late 1970s to generate high-quality routes for meter readers in a small residential neighborhood. Next, we present computerized models developed in the 1990s and then move to automatic meter reading of the late 2000s. The current practice of meter reading in the utility industry using sophisticated software is highlighted.

Historically, one of the most important applications of arc routing deals with winter maintenance of roads. In particular, when there is a danger of ice forming, vehicles are dispatched to apply an anti-icing agent such as rock salt to road surfaces. There are many factors, such as the timing of the salt spreading operation and the amount of salt that is spread on the roads, that need to be taken into account when developing arc routing models for salt spreading. We describe several models that use exact and heuristic methods to solve these types of problems. We present a detailed discussion of several applications that highlight modeling challenges and solution approaches.

We point out that the chapter by Assad and Golden [1] provides detailed background on arc routing methods and applications before 1995. The authors describe routing for meter reading and snow and ice control.

13.2 ▪ Meter reading

13.2.1 ▪ Introduction

Utility companies need to read the electric, gas, and water meters of their residential and commercial customers on a regular basis. Typically, a company sends out readers each day to individually read the meters of customers. Generating the routes of the meter readers is an arc routing problem that has received the attention of researchers and practitioners over the last 40 years. In the next five sections, we summarize key papers and their contributions from the late 1970s to the late 2000s, and describe the current practice of meter reading in the utility industry.

13.2.2 ▪ Early modeling efforts: Late 1970s

Early efforts focused on developing algorithms to find Euler cycles for small Euclidean problems with less than 50 nodes and 100 edges. These algorithms were effective at producing new solutions with fewer routes than in existing solutions.

In the city of Beersheva, meter readers from the Israeli Electric Company (IEC) are transported from the central office to each of the city's eight neighborhoods, read the meters of customers on a daily work list within a five-hour time limit (readers are walking and deadheading is permitted), and return home or to the central office. The street network is undirected, and a narrow street can be covered in a zig-zag (meander) manner. A U-turn can occur only at an intersection. In a 1979 paper, Stern and Dror [27] develop an algorithm based on a route-first cluster-second approach. First, a giant Euler cycle that covers all of the required edges (streets) in the network is generated. Second, the Euler cycle is partitioned into working routes that satisfy the five-hour time limit. They apply their algorithm to a Beersheva neighborhood with 42 nodes (intersections) and 62 edges. IEC uses 24 routes to read the meters in this neighborhood. The solution from the Stern and Dror algorithm needs only 15 routes—a savings of nearly 38%.

13.2.3 ▪ Computerized routing: 1990s

In the 1990s, geographic data became available in the form of GBF/DIME files (Geographic Base File/Dual Independent Map Encoding). There were modeling challenges because street data were not always compatible with the geographic data. Optimization algorithms, graphics, and interactive features became available for the first time in meter-reader software systems.

In the late 1980s, Southern California Gas (SOCAL) spends more than $15 million annually to read each of the 10,042 routes once a month by over 475 meter readers. SOCAL initiates a study to test computerized routing using software from DISTINCT Management Consultants (now known as RouteSmart Technologies [25]). Wunderlich et al. [32] describe the modeling effort for SOCAL. The study is motivated by three factors: (1) a four-year backlog in revising routes with lots of overtime; (2) slow development of new routes in growing service areas; and (3) union standards that call for less time to cover a street. SOCAL tests the software on a sample region with hills and level areas that is 2.5% of the total service area. The sample region has 242.5 meter-reader routes with 6,900 street segments (called centerlines, which are both sides of a street between two adjacent intersections) that are covered by walking, driving, and a combination of the two modes. An arc-partitioning algorithm (Bodin and Levy [2]) is used to cluster street segments into balanced routes (one day's work for a meter reader). The solution from the computerized routing system produces 236 routes for the sample region (6.5 fewer routes than the

SOCAL routes) with better balance and significant reductions in overtime and deadhead walking and driving times. Based on these results, more than $874,000 per year could be saved by adopting computerized routing for the entire service area. SOCAL takes delivery of a meter-reader software system from DISTINCT in 1991.

Lee and Welander [13] describe a computerized routing effort for The Metering Section of Seattle Public Utilities (SPU). SPU uses walking and driving routes to read over 174,000 water meters. On a given day, meter readers cover 15 to 22 routes. There is workload imbalance, with some routes taking less than eight hours and others taking more. There is new growth in parts of the city that require certain routes to pick up the new meters, thereby adding to the imbalance. SPU conducts a pilot project on 16,098 accounts with a variety of field conditions (e.g., hilly topography) to study the rerouting of the meter readers and decides in the fall of 1997 to proceed with the rerouting. The authors provide a detailed description of the project including the data conversion effort, preparation of the routing dataset, and the actual routing process conducted with software from RouteSmart Technologies. New routes are more balanced than old routes (i.e., number of meters is more evenly distributed on the new routes), and more meters are assigned to routes within an eight-hour workday. New routes are 16% to 46% more efficient than old routes (as measured by the number of customer accounts on the new routes).

Lee and Welander [14] provide an update on the project. Rerouting of northern Seattle is completed in June 1998. Rerouting of the southern part of the city is on hold due to several concerns about the new northern routes raised by the meter readers, including unfamiliarity with the geographical location of some routes and meter locations. In order to proceed with the project, the authors make several findings (e.g., need to incorporate the preferences of the meter readers into the routes) and recommendations (e.g., preview routes before they are implemented). They point out that technology (computerized routing) is only one piece of the rerouting project. To move the project forward requires ownership from all of the stakeholders (e.g., meter readers, analysts, management).

13.2.4 • Computerized routing: Early 2000s

In the early 2000s, a geographic information system (GIS) was combined with powerful routing algorithms (heuristics) to form a highly visual computerized system. This system could generate near-optimal routes for large service areas with thousands of customers (meters).

Utility companies have hundreds of thousands of meter reads in a month. Generating highly efficient routes for meter readers is a task that is complicated by a number of factors including balancing workload among routes, meter-reading modes (walking, driving, combination of walking and driving), high density or low density of meters in a small geographic area, amount of read time, meander (zig-zag) or nonmeander service, and natural boundaries such as highways, rivers, and lakes. In a 2002 paper, Levy, Sniezek, and Cox [16] describe how a GIS can address each of the above factors. For example, in a GIS, several layers (themes) of data such as water, major highways, and railroad tracks can be displayed in a service area. The displayed layers can then be used to select a subset of meters within the area for route planning purposes. Most importantly, algorithms (heuristics) that generate near-optimal routes can be displayed in a GIS. Combining powerful algorithms with the visual power of a GIS creates a route optimization system that allows planners to generate high-quality solutions and refine them very quickly.

13.2.5 ▪ Automatic meter reading: Late 2000s

With the widespread availability of radio frequency identification (RFID) technology in the late 2000s, an RFID tag can be attached to a meter. A truck equipped with a reading device traverses streets in the service area and collects data from the meters automatically. It is not necessary to visit each meter individually because a meter can be read from a predefined distance. In the automated meter reading (AMR) problem, the utility seeks vehicle routes that cover all meters (customers) in the service area and minimize the total distance traveled. This problem is similar to the traveling salesman problem (TSP) except that a tour does not need to visit each meter (node). Dong, Yang, and Chen [5] call this the automatic meter reading shortest tour problem (AMRSTP). They formulate a mixed integer, nonlinear program with Euclidean distances for the AMRSTP which is difficult to solve to optimality even for small- to moderate-size instances. They develop a clustering algorithm and a convex hull algorithm for generating solutions to the AMRSTP. The authors convert the AMRSTP to a TSP by introducing supernodes that can cover all customers within a certain radius. Both algorithms are tested on 19 instances with 100 to 1,000 customers in increments of 50 customers with various radii for the supernodes. Each instance is solved five times, and the total tour length is averaged. The performance of the algorithms depends on the radius and the number and distribution of customers. For example, when the radius is fixed at 10, the convex hull algorithm produces the shortest average tour length on all instances from 100 to 1,000 customers with a running time of less than one second.

With RFID technology, meter readers need to be within a radius r (generally 500 to 1,000 feet) to read meters. They no longer have to traverse every street in a neighborhood, but simply need to get close enough to read a meter. The use of RFID tags transforms a TSP over a street network to a close enough TSP (CETSP), that is, find a path that begins and ends at the depot and minimizes the distance traveled while passing within r feet of each customer (meter). Shuttleworth et al. [26] develop a two-stage process to solve the CETSP for a single meter-reader route. In the first stage, two heuristics (weighted bang for buck (WB4B), distance weighted bang for buck (DWB4B)) and two integer programs (IP1, IP2) specify a subset of street segments that are traversed by a meter reader using the radius r. In the second stage, a travel path (cycle) is generated that traverses the specified street segments. The four solution procedures are tested on a real-world data set (street network) with 150,000 customers and 16,500 street segments that are partitioned into 18 mobile AMR meter-reader routes. The partitions are created by software from RouteSmart Technologies. A computational experiment is carried out on a dense partition with 10,230 customers and 1,099 street segments (this corresponds to a single route with 97.5 miles of segments and a time duration of 9 hours and 22 minutes given by the RouteSmart sequence solver). With a 500-foot radius, the WB4B and IP2 solutions use only 470 street segments each, with a length of 64.4 miles (7 hours and 6 minutes) and 59.1 miles (7 hours and 15 minutes), respectively. These two solutions save more than 22% in time duration when compared to the single route solution. The authors also generate solutions for a sparse partition and for a radius of 350 feet. Over all 18 partitions with a 500-foot radius, WB4B, IP1, and IP2 are the best performers. They save about 15% in time and 20% in distance over the solution that traverses all 16,500 street segments.

Finally, we point out that although they do not use arc routing models or methods, the papers by Gavirneni, Clark, and Pataki [9] and Groër, Golden, and Wasil [10] describe interesting applications in meter reading. In the first paper, the authors develop integer programming models to determine the optimal locations of receivers in service areas. In this version of AMR, the meters transmit their data to receivers mounted on utility poles.

In the second paper, the authors address the problem (called the Balanced Billing Cycle Vehicle Routing Problem (BBCVRP)) that occurs when, over time, routes become inefficient and fractured with imbalanced workloads for the meter readers. This happens because a utility has not revised routes when new accounts are opened in a service area or old accounts are closed. A three-stage algorithm for solving the BBCVRP uses heuristics and integer programming to create meter-reading routes for each billing day and to assign customers to these routes. The goals are to reduce the length of meter-reading routes and to balance the workload on these routes. Computational results on four actual data sets with 3,000 to 4,000 customers show that the three-stage algorithm produces efficient and balanced routes in about one hour of computing time.

13.2.6 • The practice of computerized meter reading and related arc routing in the utility industry

RouteSmart Technologies has been a leading provider of route optimization solutions for the utility industry with significant experience working with both municipal and investor-owned utilities. In particular, over the last two decades or so, RouteSmart Technologies has helped clients develop efficient routes for meter readers. For example, the Water Bureau of the City of Portland, Oregon, used RouteSmart's computerized system to re-route 182,000 meters and achieve a 36% increase in productivity (roughly speaking, the number of meters read per day increased by 36%). RouteSmart algorithms have been created by a team of software developers with advanced degrees (mainly Ph.D.'s) in operations research. These algorithms involve a combination of clever heuristics and optimal solutions to subproblems.

The merger of Indianapolis Water with Citizens Energy Group in 2011 meant that the two companies would have to combine water and gas meter-reading routes for 600,000 customers. The full route design was carried out using the RouteSmart product. There were two major merger-related challenges. First, the gas and water meter-reading routes were separate and the associated billing days were independent. The goal was to transition smoothly from separate to integrated meter-reading routes. So, the billing days for each customer (meter) had to be reconciled. The merged utility tried to maintain the gas billing days for most customers, while changing water billing days by no more than two or three days per month, using a RouteSmart tool, in order to balance the workload over the 22 or so billing days per month. Next, RouteSmart software was applied to generate efficient daily meter-reading routes.

Most gas and electric utilities conduct natural gas leak surveys along the gas distribution system within their area of service. These surveys are mandated with a specified frequency by federal and state authorities. Typically, this involves having personnel from the utilities (carrying detection equipment) walk the natural gas pipeline routes up to the outlet of a customer's gas meter. If a leak is detected, its precise location is determined. Since natural gas is flammable, leaks must be repaired as quickly as possible.

The Colorado Springs Utilities is a four-service utility company (gas, electric, water, and wastewater). There are 190,000 gas customers. The gas system spans 2,350 miles of main lines and 2,650 miles of service lines. Certain gas services (called protected) must currently be inspected every five years; others (called unprotected) must be visited every three years. The Colorado Springs Utilities decided to use RouteSmart software to route its inspectors; the process is underway. They estimate a savings of 2,000 hours in routing time and a total cost savings of $160,000. In the future, a goal is to combine the inspection of protected and unprotected services.

An application of AMR routing in southern California was recently demonstrated,

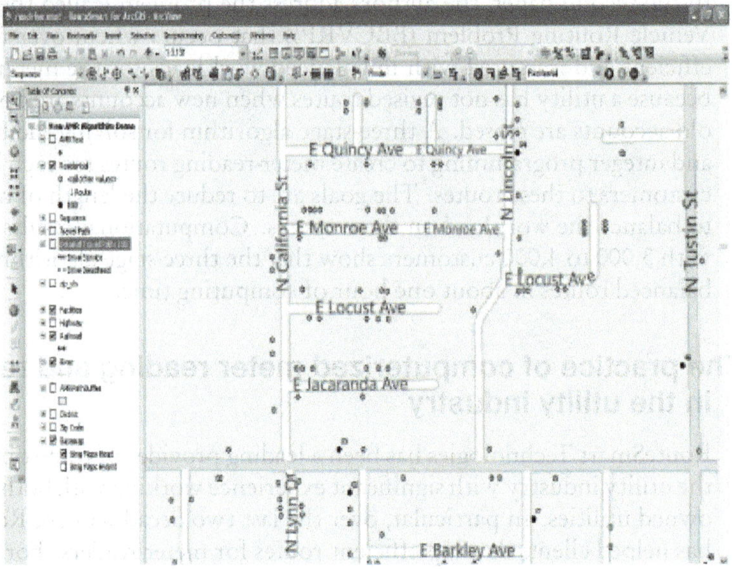

Figure 13.1. *A neighborhood on a route.*

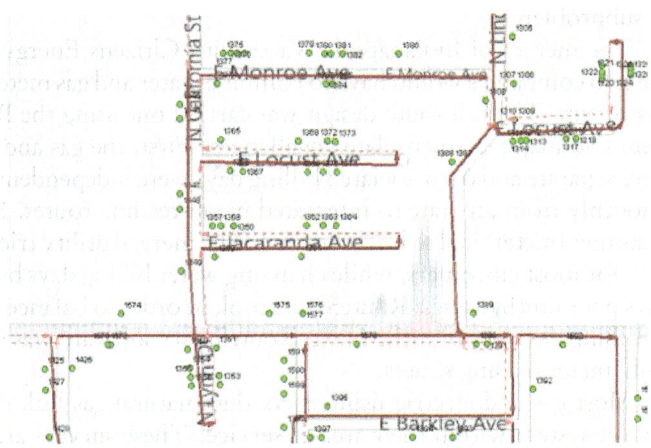

Figure 13.2. *A traditional route through a neighborhood.*

again using proprietary RouteSmart software. Given the effective read radius associated with the AMR (RFID) technology and the geography and curvature of the street network, the goal is to select a subset of streets to traverse in order to minimize total distance and deadhead mileage, while reaching all meters.

RouteSmart's AMR routing algorithm imagines a buffer (or bubble) around each street segment with a width equal to the effective read radius (also called the read range buffer). We illustrate the impact of AMR routing by focusing on a typical (small) neighborhood from southern California in which all meters must be reached or covered by the RFID

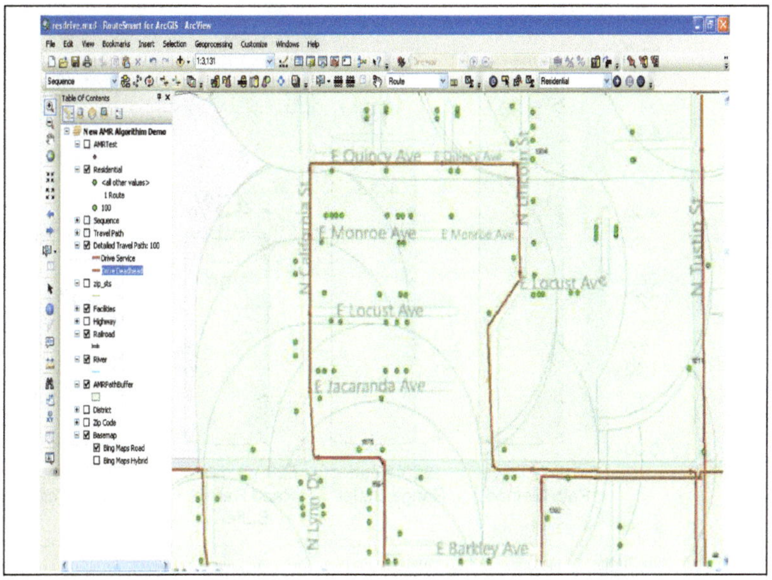

Figure 13.3. *AMR route through the same neighborhood.*

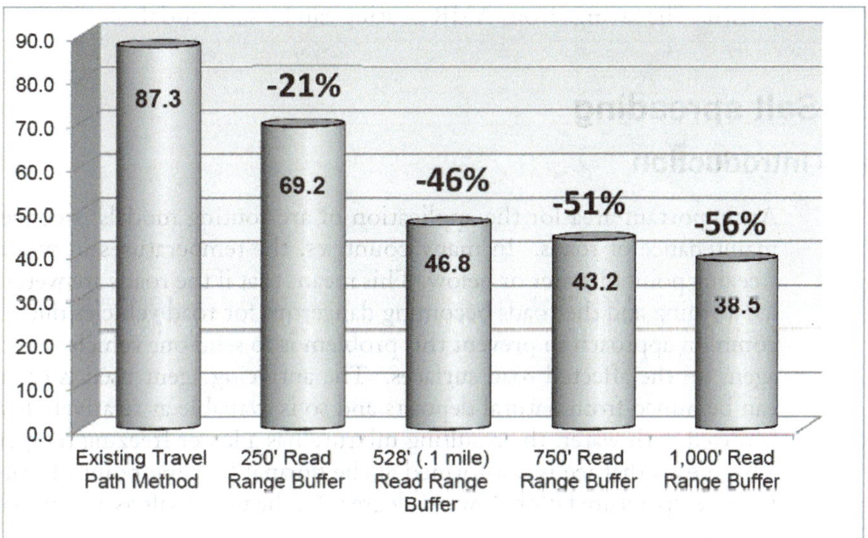

Figure 13.4. *New algorithm's impact on route miles.*

technology. In Figure 13.1, we see the streets and customer locations in this neighborhood.

In Figure 13.2, we see a traditional meter-reading route (with very limited RFID technology) in which every street and every meter must be visited. In Figure 13.3, we see a much shorter AMR route (along with the buffers and bubbles mentioned earlier, showing that every meter is covered by the bubbles from the streets on the route). In Figure 13.4, the savings in total distance as a function of effective read radius are displayed. Finally, in

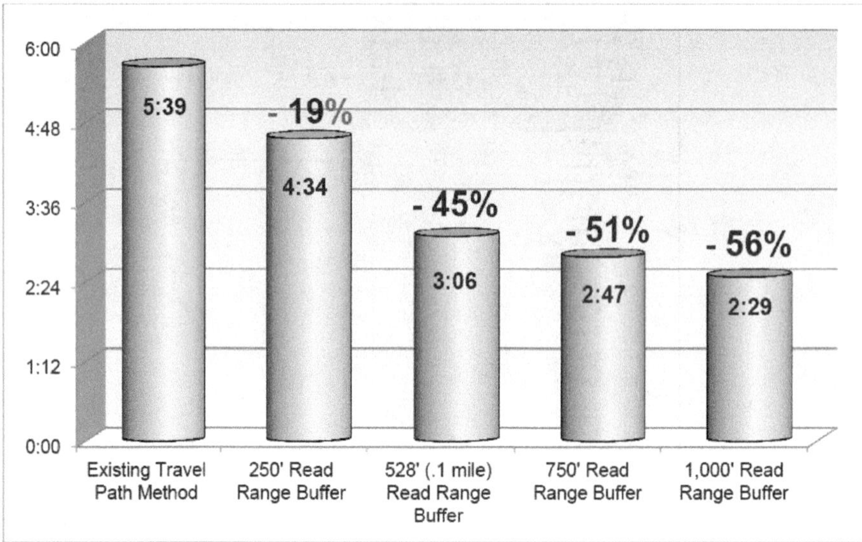

Figure 13.5. *New algorithm's impact on route time (in hours).*

Figure 13.5, the analogous savings in travel time as a function of effective read radius are shown. The savings from AMR routing can be substantial.

13.3 ▪ Salt spreading

13.3.1 ▪ Introduction

An important area for the application of arc routing models is concerned with winter maintenance of roads. In many countries, the temperatures in winter may fall to the freezing point of water or below. This means that if the roads are wet, there is a danger of ice forming and the roads becoming dangerous for road vehicles due to a lack of grip. A common approach to prevent this problem is to send out vehicles to spread an anti-icing agent on the affected road surfaces. The anti-icing agent used is often rock salt as this can be mined from natural deposits and so is available at relatively low cost. When salt is mixed with water, the resulting mixture has a lower freezing temperature than water. This means that ice is not formed at the normal freezing point of water of 0 degrees C. If the temperature falls below —5 degrees C, the use of salt as an anti-icing agent becomes less effective.

Other anti-icing agents can be used, such as calcium magnesium acetate. Alternative materials may be more or less cost effective compared to salt, but this will depend on local availability and conditions. When being used to prevent ice forming, there may be advantages in wetting the salt so that it sticks better to the road surface and will accelerate the dissolving of the salt particles. It is also possible to treat the salt with molasses, an agricultural by-product from sugar beet. The effect of this treatment is to help the salt adhere to the road surface and so enables the same anti-icing effect to be achieved with lower rates of spread.

If hard packed snow or ice has already formed on a road, then salt may also be used as a de-icing agent to help to remove this layer which makes road travel more treacherous. In some places grit, sand, or other abrasive material may also be spread on the road surface to help vehicles maintain a good grip. This practice accounts for the use of the term "winter gritting," which is used in some places to describe these operations.

The timing of the salt spreading operation is important. If a road is treated too early, then the salt may be washed away if rain is falling or blown away by the wind before the temperature drops to freezing conditions. If the road is treated too late, then ice may have already started to form and the road will be dangerous for traffic traveling on the road before it has been treated. So there are often time windows within which roads must be treated.

The amount of salt spread is also important. If too little is applied to the road for the prevailing conditions, then the resulting mixture will not have a sufficient concentration of salt to prevent freezing. On the other hand, if more salt is spread than is necessary to prevent freezing, then this will waste resources and money. Also, the salt will pollute the environment, causing damage to roadside plants and trees, and can cause an increased rate of corrosion to the metal parts of vehicles if not washed off.

In some places, conditions may require snow plowing and salt distribution to be carried out at the same time. A vehicle will be used to sweep snow off the highway and at the same time to spread salt on the swept road. In these cases, there will be a strong link with the models required for snow removal which are described in Chapter 14 of this book. However, in countries such as the United Kingdom, the climate is such that it is much more common for salt distribution to be required on its own on frosty winter mornings rather than in combined plowing and salting operations.

The vehicles used to distribute the salt on roads are normally specially designed for this operation. The salt is loaded into the back of the vehicle where it is fed to the rear by a mechanical conveyor system. The rate of spread may be controlled by the driver from the cab. The salt falls on a revolving plate which throws the salt on to the road behind the vehicle. The angle of the plate may be adjusted so that the salt from one vehicle may be able to treat both sides of the road in one pass or may treat more than one lane of a multilane highway. The speed of the vehicle will also affect the rate of spread, and vehicles will normally travel more slowly when treating roads than when they are deadheading, which means traveling along a road without treating it.

Salt distribution is an operation where it is very natural to model it using an arc routing model. In some other applications, such as postal delivery, for example, an arc routing model may be used as an alternative to a node routing model where a very large number of individual customer locations must be included. However, in salt distribution, the practical requirement is for the whole length of each required road section to be treated and so naturally fits with an arc routing modeling approach.

13.3.2 ▪ The arc routing model

The problems addressed for spreading operations are varied. In this section, there is a discussion of the issues which may be taken into account in a modeling approach and which may affect the details of the formulation of the problem and the solution approach that is used.

Various surveys have examined vehicle routing models for salt spreading–type operations which attempt to categorize the issues and problem characteristics in different ways. Perrier, Langevin, and Campbell [24] and Perrier et al. [22] provide useful sources.

The first area concerns the details of the road network over which the salt spreading operation is to be carried out. This may be available in the form of a digitized road map of some type, but it is important that relevant features of the road network are correctly recorded. The data required for each road segment, to be represented by an arc in the model, include the nodes representing the ends of the arc, the length of the arc, whether it is one way or not, and whether it is required for treatment in one of the salting categories. In addition, depending on the details of the application, other information may be needed, such as the number of lanes and factors affecting the speed of vehicles treating the road, such as the speed limit and the incline of the road. There may also be the need to indicate the number of passes required for the road to be treated (which may depend on the width of the road) or the frequency of treatments. For each node in the graph representing the road network, as well as recording the arcs that meet at that node, information may also be needed on restricted turns.

Normally detailed representations of road networks will be required such as those used in satellite navigation systems. Oversimplified representations of junctions, including one-way ramps, may result in key sections of the network being untreated or a significant underestimate of the deadheading required to treat all the required roads. For example, consider the junction illustrated in Figure 13.6. This represents a major highway in the U.K. running horizontally in the diagram with traffic keeping to the left-hand side. Four one-way slip roads or ramps allow traffic to exit or enter the highway at the junction. Two bridges carry traffic over the highway as part of the junction design allowing drivers to choose their route. If the highway and ramps require treatment with salt, a gritter approaching the junction from A must either treat the one-way ramp labeled B or continue treating the highway. If the gritter treats ramp B, it can salt other roads off the highway and then rejoin the highway by an entry ramp, but the section of highway between the exit and entry ramps labeled C will be untreated. If the gritter continues on the highway to treat the section labeled C, then the ramp labeled B is left untreated. The only way for both sections B and C to be treated is for another vehicle to approach the junction from A and treat the section of road left by the first gritter. If it is a long way from the previous junction that allows the second gritter to pass along the highway and approach the junction illustrated from A, then significant deadheading is required by the second gritter. This would not be recognized if the detail of the road layout is not captured in the network representation used in the routing model.

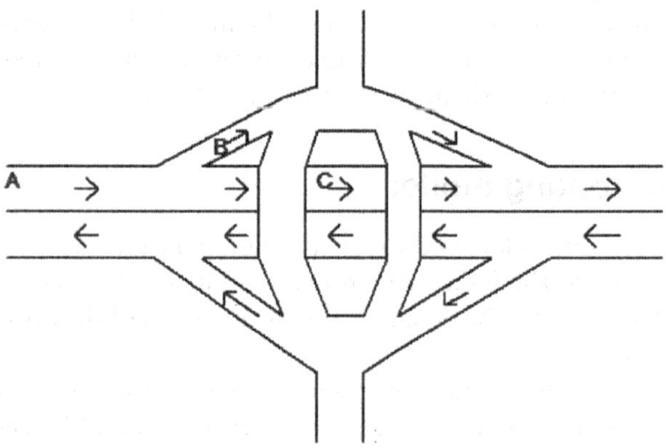

Figure 13.6. *A major road junction.*

As mentioned in the previous paragraph, the roads in a particular geographical area are normally categorized according to their requirements in terms of a spreading operation. Some roads may receive a top priority which requires them to be treated as soon as possible once the operation starts or by a particular time. Other roads may not be required for treatment, either because they are minor roads not normally included in such operations, or because other agencies are responsible for treating the road. Such roads may still be included in the graph representing the road network because it may be useful for the vehicles carrying out the spreading operation to deadhead over them. There may also be other categories for roads in connection with the salt spreading operation. For example, roads may be divided into categories referred to as deadline classes. Each category corresponds to a time limit by which the roads in that category must be treated once the operation has started. The most important roads are in categories with relatively low time limits. Such a system differs from a priority system where all roads in one priority category must be treated before any roads in a lower priority category. Using deadline classes, a better set of routes may be produced if a road in one deadline class is treated before another road in a deadline class with a lower time limit. This may occur when the road in the category with the higher time limit is one that must be used to link subsets of roads with lower time limits and returning to treat the linking road is more expensive than treating it when a gritting vehicle first travels along it.

In static models, the salt spreading categories are determined in advance and are unchanging. These may be determined by managers responsible for the condition of the roads at the beginning of a winter season and used whenever weather conditions require road treatment. In some places, thermal mapping of roads may be used to inform the appropriate categories for treatment, e.g., see Handa, Chapman, and Yao [11]. In dynamic models, the road categories could change according to changes in the weather conditions and may be informed by a system that monitors road surface temperatures.

Information will also be required on the vehicles that carry out the spreading operation. For each vehicle, details are needed of its capacity, its rate of spread when treating a road (which may depend on the road and the weather conditions), and its speed when treating a road and deadheading (which may depend on the road and the weather conditions). Vehicles may be required to start and finish at one or more depots, and there may be additional time constraints due to the availability of drivers and legal restrictions on their driving times.

The locations of salt supplies (or whatever is being spread on the roads) need to be specified. In many cases, these will correspond to the starting depots of the vehicles, but in some cases additional supplies may be available at other places which could be beneficial for vehicles to use when reloading during a long route. In many cases, these locations are fixed in advance, but some applications could include the locations of depots or refilling points as additional decision variables to be determined by the model.

Other characteristics of the problem that may be required in specifying and formulating a problem include any requirements for compactness or for routes to be nonoverlapping and times required for a vehicle to load or reload.

The decisions needed in any arc routing application for spreading will specify, for each vehicle, its route and whether it treats or deadheads any road over which it travels. Some models may also aim to combine the routing decisions with other decisions, such as the number and type of vehicles to be used or the number and location of the depots used as bases for the vehicles or points at which the vehicles can replenish their supplies.

The overall objective of an arc routing model in this application area is normally to minimize the total relevant variable costs subject to the types of constraints that have been described. The total cost may be approximately proportional to the total distance traveled, but if the number of vehicles and drivers used in the operation may vary, then there may be additional financial benefits from reducing the number of vehicles used, even if this results in a greater total distance traveled.

13.3.3 ▪ Solution methods

13.3.3.1 ▪ Exact solution methods

In general, there is little evidence of exact solution methods being used in real-life cases. In practice, the size of the problems and the number and complexity of the constraints that are important make it difficult to formulate exact models that can be solved in a reasonable amount of computing time. However, there are some models where exact approaches have been designed which are particularly relevant to the types of model and constraints found in spreading applications. One example is provided in Letchford and Eglese [15], where it is shown how a rural postman problem where the required arcs are categorized into deadline classes can be solved effectively by using a cutting-plane approach for an integer programming formulation. Another example is provided by Tagmouti, Gendreau, and Potvin [28]. In this application, the cost to treat a road depends on the time of service. This is relevant to spreading operations where weather reports indicate bad weather passing across a region and where each road has a best time for treatment with penalties if it is treated earlier or later than a preferred time window. The graph representing the road network is directed and so can be transformed to an equivalent node routing problem without increasing the size of the problem. The formulation results in a nonlinear, mixed integer program with capacity constraints and time-dependent treatment costs. The solution approach uses column generation.

13.3.3.2 ▪ Heuristic solution methods

When results are required for a real-life case, solution methods tend to rely on heuristic approaches. Early work has used relatively simple constructive heuristics, but as various metaheuristic algorithms have been developed, these have been used for salt spreading applications. Examples of different styles of heuristic approaches can be found in the literature and include simulated annealing, tabu search, genetic algorithms, and other approaches. Rather than discuss the heuristics separately here, the solution methods will be mentioned when describing particular application papers in the next section.

13.3.4 ▪ Applications

Sources that provide surveys of applications of arc routing models to spreading applications include Campbell and Langevin [4], Perrier, Langevin, and Campbell [23], Perrier, Langevin, and Campbell [24] and Perrier et al. [22] .

In this section, examples of some of these applications will be discussed to emphasize special features of this area of application and how various authors have responded to solving the challenging problems that are posed.

Eglese and Li [7] contrast a normative approach to a prescriptive approach in managing a winter gritting operation. The normative approach surveyed a sample of 31 areas in the UK and, for each area, measured the routing efficiency defined by road distance treated with salt divided by the total distance of the planned routes. Observed routing efficiencies ranged from 56% to 80%. The approach then suggested that routing efficiencies of less than 70% indicated areas where there was the possibility for improvement in the planned routes. However, the paper gives examples where a routing efficiency of 70% may not be achievable due to the large average distance between junctions and the topology of the road network. In particular, it demonstrates that road networks forming a square grid pattern may be much more efficient for salt spreading routes than a road network containing a high proportion of T-junctions where three road segments meet.

In Eglese [6], a heuristic algorithm is described for solving an arc routing problem based on a project carried out for a local authority in the UK. The project was part of a larger study considering the number and locations of depot sites needed in the area. The model considered multiple depot sites, enabling routes and costs to be determined for different scenarios in terms of the number and location of the depots. The total cost to be minimized for any scenario depended on the number of gritting vehicles required and the total distance traveled. Constraints included the maximum amount of salt carried by a vehicle, which related to the maximum length of road that could be treated and time constraints by which different categories of road should be treated. The solution approach builds on a method proposed by Male and Liebman [19] for waste collection. It consists of initially solving the unconstrained Chinese postman problem for the set of roads requiring treatment to find the additional deadheading arcs required. These are then used in creating a graph known as a cyclenode network where each node represents a cycle or circuit in the original road network and the edges of the cyclenode network link nodes where the corresponding cycles have at least one common road junction on the original road network. Depot nodes are then added with edges indicating on which cycles the depots are located. A tree rooted at a depot in the cyclenode network represents a potential vehicle route in the road network. Following a greedy heuristic to produce an initial solution, local search can be used to explore other solutions, using neighborhood moves such as removing and adding nodes from the trees in the cyclenode graph. Simulated annealing is used to guide the search to find a low cost solution. The method was able to find solutions where the number of routes needed was less than or equal to the number used in practice. However, the most significant conclusion from the study was that the number of depots required in the local authority area could be more than halved without increasing the number of gritting vehicles needed.

The approach described in Eglese [6] was appropriate for a strategic study to determine the number and location of depots. Restricting the routes considered to those made up of cycles created in the initial stage of the solution method reduced the size of the problem and gave rise to a simple neighborhood structure for the local search algorithm. However, there are disadvantages when the method is used for detailed operational routing. The approach treats all roads as edges, and this could lead to infeasible solutions if a significant number of one-way roads are present in the network. Also, there were examples where improved routes could be found that were not made up of the cycles formed in the first stage. In Li and Eglese [17], an alternative heuristic approach is described which does not have these limitations and can be used for detailed route planning. A greedy-style algorithm is proposed to produce an initial set of routes where detailed constraints on one-way roads, capacity restrictions, and time constraints are all observed. A computer program displayed the routes on a screen and then provided support for a user to modify the routes to search for improvements while satisfying the constraints. The approach enabled good sets of routes to be produced in many different situations.

Although the visual interactive approach to finding routes was very flexible, it required an experienced user and could take significant time to explore different possibilities when looking for a good set of routes. Eglese and Li [8] describe a heuristic for detailed vehicle routing suitable for applications like salt spreading where vehicle capacity and time deadline constraints must be satisfied on a mixed graph representing a road network. The heuristic algorithm proposed is called LOCSAR (LOCal Search for Arc Routing) and uses a tabu search metaheuristic. LOCSAR is able to produce improved solutions compared to the results of previous methods without the need for extensive experience and time from the user.

A practical case involving the combination of depot siting and spreader routing for the province of Antwerp, Belgium, is described in Lotan et al. [18]. The formulation of the problem takes into account the priority classes for the roads to be treated, capacity constraints for the vehicles, and time constraints for completing the treatment of the roads with salt. A first stage iteratively solves a location-routing problem on the high-priority roads which have a tree structure rooted around the ring of Antwerp in order to determine the best locations for the depots. The second stage then incorporates low-priority roads by assigning them to their closest depot, while taking into account the need for the road segments assigned to each depot to be Eulerian in order to minimize additional deadheading. The final stage constructs feasible routes and locates places where salt may be stored and vehicles refilled. This method produced improvements over the solution being used at the time. It provided useful information on the trade-offs in treating two lanes in one pass when compared to making two passes and the benefits of good locations for the refill points.

The study described in Muyldermans et al. [21] focuses on a method to design appropriate sectors for salt spreading operations. The sectors should be contiguous, balanced, and compact. Locations of the depots are given, and the objective is to design the sectors to minimize the number of vehicles required for the spreading operation. The heuristic follows a series of phases, the first of which is similar to the approach in Eglese [6] based on the method of creating cycles proposed by Male and Liebman [19] for waste collection. Cycles adjacent to depots are initially assigned to those depots, and then adjacent cycles are added in a systematic way to create an initial partial assignment of cycles to depots. In the next phase, the remaining cycles are assigned using criteria relating to sector balance, compactness, and fleet size. In the final phase, interchanges of cycles are allowed between sectors assigned to different depots to decrease the number of vehicles required, while satisfying additional constraints. A user-interactive approach can be used here to incorporate special features that are hard to quantify. The approach was used for a network from the province of Antwerp. The routes produced based on the new sectors gave cost savings of about 14% for deadheading compared to the same routing approach on the sectors in the area.

Sector design is the focus of Muyldermans, Cattrysse, and Van Oudheusden [20]. Three heuristic methods are proposed and compared for designing sectors that are suitable for operations such as salt spreading. The first heuristic builds individual edges into sectors while the second heuristic uses the cycles generated as in Muyldermans et al. [21] as building blocks for the sectors. The third heuristic uses a mixed integer linear program to assign the cycles into sectors. The three heuristics were tested on road networks from the province of Antwerp. The authors concluded that the relative effectiveness of the heuristics depended on the size of the networks and the capacity of the vehicles.

Toobaie and Haghani [31] consider how to partition a road network into sets of roads that can each be treated by a single gritting vehicle. The route for each vehicle is then found by solving an unconstrained Chinese postman problem for each partition. The formulation takes into account that some roads may require separate passes on each side of a divided highway, while others may be treated in a single pass. The objective is to minimize the number of vehicles and deadheading distance while satisfying vehicle capacity, route continuity, and workload balance constraints. The key stage of the method is to solve the minimal arc partitioning problem. This requires the original graph representing the road network to be divided into a minimum number of connected subgraphs so that the requirement for salt on the required roads in each subgraph is less than or equal to the capacity of a spreading vehicle. A genetic algorithm was proposed and tested for this problem using data from a county in Maryland, USA. The results showed improvements in the number of vehicles, deadheading distance, and the work imbalance compared to the routes used by the county at that time.

All the applications discussed so far have been for static models where all the required roads for treatment are known before the salt spreading operation starts. Indeed, in many cases, the applications have been at a strategic level where the same set of roads are assumed to require treatment on every occasion during a winter season when salt spreading is required. However, in Handa, Chapman, and Yao [11] routing requirements on a particular day are linked to Road Weather Information Systems (RWIS), which predict the temperatures of the road surfaces in a network according to the time of day or night. This leads to a dynamic problem where the set of roads requiring treatment and the appropriate amount of salt needed for each road may change for each operation and even during an operation. A memetic algorithm is used to solve the resulting routing problems, and a prototype system was tested on a road network in South Gloucestershire, England. In a subsequent paper, Handa et al. [12] considered the need to produce robust solutions where the routes did not change for every salt spreading operation. This could be an important practical issue to ensure that drivers know and are able to follow the planned routes. The solution approach ranked the routes according to when the roads on them would need treating by making use of a thermal map of the roads in the network. Routes that only contained roads that did not require treatment on relatively warmer occasions could simply be omitted unless the temperature dropped. The results of the study suggested that this approach could lead to a reduction in the total distance traveled by the vehicles by about 10%.

Cai, Liu, and Cao [3] consider the problem of designing routes for the vehicles treating the roads at the same time as finding the best depot locations. Their model includes constraints on the capacities of the vehicles and the maximum number of routes. A heuristic algorithm is proposed to solve the model based on tabu search. Testing is carried out using data from a road network in Chanchung city, China, and confirms a benefit from considering routing and depot location simultaneously rather than using a sequential approach.

The study by Tagmouti, Gendreau, and Potvin [28] has already been mentioned in the section on exact methods. However the authors show that the exact method proposed is only viable for relatively small problems, so a heuristic approach is described in Tagmouti, Gendreau, and Potvin [29]. The problem addressed is an arc routing problem where the costs of treatment depend on the time of treatment. The cost function relates to the times at which a storm passing over the area will affect particular roads and penalizes early and late treatment according to a prescribed piecewise linear function. The heuristic method proposed is based on variable neighborhood descent (VND), which can be thought of as a variant of variable neighborhood search. This method is tested on benchmark problems and compared to a heuristic for node routing problems with a similar structure. This study was extended in Tagmouti, Gendreau, and Potvin [30] to consider a dynamic version of the same problem. In this version, regular updates are received on the weather which may alter the relationships between the cost and time of treatment for any road in the area. A starting solution is obtained by using the VND heuristic based on the initial weather report. After each updated weather report has been received, the required roads that have already been treated are removed, road segments partially treated are assumed to be completed by the vehicle currently spreading salt on the surface, and other required roads yet to be treated have their cost-time service function updated. The VND is then applied to the new static problem that has been generated. Test sets are generated which illustrate the benefit of the dynamic approach, particularly when weather reports change significantly during the course of the spreading operation.

13.4 ▪ Conclusions

Meter reading and salt spreading are both operations which can naturally be represented using arc routing models. In both cases, significant operational improvements have been obtained in practical applications. Over the years, the technology used for these applications has changed. Both operations have benefited from the introduction of geographical information systems to represent the road networks over which vehicles must pass, though care is still needed in the accuracy and level of detail provided. In the case of meter reading, the introduction of RFID tags and automated meter reading has transformed the structure of the routing problem that needs to be solved. In the case of salt spreading, changes in technology have not had such a fundamental effect on the routing problem. However, links to road weather information systems may give increasing opportunities for dynamic or online routing systems to be used for spreading operations that will provide the service needed in a timely fashion.

Acknowledgments

We thank Dr. Laurence Levy at RouteSmart Technologies, Inc., for his cooperation and help with this chapter.

Bibliography

[1] A. ASSAD AND B. GOLDEN, *Arc routing methods and applications*, in Network Routing, M. Ball, T. Magnanti, C. Monma, and G. Nemhauser, eds., North-Holland, Amsterdam, 1995, pp. 375–483.

[2] L. BODIN AND L. LEVY, *The arc partitioning problem*, European Journal of Operational Research, 53 (1991), pp. 393–401.

[3] W.P. CAI, G. LIU, AND W.S. CAO, *A study of vehicle and materials depot location problems for winter road maintenance*, in Proceedings of the 9th International Conference of Chinese Transportation Professionals, ICCTP 2009, Vol. 358, Critical Issues in Transportation System Planning, Development and Management, American Society of Civil Engineers, Reston, VA, 2009, pp. 1530–1535.

[4] J.F. CAMPBELL AND A. LANGEVIN, *Roadway snow and ice control*, in Arc Routing: Theory, Solutions and Applications, M. Dror, ed., Kluwer, Boston, MA, 2000, pp. 389–418.

[5] J. DONG, N. YANG, AND M. CHEN, *Heuristic approaches for a TSP variant: The automatic meter reading shortest tour problem*, in Extending the Horizons: Advances in Computing, Optimization, and Decision Technologies, E. Baker, A. Joseph, A. Mehrotra, and M. Trick, eds., Springer, Berlin, 2007, pp. 145–163.

[6] R.W. EGLESE, *Routing winter gritting vehicles*, Discrete Applied Mathematics, 48 (1994), pp. 231–244.

[7] R.W. EGLESE AND L.Y.O. LI, *Efficient routing for winter gritting*, The Journal of the Operational Research Society, 43 (1992), pp. 1031–1034.

[8] ——, *A tabu search based heuristic for arc routing with a capacity constraint and time deadline*, in Metaheuristics: Theory and Applications, I.H. Osman and J.P. Kelly, eds., Kluwer, Boston, MA, 1996, pp. 633–649.

[9] S. GAVIRNENI, L. CLARK, AND G. PATAKI, *Schlumberger optimizes receiver location for automated meter reading*, Interfaces, 34 (2004), pp. 208–214.

[10] C. GROËR, B. GOLDEN, AND E. WASIL, *The balanced billing cycle vehicle routing problem*, Networks, 54 (2009), pp. 243–254.

[11] H. HANDA, L. CHAPMAN, AND X. YAO, *Dynamic salt routing optimization using evolutionary computation*, in Proceedings of the 2005 IEEE Congress on Evolutionary Computation, IEEE Computer Society, Piscataway, NJ, 2005, pp. 158–165.

[12] H. HANDA, D. LIN, L. CHAPMAN, AND X. YAO, *Robust solution of salting route optimisation using evolutionary computation*, in 2006 IEEE Congress on Evolutionary Computation, IEEE Computer Society, Piscataway, NJ, 2006, pp. 3098–3105.

[13] S. LEE AND B. WELANDER, *Routing Meter Readers...Leveraging an Investment in Data Conversion*, Tech. report, Esri International User Conference, 1998. Available at http://training.esri.com/bibliography/index.cfm?event=general.recorddetail &id=5448 (accessed on 19 August 2012).

[14] ——, *Routing Meter Readers...Leveraging an Investment in Data Conversion: Project Update 1998–1999*, Tech. report, Esri, 1999. Available at http://proceedings.esri.com/library/userconf/proc98/proceed/to450/pap443/ P443UPDATE. HTM (accessed on 19 August 2012).

[15] A.N. LETCHFORD AND R.W. EGLESE, *The rural postman problem with deadline classes*, European Journal of Operational Research, 105 (1998), pp. 390–400.

[16] L. LEVY, J. SNIEZEK, AND B. COX, *Utility Meter Route Management and Optimization Using GIS*, Tech. report, Electric and Gas Utilities User Group, Coeur d'Alene, Idaho, 2002. Available at http://proceedings.esri.com/library/userconf/egug2002/ presentations/meter_route_mgt.pdf (accessed on 19 August 2012).

[17] L. LI AND R.W. EGLESE, *An interactive algorithm for vehicle routing for winter gritting*, The Journal of the Operational Research Society, 47 (1996), pp. 217–228.

[18] T. LOTAN, D. CATTRYSSE, D. VAN OUDHEUSDEN, AND K.U. LEUVEN, *Winter gritting in the province of Antwerp: A combined location and routing problem*, Belgian Journal of Operations Research, Statistics and Computer Science, 36 (1996), pp. 141–157.

[19] J.W. MALE AND J.C. LIEBMAN, *Districting and routing for solid waste collection*, Journal of the Environmental Engineering Division, ASCE, 104 (1978), pp. 1–14.

[20] L. MUYLDERMANS, D. CATTRYSSE, AND D. VAN OUDHEUSDEN, *District design for arc-routing applications*, The Journal of the Operational Research Society, 54 (2003), pp. 1209–1221.

[21] L. MUYLDERMANS, D. CATTRYSSE, D. VAN OUDHEUSDEN, AND T. LOTAN, *Districting for salt spreading operations*, European Journal of Operational Research, 139 (2002), pp. 521–532.

[22] N. PERRIER, J.F. CAMPBELL, M. GENDREAU, AND A. LANGEVIN, *Vehicle routing models and algorithms for winter road spreading operations*, in Hybrid Algorithms for Service, Computing and Manufacturing Systems: Routing and Scheduling Solutions, J.R. Montoya-Torres, A.A. Juan, L.H. Huatuco, J. Faulin, and G.L. Rodriguez-Verjan, eds., Information Science Reference, Hershey, PA, 2012, pp. 15–45.

[23] N. PERRIER, A. LANGEVIN, AND J.F. CAMPBELL, *A survey of models and algorithms for winter road maintenance part I: System design for spreading and plowing*, Computers & Operations Research, 33 (2006), pp. 209–238.

[24] ——, *A survey of models and algorithms for winter road maintenance part III: Vehicle routing and depot location for spreading*, Computers & Operations Research, 34 (2007), pp. 211–257.

[25] ROUTESMART, *Routesmart Technologies*, http://www.routesmart.com (accessed on 19 August 2012).

[26] R. SHUTTLEWORTH, B. GOLDEN, S. SMITH, AND E. WASIL, *Advances in meter reading: Heuristic solution of the close enough traveling salesman problem over a street network*, in The Vehicle Routing Problem: Latest Advances and New Challenges, B. Golden, S. Raghavan, and E. Wasil, eds., Springer, Berlin, 2008, pp. 487–501.

[27] D. STERN AND M. DROR, *Routing electric meter readers*, Computers & Operations Research, 6 (1979), pp. 209–223.

[28] M. TAGMOUTI, M. GENDREAU, AND J.-Y. POTVIN, *Arc routing problems with time-dependent service costs*, European Journal of Operational Research, 181 (2007), pp. 30–39.

[29] ——, *A variable neighborhood descent heuristic for arc routing problems with time-dependent service costs*, Computers and Industrial Engineering, 59 (2010), pp. 954–963.

[30] ——, *A dynamic capacitated arc routing problem with time-dependent service costs*, Transportation Research Part C, Emerging Technologies, 19 (2011), pp. 20–28.

[31] S. TOOBAIE AND A. HAGHANI, *Minimal arc partitioning problem*, Transportation Research Record, 1882 (2004), pp. 167–175.

[32] J. WUNDERLICH, M. COLLETTE, L. LEVY, AND L. BODIN, *Scheduling meter readers for Southern California Gas Company*, Interfaces, 22 (1992), pp. 22–30.

Chapter 14

Advances in Vehicle Routing for Snow Plowing

James F. Campbell
André Langevin
Nathalie Perrier

14.1 ▪ Introduction

Snow plowing is a key activity to maintain the mobility required in modern societies during winter. Snow plowing involves the physical removal of snow from roadways, and sometimes from associated areas such as sidewalks and road shoulders, by vehicles equipped with metal blades, brooms, or blowers. This is typically the responsibility of a public sector governmental organization (e.g., a public works or transportation department of a municipality or regional government), although the actual work may be contracted out to private firms. For some historical perspectives on snow removal, including snow plowing, see Campbell and Langevin [8].

Snow plowing in a region is typically accomplished with a fleet of vehicles, usually trucks, equipped with angled fixed or movable metal blades that shear snow and ice from the roadway and then direct it to the sides of the roadway or traffic lane. Snow plowing is just one of a set of interrelated winter road maintenance operations that also include spreading of chemicals (e.g., salt) and/or abrasives on roadways, loading snow into trucks, and hauling snow to disposal sites. Snow plowing is often done by vehicles simultaneously involved in spreading chemicals to depress the freezing point on the roadway. For a comprehensive review of *operations research* (OR) methods developed for the variety of winter road maintenance operations, see Perrier, Langevin, and Campbell [28, 29, 30, 31].

Snow plowing operations generally begin once a sufficient amount of snow or ice has accumulated and continues until the precipitation ends, or until an appropriate roadway surface condition is achieved, as specified by the agencies' level of service policies. Note that other winter road maintenance operations may begin before the precipitation starts, such as anti-icing (spreading chemicals to prevent the bonding of ice to the roadway) or continue after the storm has ended in the case of snow disposal. The most common activities for snow-fighting are plowing and spreading, which are often performed for each storm event on a repetitive basis over the same routes until safe travel conditions are achieved. A variety of different vehicle types and configurations are used for snow plowing, including trucks with straight or "V"-shaped front-mounted blade plows, wing plows (to the side), under-body plows, and towed plows; motor graders; snow blowers or rotary plows; and sweepers (Boselly [4]). Snow plowing is generally limited to one traffic lane at a time for a single vehicle, in contrast to materials spreading, which can sometimes cover more than a single traffic lane, though plows are not of uniform width and teams of vehicles can be used to plow multiple lanes simultaneously in echelons.

Snow plowing is a very common winter road maintenance operation and an essential component of snow-fighting when snow and ice accumulate to a level greater than what can be handled by chemical treatments alone. It is also a very expensive activity due to the high costs for the specialized equipment and the required personnel, and the need for immediate action in response to storms. A study in Sweden (Nordin [24]) showed that snow plowing was the single most time and energy intensive road maintenance activity (not just for winter operations, but year round) in consuming 30.9% of the annual road maintenance hours and 29.7% of the annual fuel consumed for maintenance districts. Comprehensive information on costs for plowing, and more generally for snow-fighting, are difficult to document as they vary with the annual intensity of winter, and are generally incurred by the different levels of government (federal, regional, and local) responsible for maintaining the various roadways. To give an indication of the magnitude of the problems, we consider three countries for which data is available.

- In the USA, 81% of the land (over 7 million square km), 71% of the population (over 215 million people), and 75% of the roadways (over 4.8 million km) are in "snowy regions" and state and local expenditures for winter road maintenance average about \$2.3 billion each year (PIARC Technical Committee B5 Winter Service [32]).
- Japan, where some large urban areas experience extreme snowfall amounts (Sapporo, with a population of nearly 2 million has an average snowfall of about 5 meters), spends \$500 billion on snow and ice control annually (PIARC Technical Committee B5 Winter Service [32]).
- Following two severe winters in 2008–2010, the British Department for Transport undertook a "review of the resilience of England's transport systems" which found that annual expenditures for winter road maintenance activities are about \$250 million, with the total cost to the economy (including lost output, health-related costs, accidents, indirect costs, "social adversity," etc.) for an "average" winter being about \$1.5 billion (Winter Resilience Review [40]). The large majority of this expenditure is on spreading activities, treating the roads with salt, though some is for snow plowing.

From an OR perspective, the routing of snow plows (and winter road maintenance vehicles) is often viewed as a rather straightforward application for arc routing models. However, snow plowing, and other winter road maintenance operational problems, tend to be very challenging and site specific because of the wide range and dynamic nature of the operating conditions, the diverse sets of overlapping administrative jurisdictions (a single geographic region may include roadways serviced by federal, regional, and local governments, some of which may be subcontracted to private firms), operational constraints (e.g., no left turns and limited locations for U-turns), and stringent level of service constraints. See Campbell and Langevin [8] and Perrier, Langevin, and Campbell [28] for more discussion of the practical challenges associated with snow and ice removal.

Because of the inherent complexity and difficulty of snow plow vehicle routing problems, these remain an active topic of research within arc routing. Over a decade ago,

Campbell and Langevin [8] reviewed 25 papers on winter road maintenance and route optimization. At that time, the wide gap between theory and practice was noted and it was suggested that there are great opportunities for OR to contribute in the area of route optimization for winter road maintenance. More recently, Perrier, Langevin, and Campbell [31] reviewed 14 vehicle routing models for plow and/or spreader truck routing and Perrier et al. [26] reviewed and classified 26 spreader routing research papers.

Our focus in this paper is to review the latest OR research on snow plow routing and report a new case study on implementation of route optimization for snow plowing. The remainder of this paper is organized as follows. Section 2 provides background on snow plowing operations and some discussion of reasons for the complexity of real-world snow plowing. Section 3 describes recent optimization models for snow plowing, and Section 4 includes a case study on implementation of optimized plow routes. Section 5 is the conclusion with some promising areas for future research.

14.2 ▪ Plowing operations

Plowing operations consist in using vehicles to push the snow to the side of the roadways during a snow fall and in some cases after the end of the snow fall. The role of snow plowing in conjunction with deicing operations is to remove as much snow and ice as possible before applying chemicals. The snow plow routing problem consists of determining a set of plow routes to meet some objective, such that all desired road segments are serviced (plowed) and all the operational constraints are satisfied. See Perrier, Langevin, and Campbell [31] for a description of the relevant characteristics of snow plow routing problems, the operational constraints imposed on the routes, and possible objectives for the optimization process.

In plowing operations, each lane of a roadway must be serviced separately. Hence, lanes are modeled as arcs in a directed graph. The transportation network serviced by a particular agency can be viewed as "urban" or "rural," depending on whether it is required to service all road segments or only a subset of road segments, respectively. However, in practice roadways in a common geographic area may be serviced by several different agencies (and sets of plows) based on the associated maintenance jurisdictions. For example, an urban area may include federal highways as well as regional (e.g., state or provincial) roadways where the regional government is responsible for servicing only the regional roadways.

Typically, the roads of a network are partitioned into classes, based primarily on traffic volume, which induces a service hierarchy. Roads carrying the heaviest traffic are given the highest level of service in order to provide safe roads for the greatest number of motorists. This highest class of roads often includes multilane highways. Roads that access critical facilities (e.g., hospitals) or are especially hazardous (e.g., bridges, steep hills, and sharp curves) may also be given the highest priority. Roads in the next highest priority class carry medium volumes of traffic, with a lower priority class for local and residential roads having the lowest levels of traffic. In some areas, local and residential streets are not plowed except for extreme storms. Associated with each road segment is a service frequency, which is possibly dependent on the hierarchy of the network (e.g., the road segment should be plowed at least once every 2 hours). For guidelines on setting the level of service, see Blackburn et al. [3].

Also associated with each road segment is a cost, which generally represents its length, and up to three traversal times, which are possibly dependent on the vehicle type: the time required to service the road segment, the time of deadheading the road segment if it has not yet been serviced, and the time of deadheading the road segment if it has already been serviced. Deadheading occurs when a vehicle must traverse a road segment without servicing it. Since plowing operations are limited to one lane at a time, multilane road segments necessitate either multiple separate passes or tandem plow patterns in echelon formations (i.e., the multiple lanes are cleared simultaneously with a plow in each lane).

The routes must satisfy several operational constraints imposed by the level of service policies and the characteristics of the transportation network, road segments, service sectors, vehicles, and crews. The routes usually start and end at the same depot locations. In plowing operations, the route start times are dependent on the accumulation of snow and ice on the roadway surface. To balance the workload, routes are often designed to be approximately the same length or duration. This helps ensure that plowing will be completed in a timely fashion. Class continuity requires that each route services road segments with the same priority classification. Thus, if a lower-class road is included in a route servicing higher-class roads, its service level may be upgraded. In plowing and snow disposal operations, it is often desirable that both sides of a two-lane, two-way road (one lane each way) be serviced by the same vehicle in a single route.

Finally, several objectives can be considered for the routing of vehicles for snow plowing. Typical objectives are minimization of the distance covered by deadheading trips (or the deadhead travel time); minimization of the time for service completion; minimization of the penalties associated with turns; minimization of the terms penalizing the violation of some operational constraints; or any weighted combination of these objectives. Because of the hierarchy of levels of service associated with the different priority classes of roadways, a simple distance-based objective does not reflect well the goals of most operating agencies.

To illustrate the complexity of snow plowing, we describe the operations as conducted in Dolbeau-Mistassini, a relatively wide-spread municipality in Northern Quebec with an urban district and a fairly large rural section. The road network under the responsibility of Dolbeau-Mistassini is divided into commercial, residential, and rural components which induce a 3-class hierarchy. The area is also traversed by national roads whose maintenance is performed by the province of Quebec, but these national roads can be used for deadheading by the equipment of Dolbeau-Mistassini. Dolbeau-Mistassini uses eight vehicles for its plowing activities: three trucks, two graders, and three loaders. The three trucks are also used for spreading activities. There are some vehicle-road compatibility constraints that depend on the vehicle characteristics (size, speed, etc.) and the road characteristics (width and type). Also, it is preferable to use the vehicles with the highest speed to service the rural area.

There are two main difficulties when planning the plow routes. First, the plowing and the spreading routes are interlaced. At the beginning of the snowfall, the three trucks start spreading de-icers on the commercial section of the network. When the snow accumulation reaches 2.5 cm, the other five vehicles start their plowing activities, joined by the three trucks as they finish the spreading on the commercial network. The plowing of the network must be done while respecting the hierarchy and the vehicle-road compatibilities. The three trucks complete their routes by spreading the residential and rural sections of the network. Balancing the eight routes so that the entire operations are completed as soon as possible is a challenging problem. The second main difficulty is related to the need to avoid U-turns as much as possible and the prohibition of U-turns, for safety reasons, at the intersections of a municipal street and a national road. Our experience with existing arc routing algorithms indicates that they fail to generate satisfactory routes in all respects.

14.3 ▪ Vehicle routing models for plowing

Most early systems for routing snow plows relied on relatively simple constructive heuristics, manual procedures, simulation methods, or a combination of these approaches (Perrier, Langevin, and Campbell [31]). Simulation was often used to evaluate routes developed manually or with simple heuristics and to interactively develop routes (e.g., Cook and Alprin [12], Tucker and Clohan [39]). More recently, Damodaran and Krishnamurthi [14] proposed a simulation model to aid planners for the real-time planning of snow plowing and salt spreading operations where routes must be revised when weather conditions change. The simulation model uses the constructive heuristic developed by Krishnamurthi and Damodaran [22] for a hierarchical postman problem with linear precedence relation where the subgraph induced by each class may not be connected. The accuracy of the model was validated using data from the city of DeKalb, Illinois.

A number of different heuristic procedures have been proposed for the routing of vehicles for plowing operations, and Perrier, Langevin, and Campbell [31] summarize the research prior to 2004 in three categories: constructive methods, composite methods, and adaptation of metaheuristics. That review found that "all algorithms developed for the routing of vehicles for winter road maintenance are heuristics," because of the difficulty and site-specific nature of the problems. That review also documented the trend from simple constructive methods for undirected and directed versions of the *Chinese postman problem* (CPP) to more realistic models solved via local search techniques and composite heuristic methods.

In this paper we report on recent research in optimizing routing for snow plows not covered in the Perrier, Langevin, and Campbell [31] survey. The recent works show a shift from heuristic approaches to mathematical programming-based approaches, as well as efforts to include more of the real-world complexity in snow plow routing. Note that the chapter by Eglese, Golden, and Wasil in this volume [15] also reports on some developments on routing for spreading. The remainder of this section describes the new research in this area, closing with a table that summarizes these new works.

14.3.1 ▪ Mathematical programming-based methods

Golbaharan [17] studied a multi-depot snow removal routing problem with time windows after snowfall. This problem consists of designing a set of minimum cost routes for homogeneous snow plows after the snow has stopped falling while covering every required road segment exactly once within its associated time window. Every snow plow starts from a depot and returns to the same depot. The problem is formulated as a linear integer programming model. Let $G=(V,E)$ denote a network, where V is the set of nodes and E is the set of edges. The author specifies that edges represent ordinary one-way or two-way directed road segments, but no clarification is made on how road segments are converted to edges and arcs. A route r in G is defined as a sequence $(e_1,e_2,\dots,e_n)$ of n edges where the terminal node of edge e_{k-1} is the start node of edge $e_k, k=2,\dots,n$. Let D be the set of depots. For each depot $d\in D$, define R_d as the set of all feasible routes with respect to the time windows on the road segments, originating and ending at depot d. Let also $R=\cup R_d$, and define $E_p\subset E$ as the set of required road segments. For each depot $d\in D$, let y_d be a nonnegative integer variable representing the number of "extra" snow plows used at depot d, where an "extra" plow is one in addition to those available at the depot initially. Let v_d and w_d represent the number of snow plows available at depot d and the charge for using one extra snow plow in addition to those available at depot d, respectively. For each route $r\in R$, let x_r be a binary variable equal to 1 if and only if route r is in the solution and define c_r as the cost of route r. Finally, for each route $r \in R$ and for each edge $e \in E_p$, define the binary constant a_{er} equal to 1 if and only if edge e is plowed in route r.

The multi-depot snow removal routing problem with time windows after snowfall can then be formulated as follows:

$$\text{(14.1)} \qquad \text{Minimize} \qquad \sum_{r\in R} c_r x_r + \sum_{d\in D} w_d y_d$$

$$\text{(14.2)} \qquad \text{s.t.} \qquad \sum_{r\in R} a_{er} x_r = 1 \qquad (e \in E_p),$$

$$\text{(14.3)} \qquad \sum_{r\in R_d} x_r \leq v_d + y_d \qquad (d \in D),$$

$$\text{(14.4)} \qquad x_r \in \{0,1\} \qquad (r \in R),$$

$$\text{(14.5)} \qquad y_d \geq 0 \text{ and integer} \qquad (d \in D).$$

The first term of the objective function (14.1) defines the total cost of the routes, while the second term is a penalty cost for using extra snow plows. Constraints (14.2) stipulate that each required edge be covered exactly once. Constraints (14.3) are the depot fleet size constraints. Whenever the number of available vehicles at a given depot is exceeded, the objective function is penalized. Due to constraints (14.3), the problem can be regarded as a constrained set partitioning problem. Each column a_r of the constraint matrix $A = (a_r)$, $r \in R$, may be viewed as a subset, where $e \in a_r$ if and only if $a_{er} = 1$. There are no explicit constraints imposed on satisfying the time criteria on the edges, but the columns in A are feasible with regard to satisfying the time windows. The problem is then to select a set of columns, which yields a minimal cost partition of the edges in E_p fulfilling constraints (14.3). The problem is reformulated as a constrained set covering problem by redefining constraints (14.2) so that each edge in E_p is covered at least once. If an edge is covered more than once, lower cost modified versions of the column in which the edge is treated as deadheaded instead can always be added. The linear relaxation of the resulting constrained set covering problem is solved with a column generation algorithm. The constraints on the number of snow plows available at each depot and the constraints which guarantee that a required road segment is serviced are included in the master problem, while the subproblems contain the time window constraints and network flow constraints.

The master problem is solved by the dual simplex method, and the column generation is terminated when a predetermined number of columns are generated. The initial set of columns is generated with a simple constructive heuristic. The subproblem for every depot can be formulated as a shortest path problem with time windows associated with the edges in the network, and it is solved with a label-setting algorithm. The column generation algorithm is followed by a routine to obtain an integer solution to the final master problem. Two different approaches are considered: (1) a branch-and-bound algorithm with a variable reduction procedure, and (2) a greedy procedure similar to those used for set covering problems, followed by an improvement heuristic. Computational experiments are carried out on real-life data from Eskilstuna, Sweden, involving 7 depots, 21 snow plows, 362 nodes, and 814 edges, including 707 required edges. Results showed that, in general, the greedy approach is faster and more robust to achieve a good solution compared to the branch-and-bound algorithm, and that the improvement heuristic is quite useful in reducing cost about 20% below that from the greedy approach alone.

Sochor and Yu [38] proposed a solution method using subgradient optimization on a Lagrangian dual problem in combination with a greedy heuristic to solve the multi-depot snow removal routing problem with time windows after a snowfall. The problem is formulated as a constrained set covering problem of the sort reformulated by Golbaharan [17]. The solution method calculates a lower bound by applying Lagrangian relaxation to the problem. Let $\lambda, \mu \geq 0$. The maximization is over the dual variables λ and μ. The Lagrangian relaxation is

$$\begin{aligned} \text{Maximize} \quad & Z_D(\lambda,\mu) = \min_{x_r, y_d} \sum_{r \in R} c_r x_r + \sum_{d \in D} w_d y_d - \sum_{e \in E} \lambda_e \left(\sum_{r \in R} a_{er} x_r - \rho_e \right) \\ & \qquad + \sum_{d \in D} \mu_d \left(\sum_{r \in R} \delta_{dr} x_r - v_d - y_d \right) \\ \text{s.t.} \quad & x_r \in \{0,1\} \qquad (r \in R), \\ & y_d \geq 0 \text{ and integer} \qquad (d \in D), \end{aligned}$$

where ρ_e and δ_{dr} are two binary constants. For every edge $e \in E$, ρ_e is equal to 1 if and only if edge e is a required edge, and for every depot $d \in D$ and for every route $r \in R$, δ_{dr} is equal to 1 if and only if route r starts and ends at depot d. A subgradient approach is used to find near-optimal dual variables to be sent to the column generation algorithm developed by Golbaharan [17], which returns routes for the snow plows. A greedy heuristic finds a feasible solution, which gives an upper bound to the problem. The algorithm is applied iteratively until a sufficient precision β is obtained or until a given time limit is exceeded. The algorithm can be outlined as follows:

1. *Initialization.* Initialize the dual variables λ and μ. Set the upper bound $UBD^* := \infty$.

2. *Column generation.* Send the dual variables to Golbaharan's column-generation algorithm (Golbaharan [17]), and generate new columns (routes) based on this algorithm. Set the lower bound $LBD^* := 0$.

3. *Lagrangian relaxation.* Use multiplier vectors λ and μ to calculate the lower bound $ZD(\lambda; \mu)$. Update the best incumbent bound LBD^* and multipliers λ and μ.

   Update multipliers. Update multiplier vectors with the subgradient method. If the maximum number of subgradient iterations is reached or the lower bound improvement ≤ 0.01 in an iteration, then go to step 4. Otherwise, return to step 3.

4. *Greedy procedure.* Use $\overline{\lambda} = (1+\rho)\lambda^*$ and $\overline{\mu} = (1+\rho)\mu^*$, where ρ is a uniform random value in a symmetric interval around 0, to compute a heuristic solution to the problem. Update the best incumbent bound $UBD^* = \min\{UBD^*, Z\}$ and solution x^*. If the maximum number of iterations τ is reached, then go to step 5. Otherwise, repeat step 4.

5. If a sufficient precision is obtained (i.e., $UBD^* \leq \beta * LBD^*$) or the time limit is exceeded, stop. Otherwise, return to step 2.

The initial solution in step 2 consists of artificial columns which cover one edge each and have a high cost. Steps 3 and 4 are applied only on required edges. In step 3, a lower bound to the current problem is calculated and the dual multipliers are updated using a subgradient approach. Given a multiplier vector, the greedy heuristic chooses a column with minimal value and this process is repeated until either all the rows are covered or the current partial solution value becomes too expensive and exceeds the best upper bound. Once the greedy heuristic is completed and the best solution is updated, the algorithm returns to step 2 and requests new columns when the stop requirements are not met. Computational experiments were performed on the road network of Eskilstuna for different values of problem parameters. Comparisons with the column generation approach proposed by Golbaharan [17], before running the improvement procedure, showed that the algorithm produced good results at an early stage and required many fewer columns to be generated than with the approach in Golbaharan [17].

Razmara [34] addressed the *snow removal routing problem* for a nonhomogeneous fleet of snow plows *with time windows during snowfall* (SRRPTWDS). In contrast to the plow routing problem with time windows only after snowfall, the snow plows must now start the removal of snow during snowing conditions. When a time horizon for the duration of snowfall is given, the problem is called the time horizon SRRPTWDS, as opposed to the periodic case, called the periodic SRRPTWDS, where the snowfall can be thought of as continuing. The time horizon SRRPTWDS can be considered as a routing problem with dynamic time windows, while in the periodic SRRPTWDS the time windows are transformed into service frequencies. Given a time horizon and the corresponding time intervals for each road segment, the problem with time horizon dependency is formulated as a linear integer program, but no solution method is presented.

The periodic SRRPTWDS is formulated as a linear integer programming problem and solved using a Dantzig–Wolfe decomposition scheme similar to the one developed by Golbaharan [17] for solving the multi-depot snow removal routing problem with time windows after snowfall. Consider again the undirected network $G=(V,E_p\cup E_T)$, where E_p is the set of required edges and E_T is the set of auxiliary edges used only for deadheading. The edges in E_T may or may not belong to the operational district under consideration, but are added to construct a connected network. Every node in the network can be considered as a potential depot from which a route can start and end. Let R be the set of routes in G. For each route $r \in R$ and for each edge $e \in E_p$, define the binary constant a_{er} equal to 1 if and only if route r covers edge e. Associated with each route $r \in R$ is a service frequency corresponding to the number of times the route must be repeated during snowfall conditions. Obviously, the time of beginning of service of a route r, called *period*, needs to be larger than the duration d_r of the route. Also, to maintain feasibility with respect to the time restrictions on road segments, each required road segment $e \in E_p$ with maximum service time m_e must be plowed in a route with a period $p \leq m_e$. Periods fulfilling these conditions are called feasible. Then, the set of feasible periods for a route r is $P_r = \{p : d_r \leq p \leq m_e, e \in E_p\}$. A route r with $P_r \neq \emptyset$ is called a *feasible route*. Among the feasible periods for a route r, the *proper period* for the route r is $p_r = \max\{p : p \in P_r\}$. Let K be the set of vehicle types. For any type of vehicle $k \in K$, define R_k as the set of feasible routes that may be operated by a snow plow of type k. Then, for each vehicle type $k \in K$ and for each route $r \in R_k$, define C_{kr} as the cost of route r operated by vehicle of type k and computed with respect to its proper period. For each route $r \in R$ and for each type of vehicle $k \in K$, let x_{kr} be a binary variable equal to 1 if and only route r is operated by vehicle of type k. Finally, for each vehicle type $k \in K$, define y_k as the number of "extra" vehicles of type k in the network, and let v_k and w_k represent the number of available vehicles of type k and the cost of using an extra snow plow of type k, respectively. The problem is formulated similar to that in Golbaharan [17] as follows:

$$\text{(14.6)} \qquad \text{Minimize} \qquad \sum_{k\in K}\sum_{r\in R_k} C_{kr}x_{kr} + \sum_{k\in K} w_k y_k$$

$$\text{(14.7)} \qquad \text{s.t.} \qquad \sum_{k\in K}\sum_{r\in R_k} a_{er}x_{kr} \geq 1 \qquad (e \in E_p),$$

$$\text{(14.8)} \qquad \sum_{r\in R} x_{kr} \leq v_k + y_k \qquad (k \in K),$$

$$\text{(14.9)} \qquad x_{kr} \in \{0,1\} \qquad (k \in K, r \in R),$$

$$\text{(14.10)} \qquad y_k \geq 0 \text{ and integer} \qquad (k \in K).$$

The objective function (14.6) minimizes the total cost of routes and the cost of using extra vehicles. Constraints (14.7) state that each edge must be plowed at least once by a vehicle that is permitted to operate on the edge. Constraints (14.8) impose a limit on the number of vehicles that can be utilized for each type. In the decomposition scheme, the master problem is formulated as a constrained set covering problem in which the required edges must be plowed at least once, and a constraint on the number of vehicles is imposed for each vehicle type. In contrast to the master problem in Golbaharan [17], the vehicle constraints are here defined over vehicle types instead of over depots and the columns in the master problem have different types. Since the plowing routes in the undirected network G essentially correspond to connected subgraphs, the subproblems are formulated as prize-collecting connected subgraph problems. The prize-collecting connected subgraph problem is a generalization of the prize-collecting Steiner tree problem, which is itself derived from the prize-collecting traveling salesman problem introduced by Balas [1].

The linear programming relaxations of the subproblems are solved by generating valid inequalities, called connectivity cuts, and an integer solution is found heuristically. The connectivity cuts are generated by means of different procedures based on generation of labels, solving shortest path problems and/or solving maximum flow problems. In addition to connectivity cuts, a set of simple valid inequalities are derived a priori and added to the subproblems in order to strengthen the linear relaxations. At last, an integer solution to the master problem is found by the greedy search procedure developed by Golbaharan [17] and the obtained solution is then improved by means of a local search method. Tests on the Eskilstuna operational district containing 297 nodes and 443 edges showed that the best integer solution is obtained by generating 10,000 columns. The relative gap between the fractional solution of the master problem and the integer solution found by the greedy search procedure is approximately 22%, which by local search is reduced to approximately 8.5%. The number of vehicles used is also reduced by three.

In addition to the column generation algorithm, a tabu search heuristic is developed to solve the periodic SRRPTWDS. An initial solution to the tabu search is constructed with a solution method inspired by the Clarke and Wright algorithm for the vehicle routing problem [11]. Given an edge $e \in E_p$, the first phase determines the existing route, with period p as near as possible to m_e, into which this edge should be inserted. If all the routes do not have any common node with edge e, a new route consisting of edge e only is formed. This results in generating a large number of routes since no edge is used as deadheading. The second phase then tries to reduce the number of routes by applying a savings procedure that introduces shortest chains of deadheaded edges to merge the routes when yielding a positive saving. That is, route r_k and route r_l with costs c_k and c_l, respectively, are merged only if the new route determined by merging r_k and r_l is feasible and the distance savings $s_{kl} = c_k + c_l - sp_{kl} > 0$, where sp_{kl} is the length of a shortest chain between the routes. To explore solution neighborhoods, two sets of moves can be applied during the search: swapping and insertion. A $\lambda-\mu$ exchange, or swap-move, between a pair of routes A and B swaps a chain of required edges of length $l_A \leq \lambda$ on route A with a chain of required edges of length $l_B \leq \mu$ on route B. An insert-move removes a chain of required edges from one route and inserts it into another route. The new routes may include some shortest chains of deadheaded edges as the result of swapping or inserting the chains. Also, the period of a route may be updated to maintain feasibility. A move registered as tabu makes all the edges included in the move tabu. No diversification is applied to the search since the operations defining the moves result in investigating moderately remote neighbors. However, the intensification of the search is carried out by giving a higher priority to the insert moves and decreasing the length of the tabu list. The search is terminated when no improvement within a predefined number of iterations is obtained. Denote by $p_{chain(A)}$ the period of a chain of required edges *chain*(*A*), where $p_{chain(A)} = \min\{m_e : \forall e \in chain(A)\}$. The tabu search for the periodic version of the SRRPTWDS can be summarized as follows:

1. *Initialization*. Generate an initial solution s_{init} using the two-phase method. Define $f(s)$ as the objective value of a solution s. Set $s_{best} := s_{init}$ and $f(s_{best}) := f(s_{init})$. Define also s_{cur}, L, and k as the current solution, a predetermined value, $L \in Z^+$, and the number of iterations without any improvement, respectively. Set $k := 0$.

2. *Choice of the best non-tabu neighbor to* s_{cur}. Repeat the following steps for different values of λ and μ until a move is done:

   (a) Choose the values of λ and μ. Randomly select a route A. Randomly select a chain of required edges $chain(A)$ on route A with $l_A \leq \lambda$. Randomly select another route B such that $p_B \leq p_{chain(A)}$. If inserting $chain(A)$ in B is possible, then make an insert move and go the step (c). Otherwise, go to step (b).

   (b) Randomly select a chain of required edges $chain(B)$ on route B such that $p_A \leq p_{chain(B)}$ and $l_B \leq \mu$. If swapping the chains $chain(A)$ and $chain(B)$ is possible, then make the swap move and go to step (c). Otherwise, return to step (a).

   (c) Let s_{new} denote the solution obtained from the search and go to step 3.

3. If $f(s_{new}) < f(s_{best})$, then set $s_{best} := s_{new}, s_{cur} := s_{new}$ and $f(s_{best}) := f(s_{new})$, Update the tabu list, and reset the counter $k := 0$. Otherwise, set $s_{cur} := s_{new}$ and update the counter $k := k + 1$.

4. *Termination criteria*. If $k < L$, then go to step 2. Otherwise, stop.

Comparisons with the column generation algorithm on data from the operational district of Eskilstuna, Sweden, consisting of 283 nodes and 344 edges with a homogeneous vehicle fleet indicated that the tabu search heuristic obtained the best solution in less computational time.

Perrier, Langevin, and Amaya [27] proposed a formulation for the routing of vehicles for snow plowing operations in urban areas. The model incorporates a hierarchical objective, general precedence relation constraints with no assumption on class connectivity, different service and deadhead speed possibilities, separate pass requirements for multilane road segments, class upgrading possibilities, and vehicle-road segment dependencies. The hierarchical objective is very important in many practical settings to ensure that roadways of a higher priority class are all plowed before roadways of a lower class. With several priority classes, this requires coordination of several levels of routes in the solutions. To model this problem, let $G = (V, A \cup E)$ be a strongly connected mixed graph, where $V = \{v_0, v_1, \ldots, v_n\}$ is the vertex set (v_0 is the depot), $A = \{(v_i, v_j) : v_i, v_j \in V$ and $i \neq j\}$ is the arc set, and $E = \{(v_i, v_j) : v_i, v_j \in V$ and $i < j\}$ is the edge set. Let

$M = \{1, \ldots, m\}$ be the set of vehicles. For every arc and edge $(v_i, v_j) \in A \cup E$, let a_{ij}, e_{ij}, and e_{ji} be the number of traffic lanes associated with arc (v_i, v_j), edge (v_i, v_j) from v_i to v_j, and edge (v_i, v_j) from v_j to v_i, respectively. Since each lane must be plowed separately, each arc $(v_i, v_j) \in A$ is replaced by a_{ij} copies and each edge $(v_i, v_j) \in E$ is replaced by e_{ij} arcs from v_i to v_j and by e_{ji} arcs from v_j to v_i. The resulting multigraph $G' = (V, A')$ is then directed. The arc set A' is partitioned by priority class into $\{A^1, A^2, \ldots, A^K\}$ with $A^1 \cup A^2 \cup \cdots \cup A^K = A'$ and $A^i \cap A^j = \emptyset$ for i not equal to j, which induce the service hierarchy, i.e., all arcs of class A^i must be serviced before those of class A^{i+1}. Classes $1, \ldots, K-1$ include road segments having a given priority, while class K includes road segments that can be serviced anywhere in the sequence. For every class $p = 1, \ldots, K+1$, let $TMAX_p$ be a nonnegative real variable representing the service completion time of class p. Class $K+1$ is a fictitious class introduced to include the shortest travel path to the depot from the last serviced arc in class K for each vehicle.

For every vehicle $h \in M$, let $A_h \subseteq A'$ be the subset of arcs in G' that can be plowed by vehicle h. With every vehicle $h \in M$ and every arc $(v_i, v_j) \in A_h$ are associated two positive durations s_{ij}^h and d_{ij}^h for the plowing and deadheading of arc (v_i, v_j) by vehicles h, respectively. For every vehicle $h \in M$, for every arc $(v_i, v_j) \in A_h$, and for every class $p = 1, \ldots, K$, let x_{ij}^{ph} be a binary variable equal to 1 if and only if arc (v_i, v_j) is plowed in class p by vehicle h. The mathematical formulation of the problem is based on a multicommodity network flow problem to impose the connectivity of the route performed by each vehicle with supplementary variables and constraints. In this model, each commodity corresponds to a possible class-vehicle combination and shares the same directed graph $G'' = (V \cup \{v_a\}, A' \cup A_1 \cup A_2)$ constructed from G' where v_a is an artificial vertex, $A_1 = \{(v_a, v_i) : v_i \in V\}$ and $A_2 = \{(v_i, v_a) : v_i \in V\}$. The artificial vertex is introduced to link the start and end of routes for successive priority classes.

For every vehicle $h \in M$, for every arc $(v_1, v_j) \in A_h \cup A_1 \cup A_2$, and for every class $p = 1, \ldots, K+1$, let y_{ij}^{ph} be a nonnegative integer variable representing the number of times arc (v_i, v_j) is traversed (while servicing or deadheading) in class p by vehicle h. All routes must not contain any disconnected subtours. To model this, flow variables representing fictitious units flowing through the graph G'' are introduced. For every vehicle $h \in M$, for every arc $(v_i, v_j) \in A_h \cup A_1 \cup A_2$, and for every class $p = 1, \ldots, K+1$, let w_{ij}^{ph} be a nonnegative real variable representing the number of fictitious units of flow used on arc (v_i, v_j) associated with class p and vehicle h. Finally, for every class $p = 1, \ldots, K+1$ and for every vehicle $h \in M$, let t_p^h be a nonnegative real variable representing the service completion time of class p on route h. The t_p^h variables are used to clarify the formulation and the interpretation of results. The basic model for the problem of vehicle routing for urban snow plowing operations can be stated as follows:

$$\text{(14.11)} \qquad \text{Minimize} \quad \sum_{p=1}^{K+1} M_p \times TMAX_p$$

$$\text{(14.12)} \qquad \text{s.t.} \quad TMAX_p \geq t_p^h \quad (p = 1, \ldots, K+1, h \in M),$$

$$\text{(14.13)} \qquad t_p^h = t_{p-1}^h + \sum_{(v_i, v_j) \in A_h} (s_{ij}^h x_{ij}^{ph} + d_{ij}^h (y_{ij}^{ph} - 1)) \quad (p = 1, \ldots, K+1, h \in M),$$

$$t_0^h = 0 \quad (h \in M), \tag{14.14}$$

$$\sum_{\substack{h \in M \\ (v_i,v_j)\in A_h}} \sum_{p=1}^{k} x_{ij}^{ph} = 1 \quad ((v_i,v_j)\in A^k, k = 1,\dots,K-1), \tag{14.15}$$

$$\sum_{\substack{h \in M \\ (v_i,v_j)\in A_h}} \sum_{p=1}^{K+1} x_{ij}^{ph} = 1 \quad ((v_i,v_j)\in A^K), \tag{14.16}$$

$$\sum_{(v_i,v_j)\in A_h \cup A_1 \cup A_2} y_{ij}^{ph} = \sum_{(v_i,v_j)\in A_h \cup A_1 \cup A_2} y_{ji}^{ph} \quad (v_i \in V \cup \{v_a\}, p = 1, \dots, K+1, h \in M), \tag{14.17}$$

$$y_{ij}^{ph} \geq x_{ij}^{ph} \quad ((v_i,v_j)\in A_h, p = 1,\dots,K, h \in M), \tag{14.18}$$

$$\sum_{(v_i,v_j)\in A' \cup A_1 \cup A_2} w_{ij}^{ph} = \sum_{(v_j,v_i)\in A' \cup A_1 \cup A_2} w_{ji}^{ph} \quad (v_i \in V \cup \{v_a\}, p = 1, \dots, K+1, h \in M), \tag{14.19}$$

$$y_{ij}^{ph} \leq w_{ij}^{ph} \leq |A'| y_{ij}^{ph} \quad ((v_i,v_j)\in A_h \cup A_1, p = 1, \dots, K+1, h \in M), \tag{14.20}$$

$$y_{ij}^{ph} \leq w_{ia}^{ph} \quad ((v_i,v_j)\in A_h, p = 1,\dots,K+1, h \in M), \tag{14.21}$$

$$\sum_{v_i \in V} y_{ai}^{ph} = 1 \quad (p = 1,\dots,K+1, h \in M), \tag{14.22}$$

$$\sum_{v_i \in V} y_{ia}^{ph} = 1 \quad (p = 1,\dots,K+1, h \in M), \tag{14.23}$$

$$y_{ia}^{ph} = y_{ai}^{p+1,h} \quad (v_i \in V, p = 1,\dots,K, h \in M), \tag{14.24}$$

$$y_{a0}^{1h} = 1 \quad (h \in M), \tag{14.25}$$

$$y_{0a}^{K+1,h} = 1 \quad (h \in M), \tag{14.26}$$

$$x_{ij}^{ph} \in \{0,1\} \quad ((v_i,v_j)\in A_h, p = 1,\dots,K, h \in M), \tag{14.27}$$

$$y_{ij}^{ph} \geq 0 \text{ and integer} \quad ((v_i,v_j)\in A_h \cup A_1 \cup A_2, p = 1, \dots, K+1, h \in M), \tag{14.28}$$

$$w_{ij}^{ph} \geq 0 \quad ((v_i,v_j)\in A_h \cup A_1 \cup A_2, p = 1, \dots, K+1, h \in M), \tag{14.29}$$

$$TMAX_p \geq 0 \quad (p = 1, \ldots, K+1), \tag{14.30}$$

$$t_p^h \geq 0 \quad (p = 0, \ldots, K+1), \tag{14.31}$$

where $M_1 >> M_2 >> \cdots >> M_{K+1} = 1$. The objective function (14.11) minimizes the time to plow roads in the first priority class, then the completion time of the second class, and so on. As highlighted by Cabral et al. [5], the notation ">>" is used to designate that in any feasible solution,

$$M_p \times TMAX_p > \sum_{k=p+1}^{K+1} M_k \times TMAX_k$$

must be satisfied for $p = 1, \ldots, K$. Constraints (14.12) state that the maximum completion time of a given priority class must be greater than or equal to the completion time of this class on any route. Constraints (14.13) and (14.14) define the service completion time of each priority class on each route and the start time of each route, respectively. Constraints (14.15) and (14.16) ensure that each arc of a given priority class is plowed either in this class or in any of the classes of higher priority by exactly one eligible vehicle and that all other arcs are serviced by exactly one eligible vehicle, respectively. A vehicle is *eligible* for a certain arc if it can plow or deadhead this arc. Constraints (14.17) ensure route continuity for each possible priority class-vehicle combination. Constraints (14.18) state that an arc is plowed by an eligible vehicle in a given priority class only if it is traversed by the same vehicle in the same class. Flow conservation at every node for each priority class and for each vehicle is imposed by constraints (14.19). Constraints (14.20) ensure that the flow on every arc associated with a priority class and an eligible vehicle is positive if and only if this arc is traversed (while plowing or deadheading) in that class by that vehicle. Constraints (14.21) ensure that each partial route associated with a priority class and a vehicle does not contain any disconnected subtours. Constraints (14.22) and (14.23) require that each priority class-vehicle combination be associated with exactly two vertices of G: one at which the route must start service, called the *start vertex*, and one at which the route must end service, called the *end vertex*, respectively. Constraints (14.24) ensure that the end vertex associated with a priority class and a vehicle corresponds to the start vertex associated with the next lower class and the same vehicle. Constraints (14.25) and (14.26) require that each route starts and ends at the depot location, respectively. A feasible plow route consists of a sequence of partial routes for priority classes 1, 2, 3, etc., where the partial route for plow h in priority class k begins at time t_{k-1}^h and ends at time t_k^h. A schematic representation of a feasible route h in G'' is illustrated in Figure 14.1. Recall that the partial route associated with class $K+1$ corresponds to the shortest path from the last serviced arc by vehicle h in class K to the depot v_0.

Two solution approaches are proposed to solve the model. The first method, called the *parallel algorithm*, constructs several routes in parallel by sequentially solving a *multiple vehicle rural postman problem* (m-RPP) with vehicle-road segment dependencies, turn restrictions, and load balancing constraints for each priority class, considering all traversed arcs as already plowed. The m-RPP consists of designing a set of m vehicle routes of least total cost, such that each route starts and ends at the depot and each required arc appears in at least one route and is serviced by exactly one vehicle. For each priority class, the m-RPP is formulated as a linear, mixed integer program. The second approach, called the *cluster-first, route-second algorithm*, first partitions the arcs into clusters, each having approximately the same workload by adapting a technique proposed by Benavent et al. [2] based on solving a generalized assignment problem. A vehicle route is then constructed

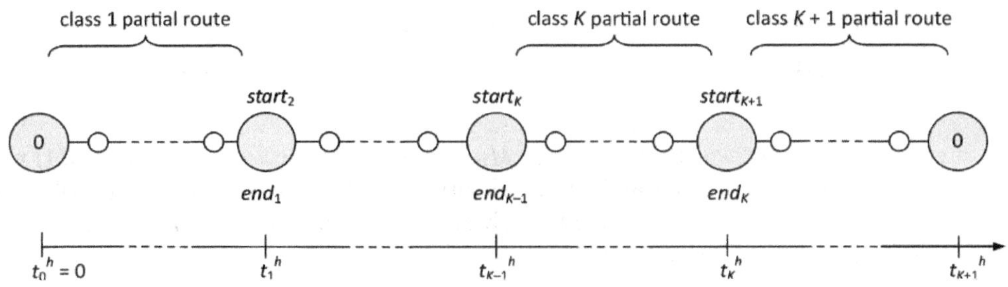

Figure 14.1. *Schematic representation of a feasible route h in G″, from Perrier, Langevin, and Amaya [27].*

in each cluster through the solution of a *hierarchical rural postman problem* (HRPP) with class upgrading possibilities, vehicle-road segment dependencies, and turn restrictions. For each cluster, the HRPP is formulated as a linear, mixed integer program. All linear integer models were solved with CPLEX. The two solution approaches were tested on data from the road network of Dieppe, Canada, involving 462 vertices and 1234 arcs partitioned into three classes to be serviced by eight vehicles of three different types. Results indicated that the two methods produced routes that dominate the existing routing plan of City of Dieppe. Solutions were obtained in a few hours of computing time, which was deemed satisfactory since the problem needs only be solved once every winter season.

To address the dynamic nature of winter maintenance operations in which fleet and other resources must be directed in real time to where they are most needed in response to equipment breakdowns or weather changes, Fu, Trudel, and Kim [16] formulated a winter road maintenance scheduling problem that can be used to schedule snow plows in real time. Given a fleet of plowing vehicles and a set of predefined maintenance routes, the problem consists of developing an operations plan for the available service vehicles that specifies route assignment, service type, and the corresponding start time. The scheduling model takes into account both operating costs and level of service requirements, as well as road network topology, road classes, and weather forecasts. To capture the dynamic nature of the problem, a planning horizon is divided into a set of uniform time intervals. These intervals represent the shortest intervals which can be scheduled or during which any operation can take place. Let I be the set of road segments on a road network. For each road segment $i \in I$, define l_i, λ_i, and ϕ_i as the length of road segment i, expressed as the number of time intervals required to service it or to deadhead it, the weight factor ascribing "importance" values to road segment i, and the minimum level of service for road segment i, represented by the maximum allowable snow accumulation, respectively. The weight factors λ_i allow modeling of different priority classes of roads. Let K be the set of predetermined maintenance routes, and for each route $k \in K$, define L_k as the length of route k, expressed as the number of time intervals required to traverse it. For each route $k \in K$ and for each road segment $i \in I$, let r_{ki} represent the ordinal value of road segment i's position in route k and let $R = \{r_{ki}\}$. Define T as the set of time periods. For each route $k \in K$ and for each time period $t \in T$, let x_{kt} be a binary variable equal to 1 if and only if a service vehicle is to depart on route k at the beginning of time interval t. Let $X = \{x_{kt}\}$. For each road segment $i \in I$ and for each time period $t \in T$, let y_{it} and y'_{it} be two intermediate nonnegative integer variables representing the number of vehicles on road segment i at time interval t and the number of vehicles that are actually servicing road segment i at time interval t, respectively. The y_{it} variables are used to reflect the fleet size restriction, and the y'_{it} variables are used to model the removal of snow. Let $Y = \{y_{it}\}$ and $Y' = \{y'_{it}\}$. The values of the y_{it} and y'_{it} variables can be determined as linear functions of scheduling decisions (X), route-link incidental relationship (R), and link service speed or time (l). That is, $Y = f(X,R,l)$ and $Y' = f'(X,R,l)$, where f and f' are linear functions. For each road segment $i \in I$ and for each time period $t \in T$, define also c_{it} as an intermediate nonnegative real variable representing the estimated pavement condition of road segment i at the end of time interval t, and let w_{it} represent the additional snow accumulation on road segment i caused by snowfall at time interval t. The value of the c_{it} variables is determined recursively as a function of the condition level in the previous time interval, plus any weather-related changes, minus any changes due to plowing. That is, $c_{it} = \max\{0, c_{i,t-1} + w_{it} - \Delta . y'_{it}\}$, where Δ represents the maximum depth of snow that can be cleared in a single pass of a vehicle, and c_{i0} is the initial road surface condition at the beginning of time interval 0. Finally, define d as the size of the vehicle fleet. The winter road maintenance scheduling problem can be formulated as the following linear, mixed integer program:

$$\text{(14.32)} \quad \text{Minimize} \quad \sum_{i\in I}\sum_{t\in T} \lambda_i c_{it} + \beta \sum_{k\in K}\sum_{t\in T} L_k x_{kt}$$

$$\text{(14.33)} \quad \text{s.t.} \quad c_{it} = \max\left\{0, c_{i,t-1} + w_{it} - \Delta \cdot y'_{it}\right\} \quad (i \in I, t \in T),$$

$$\text{(14.34)} \quad c_{it} \le \phi_i \quad (i \in I, t \in T),$$

$$\text{(14.35)} \quad \sum_{i\in I} y_{it} \le d \quad (t \in T),$$

$$\text{(14.36)} \quad Y = f(X,R,l),$$

$$\text{(14.37)} \quad Y' = f'(X,R,l).$$

The objective function (14.32) minimizes a weighted total of the negative impact of service (via c_{it}) and the cost of providing service. The first term of the objective function represents the total road condition index or snow depth adjusted by road class weighting factor. The second term represents a function of service kilometers, where β is a factor assigning relative importance to these two terms as well as converting the two terms into a common unit. Constraints (14.33) represent the road condition transition based on a recursive formula described above. Constraints (14.34) and (14.35) enforce service level and fleet size restrictions, respectively. Constraints (14.36) and (14.37) define the location indices Y and Y' from the values of X, R, and l.

Model (14.32)–(14.37) is solved using a branch-and-bound method with bounds computed using the simplex algorithm. The model was applied to a realistic case based on the winter maintenance plan of the City of Waterloo, Ontario, Canada. Four weather forecasting scenarios were considered, each assuming the same total amount of snowfall over the treatment area but a different level of forecasting accuracy. In particular, the *time and spatially accurate* scenario corresponds to a perfect forecast of the storm where perfect information is assumed on the snowfall rates for each road segment at every time interval. The *time and spatially averaged* scenario represents the least accurate weather forecast where a single snowfall amount is assumed for the entire storm duration for all roads. Good performance was obtained, as measured by the 28% predicted reduction in overall snow accumulation throughout the test period when the time and spatially accurate scenario is compared with the time and spatially averaged forecast. This result illustrates the benefit of responding to highly accurate real-time weather information for winter road maintenance operations.

Kuchera and Gupta [23] proposed a model and a heuristic method for finding plowing and spreading routes with maximum route duration, vehicle capacity, and both-sides service constraints. Each vehicle must start and end a route at the depot. In each route, a road segment can be deadheaded or serviced, but not both. This work was motivated by the street network in St. Louis County, Minnesota, which includes many dead ends and loops that require additional time to turn around a snow plow. To model this, the time to perform each turnaround must be added to the deadhead time. The objective function gives priority to high-traffic roads by maximizing the sum of the product of road length and the *annual average daily traffic* (AADT) count on the road segment. The objective function also included a penalty for each time a vehicle must perform a turnaround. Let $G=(V,A)$ be a directed network with counterpart arcs in opposite directions between intersection nodes, where $V=\{v_0,v_1,\ldots,v_n\}$ is the vertex set and $A=\{(v_i,v_j):v_i,v_j\in V$ and $i\neq j\}$ is the arc set. For each arc $(v_i,v_j)\in A$, define a_{ij}, d_{ij}, ss_{ij}, and ds_{ij} as the AADT count of arc (v_i,v_j), the length of arc (v_i,v_j) (in miles), the speed of a vehicle servicing arc (v_i,v_j), and the speed of a vehicle deadheading arc (v_i,v_j), respectively. For each arc $(v_i,v_j)\in A$, define the binary constant r_{ij} equal to 1 if and only arc (v_i,v_j) is an existing road segment in graph G. For each arc $(v_i,v_j)\in A$, let x_{ij} and y_{ij} be two binary variables equal to 1 if and only if arc (v_i,v_j) is serviced and deadheaded by a vehicle, respectively. Also, for each arc $(v_i,v_j)\in A$, let t_{ij} and b_{ij} be two binary variables equal to 1 if and only if a vehicle turns around at node j on arc (v_i,v_j) due to a dead end (or a constraint that forces the vehicle to turn around) and due to a three-node loop, respectively. Finally, define δ, c, w, W, and K as the average time required by a vehicle to perform a turnaround, the cost associated with a turnaround, the sand/salt application rate (in pounds per lane mile), the spreader sanding/salting capacity (in pounds), and the maximum duration of a route (in hours). The formulation is given next.

$$\text{(14.38)} \quad \text{Maximize} \sum_{(v_i,v_j)\in A}\left(d_{ij}a_{ij}x_{ij}-c\cdot t_{ij}-\frac{1}{6}c\cdot b_{ij}\right)$$

$$\text{(14.39)} \quad \text{s.t.} \sum_{(v_i,v_j)\in A}\left(x_{ij}\frac{d_{ij}}{ss_{ij}}+y_{ij}\frac{d_{ij}}{ds_{ij}} +\delta\cdot t_{ij}+\frac{1}{6}\delta\cdot b_{ij}\right)\leq K,$$

$$\text{(14.40)} \quad t_{ij}\leq x_{ij}+y_{ij} \quad ((v_i,v_j)\in A),$$

$$\text{(14.41)} \quad t_{ij}\leq x_{ji}+y_{ji} \quad ((v_i,v_j)\in A),$$

$$\text{(14.42)} \quad \left(x_{ij}+y_{ij}\right)-\sum_{(v_j,v_k)\in A} r_{jk}\left(x_{jk}+y_{jk}\right)\leq t_{ij} \quad ((v_i,v_j)\in A),$$

$$\text{(14.43)} \quad t_{ij}+t_{ji}\leq 1 \quad ((v_i,v_j)\in A),$$

$$\text{(14.44)} \quad \frac{1}{6}\Big[\left(x_{ij}+y_{ij}\right)+\left(x_{ji}+y_{ji}\right)+\left(x_{jk}+y_{jk}\right) +\left(x_{kj}+y_{kj}\right)+(x_{ki}+y_{ki})+(x_{ik}+y_{ik})\Big]\leq b_{ij} \quad ((v_i,v_j),(v_j,v_k),(v_k,v_i)\in A),$$

$$(14.45)\qquad \sum_{v_j \in V:(v_i,v_j)\in A} x_{ij} + y_{ij} = \sum_{v_j \in V:(v_j,v_i)\in A} x_{ji} + y_{ji} \qquad (v_i \in V),$$

$$(14.46)\qquad \sum_{(v_i,v_j)\in A} w \cdot d_{ij} x_{ij} \leq W,$$

$$(14.47)\qquad x_{ij} + y_{ij} \leq 1 \qquad ((v_i,v_j)\in A),$$

$$(14.48)\qquad x_{ij} = x_{ji} \qquad ((v_i,v_j)\in A),$$

$$(14.49)\qquad x_{0m} + y_{0m} = 1,$$

$$(14.50)\qquad x_{m0} + y_{m0} = 1,$$

$$(14.51)\qquad x_{ij}, y_{ij}, t_{ij}, b_{ij} \in \{0,1\} \qquad ((v_i,v_j)\in A).$$

The objective function (14.38) effectively gives higher priority to roads with higher AADT counts and longer lengths, and it penalizes turnarounds. The third term in the objective function is multiplied by 1/6 since six variables b_{ij} are implied for each three-node loop in which a turnaround occurs. Constraint (14.39) enforces time limitations accounting for turning the plow around. Constraints (14.40)–(14.44) account for turnarounds. Constraints (14.40)–(14.41) state that a vehicle must travel both arcs (v_i, v_j) and (v_j, v_i) for a turnaround to be counted. Constraints (14.42) state that if a vehicle travels arc (v_i, v_j) and has no other option at node j except to travel arc (v_j, v_i), then a turnaround is counted. Constraints (14.43) avoid single-road subtours occurring when turnarounds are performed at each end of a single road segment. Constraints (14.44) state that if a vehicle plows (deadheads) one side of the street in a three-node loop, it must turn around and plow (deadhead) the other side of the street. Flow conservation equations must be satisfied for each node via constraints (14.45). Constraint (14.46) ensures that the vehicle capacity is not exceeded. Constraints (14.47) ensure that a vehicle can either service or deadhead a road segment, but not both. Constraints (14.48) ensure that if one side of the street is serviced, then the other side of the street must also be serviced. Finally, constraints (14.49) and (14.50) ensure that a vehicle travels from the depot (node v_0) to the beginning of the route (node m) and returns to the depot along this same road segment.

The heuristic solution method consists in sequentially solving model (14.38)–(14.51) for each route until all arcs in A are serviced. For the first route, all arcs in A are available to be selected for plowing and spreading. If a second route is required, then x_{ij} is set to 0 for those arcs that were already serviced in the first route, and so on until all arcs in A are serviced. If a route contains subtours, a heuristic is used to eliminate them by representing each subtour as a node and solving the underlying traveling salesman problem. The accuracy of the heuristic method was validated using road and weather data from St. Louis County. The heuristic was embedded in a decision support tool to assist planners in determining the sequence in which to service road segments within a route, the time required to service each route, the number of vehicle operators needed, and the amount of sand/salt required. In a related report (Gupta et al. [19]), the authors describe a decision support tool to optimize workforce deployment decisions under different storm conditions, including storm uncertainty.

The research described up to this point has focused primarily on developing routes for plows, independently from other winter road maintenance decisions. However, there are strong interactions between snow plow routing and several other decisions, including the design of service sectors, the location of disposal sites and depots, and the routing of vehicles for spreading, loading, and hauling operations. (See Perrier, Langevin, and Campbell [28, 30, 31].) Models that take all these aspects of winter road maintenance into consideration become extremely complex, if not simply intractable. The traditional approach has thus been to solve these interdependent problems separately or sequentially, with the plow routes being among the latter decisions. Obviously, a sequential approach may lead to suboptimal decisions at all planning levels.

In an effort to integrate several different decisions, Jang, Noble, and Hutsel [20] proposed a formulation and a solution method for a combined depot location, sector design, spreading and plowing route design, fleet configuration, and vehicle scheduling problem. The model, which assumes a heterogeneous fleet, takes into account road class service frequency and spreader capacity constraints. Service routes must be established exclusively for each class, but a vehicle can service multiple routes in different classes with different service frequencies. The objective considered is to minimize the number of working vehicles. Let $G = (V,A)$ be a directed graph where $V = \{v_1, v_2, \ldots, v_n\}$ is the node set and $A = \{(v_i, v_j) : v_i, v_j \in V, i \neq j\}$ is the arc set. In addition, let $D \subset V$ be the set of potential depot locations and let $R \subset A$ represent the set of required arcs. With every arc $(v_i, v_j) \in R$ are associated a required amount of material to spread q_{ij}, a service time t'_{ij}, a deadheading time $t_{ij} < t'_{ij}$, and a service frequency per 12 hours f_{ij}. Define also K and H as the sets of routes and vehicles, respectively. For every route $k \in K$, let $f_k = \max\{f_{ij} : (v_i, v_j) \in K\}$ be the service frequency of route k for 12 hours. Each vehicle $h \in H$ has the material capacity to spread Q_h. For every arc $(v_i, v_j) \in R$ and for every route $k \in K$, let x_{ijk} and y_{ijk} be two binary variables equal to 1 if and only arc (v_i, v_j) is serviced and deadheaded in route k, respectively. For every vehicle $h \in H$, let z_h be a binary variable equal to 1 if and only if vehicle h is used. For every route $k \in K$ and for every vehicle $h \in H$, let u_{kh} be a binary variable equal to 1 if and only if vehicle h serves route k. Also, for every depot $d \in D$, let w_d be a binary variable equal to 1 if and only if location d is selected as a depot. For every vehicle $h \in H$ and for every depot $d \in D$, let s_{hd} be a binary variable equal to 1 if and only if vehicle h is assigned to depot d. Finally, let t be the time required to refill a vehicle. The formulation is as follows:

$$\text{(14.52)} \quad \text{Minimize} \quad \sum_{h \in H} z_h$$

$$\text{(14.53)} \quad \text{s.t.} \sum_{j:(v_j,v_i) \in A} (x_{jik} + y_{jik}) - \sum_{j:(v_i,v_j) \in A} (x_{ijk} + y_{ijk}) = 0 \quad (k \in K, v_i \in V),$$

$$\text{(14.54)} \quad \sum_{k \in K} x_{ijk} = 1 \quad ((v_i, v_j) \in R),$$

$$\text{(14.55)} \quad \sum_{(v_i,v_j) \in R} q_{ij} x_{ijk} \le \sum_{h \in H} Q_h u_{kh} \quad (k \in K),$$

$$\text{(14.56)} \quad \sum_{j:(v_d,v_j) \in A} (x_{djk} + y_{djk}) \ge u_{kh} + s_{hd} - 1 \quad (k \in K, h \in H, d \in D),$$

$$\text{(14.57)} \quad \sum_{i:(v_i,v_d) \in A} (x_{idj} + y_{idk}) \ge u_{kh} + s_{hd} - 1 \quad (k \in K, h \in H, d \in D),$$

$$\text{(14.58)} \quad \sum_{v_i,v_j \in S} (x_{ijk} + y_{ijk}) \le |S| - 1 + n^2 \theta_k^S + n^2 (2 - u_{kh} - s_{hd}) \quad (k \in K, h \in H, d \in D,$$

$$S \subset V\setminus\{d\}, S \neq \emptyset),$$

$$\sum_{v_i \in S, v_j \notin S} (x_{ijk} + y_{ijk}) \tag{14.59}$$

$$\geq 1 - \phi_k^S - n^2(2 - u_{kh} - s_{hd}) \quad (k \in K, h \in H, d \in D, \; S \subset V\setminus\{d\}, S \neq \emptyset),$$

$$\theta_k^S + \phi_k^S \leq 1,\ \theta_k^S, \phi_k^S \in \{0,1\} \quad (k \in K, d \in D, \; S \subset V\setminus\{d\}, S \neq \emptyset), \tag{14.60}$$

$$\sum_{h \in H} u_{kh} = 1 \quad (k \in K), \tag{14.61}$$

$$u_{kh} \leq z_h \quad (k \in K, h \in H), \tag{14.62}$$

$$f_k \geq f_{ij} x_{ijk} \quad (k \in K, (v_i, v_j) \in R), \tag{14.63}$$

$$\sum_{k \in K, (v_i, v_j) \subset A} f_k \Big(t'_{ij} x_{ijk} \tag{14.64}$$

$$+ t_{ij} y_{ijk} + t \Big) u_{kh} \leq 12 \quad (h \in H),$$

$$\sum_{d \in D} w_d = M, \tag{14.65}$$

$$\sum_{d \in D} s_{hd} = z_h \quad (h \in H), \tag{14.66}$$

$$s_{hd} \leq w_d \quad (h \in H, d \in D), \tag{14.67}$$

$$x_{ijk}, y_{ijk}, u_{kh}, s_{hd}, z_h, w_d \in \{0,1\} \quad (k \in K, h \in H, d \in D, (v_i, v_j) \in A). \tag{14.68}$$

The objective function (14.52) minimizes the number of vehicles used. Constraints (14.53)–(14.60) are extensions of the capacitated arc routing problem. Constraints (14.53) are flow conservation equations for each route. Constraints (14.54) ensure that each required arc is serviced. Constraints (14.55) guarantee that the vehicle capacity is never exceeded. Constraints (14.56) and (14.57) force all routes to start and end at the associated depot. Constraints (14.58)–(14.60) ensure that the solution does not contain any subtours, where θ_k^S and ϕ_k^S are binary variables defined in (14.60). Constraints (14.61) ensure that each route is serviced by exactly one vehicle. Constraints (14.62) ensure that each route is assigned to a vehicle that is used. Constraints (14.63) state that the service frequency of a route is determined by the maximum service frequency of any arc in the route. Constraints (14.64) allow vehicles to serve multiple routes as long as the service frequency of each route is satisfied. Constraint (14.65) states that exactly M depots are to be located. Constraints (14.66) require each vehicle used to be assigned to exactly one depot. Finally, constraints (14.67) ensure that each vehicle is assigned to a depot that is selected.

This is a complex formulation that includes elements of capacitated arc routing (constraints (14.53)–(14.60)) and facility location with heterogeneous vehicles, multiple routes, and time constraints, as well as nonlinear constraints (constraints (14.64)) linking vehicles to multiple routes. Therefore, the authors proposed an iterative approach to solve the model with four stages:

1. *Location of depots and assignment of arcs to sectors.*

2. *Construction of initial routes.*

3. *Improvement of routes and assignments of arcs to sectors.*

4. *Configuration of fleet and scheduling.*

From stage 4 the solution procedure loops back to stages 3, 2, and 1 iteratively to improve the solution by changing the number and locations of depots, as well as the assignments of arcs to sectors, the number and design of routes, and the number and schedules of the vehicles. Each stage is described briefly in turn.

The first stage to locate depots and assign arcs to sectors (i.e., depots) includes two procedures. Initially, this is accomplished via an integer programming formulation (a p-median), but in subsequent iterations a greedy add-drop heuristic is used to locate depots. For details for the greedy procedure, see Jang, Noble, and Hutsel [20].

In stage 2 to construct the initial routes, each priority class in each sector is treated separately. Initial routes are constructed by simply dividing a Chinese postman tour into feasible routes. This provides an upper bound on the number of feasible routes required. Some of these initial routes are then combined in subsequent iterations to reduce the total number of routes. Stage 3 includes sequential application of intra-route and inter-route improvement procedures, first within each sector and then across sectors. Throughout the improvements, both the capacity and the service frequency constraints should be met. The improvement heuristics consider moving and exchanging sets of arcs, extending the single arc movement and exchange heuristics developed by Qiao [33]. See Jang, Noble, and Hutsel [20] for details on the implementation of the improvement heuristics.

In stage 4 to determine the fleet configuration and schedules for the vehicles, the routes for each depot are treated separately. The goal is to minimize the number of vehicles required, so the key is to effectively use a vehicle to service multiple routes while obeying the specified service frequencies on all routes. The earlier stages in the algorithm determine the routes for each depot, so the associated service times, service frequencies, and amounts of materials to spread are known. Recall that vehicles are involved in plowing and spreading in this model. The fleet configuration and scheduling problem can then be formulated as an integer programming problem.

Let I be the set of feasible routes and J be the set of available heterogeneous vehicles. For each vehicle $j \in J$, let y_j be a binary variable equal to 1 if and only if vehicle j is used, and let also Q_j represent the maximum amount of material that a heterogeneous vehicle j can spread. Let t be the time to refill a truck with materials at a depot. For each route $i \in I$ and for each vehicle $j \in J$, let x_{ij} be a binary variable equal to 1 if and only if route i is served by vehicle j. For each route $i \in I$, define f_i, q_i, and s_i as the service frequency of route i, the amount of material to spread on route i, and the service duration of route i, respectively. The fleet configuration and vehicle scheduling are determined by solving the following 0-1 integer program:

$$\text{Minimize} \quad \sum_{j \in J} y_j \tag{14.69}$$

$$\text{s.t.} \quad \sum_{j \in J} x_{ij} = 1 \qquad (i \in I), \tag{14.70}$$

$$\sum_{j \in J} Q_j x_{ij} \geq q_i \qquad (i \in I), \tag{14.71}$$

$$\sum_{i \in I} f_i(s_i + t)x_{ij} \leq 12 \qquad (j \in J), \tag{14.72}$$

$$x_{ij} \leq y_j \quad (i \in I, j \in J), \tag{14.73}$$

$$x_{ij} \in \{0,1\} \quad (i \in I, j \in J), \tag{14.74}$$

$$y_j \in \{0,1\} \quad (j \in J). \tag{14.75}$$

The objective function (14.69) minimizes the number of vehicles used. Constraints (14.70) assign each route to exactly one vehicle. Constraints (14.71) ensure that the demand of each route is satisfied. Constraints (14.72) allow each vehicle to serve multiple routes while satisfying the service frequency over a 12-hour period. Constraints (14.73) ensure that a route is assigned to only an operating vehicle.

The four-stage heuristic was applied to the state highway road network in Boone County, Missouri, which includes 1030 lane-miles in three priority classes serviced by the *Missouri Department of Transportation* (MoDOT). Results indicated that MoDOT could achieve the same high level of service with up to 30.4% fewer snow removal vehicles. Tests were also conducted for closing one of the existing depots and for using fewer optimally located depots, and the results further reduced the number of vehicles required, while slightly improving the service frequencies. A similar model was also applied to MoDOT's St. Louis District, which includes about 6000 lane-miles (Jang, Noble, and Nemmers [21]). Optimized routes were implemented during the 2011–2012 winter, but results from field testing the new routes are not available due to the limited snowfall that winter.

14.3.2 ▪ Metaheuristics

Dali [13] proposed a sequential constructive heuristic for designing snow plow routes in a multidepot network so as to minimize the total deadhead distance, while satisfying some side constraints such as service continuity, both-sides service, vehicle capacity, and maximum time for service completion. The problem is modeled as a capacitated arc routing problem. Consider a directed graph $G=(V,A)$, where $V=\{v_0,v_1,\ldots,v_n\}$ is the vertex set, with v_0 being the depot, and $A=\{(v_i,v_j): v_i,v_j \in V \text{ and } i \neq j\}$ is the arc set, and let $R \subseteq A$ be the subset of required arcs. The heuristic can be outlined as follows.

Repeat the following steps until all required arcs are serviced:

1. Determine the nonserviced required arc $(v_i,v_j) \in R$ that is farthest from the depot v_0.

2. Create a feasible vehicle route consisting of a shortest path of nonserviced required arcs between v_0 and v_i, the required arc (v_i,v_j), and a shortest path of nonserviced required arcs between v_j and v_0.

3. Select a nonserviced required arc that induces the smallest detour, and insert this arc into the route.

4. Repeat step 3 as long as the vehicle capacity and the maximum service completion time constraints permit.

This heuristic combines the ideas contained in the parallel-insert algorithm developed by Chapleau et al. [9] for the capacitated arc routing problem, except that the routes are constructed one by one. Steps 1 and 2 generate a vehicle route by determining the farthest required arc from the depot. Then, in steps 3 and 4, the nearest nonserviced required arcs from the route are sequentially inserted into the route as long as it remains feasible. The route construction process is repeated until all required arcs are serviced. The author did not, however, provide numerical results.

Salazar-Aguilar, Langevin, and Laporte [36] introduce a synchronized arc routing problem for snow plowing operations. In this problem, routes must be designed in such a way that street segments with two or more lanes in the same direction are plowed simultaneously by synchronized vehicles in a *tandem operation*. The authors provide a mathematical nonlinear integer programming formulation. Since the model is intractable even for small size instances, a local search heuristic based on the *Adaptive Large Neighborhood Search* (ALNS) metaheuristic (Ropke and Pisinger [35]) is developed. The heuristic constructs an initial set of feasible routes, and then attempts to improve them by means of different destroy/repair operators. The choice of operators is controlled dynamically according to their past performance. Each operator has a weight which controls how often it is used during the search, and the weights are adjusted as the search progresses so that the heuristic adapts to the instance at hand and to the state of the search. The initial set of routes is constructed while ensuring the synchronization on multilane roads. For the improvement phase, five destroy/repair operators have been developed:

- $N1$: *Best arc sequence removal best insertion*: The aim of this operator is to reduce the makespan. A sequence of arcs is randomly chosen and removed from the longest route. It is then inserted into another route which traverses the maximum number of arcs belonging to the sequence. The insertion point is after the arc having the shortest path from the initial arc of the sequence.

- $N2$: *Random arc sequence removal-insertion*: The goal of this operator is to diversify the search. A sequence of arcs is removed from the longest route and randomly inserted into another route.

- $N3$: *Best interchange of arc status (serviced-traversed)*: This operator attempts to intensify the search process. An arc (i,j) serviced by the longest route P_d is randomly chosen. If another route uses this arc just for traversal, the statuses of this arc in these two routes are interchanged. Route P_d is reoptimized by computing shortest paths in the links affected by the change of status of arc (i,j).

- $N4$: *Worst interchange of arc status (traversed-serviced)*: This operator attempts to diversify the search. The route with the shortest duration is chosen, and an arc (i,j) which is only traversed by the route is randomly selected. Another route P_d servicing (i,j) is identified, and an interchange of statuses in the arc (i,j) is carried out in both routes. As in $N3$, P_d is reoptimized by computing shortest paths in the links affected by the change of status of arc (i,j).

- $N5$: *First-traversed first-serviced*: The aim of this operator is to intensify the search process. All routes in the current solution P are reoptimized. Since each arc can be traversed several times by the same route but should be serviced only once, this procedure is applied to each route P_d in P in such a way that the arcs that are both serviced and traversed by P_d are reordered in an attempt to reduce the makespan. Thus, each arc that is both serviced and traversed by P_d will be forced to be serviced the first time it is traversed in route P_d.

The performance of the proposed ALNS procedure was evaluated on large artificial and real instances. The average improvement yielded by the improvement phase of the algorithm is significant, which is an indication of the efficiency of this phase.

Table 14.1 summarizes the snow plow routing research described in the previous sections.

Table 14.1. *Characteristics of recent plow routing models.*

Authors	Problem type	Planning level	Problem characteristics	Objective function	Model structure	Solution method
Golbaharan [17]	Plow routing	Operational	Multi-depots, time windows	Min route cost plus penalty for extra vehicles used	Set partitioning problem	Column generation
Sochor and Yu [38]	Plow routing	Operational	Multi-depots	Min route cost plus penalty for extra vehicles used	Set covering problem	Column generation
Razmara [34]	Plow routing	Operational	Time windows, service frequencies, heterogeneous fleet	Min route cost plus penalty for extra vehicles used	Set partitioning problem	Column generation and tabu search
Perrier, Langevin, and Amaya [27]	Plow routing	Tactical	Service hierarchy (general precedence relation), separate passes for multilane roads, class upgrading, vehicle-road segment dependency, turn restrictions, load balancing	Min hierarchical objective	Linear MIP	Optimization-based heuristics
Fu, Trudel, and Kim [16]	Plow scheduling	Real-time	Service hierarchy, minimal service level, maximum number of vehicles	Min total weighted service kilometers and negative impact of service	Linear MIP	Branch and bound
Kuchera and Gupta [23]	Plow and spreader routing	Operational	Vehicle capacity, service hierarchy, maximum route duration, both-sides service	Max compliance with the service hierarchy and turn restrictions	Linear 0-1 IP	Optimization-based method
Jang, Noble, and Hutsel [20]	Combined depot location, sector design, plow and spreader routing, vehicle scheduling, and fleet configuration	Strategic and operational	Service hierarchy, heterogeneous fleet, vehicle capacities, service frequencies, multiple routes per vehicle	Min number of vehicles used	Capacitated arc routing problem	Optimization-based composite method
Dali [13]	Plow routing	Operational	Service continuity, multiple depots, both-sides service, vehicle capacity, maximum time for service completion	Min deadhead distance	Capacitated arc routing problem	Constructive heuristic
Salazar-Aguilar, Langevin, and Laporte [36]	Plow routing	Operational	Service continuity, tandem operations, vehicle capacity, maximum time for service completion	Min completion time	Linear MIP	Metaheuristic (adaptive large neighborhood search)

14.4 ▪ Case study of implementation of "optimized" plow routes

This section presents a case study of implementation of route optimization for snow plowing in Centennial, Colorado, for which the city received the 2012 Achievement Award for Transportation Operations from the Institute for Transportation Engineers. Throughout this section, "optimized" is used to refer to the new and improved plow routes developed for Centennial, which is not to imply that these are proven mathematically to be the best in regards to any particular metric.

This case documents the benefits of route optimization software combined with GPS-based vehicle location and route guidance systems. Centennial is a city with a population of about 100,000 and an area of 27.9 square miles in the southern part of the Denver Metropolitan Statistical Area. Details for this case are derived from Callen [6, 7], Sedlak [37], and City of Centennial [10]. Centennial includes approximately 470 lane-miles of roadways plowed for every storm, including some multilane roads. Public works service in Centennial are provided by CH2M HILL via a public-private partnership, and private firms are contracted for snow plowing. Plows generally clear a 10-foot wide path, although traffic lanes may be wider. Snow is plowed from the travel lanes during a storm, though the turn lanes and paved medians may be plowed after the storm. Both 1-axle and 2-axle trucks with spreaders are used in snow removal operations. State highways in Centennial are serviced by the *Colorado Department of Transportation* (CDOT). For most storms snow is plowed only from the major roadways, which are classified in two priority classes. Priority 1 roadways consist of major arterial streets, which are generally located on a one-mile grid and have four traffic lanes. Some roadways needed to access emergency services are also in priority 1. Priority 2 roadways include the minor arterial and major collector streets that link the major arterials in priority 1 with the residential streets. Residential streets are generally not plowed, though in the event of a severe snowfall they can be plowed (at the discretion of the city). The roadways to plow in Centennial include approximately 265 lane-miles classified as priority 1 and 203 lanes-miles classified as priority 2. The objective for snow plowing in Centennial is to complete the plowing of all roads as quickly as possible, with the priority 1 roads completed before beginning the priority 2 roads. More specifically, the primary objective in designing routes was to ensure that the routes in each priority class could be completed at about the same time. Secondary objectives were to minimize the aggregate completion time to plow all routes and the aggregate mileage. Reducing the number of plows required was not considered. Prior to the route optimization study, Centennial used 10 plows to service 10 priority 1 "routes" and 10 priority 2 "routes," where these routes were specified only as a collection of roadways to plow without a specified sequence for plowing. Drivers were free to service the routes in whatever order they deemed best.

The route optimization study was undertaken by C2Logix, Inc., using the FleetRoute™ software (http://www.c2logix.com/software/fleet-software-fleetroute), in conjunction with considerable pre- and postprocessing to account for practical aspects not handled automatically by the software. One difficulty in route design was to accommodate U-turns, which were prohibited in many places, but feasible in certain locations (e.g., some parking lots). Other routing preferences used in the route design include favoring turning right and going straight at intersections, and discouraging turning left. Another difficulty for plow routing in Centennial derived from the need to obey the hierarchy in clearing the two priority classes with a subsidiary goal of minimizing deadheading travel between the priority 1 and 2 routes and from the depot to the beginning of the route. Keys to developing improved routes were (i) balancing of workload across the routes, (ii) replacing the 10 priority 1 routes (of 4 lanes) served by a single plow each, with five priority 1 routes served by tandem plows, and (iii) linking the end of the priority 1 routes with the start of the priority 2 routes. Thus, with the improved routes, each pair of plows from a priority 1 route would split apart to start two priority 2 routes, and then at the end of those they would rejoin to start the priority 1 route in tandem again. Balancing the routes and linking the starting and ending points proved crucial to improve service and reduce costs.The plow speeds for modeling were set at 13.5 mph when plowing, based on field data, and the posted speed limit for the roadway (or 25 mph if no speed limit was posted) when not plowing.

Some drivers were initially resistant to the route optimization study and expressed doubts that a "computer" could tell them what to do or improve on their own route plans. Therefore, in addition to finding improved routes, with the help of FleetRoute™, the use of guided-driver technology was essential for success. The new routes were programmed into Garmin GPS units, so that the drivers could be provided with real-time directions to find and follow the routes. This required some customized programming of the GPS units, but the result was essential in providing (i) visual displays and voice directions of the new routes, (ii) real-time location-based directions to rejoin the route should a driver get off route, and (iii) assurance that all roadways in the route were actually covered. These benefits were especially important for new drivers and replacement drivers, who were not as familiar with a particular route or region.

The first step to assess the potential benefits from the route optimization study was to compare the modeled performance of the improved routes with the performance prior to the route optimization study. Unfortunately, the actual routes being followed were not known, since each driver was free to decide to how to service the roadways in their "route." Furthermore, the routes (really just collections of roadways) specified by the city were not exactly the same as those being used by the present contractor, so two sets of "before" routes were addressed: (i) "city routes" were those collections of roadways specified by the city, but serviced in a sequence as determined using the FleetRoute™ software; and (ii) "contractor routes" were those collections of roadways being serviced by the contractor, but serviced in a sequence as determined using the FleetRoute™ software. Because the software was used to determine the travel paths and travel times, the performance for the city routes and contractor routes may overstate actual performance by the drivers (i.e., the actual times to service the roadways may be greater). Table 14.2 provides some details on the city routes, current routes, and "optimized" routes. The first four columns of results provide the times (in minutes) for traveling from the depot to the start of the priority 1 routes, the time to complete the priority 1 routes, the time to travel between the priority 1 and 2 routes, including the time to return from the priority 2 route to the start of the priority 1 route to begin servicing it again, and the time to complete the priority 2 routes, respectively. The last two columns report the total time for all routes and the total mileage. The table is divided horizontally into sections for each of the three route types, with a summary of the changes from the contractor routes to the "optimized" routes in the last section. For each route type (or the summary) the table reports the total for all 10 vehicles, the range, and the maximum for a single vehicle.

These results show that the city routes and contractor routes are quite similar in performance, with the contractor routes being a little shorter (18 minutes out of 38.7 hours) and better balanced, but still having a range of over 2 hours (137 minutes) and a maximum of over 5.5 hours. The routes from the optimization study are much better in reducing variability, though the total time is only 95 minutes (4%) less than with the contractor routes. However, the range for these new routes is only 52 minutes compared to 137 minutes for the contractor routes (a 62% decrease), and the longest of these new routes is only 252 minutes compared to 332 minutes for the longest contractor route (and 354 minutes for the longest city route). Thus, the completion time for all routes is reduced by 80 minutes or 24% with the "optimized" routes, although the total savings in miles and time are

Table 14.2. *Centennial route lengths before and after the route optimization study (all times are in minutes).*

		Time from facility to start of priority 1	Time for priority 1	Time between priority 1 and priority 2	Time for priority 2	Total time	Total mileage
City routes	Total	110.7	1140.9	46.1	1040.3	2339	645
	Range	21.0	110.9	10.4	125.0	187	59
	Max	22.2	177.6	12.5	162.4	354	104
Contractor routes	Total	110.1	1125.3	47.5	1039.0	2321	618
	Range	16.8	144.5	5.5	71.1	137	46
	Max	18.7	188.3	7.0	138.9	332	96
"Optimized" routes	Total	81.2	1042.5	42.2	1059.2	2226	587
	Range	30.3	41.6	11.2	25.5	52	14
	Max	15.1	117.8	11.8	117.8	252	67
Savings	Total	29.0	82.8	5.3	−20.2	95	31
	Range	−13.5	102.9	−5.7	45.6	85	32
	Max	3.6	70.5	−4.8	21.1	80	29

only 5% and 4%, respectively. Note that the majority of the savings derive from improved priority 1 routes, both in reducing the length and in better orienting the routes relative to the depot; the total time for all 10 "optimized" priority 2 routes is actually longer than the time for the priority 2 contractor routes, though the maximum time and range are much smaller in the new routes.

The new optimized routes were implemented in the 2011–2012 winter, and the benefits in Centennial were documented from several storms. For a 12-inch snowstorm prior to the route optimization study on December 21, 2011, the priority 1 and 2 routes were completed in 8 hours, which was deemed to be typical performance. In a 15-inch storm on February 2, 2012, using the new "optimized" routes, the completion time for the priority 1 and 2 routes was reduced 28% to 5.5 hours. In a subsequent smaller storm, the time was reduced to 4.5 hours. These benefits translate into the need for less equipment to achieve the targeted level of service or the ability to provide a higher level of service with a given fleet. These benefits were achieved not just as a result of the new improved routes, but as a combined result of route "optimization" and route guidance technology: Route "optimization" can help equalize the workload and provide improved travel paths, but route guidance technology is needed to help ensure that the plows actually follow the desired travel paths. Further, beyond what the route optimization study and the technology provided, the successful implementation in Centennial required commitment and involvement of the staff across the organization, from managers down to drivers, in finalizing the routes. Deep involvement of the City staff was essential in allowing the users to take ownership of the route optimization process, including the use of route guidance technology, and realize the maximum potential for savings.

14.5 ▪ Conclusion

This chapter reviews recent research on snow plow route optimization and documents a trend in the recent advances to mathematical-based optimization and to models that include greater real-world complexity. Interesting and important recent research directions also include adapting new metaheuristics, such as Adaptive Large Neighborhood Search, and developing models that integrate several distinct problems into a more comprehensive composite model for winter road maintenance. While the last decade has witnessed some impressive progress in snow plow routing research, and "in practice, vehicle routing may be the single biggest success story in operations research" (Golden, Raghavan, and Wasil [18]), similar progress has not occurred in implementing route optimization for winter road maintenance, including snow plowing. It does appear that models are generally still not comprehensive enough to consider all that needs to be included, and the mathematical programming-based models do not have the ease of use required by operating personnel. In many agencies, plow routes are still often created manually, or on the fly by the drivers, even though software for snow plow route "optimization" has existed for decades. This lack of use of such software for snow removal seems to be due to several reasons: (i) shortcomings of the software for plow route optimization, (ii) a lack of understanding and comfort with optimization software among public works personnel, (iii) difficulties in budgeting for software and route optimization studies, as opposed to for equipment and materials, (iv) challenging organization structures where plowing is often contracted to private firms, and (v) the difficulty of assessing the return on investment for route optimization. Both (iii) and (v) are especially affected by the inconsistent nature of winter operations from year to year due to the unpredictable and varying severity of winter.

The shortcomings of many arc routing software packages with regard to snow plow route optimization stem from the complexity of winter road maintenance operations and the nature of the real-world objectives, especially with hierarchies of levels of service and time-based objectives. One particular area of complexity for snow plow routing is the need to properly model road intersections and interchanges, where many turn lanes may need to be added to the mathematical networks (electronic maps). This can generate considerable deadheading in solutions. Because many commercial software packages for arc routing do not address the complexities required for snow removal operations, implementable solutions generally require extensive and expensive customization of routes. For example, the Ohio DOT undertook a program to use a snow removal route optimization software package about a decade ago (Ohio Department of Transportation [25]). However, after careful testing they found that the software "had limitations that prevented it from providing real-life routing scenarios," especially in handling complex interchanges and ramps, and was "extremely time-consuming and cost prohibitive." As a consequence the program was discontinued. From a software vendor's perspective, the fragmented nature of public works operations and the seasonal nature of snow removal activities suggest a smaller market and/or a more difficult market to penetrate than for some other arc routing applications (e.g., waste collection, street sweeping, postal operations) where single organizations have very large problems and operations are performed on a more consistent basis. Yet, in aggregate the savings from optimizing winter road maintenance routes seems substantial.

A final important reason for the lack of implementation of optimized snow plow routes is that optimizing routes is only part of the puzzle and managing the process of snow removal to ensure drivers follow the optimized routes is equally important. A key component is the capability to easily and effectively communicate the plow routes to the drivers, often under difficult operating conditions such as during storms, in traffic, amidst parked or abandoned vehicles, and at night. *Automated vehicle location* (AVL) technologies coupled with optimized routes allows drivers to receive real-time driving directions, without needing to consult maps or memorize routes, as well as real-time directions to return to a route should the driver travel off-route, either unintentionally or due to unanticipated conditions. Thus, the use of guided-driver technology may be an essential tool to leverage the value of the route optimization for winter road maintenance. But success also depends on the organizational dynamics, and involvement and buy-in by all stakeholders is essential to exploit the benefits from route optimization. We hope that the recent Transportation Achievement Award from the Institute of Transportation Engineers for the implementation of snow plow route optimization in Centennial, Colorado, signals a growing recognition of both the benefits of route "optimization" for winter road maintenance and the ability to handle real-world complexities. In summary, while the contributions to the practice of route optimization for snow plowing have been slow to come, we do believe that new models that are more comprehensive and more attuned to the practical constraints of winter operations will be useful in practice, when coupled with the appropriate technology and attention to organizational dynamics.

Acknowledgments

The authors would like to thank Monty Sedlak and Kevin Callen for their valuable comments and insights on the case study.

Bibliography

[1] E. BALAS, *The prize collecting traveling salesman problem*, Networks, 19 (1989),

pp. 621–636.

[2] E. BENAVENT, A. CAMPOS, A. CORBERÁN, AND E. MOTA, *The capacitated arc routing problem: A heuristic algorithm*, Qüestiió, 14 (1990), pp. 107–122.

[3] R.R. BLACKBURN, K.M. BAUER, D.E. AMSLER SR., S.E. BOSELLY, AND A.D. MCELROY, *Snow and ice control: Guidelines for materials and methods*, Tech. report, Transportation Research Board, Washington, DC, 2004.

[4] S.E. BOSELLY, *Update of the AASHTO guide for snow and ice control*, Tech. Report prepared as part of NCHRP Project 20-7, Task 250, National Cooperative Highway Research Program, Transportation Research Board, December 2008.

[5] E.A. CABRAL, M. GENDREAU, G. GHIANI, AND G. LAPORTE, *Solving the hierarchical Chinese postman problem as a rural postman problem*, European Journal of Operational Research, 155 (2004), pp. 44–50.

[6] K. CALLEN, *C2Logix FleetRoute Snow Plow Routing Set-up and Analysis for the City of Centennial, CO* (2011).

[7] ——, Private communication (2012).

[8] J.F. CAMPBELL AND A. LANGEVIN, *Roadway snow and ice control*, in Arc Routing: Theory, Solutions and Applications, M. Dror, ed., Kluwer, Boston, MA, 2000, pp. 389–418.

[9] L. CHAPLEAU, J.A. FERLAND, G. LAPALME, AND J.-M. ROUSSEAU, *A parallel insert method for the capacitated arc routing problem*, Operations Research Letters, 3 (1984), pp. 95–99.

[10] CITY OF CENTENNIAL, *Snow and ice control plan, Updated*, Tech. report, Centennial, CO, December 2012.

[11] G. CLARKE AND J. WRIGHT, *Scheduling of vehicles from a central depot to a number of delivery points*, Operations Research, 12 (1964), pp. 568–581.

[12] T.M. COOK AND B.S. ALPRIN, *Snow and ice removal in an urban environment*, Management Science, 23 (1976), pp. 227–234.

[13] Z. DALI, *Optimization of vehicle routing for plowing and snow disposal*, in Proceedings of the 9th International Conference of Chinese Transportation Professionals, ICCTP 2009, Critical Issues in Transportation System Planning, Development, and Management 358, 2009, pp. 2738–2744.

[14] P. DAMODARAN AND M. KRISHNAMURTHI, *A continuous simulation model for snow removal*, International Journal of Industrial Engineering, 12 (2005), pp. 189–199.

[15] R. EGLESE, B. GOLDEN, AND E. WASIL, *Route optimization for meter reading and salt spreading*, in Arc Routing: Problems, Methods, and Applications, A. Corberan and G. Laporte, eds., MOS-SIAM Series on Optimization 20, SIAM, Philadelphia, 2014.

[16] L. FU, M. TRUDEL, AND V. KIM, *Optimizing winter road maintenance operations under real-time information*, European Journal of Operational Research, 196 (2009), pp. 332–341.

[17] N. GOLBAHARAN, *An Application of Optimization to the Snow Removal Problem: A Column Generation Approach, Thesis No. 886*, Ph.D. thesis, Linköping University, Linköping, Sweden, 2001.

[18] B.L. GOLDEN, S. RAGHAVAN, AND E.A. WASIL, *Preface*, in The Vehicle Routing Problem: Latest Advances and New Challenges, Springer, New York, 2008.

[19] D. GUPTA, E. TOKAR-ERDEMIR, D. KUCHERA, A.K. MANNAVA, AND W. XIONG, *Optimal workforce planning and shift scheduling for snow and ice removal*, Tech. Report MN/RC 2011-03, Minnesota Department of Transportation Research Services Section, St. Paul, MN, 2010.

[20] W. JANG, J.S. NOBLE, AND T. HUTSEL, *An integrated model to solve the winter asset and road maintenance problem*, IIE Transactions, 42 (2010), pp. 675–689.

[21] W. JANG, J. NOBLE, AND C. NEMMERS, *Optimizing winter/snow removal operations in MoDOT St. Louis District—Includes outcome based evaluation of operations*, final report CMR 12-007, Missouri Department of Transportation, October 2011.

[22] M. KRISHNAMURTHI AND P. DAMODARAN, *A modified postman tour heuristic for efficient snow removal planning*, in Proceedings of the 7th Industrial Engineering Research Conference, Banff, Canada, 1998.

[23] D. KUCHERA AND D. GUPTA, *A decision support tool for optimizing winter maintenance operations*, in Transportation Research Board 88th Annual Meeting, 2009, no. 09-3244.

[24] L. NORDIN, *Energy Use within Road Maintenance Operations with Potentials for Increased Efficiency*, Licentiate thesis A135, University of Gothenburg, Gothenburg, Sweden, 2011.

[25] OHIO DEPARTMENT OF TRANSPORTATION, *Snow & Ice Practices*, Division of Operations—Office of Maintenance Administration, March 2011.

[26] N. PERRIER, J.F. CAMPBELL, A. LANGEVIN, AND M. GENDREAU, *Vehicle routing models and algorithms for winter road spreading operations*, in Hybrid Algorithms for Service, Computing and Manufacturing Systems: Routing and Scheduling Solutions, J.R. Montoya-Torres, A.A. Juan, J. Huaccho Huatuco, G.L. Rodriguez-Verjan, and J. Faulin, eds., IGI Global, Hershey, PA, 2012, pp. 15–45.

[27] N. PERRIER, A. LANGEVIN, AND C.A. AMAYA, *Vehicle routing for urban snow plowing operations*, Transportation Science, 42 (2008), pp. 44–56.

[28] N. PERRIER, A. LANGEVIN, AND J.F. CAMPBELL, *A survey of models and algorithms for winter road maintenance. Part I: System design for spreading and plowing*, Computers & Operations Research, 33 (2006a), pp. 209–238.

[29] ——, *A survey of models and algorithms for winter road maintenance. Part II: System design for snow disposal*, Computers & Operations Research, 33 (2006b), pp. 239–262.

[30] ——, *A survey of models and algorithms for winter road maintenance. Part III: Vehicle routing and depot location for spreading*, Computers & Operations Research, 34 (2007a), pp. 211–257.

[31] ——, *A survey of models and algorithms for winter road maintenance. Part IV: Vehicle routing and fleet sizing for plowing and snow disposal*, Computers & Operations Research, 34 (2007b), pp. 258–294.

[32] PIARC TECHNICAL COMMITTEE B5 WINTER SERVICE, *Snow and Ice Databook 2010*, PIARC, Quebec, Canada, 2010.

[33] H. QIAO, *Capacitated Rural Directed Arc Routing Problem: Algorithms and Applications*, Master's thesis, University of Maryland, College Park, MD, 1999.

[34] G. RAZMARA, *Snow Removal Routing Problems: Theory and Applications*, Ph.D. thesis, Linköping University, Linköping Studies in Science and Technology, Linköping, Sweden, 2004.

[35] S. ROPKE AND D. PISINGER, *An adaptive large neighbourhood search heuristic for the pickup and delivery problem with time windows*, Transportation Science, 40 (2006), pp. 455–472.

[36] M.A. SALAZAR-AGUILAR, A. LANGEVIN, AND G. LAPORTE, *Synchronized arc routing for snow plowing operations*, Computers & Operations Research, 39 (2012), pp. 1432–1440.

[37] M. SEDLAK, Private communication (2012).

[38] J. SOCHOR AND G. YU, *A heuristic method for routing snowplows after snowfall*, Tech. report, Linköping Institute of Technology, Linköping, Sweden, 2004.

[39] W.B. TUCKER AND G.M. CLOHAN, *Computer simulation of urban snow removal*, in Snow Removal and Ice Control Research. Special report no. 185, Transportation Research Board, Washington, DC, 1979, pp. 293–302.

[40] WINTER RESILIENCE REVIEW, *The Resilience of England's Transport Systems in Winter*, final report, Department for Transport, United Kingdom, October 2010.

Chapter 15
Routing in Waste Collection Applications

Gianpaolo Ghiani
Cândida Mourão
Leonor Pinto
Daniele Vigo

15.1 • Introduction

Today, municipalities all over the world face a great number of problems related to waste management. In fact, in addition to the large and increasing amount of solid waste generated each year, a rising public concern for environmental preservation makes *Solid Waste Management* (SWM) one of today's main issues for public decision makers.

For planning purposes, an SWM system can be decoupled into two main subsystems: a regional and a municipal management system. The regional management system is in charge of the overall SWM organization, including the facilities network for waste treatment and disposal (e.g., incinerators, landfills, transfer stations) and the specific traffic regulations for accessibility. The municipal management system is responsible for the transportation of the various types of waste from their source of production to either a treatment plant or a final disposal site. This involves collecting a set of of containers, also called bins, which are spread over the territory. It also includes decisions on the type of waste to be collected and the means to be used, sizing and characteristics of the collection points, fleets and crews, the collection services frequency, crew scheduling, and vehicle routing problems.

Some of these problems have already been examined and categorized by Marks and Liebman [40], who identified some relevant issues to be addressed by operations research methodologies. Since this seminal study, both real-world waste management systems and optimization methodologies and tools have undergone a considerable evolution, as reported in the recent survey on SWM strategic and tactical issues by Ghiani et al. [30]. Examples of new and increasingly important subjects are waste separation, which requires proper definition of waste types and characteristics, and green logistics, which measures the environmental impact of the collection and treatment systems. The latter, for example, may aim to reduce energy consumption, waste production, and pollutant emissions. From the 1970s waste management systems started to appear within operation research applications, with the sanitation departments of large cities in developed countries, such as New York or Washington, DC, noticing the need for optimization. Since Beltrami and Bodin [10], several studies have been carried out, leading to the design of a computerized system by Bodin et al. [15]. Recent surveys on routing studies with waste collection applications include Golden, Assad, and Wasil [31], Kim, Kim, and Sahoo [37], Francis, Smilowitz, and Tzur [25], Wøhlk [59], Corberán and Prins [21], and Benjamin [11].

The focus of this chapter is the routing of vehicles in municipal waste systems. This problem poses complex management challenges and requires a large amount of resources. Vehicles start and end their service route at one or more depots, with the waste usually dumped at a landfill or a treatment plant, which must be visited during the route. Waste containers are generally concentrated at a set of collection sites, with either a small number of large containers or individual smaller bins spread over the streets of a neighborhood. Each site is characterized by a specific rate at which the various types of waste are left. Such rates are random in nature and exhibit both long-term (yearly) and short-term (weekly) patterns. Examples of parameters that must be estimated are the time needed to dump a bin into the truck or the time to unload at the landfill.

More precisely, the solid waste collection problem consists of designing a set of vehicle routes such that

- every route starts and ends at a depot, and the vehicle must arrive empty to the last depot in the event that it does not coincide with a landfill or a treatment site;
- every bin is emptied before it overflows, with a minimum frequency depending on the type of waste and on the season (e.g., food or organic waste might be collected more often in summer than in winter for sanitary reasons), and following a specific pattern (e.g., bins should be always emptied on the same weekday);
- on every route the total amount of waste collected between the depot and the first visit to a landfill or treatment site, or between two successive visits to a landfill or treatment site, must not exceed the vehicle capacity;
- the duration of any route does not exceed a maximum shift length, in accordance with local rules and regulations.

Further operational constraints often occur in practice, namely,

- recyclable materials (glass, paper, metal, plastic, textiles, etc.) may be collected together by vehicles with multiple compartments;
- when landfills and treatment plants are far from the municipalities, the waste may be carried to some transshipment sites, consolidated, and then transported to its final destination by large trucks, which usually imposes additional synchronization requirements between the two phases;
- the shape of the vehicle routes might be constrained (e.g., routes may be required to replicate the subdivision of the municipality, or to avoid overlapping);
- some vehicles may be forbidden to traverse some streets (e.g., large vehicles may be banned from narrow streets in historical centers);
- collection sites might not be visited on specific days;

- some streets must be serviced during prescribed time windows (e.g., at night to avoid traffic congestion or during the day to avoid noise in residential areas);
- the opening hours of disposal and treatment sites must be taken into account;
- some maneuvers (e.g., U-turns) might be prohibited at some intersections.

The waste collection problem described above can be conceptually split into two main components: (i) determination of the collection days for each location and each type of waste; and (ii) definition of the vehicle routes, once the collection days have been determined. With the purpose of keeping the operations simple, the first issue is generally addressed by defining a periodic schedule. The second issue can be modeled as a node or arc routing problem.

Node Routing (NR) is appropriate to handle problems with large containers located at a relatively small number of individual locations, as is the case of hazardous materials (e.g., industrial or hospital waste), recyclables (e.g., glass, paper or plastic), commercial waste, or even household collection whenever large containers are used (e.g., underground containers). *Arc Routing* (AR) deals with continuous collection, which applies to some household refuse collection cases. The objective of the optimization is the minimization of the overall cost, which is essentially dependent on the number of routes and the total distance traveled by the vehicles.

General Routing Problems (GRP), which consider both node and arc demands, have also been proposed within many applications. Notwithstanding, in waste collection, the use of different vehicles usually enables the separation between arc and node routing case studies. This is reflected by the fact that there are only a few general routing applications in waste collection, such as Álvarez-Valdés et al. [1], Corberán, Mejía, and Sanchis [20], and Prins and Bouchenoua [48].

The aim of this chapter is to highlight approaches constructed to deal with real-world problems. The most promising approach (NR or AR) is determined by the underlying nature of each case.

The remainder of the chapter is organized as follows. In Section 2, node routing waste collection studies are surveyed, while Section 3 is devoted to arc routing approaches. In Section 4, an in-depth description of a case study in Portugal is provided. Finally, in Section 5 future research topics are presented.

15.2 ▪ Node routing and waste collection

As outlined in the previous section, there are a number of waste collection problems in which node routing modeling is more suitable than arc routing modeling. This is the case when the collection takes place at large bins located at a relatively small set of sites, or if the containers are scattered over a wide (rural) area.

Despite the practical relevance of node routing modeling, only a few studies have been devoted to this approach. One of the first papers is by Beltrami and Bodin [10], who develop a cluster-first route-second approach to solve a refuse collection problem in New York. More recently, Tung and Pinnoi [58] adapt the well-known I1 insertion algorithm of Solomon [55], for the *Vehicle Routing Problem* (VRP) *with Time Windows*, to a waste collection problem in Hanoi, Vietnam. The initial solution is then refined through a local search procedure using Or-opt and 2-opt* neighborhoods. The proposed method is able to achieve considerable savings with respect to the existing manual solutions, both in terms of the vehicles used and route efficiency.

Research in this field is now beginning to take into account the fact that collection does not take place every day. Indeed, for each pick-up point a collection frequency is specified. Such a frequency can be, for instance, equal to 1, 2, or 3 visits within a planning horizon spanning 5 to 7 days for undifferentiated urban waste and 2 to 4 weeks for recyclable or industrial waste collection. In rural areas the frequencies can be even lower, e.g., a few visits per month. The collection frequency is dictated by the waste generation rate as well as by hygiene requirements. Angelelli and Speranza [3] were among the first to address the periodic nature of the problem with a node routing approach. They study two real-world cases, one in the Trompia Valley, Italy, and the other in Antwerp, Belgium, with heterogeneous vehicles and the presence of intermediate transhipment facilities. A *Tabu Search* (TS) proposed by the same authors for the *Periodic VRP* (PVRP) *with Intermediate Facilities* (PVRP-IF) is applied (see Angelelli and Speranza [4]). To speed up the search, which examines 343 and 180 collection sites, respectively, groups of close sites sharing the same service requirements are clustered into macro-points. The algorithm assigns a random collection schedule to the points and builds an initial schedule accordingly. Then, a TS is developed by using a neighborhood defined by exchanges of containers among routes, changes in their visit schedules, and the merge/split of routes. Infeasible solutions are handled by means of appropriate penalties in the objective function. Several simulations are run considering different types of vehicles (i.e., side-loader with fixed or with demountable body vehicles), different shift durations, and the presence of unloading facilities inside or near the collection area. The relevant cost savings are reported.

A similar PVRP model is proposed by Baptista, Oliveira, and Zúquete [7] who describe a heuristic approach for a waste paper collection problem in Almada, Portugal. The problem involved two vehicles and 59 collection points over a time horizon of 20 days. The objective was the maximization of the net revenue. The authors define an integer linear programming model and a heuristic algorithm inspired by that of Christofides and Beasley [17] for the PVRP. Savings in collection costs, derived from implementing the computed solution, were significant enough to obtain an overall positive net income.

Another example of a PVRP application is provided by Teixeira, Antunes, and Sousa [57] who present a case study related to the collection of three recyclable waste materials in Aveiro and Coimbra, Portugal. The large size of the instances (two depots and thousands of points to be visited in a time horizon of 20 working days) motivated the authors to adopt a three-phase heuristic approach. In the first phase, the collection area is subdivided into five zones that are planned separately. The aim of the zoning procedure is to group the municipalities into balanced areas. Then, in the second phase, the type of waste to be collected on each day and in each zone is identified by distributing, as regularly as possible, the collection activities along time. Finally, the collection routes are constructed by an insertion heuristic. The computational results showed distance reductions even with increased collection frequencies.

Similar experiments and results have been reported by Sahoo et al. [49] and by Kim, Kim, and Sahoo [37], who discuss real-world applications in the United States, characterized by multiple depots, multiple trips per shift, and several operational soft and hard constraints (such as route geographic compactness, load balancing between routes, and drivers' lunch breaks).

Aringhieri et al. [6] study the collection and disposal of bulky recyclable waste in Perugia. Besides the pick-up of full containers, dumping operations at plants as well as the delivery of empty containers are also considered. The problem is modeled as a directed VRP and solved through a local search based algorithm previously developed.

Nuortio et al. [46] are among the first to report the use of modern metaheuristics to solve waste collection problems. In particular, they make use of a *Guided Variable Neigh-*

borhood Thresholding procedure, proposed by Kytöjoki, Nuortio, and Bräysy [38], to a waste collection problem modeled as a *PVRP with Time Windows*. Very large instances with around 14,000 collections points were considered. Remarkable savings, sometimes larger than 40% were achieved. More recently, Benjamin and Beasley [12, 13] have proposed several metaheuristic approaches, including TS and *Variable Neighborhood Search* (VNS), to solve a commercial waste collection problem similar to the one considered by Kim, Kim, and Sahoo [37]. Hemmelmayr et al. [34] use a VNS to solve a PVRP-IF that models the collection of several separate waste types (glass, metal, plastic, and paper) in Austria.

Metaheuristics in SWM routing is also the option of Karadimas, Papatzelou, and Loumos [36], who embedded an ant colony method in a *Geographic Information System* (GIS) to optimize the collection of urban solid waste in the municipality of Athens. An adaptive large neighborhood search algorithm allowed Buhrkal, Larsen, and Ropke [16] to outperform the results of [12] on the instances proposed by [37]. Motivated by real cases from a Danish company, the problem in [16] is modeled as a VRP with time windows and driver breaks. Real-world data from the company are also used for the validation of the approach. Bing et al. [14] present a VRP model and a tabu search heuristic to pursue the "eco-efficiency" using real data of plastic waste collection in the Dutch municipality of Wageningen.

Recent efforts have been devoted to incorporating the determination of the collection frequencies and the allocation of bins to the collection sites into route collection design. In this line of research, Hemmelmayr et al. [35] study the design of a collection service by integrating both bin allocation and routing. They develop several integer linear programming models for the optimal bin allocation, examining different scenarios and incorporating them into a VNS for route design. Their extensive computational testing on instances from the literature and on real-world data showed the benefit of an integrated heuristic with respect to a classical hierarchical approach.

Finally, Muyldermans and Pang [45] have studied the simultaneous collection of several types of waste materials with multi-compartment vehicles. A *Local Search* procedure with well-known moves, and efficient neighborhood lists, is combined with a *Guided Local Search* metaheuristic. Computational results compared both multi-compartment and separate collection approaches.

15.3 ▪ Arc routing and waste collection

Waste collection can be conveniently modeled as an *Arc Routing Problem* (ARP) whenever the waste is in small containers located almost continuously along a street. This is the case of many household waste collection problems in urban areas. Under these assumptions, collection can be modeled as some variant of the classical *Capacitated ARP* (CARP) on an undirected, directed, or mixed graph. We focus on some of the most relevant studies.

Beltrami and Bodin [10] model a street sweeping problem as a multiple *Chinese Postman Problem* with time limit constraints. Feasible solutions are achieved by two heuristic methods. In the first one, a single giant route traversing all streets is initially determined. Then, this tour is divided into feasible trips with respect to a given time limit. The second heuristic starts with a sectorization procedure to identify small zones compatible with the service of one vehicle. Later, a set of feasible routes is determined sequentially. This preliminary work helped in the design of one of the very first computerized sanitation and scheduling systems by Bodin et al. [15]. A GIS for the town of Oyster Bay, New York, was used to design efficient collection routes and to assess alternative collection schemes under a variety of policies.

Using data from the solid waste management system in Cleveland, Ohio, Clark, and Lee [18] develop a simulation model to compare alternative heuristic approaches to waste collection, including waiting times at dump-sites, and the potential use of transfer stations. The authors describe the data gathering phase and the relationships between some parameters that are adjusted by using regression techniques. They also establish productivity measures for solid waste collection systems, and decision-making tools so that trade-offs among critical investments may be better evaluated. As a consequence, a severe cutback in the number of collection routes and in personnel led to an impressive reduction in the annual budget.

Other relevant case studies have been carried out by Mourão and Almeida [42] and Mourão and Amado [43], who examine waste collection on randomly generated networks that replicate the Lisbon city map. The first study defines the problem on directed graphs, with a dump-site far from both the depot and the collection district. The papers present two lower bounds and a three-phase heuristic that transforms one of the lower bound solutions into a feasible one. Lower bounds are derived from a specific problem formulation, and are computed very fast as they are based on the resolution of simple problems, such as the transportation problem. In [43] a heuristic is used to first decide the direction an edge should be served, with the result that the mixed networks can then be treated as directed ones. Then, a method is proposed to determine feasible solutions with some specific characteristics, as load balanced trips and with a few intersections. In fact, in this type of application, it is usually desirable that different vehicle routes should have as few common streets as possible and that the collection workload is fairly distributed among the crews. The results are competitive against other known CARP heuristics with better bounds and a more balanced set of trips.

Maniezzo [39] has modeled an urban solid waste collection as a CARP on a directed graph, named *Directed CARP* (DCARP). The aim is to design vehicle routes that minimize the total distance required to collect all bins, while obeying a vehicle's capacity and bin compatibility constraints. Forbidden turns and one-way streets are also considered. Large networks are needed to represent the municipalities of towns with about 100,000 inhabitants. First, each DCARP instance is transformed into a node routing equivalent one, resulting in a very large graph (with hundreds or thousands of nodes). The size of the resulting instances requires simple heuristics based on local search procedures. These metaheuristics are tested on the large real-world instances and on some benchmark data with competitive results. A *Decision Support System* (DSS) with ESRI ArcView interface (see www.esri.com) allows the author to work on road maps and to display the computational results so that they can be easily used by service operators.

Real-world road network constraints, such as undesirable or forbidden turns, have also been analyzed by Coutinho-Rodrigues, Rodrigues, and Clímaco [22], who examine waste collection problems in five cities in Portugal—Figueira da Foz, Leiria, Aveiro, Viseu, and Covilhã. The authors produced an easy-to-use software package, which constructs feasible routes and graphically represents them on the network. This facilitates the practical implementation. Several CARP heuristics are compared, and the well-known *Path Scanning* algorithm of Golden, DeArmon, and Baker [32] is adapted to handle additional features such as a heterogeneous fleet, mixed networks, forbidden or penalized turns (usually left or U-turns), and time limit constraints. The reported savings on the total distance traveled were about 49%. The control of turns in a waste collection case study was previously considered by McBride [41], where an assignment problem is solved to assign each "entrance" in a node to a unique exit, thus avoiding bad turns.

Ghiani et al. [26] present a study of a waste collection problem in Castrovillari, a small town in southern Italy with a population of around 25,000 inhabitants spread over a wide area. In this case, large vehicles were not allowed to traverse the narrow streets located in historical sites. In addition, some streets had to be serviced at night to avoid traffic congestion, thus requiring the use of time windows. Finally, some vehicles were not able to unload some bins (two types of bins were in use). All trucks were based at a depot, located in the central area, and had to dump in a sanitary landfill site on the outskirts of the town. The resulting instance is defined over a mixed multigraph with 5273 nodes, 5292 edges, and 584 arcs. The authors developed a computerized system, based on a cluster-first route-second approach, which avoided overtime in the service and reduced the total costs by about 8%. Computational tests confirm the robustness of the solutions in response to changes in the amount of waste produced.

Belenguer et al. [9] have developed heuristics and lower bounds for the *Mixed CARP* (MCARP). Feasible solutions incorporating forbidden turns or windy edges (i.e., with asymmetric traversal costs) are generated by three constructive heuristics and a memetic algorithm. A linear formulation reinforced with valid inequalities permits the use of a cutting-plane algorithm to find tight lower bound values. These procedures provided very good results on randomly generated waste collection problems. Lower bounds in [9] are compared with those of Gouveia, Mourão, and Pinto [33], who propose a compact formulation and a relaxation for the mixed CARP associated with waste collection in Lisbon, Portugal. The models used flow variables to guarantee the solutions' connectivity with a polynomial number of constraints. The best-known lower bounds for some benchmark instances were improved in [33] with a relaxation able to produce very good bounds in fairly small computation times.

Bautista, Fernández, and Pereira [8] have presented a study carried out in the municipality of Sant Boi de Llobregat, within the metropolitan area of Barcelona, Spain. This municipality, with over 80,000 inhabitants, is represented by a network with 220 required edges and 459 arcs, inducing a subgraph with 72 strongly connected components. Moreover, the map of the city contains 4150 turns, 1316 of which are forbidden ones (76% are U-turns). The mixed ARP is modeled as a node routing that accounts for the forbidden turns. Two ant colony heuristics, which combine constructive and local search methods exploring various neighborhoods, are discussed. One constructive method has been developed for each of the two metaheuristics: a nearest neighbor and a nearest insertion, where forbidden turns are avoided by allowing the visit to only one node per cluster. The existing waste routes in the municipality were considerably improved, with a total length reduction of around 35%. The software application, named SIRUS, provided a graphical tool to help the municipality managers in several tasks, such as the location of collection areas, allocation of bins, and the design of routes.

Recently, Santos, Coutinho-Rodrigues, and Antunes [50] have presented a user-friendly, *web-based Spatial DSS* (wSDSS) to design multiple feasible routes for urban waste collection in Coimbra, Portugal. Due to the spatial nature of these problems, the wSDSS requires a GIS, thus permitting high-speed interactive systems to compare alternatives in real time. In addition to the previously described issues (i.e., mixed networks, forbidden turns, and vehicle capacity), this work also considers time windows and time limit constraints. A CARP heuristic approach and an ant colony metaheuristic are adapted to include the practical constraints. The wSDSS combines optimization methods with Google Maps, resulting in an easy-to-use tool that can be used for what-if analysis related to changes in the input data.

Other CARP generalizations are frequently reported to deal with waste collection case studies, as illustrated next. Ghiani, Improta, and Laporte [28] introduced a variant to the *CARP with multiple Intermediate Facilities* (CARPIF) as dump-sites: a key feature in many waste collection problems. Lower bounds, based on a relaxation defined for the so-called deadheading variables, and two heuristics are proposed. One transforms a feasible solution to the *Rural Postman Problem* into a feasible one for the CARPIF by optimally introducing IFs. The second one uses a graph modification where a feasible CARPIF solution can be found by means of a CARP heuristic. Three sets of benchmark instances were adapted to the CARPIF, and good gap values were reported.

A CARPIF generalization with tour *Length* (L) constraints (CLARPIF) was later addressed by Ghiani et al. [29]. The new ant colony optimization procedure here developed provides substantial improvements over the constructive partitioning and TS methods by Ghiani et al. [27]. The CLARPIF is also the focus of Polacek et al. [47], who have developed a simple and robust VNS that outperforms the known methods. The VNS is simple as it uses only one exchange operator in the shaking phase, and a simple inverting operator, for the local search. Its robustness is due to its ability to solve different problem classes efficiently and to the apparent independence between the quality of the initial and the final solutions obtained.

Del Pia and Filippi [24] adapt a *CARP with multiple Mobile Dump-sites* (CARP-MD) to solve waste collection in Due Carrare, a small town located in northern Italy. The mobile dump-site feature stems from the use of two types of vehicles: a small size and a large size. The smaller ones collect the waste along the streets and must unload it into the larger ones, at some point of their tour. Both types of vehicles must therefore meet along their routes, and the larger ones are thus called the mobile dump-sites. A *Local Search* heuristic is applied with significant reduction in the duration of the routes of the case study. Moreover, the authors claim that solutions seem to be robust to variations in the waste to be collected, although in this case study there was substantial stability from week to week.

Amposah and Salhi [2] have studied a waste collection problem in Accra, Ghana, where, as often happens in developing countries, high temperatures contribute to faster decomposition of the waste, and resources are not enough to ensure that all waste is collected every day. An additional concern is the specific characteristics of the waste in Accra, which may include, for instance, toxic components. The authors model the problem as a variant of the CARP in which the objective is to optimize two distinct performance indicators: the minimization of the total distance traveled and the minimization of the uncollected waste (i.e., the inconvenience due to smell). Note that the latter is time dependent, as leaving uncollected waste for a longer time leads to a worse environmental situation. The presence of several dump-sites as well as that of directed and undirected links are modeled. A constructive heuristic, taking into account both the environmental aspects and the total distance, is used. A look-ahead strategy that tries to assess the impact of the current choices in the near future is also developed to avoid bad feasible solutions generated by the constructive method. The algorithm produced several good quality solutions that were favorably considered by the local authorities.

Finally, among the long-term decisions in waste collection systems, we consider the sectorization or partition problem. In this problem, the aim is to partition the service of large areas into smaller ones that are capable of being serviced by a single vehicle. Such a waste collection problem was the focus of Mourão, Nunes, and Prins [44]. The street network was partitioned into a number of sectors and a set of minimum cost vehicle trips was built in each sector. Two-phase heuristics—where sectors were built in a first phase—and mixed CARPs to identify the routes were then solved in a second phase were proposed. Moreover, a best insertion method which simultaneously builds sectors and trips was also presented. The heuristics considered forbidden turns during construction of the trips. In addition to solution cost, evaluation criteria such as imbalance, diameter, and dispersion measures are used to compare the algorithms, which were designed to promote workload balance and compactness as well as contiguity of the trips. Computational results on benchmark networks were quite promising, revealing small gap values in a few seconds.

Sniezek et al. [54] have studied a problem in Philadelphia, USA, where the vehicle fleet is heterogeneous, and additional constraints are needed to forbid some vehicles from servicing or traversing certain streets. Their problem is a *CARP with Vehicle-Site-Dependencies* (CARP-VSD). Besides the time minimization objective, the authors aimed to find a balanced and noninterlacing partition. The latter is a partition over the arcs, such that the resulting trips from each partition define a set of balanced trips that overlap as little as possible. The routes are produced by the RouteSmart software system (www.routesmart.com) incorporated within a GIS. The algorithm was implemented in Philadelphia with success, as the generated routes were more effective than existing ones (better balanced and with less intersections), thus being better accepted by the crews. Further developments concerning the Philadelphia case study are due to Sniezek and Bodin [53].

15.4 ▪ Modeling and solving a real-world problem

To illustrate the practical issues arising in real-world applications, we describe a new case study where a hybrid heuristic is developed to optimize the routes of the refuse collecting vehicles in Seixal, one of the 18 municipalities of the Lisbon Metropolitan Area in Portugal. Since the approach combines two integer programming models with a heuristic method to fix the values of some variables, it can be considered a *matheuristic*. The models—one which is valid for the MCARP, and the other, which is one of its relaxations—were proposed by Gouveia, Mourão, and Pinto [33]. The heuristic starts by solving the relaxed model with CPLEX (see www.ibm.com). Then, using the graph induced by that solution, some services are heuristically assigned to specific trips. Finally, the exact model is applied to produce a set of feasible trips. Since a subproblem is solved where the number of variables is dramatically reduced, the exact model is able to produce feasible solutions for large instances.

The models are defined over a mixed graph, in which the nodes represent the intersections and the *links* (i.e., arcs or edges) represent the streets. Some of the links are associated with streets where waste must be collected and make up the so-called *required links* or *tasks*. *Arc-tasks* are one-way streets or large two-way streets with no zigzag collection, while *edge-tasks* stand for two-way streets when zigzag collection is allowed. In this case study, vehicle trips must respect time limits only (vehicle capacity is not relevant and demands are used for balancing purposes). The model, denoted as $F1$, uses the following notation:

- $G=(V,A)$ is a directed graph resulting from the initial mixed graph. V is the vertex set, and A the set of arcs, where each initial edge is replaced by two opposite arcs.
- $0 \in V$ is the depot node, where each vehicle trip must start and end. A copy of the depot may be used as an intermediate node.
- A_R and E_R are the sets of required arcs and edges, respectively, in the original mixed graph.
- $R \subseteq A$ is the set of required arcs in A; hence, $|R|=|A_R|+2|E_R|$.
- $\mathcal{K}$ is the set of identical vehicles, and $|\mathcal{K}|$ is the number of trips, as all vehicles are used.

- T_{max} stands for the time limit per trip.
- $t^d_{ij}, t^s_{ij}, q_{ij}$ represent the deadheading time, the service time, and the demand of task (i,j), respectively.
- $\overline{Q} = \frac{Q_T}{|\mathcal{K}|}$ is the trip average waste, where Q_T is the total waste.
- β is a given percentage so that $\beta\overline{Q}$ is the minimum service per trip.

Variables are divided into three sets: (i) the service variables, x^k_{ij}, which take value 1 if $(i,j) \in R$ is served by vehicle k, and 0 otherwise; (ii) the deadheadings, y^k_{ij}, are the number of times that the vehicle k deadheads arc $(i,j) \in A$; and (iii) the flow variables, f^k_{ij} $\forall (i,j) \in A$, related with the remaining time in (sub)trip k. The following model formulates the household waste collection problem under analysis.

$$\text{(15.1)} \quad \text{(F1)} \quad \min \sum_{k \in \mathcal{K}} \left(\sum_{(i,j) \in A} t^d_{ij} y^k_{ij} \right)$$

$$\text{(15.2)} \quad \text{s.t.} \sum_{j:(i,j)\in A} y^k_{ij} + \sum_{j:(i,j)\in R} x^k_{ij} = \sum_{j:(j,i)\in A} y^k_{ji} + \sum_{j:(j,i)\in R} x^k_{ji} \qquad \forall i \in V,\ \forall k \in \mathcal{K},$$

$$\text{(15.3)} \quad \sum_{k \in \mathcal{K}} x^k_{ij} = 1 \qquad \forall (i,j) \in A_R,$$

$$\text{(15.4)} \quad \sum_{k \in \mathcal{K}} \left(x^k_{ij} + x^k_{ji} \right) = 1 \qquad \forall (i,j) \in E_R,$$

$$\text{(15.5)} \quad \sum_{j:(0,j)\in A} y^k_{0j} + \sum_{j:(0,j)\in R} x^k_{0j} = 1 \qquad \forall k \in \mathcal{K},$$

$$\text{(15.6)} \quad \sum_{j:(j,i)\in A} f^k_{ji} - \sum_{j:(i,j)\in A} f^k_{ij} = \sum_{j:(j,i)\in R} t^s_{ji} x^k_{ji} + \sum_{j:(j,i)\in A} t^d_{ji} y^k_{ji} \qquad \forall i \in V,\ \forall k \in \mathcal{K},$$

$$\text{(15.7)} \quad \sum_{j:(0,j)\in A} f^k_{0j} = \sum_{(i,j)\in R} t^s_{ij} x^k_{ij} + \sum_{(i,j)\in A} t^d_{ij} y^k_{ij} \qquad \forall k \in \mathcal{K},$$

$$\text{(15.8)} \quad \sum_{i:(i,0)\in A} f^k_{i0} = \sum_{i:(i,0)\in R} t^s_{i0} x^k_{i0} + \sum_{i:(i,0)\in R} t^d_{i0} y^k_{i0} \qquad \forall k \in \mathcal{K},$$

$$\text{(15.9)} \quad f^k_{ij} \le T_{max}(x^k_{ij} + y^k_{ij}) \qquad \forall (i,j) \in A,\ \forall k \in \mathcal{K},$$

$$\text{(15.10)} \quad \sum_{(i,j)\in R} q_{ij} x^k_{ij} \ge \beta\overline{Q} \qquad \forall k \in \mathcal{K},$$

$$\text{(15.11)} \quad x^k_{ij} \in \{0,1\} \qquad \forall (i,j) \in R,\ \forall k \in \mathcal{K},$$

$$\text{(15.12)} \quad f^k_{ij} \ge 0 \qquad \forall (i,j) \in A,\ \forall k \in \mathcal{K},$$

$$\text{(15.13)} \quad y^k_{ij} \ge 0 \quad \text{integer} \qquad \forall (i,j) \in A,\ \forall k \in \mathcal{K}.$$

Having a fixed service time, the objective function (15.1) minimizes the total deadheading time. Very similar to the ones in Gouveia, Mourão, and Pinto [33], constraints (15.2) to (15.9) ensure the formation of feasible trips, within the time limit. In addition,

(15.10) are needed to impose a minimum service per trip, a major request from the Seixal municipality.

Lower bounds on flow variables are also set to reduce the feasible region of this valid model, namely,

$$f_{ij}^k \geq t_{ij}^d y_{ij}^k \qquad \forall (i,j) \in A \setminus R, \quad \forall k \in \mathcal{K}, \tag{15.14}$$

$$f_{ij}^k \geq t_{ij}^s x_{ij}^k \qquad \forall (i,j) \in R, \quad \forall k \in \mathcal{K}. \tag{15.15}$$

Henceforth, model $F1$ includes constraints (15.2) to (15.15) and has a polynomial number of variables and constraints. However, it turned out to be not solvable for the instances size of the case study, which includes up to 370 required links (see Table 15.1). Therefore, we considered a relaxation of $F1$, denoted by $F2$, in which the variables are aggregate over the set of trips (see [33] for more details). $F2$ is easier to solve. Starting from the solution of the relaxed problem, the heuristic proceeds by fixing some variables and then solving the original model for a smaller instance, as detailed in Algorithm 15.1. The distances between two tasks or between the depot and a task are based on lengths defined by

$$\text{length}(a) = \begin{cases} \text{service time} & \text{if } a \text{ is an arc task,} \\ \text{deadheading time} & \text{otherwise.} \end{cases} \tag{15.16}$$

One advantage of this procedure is that it allows the introduction of additional requirements that arise in the practical application. This helps to produce solutions that are more acceptable for end users, and thus likely to be implemented. For example, a set of tasks may be forced to be assigned to a specific vehicle. Next, we detail the Portuguese case study and analyze the computational results.

The total area of Seixal is 94 Km^2, and the population is larger than 150,000 inhabitants (in accordance with 2011 census; see the Câmara Municipal do Seixal (CMS) web page, www.cm-seixal.pt/cmseixal/seixal /territorio). The municipality is divided into six *freguesias*, as Figure 15.1 shows (namely, Aldeia de Paio Pires, Amora, Arrentela, Corroios, Fernão Ferro, Seixal), and is located in the Lisbon Metropolitan Area, in the Setúbal District.

In Seixal, several residential areas include individual townhouses, where a door-to-door waste collection system has been implemented. These areas are spread all over the municipality, as seen in the map of Figure 15.1. The department in charge of street cleaning opted for this type of pickup both for population comfort and for reasons of hygiene. In fact, in areas with several individual townhouses individual containers prove to be a better choice, whereas collective bins lead to an undesirable accumulation of residues in the surroundings. This study, which is part of the project PTDC/EGE-GES/121406/2010 supported by FCT (Fundação para a Ciência e a Tecnologia, Portugal), arose to analyze and propose an improvement of the current system in the municipality. Currently, the collection routes are constructed manually. The Seixal Waste Division established some specific trial zones, each one to be collected by only one vehicle. One of these manual routes is illustrated in Figure 15.2, which also depicts the townhouses and the street network.

The data collected from a GIS were used to generate real instances. The historical data of the waste collected per trip from past years were used to compute the parameters which were assigned to the links, namely, the amount of refuse per street, the service, and deadheading times. Each instance corresponds to a particular region that includes different contiguous trial zones, established by the Waste Division. To overcome limitations

Algorithm 15.1 Hybrid Heuristic

```
Read instance; α; β
{Initialization}
Mark all tasks as nonserved
Solve the aggregate model, (F2), and let G2 = (V2, A2) be the graph induced by (F2) solution
∀u, v ∈ V2 compute l_uv as the length of the shortest path from u to v using (15.16)
{Initialization of |𝒳| trips}
for x = 1 to |𝒳| do
    Identify the farthest nonserved task, a1 = (u1, v1), both from the depot and from the previous trip first
    marked task, according to lengths l
    Mark task a1 as served
    x^x_a1 ← 1
    Q_x ← q_a1
    {Filing trips by fixing the services reaching α% of vehicles capacity in trip x}
    end ← false
    UT ← {nonserved tasks in the shortest paths between the depot and terminal nodes of a1}
    repeat
        Let a2 ∈ UT be the farthest task from the depot node
        if Q_x + q_a2 ≤ α Q̄ then
            Mark task a2 as served
            x^x_a2 ← 1
            Compute Q_x ← Q_x + q_a2
            UT ← UT \ {a2}
        else
            end ← true
        end if
    until Q_x ≥ α Q̄ or end
end for
end ← false
UT ← {nonserved tasks}
repeat
    Let a ∈ UT be the farthest task from the depot node
    UT ← UT \ {a}
    Let x be the closest trip from a
    if Q_x + q_a ≤ β Q̄ then
        Q_x ← Q_x + q_a
        Mark task a as served
        x^x_a ← 1
    else
        if UT = ∅ then
            end ← true
        end if
    end if
until Q_x ≥ β Q̄ or end
{Solving the valid model}
Consider the fixed variables, and solve the valid model F1 to get a feasible solution
```

of the data, we joined zones, which usually results in harder instances to solve. It may also produce more feasible solutions and trips recognized by the practitioners as good, in reasonable computational times.

The data set is formed by nine instances, named seix1–seix9, as displayed in Table 15.1 (column 1). The network sizes vary, with 148–361 nodes and 284–713 links (columns 2–3). The percentage of required arcs is between 9% and 24% and that of required edges is between 22% and 55%. A time limit T_{max} of 7 hours was considered, and the number of available vehicles varies from 2 to 5 (column 6).

The heuristic is evaluated against lower bound and upper bound values. Two lower bounds are achieved through CPLEX within a time limit of one hour for $F1$ and half an

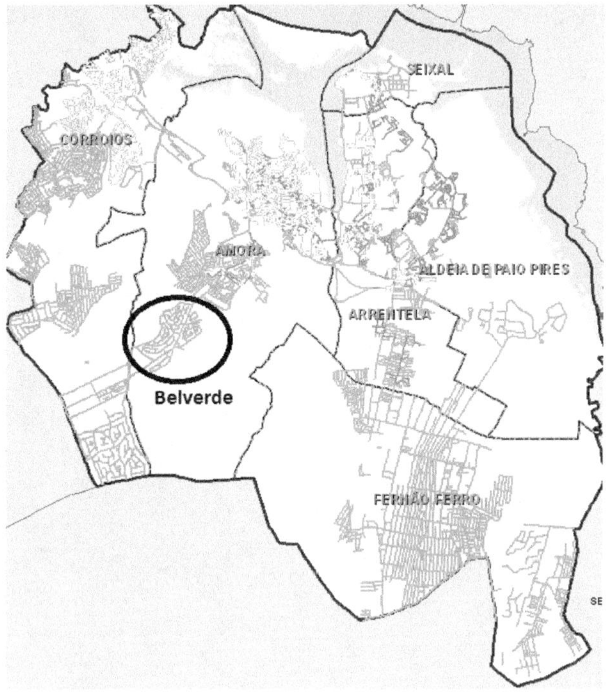

Figure 15.1. *The Seixal map. Courtesy of CMS.*

hour for $F2$. The best of them, usually the former, is considered as the lower bound value. Upper bounds are produced by solving $F1$ with CPLEX, with a one-hour time limit. For some instances, CPLEX was not able to find feasible solutions.

The last three columns in Table 15.1 show the heuristic results. Column 7 gives the percentage gap between the matheuristic value and the best lower bound given by CPLEX, while Column 8 compares it with the upper bound. In this column, a negative value results from a better hybrid heuristic solution, and an (a) means that no upper bound was found with $F1$. The last column depicts the computational times (in seconds). The results were obtained with CPLEX 12.1, on a machine with two AMD Opteron 6172 processors (24 cores), 2.1GHz (64 GB RAM).

With the proposed heuristic, we were able to automatically produce feasible solutions for all tested instances. Moreover, in 6 out of 9 cases (corresponding to negative values and (a)) we obtained better values than the CPLEX $F1$ upper bound (Column 8). Model $F1$ finds the optimum value for the three smallest instances, seix1–seix3, which coincides with the $F2$ values. In terms of waste collected the heuristic routes are as balanced as those of the original $F1$ model, which is an important issue from the real application. This requirement makes the $F1$ model harder than the CARP. Although large, the gap values, from Column 7, may reflect that the lower bounds, produced by CPLEX from the relaxed model, which aggregates trips and does not consider the balance constraint, are far from the optimum. The computational times vary from 14 s to the time limit of one hour, with an average value around half an hour. Almost all the time is consumed by CPLEX to solve the two integer models, each one with a time limit of half an hour.

One of the major concerns when dealing with applications is to construct useful tools to support decision makers. In this case study, which is still an ongoing project, the solutions produced so far have already been recognized as good by the end users.

Figure 15.2. *The Belverde route. Courtesy of CMS.*

Table 15.1. *Computational results: Municipality of Seixal.*

	Characteristics					Heuristic solution		
Name	$\|V\|$	$\|A \cup E\|$	$\|A_R\|$	$\|E_R\|$	$\mathcal{K}$	% gap LB	% gap UB	time (s)
seix1	148	284	63	119	3	12	12	538
seix2	167	318	77	70	2	2	2	14
seix3	179	350	43	120	2	52	52	523
seix4	257	437	47	235	3	42	−10	2098
seix5	286	472	56	259	3	28	(a)	2388
seix6	292	589	53	242	3	85	−11	2061
seix7	305	546	95	245	5	34	(a)	2270
seix8	319	559	99	268	5	45	(a)	3606
seix9	361	713	69	291	3	13	−28	3601

(a) CPLEX did not find an $F1$ feasible solution within one hour.

15.5 ▪ Waste collection: What's next?

Several methods developed to design waste collection vehicle tours were addressed in this chapter. Both node and arc routing approaches incorporating many specific characteristics of this problem class have been developed and successfully applied to case studies. Future research involving real-world waste collection must consider three groups of concerns: (i) methodological innovation; (ii) consideration of new issues; (iii) development of decision support systems.

Regarding the methodologies to be used for the solution of the main problems studied so far, considerable work still needs to be done. Due to the complexity of the practical constraints and the size of the instances to be solved, most of the existing approaches rely on rather basic solution methodologies and relatively few metaheuristics are employed. Hence, it will be appropriate to follow the evolution of VRP or ARP studies by extending the most successful algorithms to SWM applications. In particular, it seems very promising to mix exact results with heuristic approaches, leading to so-called matheuristic methods as described in Section 15.4.

Great research opportunities are also related to the inclusion of additional real problem characteristics into existing models, following the evolution that led to the introduction of the extended CARP (see, e.g., Belenguer et al. [9], Mourão and Almeida [42], Mourão and Amado [43]), the CLARPIF (Ghiani, Improta, and Laporte [28]), the CARP-MD (Del Pia and Filipi [24]), the CARP-VSD (Sniezek et al. [54]), and the co-collection VRP (Muyldermans and Pang [45]), which may be applied to the collection of different recyclables using vehicles with separate compartments.

Nowadays, in many countries, and in particular in the developing ones, the responsibility for street cleaning and waste collection has been transferred from municipalities to private companies. This brings with it the redefinition of the objective functions to be considered in waste collection optimization, which may shift from cost minimization to profit maximization. For instance, Aráoz, Fernández, and Franquesa [5] and Corberán et al. [19] study a cluster prize-collecting ARP, where the demand edges are clustered, and either all or none of the services of a cluster should be performed. It was possible to formulate waste collection problems in cities of Minnesota, USA, or in private companies of Buenos Aires, Argentina, this way. Another important aspect, studied by Wu et al. [60], defines objective cost functions which accommodate the effects of economies-of-scale for planning waste management systems in the region of Hamilton, Ontario, Canada. In addition, several complicating constraints should also be considered: forbidden or undesirable turns, narrow streets in old city centers, busy streets during rush hours, time windows for collection, and location of landfills or other facilities where the vehicle may need to go more than once to be emptied.

Green logistics may also become a new relevant challenge within waste collection applications. Sbihi and Eglese [52] related green logistics to waste management, where environment pollution is also taken in consideration. The authors claim that "activities concerning the transport of waste materials are clearly part of the green logistics agenda," given that obtaining trips with smaller total distances or using energy efficient vehicles offer environmental benefits such as energy and emission savings. These benefits are difficult to measure and are usually not considered explicitly, although it is clear that better trips lead to a better carbon footprint. Tavares et al. [56] incorporate in a SWM 3D GIS collection modeling a tool to measure fuel consumption and vehicle emissions, taking into account road slopes and vehicle weight. This 3D system is succefully applied in Cabo Verde.

DSS innovation seems to be an inevitable need in waste management case studies. In fact, to apply methodologies to a real case study, large quantities of real-world data must be easily managed and updated by programs. GIS have been used for routing sanitation vehicles by Bodin et al. [15] and Dallaire [23] since the end of the 1980s. Considerable innovation in GIS technologies has enable the design of effective prototypes of SWM DSSs as reported in many papers. However, careful design of GIS interfaces, ease of data entry, solution modification, and validation are still important issues for a wider diffusion of routing optimization in SWM practice. Examples of modern DSS are those developed by Santos, Coutinho-Rodrigues, and Current [51] and Santos, Coutinho-Rodrigues, and Antunes [50], who used Google Maps to solve the waste collection in Coimbra, Portugal. In this field, we should also mention the important resource for both public administrations and researchers represented by public domain GIS and digital map applications, which constitute viable alternatives to commercial ones for realistic applications.

Given the practical relevance of SWM applications, it is surprising that relatively few commercial packages explicitly tailored for this area exist on the market. We report some of the most notable exceptions. Optrak Waste Management software[6] has been used in Lisbon, Portugal. RouteSmart[7] and GeoRoute[8] software have been successfully applied to solve waste collection problems in various North American towns as reported by Sniezek et al. [54] and by Golden, Assad, and Wasil [31] among others. Caliper Corporation (USA) commercializes an add-in to TransCAD, a transportation planning software, which is able to route solid waste collection vehicles.[9] Finally, the Italian company Optit[10] has developed OptiRoute, which is a routing software based on a public domain GIS and specifically developed for urban and rural solid waste collection which was used in several applications in Italy.

Acknowledgments

Thanks are due to Mafalda Afonso for all the work with the Seixal data gathering. M.C. Mourão and L.S. Pinto wish to thank Câmara Municipal do Seixal and Fundação para a Ciência e Tecnologia (projects PEsT-OE/EGE/UI0491; PEsT-OE/ MAT/UI0152; PTDC/ EGE-GES/121406) for their support. Daniele Vigo gratefully acknowledges the support of Ministero dell'Istruzione, dell'Università e della Ricerca, Italy.

Bibliography

[1] R. ÁLVAREZ-VALDÉS, E. BENAVENT, V. CAMPOS, A. CORBERÁN, E. MOTA, J. TAMARIT, AND V. VALLS, *Arc: A computarized system for urban garbage collection*, Top, 1 (1993), pp. 89–105.

[2] S.K. AMPONSAH AND S. SALHI, *The investigation of a class of capacitated arc routing problems: The collection of garbage in developing countries*, Waste Management, 24 (2004), pp. 711–721.

[3] E. ANGELELLI AND M.G. SPERANZA, *The application of a vehicle routing model to a waste-collection problem: Two case studies*, The Journal of the Operational Research Society, 53 (2002), pp. 944–952.

[4] ——, *The periodic vehicle routing with intermediated facilities*, European Journal of Operational Research, 137 (2002), pp. 233–247.

[6] optrak.co.uk/content/optrak-waste-management
[7] www.routesmart.com
[8] www.giro.ca/en/products /georoute/index.htm
[9] www.caliper.com/Press/pr990722.htm
[10] www.optit.net

[5] J. ARÁOZ, E. FERNÁNDEZ, AND C. FRANQUESA, *GRASP and path relinking for the clustered prize-collecting arc routing problem*, Journal of Heuristics, online (2011), DOI 10.1007/s10732-011-9183-1.

[6] R. ARINGHIERI, M. BRUGLIERI, F. MALUCELLI, AND M. NONATO, *An asymmetric vehicle routing problem arising in the collection and disposal of special waste*, Electronic Notes in Discrete Mathematics, 17 (2004), pp. 41–47.

[7] S. BAPTISTA, R.C. OLIVEIRA, AND E. ZÚQUETE, *A period vehicle routing case study*, European Journal of Operational Research, 139 (2002), pp. 220–229.

[8] J. BAUTISTA, E. FERNÁNDEZ, AND J. PEREIRA, *Solving an urban waste collection problem using ants heuristics*, Computers & Operations Research, 35 (2008), pp. 3020–3033.

[9] J.M. BELENGUER, E. BENAVENT, P. LACOMME, AND C. PRINS, *Lower and upper bounds for the mixed capacitated arc routing problem*, Computers & Operations Research, 33 (2006), pp. 63–83.

[10] E. BELTRAMI AND L.D. BODIN, *Networks and vehicle routing for municipal waste collection*, Networks, 4 (1974), pp. 65–94.

[11] A.M. BENJAMIN, *Metaheuristics for the Waste Collection Vehicle Routing Problem with Time Windows*, Ph.D. thesis, Department of Mathematical Sciences, Brunel University, Uxbridge, UK, 2011.

[12] A.M. BENJAMIN AND J.E. BEASLEY, *Metaheuristics for the waste collection vehicle routing problem with time windows, driver rest period and multiple disposal facilities*, Computers & Operations Research, 37 (2010), pp. 2270–2280.

[13] ——, *Metaheuristics with disposal facility positioning for the waste collection vrp with time windows*, Optimization Letters, 7 (2013), pp. 1433–1449.

[14] X. BING, M. KEIZER, J.M. BLOEMHOF-RUWAARD, AND J.G.A.J. VAN DER VORST, *Vehicle routing for the eco-efficient collection of household plastic waste*, Waste Management, 34 (2014), pp. 719–729.

[15] L.D. BODIN, G. FAGIN, R. WELEBNY, AND J. GREENBERG, *The design of a computarized sanitation vehicle routing and scheduling system for the town of Oyster Bay, New York*, Computers & Operations Research, 16 (1989), pp. 45–54.

[16] K. BUHRKAL, A. LARSEN, AND S. ROPKE, *The waste collection vehicle routing problem with time windows in a city logistics context*, Procedia Social and Behavioral Sciences, 39 (2012), pp. 241–254.

[17] N. CHRISTOFIDES AND J.E. BEASLEY, *The period vehicle routing*, Networks, 14 (1984), pp. 237–256.

[18] R.M. CLARK AND J.C.H. LEE, JR., *Systems planning for solid waste collection*, Computers & Operations Research, 3 (1976), pp. 157–173.

[19] A. CORBERÁN, E. FERNÁNDEZ, C. FRANQUESA, AND J. SANCHIS, *The windy clustered prize-collecting arc routing problem*, Transportation Science, 45 (2011), pp. 317–334.

[20] A. CORBERÁN, G. MEJÍA, AND J.M. SANCHIS, *New results on the mixed general routing problem*, Operations Research, 53 (2005), pp. 363–376.

[21] A. CORBERÁN AND C. PRINS, *Recent results on arc routing problems: An annotated bibliography*, Networks, 56 (2010), pp. 50–69.

[22] J. COUTINHO-RODRIGUES, N. RODRIGUES, AND CLÍMACO, *Solving an urban routing problem using heuristics: A successful case study*, Belgian Journal of Operations Research, Statistics and Computer Science, 33 (1993), pp. 19–32.

[23] G. DALLAIRE, *How cities are using GIS for route optimization*, MSW Management, May/June (1996), pp. 74–79.

[24] A. DEL PIA AND C. FILIPI, *A variable neighborhood descent algorithm for a real waste collection problem with mobile depots*, International Transactions in Operational Research, 13 (2006), pp. 125–141.

[25] P.M. FRANCIS, K.R. SMILOWITZ, AND M. TZUR, *The period vehicle routing problem and its extensions*, in The Vehicle Routing Problem: Latest Advances and New Challenges, B.L. Golden, S. Raghavan, and E.A. Wasil, eds., Springer, Boston, 2008, pp. 73–102.

[26] G. GHIANI, F. GUERRIERO, G. IMPROTA, AND R. MUSMANNO, *Waste collection in southern Italy: Solution of a real-life arc routing problem*, International Transactions in Operational Research, 12 (2005), pp. 135–144.

[27] G. GHIANI, F. GUERRIERO, G. LAPORTE, AND R. MUSMANNO, *Tabu search heuristics for the arc routing problem with intermediate facilities under capacity and length restrictions*, Journal of Mathematical Modelling and Algorithms, 3 (2004), pp. 209–223.

[28] G. GHIANI, G. IMPROTA, AND G. LAPORTE, *The capacitated arc routing problem with intermediate facilities*, Networks, 37 (2001), pp. 134–143.

[29] G. GHIANI, D. LAGANÀ, G. LAPORTE, AND F. MARI, *Ant colony optimization for the arc routing problem with intermediate facilities under capacity and length restrictions*, Journal of Heuristics, 16 (2010), pp. 211–233.

[30] G. GHIANI, D. LAGANÀ, E. MANNI, AND D. VIGO, *Operations research in solid waste management: A survey of strategic and tactical issues*, Computers & Operations Research, 44 (2014), pp. 22–32.

[31] B.L. GOLDEN, A.A. ASSAD, AND E.A. WASIL, *Routing vehicles in the real world: Applications in the solid waste, beverage, food, dairy, and newspaper industries*, in The Vehicle Routing Problem, P. Toth and D. Vigo, eds., SIAM, Philadelphia, 2002, pp. 245–286.

[32] B.L. GOLDEN, J. DEARMON, AND E.K. BAKER, *Computational experiments with algorithms for a class of routing problems*, Computers & Operations Research, 10 (1983), pp. 47–59.

[33] L. GOUVEIA, M.C. MOURÃO, AND L.S. PINTO, *Lower bounds for the mixed capacitated arc routing problem*, Computers & Operations Research, 37 (2010), pp. 692–699.

[34] V.C. HEMMELMAYR, K.F. DOERNER, R.F. HARTL, AND S. RATH, *A heuristic solution method for node routing based solid waste collection problems*, Journal of Heuristics, 19 (2013), pp. 129–156.

[35] V.C. HEMMELMAYR, K.F. DOERNER, R.F. HARTL, AND D. VIGO, *Models and algorithms for the integrated planning of bin allocation and vehicle routing in solid waste management*, Transportation Science, 48 (2014), pp. 103–120.

[36] N.V. KARADIMAS, K. PAPATZELOU, AND V.G. LOUMOS, *Optimal solid waste collection routes identified by the ant colony system algorithm*, Waste Management Research, 25 (2007), pp. 139–147.

[37] B.I. KIM, S. KIM, AND S. SAHOO, *Waste collection vehicle routing problem with time windows*, Computers & Operations Research, 33 (2006), pp. 3624–3642.

[38] J. KYTÖJOKI, T. NUORTIO, AND O. BRÄYSY, *Two efficient metaheuristics for huge scale vehicle routing problems*, working paper, Department of Environmental Sciences, University of Kuopio, Kuopio, Finland, 2004.

[39] V. MANIEZZO, *Algorithms for large directed CARP instances: Urban solid waste collection operational support*, Technical report, Department of Computer Science, University of Bologna, Bologna, Italy, 2004.

[40] D.H. MARKS AND J.C. LIEBMAN, *Mathematical analysis of solid waste collection*, Public Health Service Publication 2104, US Department of Health, Education, and Welfare, 1970.

[41] R. MCBRIDE, *Controlling left and u-turns in the routing of refuse collection vehicles*, Computers & Operations Research, 9 (1982), pp. 145–152.

[42] M.C. MOURÃO AND M.T. ALMEIDA, *Lower-bounding and heuristic methods for a refuse collection vehicle routing problem*, European Journal of Operational Research, 121 (2000), pp. 420–434.

[43] M.C. MOURÃO AND L. AMADO, *Heuristic method for a mixed capacitated arc routing problem: A refuse collection application*, European Journal of Operational Research, 160 (2005), pp. 139–153.

[44] M.C. MOURÃO, A.C. NUNES, AND C. PRINS, *Heuristic methods for the sectoring arc routing problem*, European Journal of Operational Research, 196 (2009), pp. 856–868.

[45] L. MUYLDERMANS AND G. PANG, *On the benefits of co-collection: Experiments with a multi-compartment vehicle routing algorithm*, European Journal of Operational Research, 206 (2010), pp. 93–103.

[46] T. NUORTIO, J. KYTÖJOKI, H. NISKA, AND O. BRÄYSY, *Improved route planning and scheduling of waste collection and transport*, Expert Systems with Applications, 30 (2006), pp. 223–232.

[47] M. POLACEK, K.F. DOERNER, R.F. HARTL, AND V. MANIEZZO, *A variable neighborhood search for the capacitated arc routing problem with intermediate facilities*, Journal of Heuristics, 14 (2008), pp. 405–423.

[48] C. PRINS AND S. BOUCHENOUA, *A memetic algorithm solving the VRP, the CARP and general routing problems with nodes, edges and arcs*, Studies in Fuzziness and Soft Computing, 166 (2005), pp. 65–85.

[49] S. SAHOO, S. KIM, B.-I. KIM, B. KRAAS, AND JR.A. POPOV, *Routing optimization for waste management*, Interfaces, 35 (2005), pp. 24–36.

[50] L. SANTOS, J. COUTINHO-RODRIGUES, AND C.H. ANTUNES, *A web spatial decision support system for vehicle routing using google maps*, Decision Support Systems, 51 (2011), pp. 1–9.

[51] L. SANTOS, J. COUTINHO-RODRIGUES, AND J. CURRENT, *Implementing a multi-vehicle multi-route spatial decision support system for efficient trash collection in Portugal*, Transportation Research A, 42 (2008), pp. 922–934.

[52] A. SBIHI AND R.W. EGLESE, *Combinatorial optimization and green logistics*, 4OR, 5 (2007), pp. 99–116.

[53] J. SNIEZEK AND L. BODIN, *Using mixed integer programming for solving the capacitated arc routing problem with vehicle/site dependencies with an application to the routing of residential sanitation collection vehicles*, Annals of Operations Research, 144 (2006), pp. 33–58.

[54] J. SNIEZEK, L. BODIN, L. LEVY, AND M. BALL, *The capacitated arc routing problem with vehicle/site dependencies: The Philadelphia experience*, in The Vehicle Routing Problem, P. Toth and D. Vigo, eds., SIAM Monographs on Discrete Mathematics and Its Applications 9, Philadelphia, 2002, pp. 287–308.

[55] M.M. SOLOMON, *Algorithms for the vehicle routing and scheduling problems with time windows constraints*, Operations Research, 35 (1987), pp. 254–262.

[56] G. TAVARES, Z. ZSIGRAIOVA, V. SEMIAO, AND M.G. CARVALHOA, *Optimisation of msw collection routes for minimum fuel consumption using 3D GIS modelling*, Waste Management, 29 (2009), pp. 1176–1185.

[57] J. TEIXEIRA, A.P. ANTUNES, AND J.P. SOUSA, *Recyclable waste collection planning: A case study*, European Journal of Operational Research, 158 (2004), pp. 543–554.

[58] D.V. TUNG AND A. PINNOI, *Vehicle routing-scheduling for waste collection in Hanoi*, European Journal of Operational Research, 125 (2000), pp. 449–468.

[59] S. WØHLK, *A decade of capacitated arc routing*, in The Vehicle Routing Problem: Latest Advances and New Challenges, B.L. Golden, S. Raghavan, and E.A. Wasil, eds., Springer, Boston, 2008, pp. 29–48.

[60] X.Y. WU, G.H. HUANG, L. LIU, AND J.B. LI, *An interval nonlinear program for the planning of waste management systems with economies-of-scale effects: A case study for the region of Hamilton, Ontario, Canada*, European Journal of Operational Research, 171 (2006), pp. 349–372.

Chapter 16

Arc Routing Applications in Newspaper Delivery

Geir Hasle

16.1 ▪ Introduction

Periodical news publications have been around since Julius Caesar, at the time in the form of government bulletins that were carved in metal or stone. In imperial China, "imperial bulletins" were periodically published for bureaucrats, probably as early as the 3rd century AD. Newspapers proper that were targeted at the general public and had a wide coverage of topics first appeared in the early 17th century in Europe. The printing press of Gutenberg and its further development was of course an important underpinning technology. They soon spread to major cities and to overseas colonies. Newspapers appeared in North America in the early 18th century. The industrial revolution brought several technological innovations that reduced the printing price. The newspaper communication medium soon covered most areas of the globe.

There are several types of newspapers, depending on the range of topics, the frequency of publication, and the geographical target area for readers. They range from national and international newspapers with a daily frequency and a circulation in the millions, to weekly local papers focused on a specific topic with a circulation of a few hundred. Some have several editions per day; some have focused editions targeted at different geographical areas. Some newspapers are only sold through retailers; some are also regularly delivered directly to household and business readers through subscription.

The World Association of Newspapers and News Publishers (WAN-IFRA) is the global organization of the world's newspapers and news publishers. It represents more than 18,000 publications, 15,000 online sites, and over 3,000 companies in more than 120 countries. WAN-IFRA has published a yearly report on world press trends since 1988. The following information on statistics and trends are taken from their 2010 and 2011 reports, and we refer to [48] and [49] for details.

In 2011, the global newspaper publishing industry produced revenue of about USD 160 billion annually. Over three billion people, or 72% of literate adults worldwide, read a newspaper regularly. The global daily print newspaper circulation was around 520 million, with a total of some 15,000 titles. The global circulation reached its peak in the 1990s, and it dropped by some 4% between 2008 and 2011. There are, however, large differences in trends between regions. Circulation continues to rise in Asia, but declines rapidly in mature markets in the West. The latter is due to a combination of deep technological, structural, economic, and cultural changes.

The major sources of revenue for modern newspapers are advertisements, sales through retailers, and subscriptions. For many years, the newspaper industry has struggled hard to meet the competition from other media and the Internet. Media consumption patterns vary widely between regions and countries, but there is a general trend towards less time spent with newspapers compared with other media, with a resulting decline in revenues. Readers have more choices, and are less faithful to newspapers. The Web and new platforms such as smartphones and tablet PCs are attractive alternatives. In many parts of the world, the younger generation by and large does not read news on paper. The newspaper industry has answered by publishing through the new digital channels, but a good business model remains hard to find. The printed circulation decline is more than compensated by the increase in digital reading, but still, advertisement revenues are dropping. Also, the new social media are changing the ways in which media content is produced and disseminated, and newspaper companies have not yet figured out how to adapt. In this climate, it is not surprising that large parts of the global newspaper business have been forced to cut costs, with fewer titles, slimmer newsrooms, and fewer editions as results.

There are large variations between countries, but, on average, printing and distribution make up roughly 2/3 of the costs of running a newspaper. Distribution costs are typically 1/3 of cover price, and profits are typically slim, at best. It seems attractive to abandon newsprint and "go all in" for digital publishing, but the lower relative advertisement revenues constitute a counteracting force. Some American newspaper executives project that in five years many newspapers will print only the Sunday edition or perhaps two or three days per week. An alternative way of cutting costs is to improve coordination in logistics, e.g., through outsourcing and consolidation of printing and distribution between newspaper companies, and through software-tool-based optimization [12]. Another way of counteracting the decline of subscription and advertisement income is to create new revenues by utilizing the sophisticated supply chain to distribute other goods. This opportunity has been taken by many companies. It introduces even more variability and demands for agile supply chain management.

With the above introduction as a backdrop, the goal of this chapter is to give an up-to-date account of arc routing applications in the newspaper business. Such applications are found in the "last mile" distribution of subscription newspapers. In the literature, this type of distribution is mentioned as a good example of arc routing, along with postal delivery and municipal waste collection. In Section 16.4, we shall criticize this view, and advocate a more general and more adequate model, "The Mixed Capacitated General Routing Problem" that is discussed in detail in Section 16.5. First, in Section 16.2, we give the necessary background on the newspaper supply chain, with focus on the "last mile." A literature survey follows in Section 16.3. Section 16.6 describes characteristics of routing problems in carrier delivery, whereas Section 16.7 is devoted to a case study where we describe the development of a Web-based route design and revision system. We finish the chapter with summary, conclusions, and prospects for the future, including important research topics.

16.2 ▪ Logistics for newspaper distribution

As accentuated by the line "Who wants yesterday's papers?" in a popular rock tune by the Rolling Stones from 1967 [29], newspapers are highly perishable goods.[11] For a major, daily newspaper, large volumes of newsprint are processed in printing presses, and then packaged, loaded on to vehicles, and distributed from the print works to readers, typically in a multi-echelon distribution chain. Either a newspaper copy is delivered to the home of the reader via subscription, or the reader buys it from a retailer. The final delivery to subscribers is made by carriers, either on foot, by bicycle, or by car.

Newspaper logistics are tightly constrained by deadlines. Subscribers need their morning paper before they go to work. Typically, the newspaper guarantees to place a copy on the reader's doormat before 6 AM. On the other hand, the demand for up-to-date news creates a pressure in the other direction: The editor would like to push the deadline as close to delivery as possible. In a time span of a few hours, a large number of papers are printed, packed, and distributed to retailers and subscribers. Some newspapers have several editions per day, which may contribute to added flexibility, but still, a highly time and cost efficient supply chain is necessary. The guaranteed delivery time has an effect on printing, packaging, and distribution sequences: Newspapers destined for far away districts may need to go first. The widespread use of the Internet and other electronic news media has started to change the role of the paper-based newspaper from being a major supplier of the latest news to being a provider of background and in-depth content. Hence, deadlines may become more liberal, with opportunities for more efficient logistics. Nevertheless, the newspaper supply chain is severely time constrained.

A modern, efficient newspaper supply chain is not only suitable for delivery of newsprint and direct marketing inserts. As alluded to above, this fact has been utilized by newspapers and distribution companies to add much needed revenues. Many side products are suitable, but in practice, printed products such as books, magazines, and direct mail are most common. Again, the unpredictable patterns of product mix increase dynamics. The idea of static distribution routes that are valid for several months does not go well with this distribution of multiple products with unpredictable demand to a large share of households in a given area. In principle, new, optimized routes should be designed every day based on the actual products and volumes. However, there are conservative forces. The traditional newspaper carrier knows the route by heart and is often well known by the subscribers. She or he has a carrier's book, as illustrated in Figure 16.1, with all necessary delivery information that is not well suited for dynamics, communication, and interaction. The newspaper/distribution company may want to have a routing plan that covers all households in a given area, for instance, for direct mail distribution. We shall return to these issues in Section 16.6.

When we disregard content production, the supply chain for paper newspapers largely consists of printing, packaging, first-stage delivery, and final delivery to retailers and subscribers. The time pressure, the volumes involved, and the fact that the paper must be delivered to the doormats of the subscribers, together require a sophisticated supply chain. The economic pressures and increasing dynamics require an optimized supply chain design and a high degree of operational coordination, in addition to outsourcing and consolidation of logistics between companies at the strategic level as described above.

In a modern newspaper distribution supply chain, several newspapers share the same printing and distribution resources. Newspapers with a broad geographical cover may be printed in several, geographically dispersed print works. The editorial content is produced in the newsroom and sent just before (or maybe right after) the deadline to the

[11] We here disregard the fact that perished newspapers have secondary utility, e.g., as wrapping for fish'n'chips.

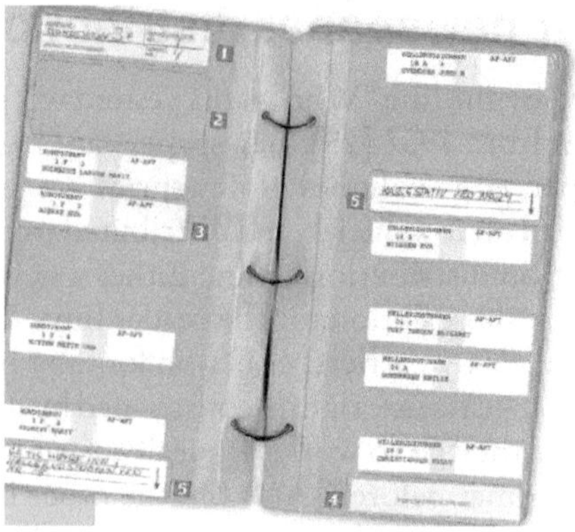

Figure 16.1. *A traditional carrier's book with paper strip delivery information. Courtesy of Aftenposten.*

allocated print work(s). The newspaper is printed, inserts are possibly added, and newspapers are bundled into packages with drop-off information on packing slips. For subscription newspapers that utilize the traditional carrier's book illustrated in Figure 16.1, route changes are printed on the packing slip. Information on new subscribers appears as new paper strips that the carrier must insert in the right places in her carrier's book to maintain a good route sequence.[12]

Newspaper bundles are conveyed to loading ramps, where they are loaded onto lorries for first echelon transportation. As space for inventory is limited, the printing sequence for multiple products or editions and vehicle dispatch are closely related. Often, loading ramps are scarce and constitute a bottleneck. Depending on volumes and geographical spread, the transportation to retailers and drop-off points for subscription newspapers is performed with a multiple echelon distribution chain. In general, smaller vehicles are used for the final delivery echelon. At transit points, reloading of different products from several print works takes place.

For nonsubscription newspapers, the distribution chain ends with delivery to retailers. However, reverse logistics may be important in the nonsubscription supply chain. Normally, retailers only pay for sold copies. Many newspapers require that unsold copies be returned. If sales prognoses are not good, large volumes of paper are transported to central facilities where the copies are counted and then forwarded for recycling. The return and centralized counting of unsold newspaper copies may seem unnecessary, as modern cash registers can provide accurate sales information, but this is unfortunately also a matter of trust.

For subscription newspapers, the final stage of the supply chain is the delivery from drop points to the mailbox or doormat of each subscriber. For a morning newspaper, bundles marked with route information are delivered to drop-off points, say, before 3:30 AM. Carrier routes are based on various transportation modes:

- pedestrian routes, with trolley,
- bicycle routes,
- driven routes, normally by car.

[12]This operation may be cumbersome, and even painful, for instance at 3AM in a blizzard on a winter morning in Norway.

There are cultural differences regarding the modes that are used, but pedestrian routes and car routes are common. Pedestrian routes are only used in dense urban or suburban districts where the total walking time is comparable to the total service time of the route. Bicycle routes operated by youngsters that were typical in many parts of the world have become less frequent, as evening subscription newspapers have become more scarce and virtually disappeared in many countries. Partly due to the fact that several titles now use the same distribution chain, and products other than newspapers have also been included, routes have become longer. Morning carrier routes are often designed to take two hours. Driven routes require a certificate, and in most countries, child labor regulations do not permit adolescents as morning paper carriers. Although the salary may not be top-notch, the typical carrier workforce now has a heterogeneous composition of adults with different sex, background, and age. For some, newspaper delivery is their sole occupation and they may operate more than one route.

The carrier picks up her printed media product bundles (possibly consisting of different newspaper titles, magazines, books, direct marketing material) at the drop-off point, and prepares delivery by unpacking the bundles and, possibly, adding inserts to newspapers. Driving carriers load the products into their vehicle, and pedestrian carriers fill their trolley. The carrier takes off to visit all subscribers on the route. Carriers on foot will normally return to the drop-off point to park the trolley (Figure 16.2, left). Driven routes may be open, if the carrier uses her own car and is only paid until the final subscriber. Depending on the type of vehicle, drivers may deliver products directly to the subscriber's mailbox. It is normal that drivers sometimes have to get out of the car and make a small pedestrian loop to serve a number of customers, for instance, in a block of flats. Similarly, the pedestrian carrier will often leave the trolley with an appropriate number of product copies and make a subtour (Figure 16.2, right). Particularly in urban areas, carriers need a large bunch of keys to enter blocks. Figure 16.3 illustrates several types of newspaper delivery to households.

Carriers on foot may of course meander, i.e., zigzag between both sides of a street, if this is time-saving. Driving carriers may do the same for appropriate streets, depending on the type of vehicle and driving regulations. Pedestrian carriers will use pedestrian roads and shortcuts that are not necessarily included in commercial road network data. A shortcut through a backyard may require a key that is only available for one carrier. In hilly areas, a pedestrian route should preferably not start uphill when the trolley is fully loaded. A carrier routing plan should have a certain level of "visual beauty," i.e., users have a preference for routing plans that look nice on the map. This beauty contest normally involves two subcriteria: compact routes and/or nonoverlapping routes. The motivation for the preferred beauty may seem unclear. It is partly connected with a user belief that plans with compact routes are more efficient, which is only partly true. However, it is clearly undesirable that several carriers run around in the same area with the risk that neighboring subscribers receive their newspaper at different times. These are real-life aspects that make carrier route design more challenging.

There are obvious similarities between postal delivery and newspaper delivery. In fact, some newspapers are delivered by the local postal service, and some modern newspaper distributers aim at adding postal mail as one of their product types but are prevented by postal monopoly. Less obvious is maybe the similarity to municipal garbage collection, because one is pickup and the other delivery. One similarity is that routes must visit virtually all households in the area. The time pressure may not be as hard in garbage collection as in newspaper delivery, although it is normally preferable that the routes finish early in the morning. The periodic nature of the garbage collection routing problem, where the selection of weekdays is an important characteristic, is not found in newspaper delivery.

Figure 16.2. *Pedestrian newspaper delivery with trolley. Left: Walking between delivery modules/returning to drop-off point. Right: Subtour without trolley. Courtesy of Aftenposten.*

Figure 16.3. *Pedestrian delivery. Left: Rack. Middle: Newspaper box. Right: Doormat. Courtesy of Aftenposten.*

Rather, individual routes for each weekday are typically preferable in newspaper delivery due to volume variation, and even routes designed for each individual day without repetition. A full distribution routing plan is often needed by the newspaper (distribution) company, for instance, to accommodate distribution of direct marketing leaflets.

It should be clear from the above description that the newspaper distribution chain presents complex and economically important design and coordination challenges. Purely manual decision-making is not adequate even for smaller newspapers. There is a high cost-cutting potential from the use of optimization-based supply-chain and routing software (see, for instance, [12] for a survey of routing software). Several providers of routing software tools specifically target newspaper distribution applications, also for the last mile carrier delivery, and there are many examples of successfully deployed vehicle routing software tools. In Section 16.7 we present a specific case, namely, the development and exploitation of a Web-based system for automatic construction and revision of carrier routes for last mile delivery of newspapers and other paper based media products. Before that, we give a brief survey of the routing literature for newspaper applications since 1996.

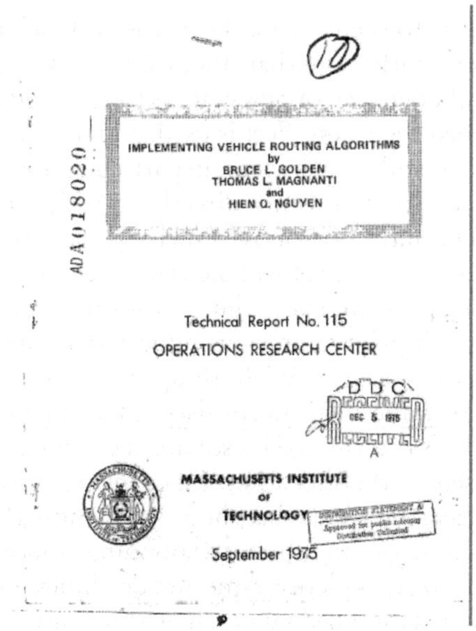
IMPLEMENTING VEHICLE ROUTING ALGORITHMS
by
BRUCE L. GOLDEN
THOMAS L. MAGNANTI
and
HIEN Q. NGUYEN

Technical Report No. 115
OPERATIONS RESEARCH CENTER

MASSACHUSETTS INSTITUTE
OF
TECHNOLOGY

September 1975

Figure 16.4. *An early scientific publication on routing in the newspaper industry [21].*

16.3 ▪ Literature survey

For a detailed survey of the early (until 1996) literature on NDP variants, we refer to [20]. The survey contains nine papers, ranging in time from the first (as far as we know) NDP paper of Golden, Magnanti, and Nguyen [22] to three papers from 1996. The nine papers that are described focus on the first part of the supply chain: distribution from the print works to drop-off points. Two papers consider a multiechelon supply chain, but none consider the last mile carrier delivery. Hence, none of the papers use an arc routing formulation.

Below, we provide an updated survey of the NDP literature since 1996. Whereas there is a substantial VRP literature for routing in waste management and postal mail delivery, there is little on routing for newspaper applications. As in the survey found in [20], we have found less than 10 relevant papers. Again, these focus on the first part of the supply chain; hence, none use an arc routing formulation.

Van Buer, Woodruff, and Olson [13] study tight coordination of production and distribution. Their case problem is the first-echelon distribution for a concrete medium-sized newspaper, in a setting where printing and distribution are tightly coupled since there is no room for finished goods inventories.[13] The newspaper has several editions targeted at different geographical areas. There is a classical tension between production and transportation concerns; i.e., the sequence determined by minimization of setup times in the press does not necessarily coincide with a good dispatching sequence based on geography. The authors give a mathematical formulation, and propose a heuristic solution method that is partly designed from comparative investigation on instances from the newspaper company. The authors conclude that reusing vehicles by allowing multiple routes per vehicle is the most important way to achieve low-cost solutions. Through this remedy, costs may be reduced dramatically.

Song et al. [45] focus on the first echelon of a two-echelon distribution chain from

[13]As mentioned above, this is typical.

the print works to the readers. Their case is Chosun-Ilbo, one of the major newspapers in South Korea with more than 100,000 readers and multiple editions. The first echelon consists of bulk delivery of newspapers to so-called agents, of which there may be several hundred. A heterogeneous fleet is used, and several deliveries (split delivery) may be used to fulfill the demand of each agent, in particular for agents that are close to the print works. The second echelon that is not directly considered is home delivery from each agent by small cars. The authors propose a method for solving the combined factory scheduling, allocation, and routing problem, i.e., the integrated problem of determining the sequence of editions at each print shop, the allocation of agents to individual print shops, and the routes from each print shop to its allocated agents. There are several complicating factors, such as sequence-dependent setup times and incompatibilities between vehicles and agents. The solution method consists of two main stages. First, the allocation of agents to print works is determined by solving a generalized assignment problem, using a regret distance. Based on the allocation, the printing schedules are determined heuristically. In the second stage, the agents that should have more than one delivery are determined (maximum two deliveries), and the corresponding volumes are calculated, both with heuristics that take printing completion time and deadlines into account. For the subsequent calculation of routes, a modified savings heuristic is used. Dispatch sequence at the print works is determined by an urgency-based heuristic. The paper reports from an empirical study on the Chosun-Ilbo operation with three print works and 400 agents in the Seoul area, for two specific days. Compared to the actual plans, the proposed method reduced costs and delays by 15% and 40%, respectively.

In their 2007 paper [15], Cunha and Mutarelli develop a MILP model for a concrete media product logistics problem that has some similarities with the Chosun-Ilbo study in [45]. The Brazilian weekly magazine Época has a circulation of some 400,000. The goal is to determine the number and location of the print works to be used, which destinations should be assigned to each selected works, the production sequence, and the mode of transportation, either airplane or truck. Savings of 7.1% of the total costs are reported, equivalent to more than USD 600,000 annually.

Russell, Chiang, and Zepeda also present research on combined production and distribution in the newspaper supply chain in their 2008 paper [42]. As for the three papers described above, the authors focus on the first part of the logistics chain, and consider the coordination and synchronization of production and delivery for multiproduct newspapers to bulk delivery locations. They define the problem as an Open VRP with Time Windows and Zoning Constraints (OVRPTWZC), and report significant improvement relative to current practice at the time for the Tulsa World, a midsize newspaper operation with a circulation of about 150,000. The authors do not consider the problem of production scheduling, but the print sequence considers geography, and the loading of vehicles naturally follows this sequence. As different geographical zones receive different advertising inserts, a zoning constraint is imposed on the routes. The proposed metaheuristic solution method first employs a parallel insertion heuristic with a tabu search improvement procedure that is invoked 10 times during construction. Iterative improvement with insert, delete, and swap operators and tabu search with a frequency-based diversification strategy and dynamic tabu tenure then follow. Savings related to fewer vehicles and reduction of total distance relative to manually generated routes that amount to some USD 70,000 are reported.

The same authors, reinforced with Xiaojing Xu, examine larger parts of the supply chain for a midsize US newspaper with both city and state editions in the Midwest that delivers to households and businesses in over 200 zip codes. They define the problem as an integrated newspaper production and distribution supply-chain management problem. The applications studied in [14] cover 2-echelon (city edition) and 3-echelon (state edition) distribution from the print works. They extend the work from the 2008 paper by presenting a 3-index mathematical formulation of the OVRPTWZC, and include uncertainty that is handled by simulation.

In [41], Russell reports on additional work on the newspaper production and distribution supply chain based on the case of the Tulsa World. The problem is still modeled as an OVRPTWZC. Constraint programming is used for modeling and solving it.

A similar problem is considered by Böhnlein, Gahm, and Tuma in [8]: The design of vehicle routes for first-echelon transportation from newspaper print shops that have to start directly after print finalization in order to meet deadlines. The variant they study allows multiple products (editions) per vehicle, and route planning must consider completion times of editions, as these may have a geographical focus. This problem is modeled as a so-called VRPTW with Cluster-Dependent start times. The authors propose a hybrid of Ant Colony Optimization and Tabu Search, and report some 20% savings for a large German newspaper with around 1,400 drop points. Böhnlein, Schweiger, and Tuma propose a multiagent solution method for a dynamic variant of the same problem in a later paper [9].

Again, in a paper by Eraslan and Derya [17], the transportation of newspapers from the print works to news agents is investigated using a VRP model. The first echelon routes for vehicles of a leading newspaper distributor company in Turkey are examined. An integer linear programming model is proposed. Scenarios with up to 55 newsagents are solved with a commercial MILP solver within reasonable computing times. It is reported that the proposed model reduced the delivery cost by 21% on average when compared to the current manual planning method.

In their TRISTAN VII paper [2], Archetti, Doerner, and Tricoire introduce the so-called Free Newspaper Delivery Problem (FNDP), i.e., the distribution of free newspapers from print works to news racks with given capacity and consumption profile at underground, bus, and tramway stations. The problem includes time windows, coordination of production and distribution, and inventory management. The goal is hierarchical: The primary objective is to minimize the time period at which all stations are subject to stockout, while the secondary objective is to minimize the total number of trips for a homogeneous fleet of vehicles, where each vehicle may take several trips. Once stock-out happens at a given customer, there will be no further replenishment. The authors give a mathematical formulation and propose solution techniques for the FNDP, including a decomposition method, heuristic, and exact approaches for the subproblems, as well as a hybrid method. Computational experiments were made on instances inspired by real data from a free newspaper in Vienna. Results show that the hybrid approach yields the best solutions.

16.4 ▪ Newspaper carrier delivery: Node or arc routing?

Our search for specific literature on newspaper routing did not find any relevant scientific papers on the last mile carrier delivery. In contrast, several introductory sections in technical and general papers on the Arc Routing Problem (ARP) briefly mention carrier newspaper delivery, last mile postal mail delivery, and garbage collection as applications where variants of the ARP would be an adequate model.

In a chapter in the *Handbook of Transportation Science* [7], Bodin, Maniezzo, and Mingozzi discuss so-called Street Routing and scheduling Problems (SRPs). Examples include routing of residential and containerized sanitation vehicles, the scheduling of meter readers, the routing of field service operations for public utilities, the routing of vehicles for other traditional local pickup and delivery operations, and the scheduling of vehicles for telephone book and newspaper deliveries. In street routing, the locations to be serviced are assigned to the street segments or intersections of a digital street network database. SRP occur primarily in local delivery operations. The authors provide detailed information on the features and characteristics of SRP and how these problems differ from the more traditional VRP: Regarding models and solution methods, they focus on node-based variants such as CVRP, VRPTW, and VRPB. They also outline a cluster-first route-second generic algorithm for solving SRP, under certain assumptions.

The VRP literature has been dichotomized between node routing problems, first defined by Dantzig and Ramser in 1959 [16], and arc routing problems, introduced by Kwan Mei-Ko in 1962 [33]. The equivalent of the basic, capacitated node routing problem (CVRP) is the Capacitated Arc Routing Problem (CARP) first defined by Golden and Wong in 1981 [23]. Although the General Routing Problem (GRP), a generalization of the TSP where one must visit both nodes and edges in the graph, was defined by Orloff in 1974 [37], its capacitated and multi-vehicle generalizations have not been much studied.

Some applications are naturally modeled as arc routing problems because the demand is fundamentally defined on arcs or edges in a transportation network. Prime examples are street sweeping, gritting, and snow clearing. However, the arc routing model has also been advocated in the literature for problems where the demand is located at nodes, for instance, distribution of subscription newspapers to households, postal mail delivery, and pickup of household waste, particularly in urban areas. In real-life cases, there are often thousands or tens of thousands of points to be serviced along a subset of all road links in the planning area. Such cases may be formulated as CARPs on the road links with demand, typically with a drastic reduction of problem size. An arc routing problem can be converted into an equivalent node routing problem, but at the expense of the size of the instances to be solved.

Capacitated and multivehicle generalizations of the GRP were introduced in a 1995 paper by Pandi and Muralidharan [38] and later in a paper by Gutierrez, Soler, and Hervaz [25]. The problem was denoted the Capacitated General Routing Problem on Mixed Graphs (CGRP, or CGRP-m) at the time. Later, the Mixed Capacitated General Routing Problem (MCGRP) has become a more standard name for it. In their 2004 paper [40], Prins and Bouchenoua motivate further studies of the MCGRP. They call it the Node, Edge, and Arc Routing Problem (NEARP) and state that "Despite the success of metaheuristics for the VRP and the CARP, it is clear that these two problems cannot formalize the requirements of many real-world scenarios." Their example is urban waste collection, where most demand may adequately be modeled on street segments, but there may also be demand located at points, for instance, at supermarkets. The MCGRP (or NEARP) is a generalization of the CVRP and the CARP, which again may be viewed as a capacitated extension of the General Routing Problem [37]. They propose a memetic algorithm for the MCGRP and investigate it empirically on standard CVRP and CARP instances from the literature. The authors also create an MCGRP benchmark set consisting of 23 grid-based test cases, the so-called CBMix-instances with between 20 and 212 required tasks, and provide experimental results for their proposed algorithm.

We would like to enhance the motivation for the MCGRP and further emphasize its importance to practice. The arc routing model for node-based demand cases such as subscription newspaper delivery is based on an underlying idea of abstraction. Some form of abstraction may be necessary to contain the computational complexity resulting from a large number of demand points in industrial routing. The assumption that all point-based demands can be aggregate into edges or arcs may be crude in practice. It may lead to solutions that are unnecessarily costly, as partial traversal of edges is not possible. In industry, a route planning task may cover areas that have a mixture of urban, suburban, and rural parts where many demand points will be far apart and aggregation would impose unnecessary constraints on visit sequences.

A more sophisticated type of abstraction is aggregation of demand based on the underlying transportation network topology. Such aggregation procedures must also take capacity, time, and travel restrictions into consideration to avoid aggregation that would lead to impractical or low quality plans. In general, such procedures will produce an MCGRP instance with a combination of demands on arcs, edges, and nodes. The generalization of CVRP, GRP, and CARP was, therefore, an important step to eliminate the arc/node routing dichotomy and thus enable the modeling of the continuum of node and arc routing problems needed for representational adequacy in real-life situations.

16.5 ▪ The mixed capacitated general routing problem

The MCGRP is defined on a connected multigraph $G=(N,E,A)$, where N is the set of nodes, E is the set of undirected edges, and A is the set of directed arcs. Let c_e denote the non-negative traversal cost for $e \in E \cup A$, also known as deadheading cost, i.e., the cost for traversing the edge/arc without servicing it. The traversal cost is zero for nodes. Let $N_R \subseteq N$ be the set of required nodes, and let q_i denote the demand and s_i the servicing cost of node $i \in N_R$. Similarly, let E_R and A_R be the set of required edges and arcs, respectively, and let q_e and s_e denote the demand and service cost of $e \in E_R \cup A_R$. A fleet of identical vehicles each with capacity Q is initially located at a special depot node. It is assumed that the size of the fleet is unbounded.

The goal is to identify a number of tours for the vehicles such that

1. every node $i \in N_R$, every edge $e \in E_R$, and every arc $e \in A_R$ is serviced by exactly one vehicle,
2. the sum of demands serviced by each vehicle does not exceed Q, and
3. the total cost of the tours is minimized.

We note that for the basic MCGRP, the total service cost is constant over all feasible solutions, so it is sufficient to minimize the sum of traversal costs.

As mentioned above, studies of the MCGRP are scarce in the literature. The first we know of is the paper by Pandi and Muralidharan from 1995 [38]. They address a generalized version of the MCGRP, i.e., routing a given heterogeneous set of vehicles over specified segments and nodes of a street network, under a route duration constraint. The problem is denoted the Capacitated General Routing Problem (CGRP). The authors formally define the CGRP and design a heuristic for solving it. They generate random test instances inspired from curb-side waste collection in residential areas on a network with 50 nodes and 100 links. They also investigate the proposed method on random instances of the Capacitated Chinese Postman Problem for which they had two lower bound procedures. The authors integrated their solver in a microcomputer-based interactive route planning system.

In [25], the homogeneous fleet version of the CGRP studied by Pandi and Muralidharan is investigated by Gutierrez, Soler, and Hervaz. The number of vehicles is given. They call the problem CGRP-m and devise a heuristic that is investigated on a set of 28 instances with between 20 and 50 vertices, 25 and 97 edges, and 0 and 16 arcs. They compare their heuristic with the heuristic proposed by Pandi and Muralidharan and conclude that their heuristic is substantially better for the homogeneous fleet version of the problem.

In [4], Bach, Hasle, and Wøhlk provide the first combinatorial lower bound procedure for the MCGRP. They also present two new benchmarks: BHW, consisting of 20 instances based on well-known CARP instances with between 29 and 410 required tasks; and DI-NEARP with 24 instances that are based on 6 real cases from last mile newspaper delivery. The number of required tasks varies between 240 and 833. They report numerical results for the proposed lower bounds for the three benchmarks, and also the best known upper bounds from the literature. Their only source of upper bounds for the BHW and DI-NEARP benchmarks was a technical report by Hasle et al. [27] that describes results from experiments on MCGRP test instances using SINTEF's industrial VRP solver Spider [26, 44]. This technical report provides new best-known results for the CBMix benchmark and the first upper bounds for the BHW and DI-NEARP benchmarks. For three of the BHW instances, the gap is closed: BHW2, BHW4, and BHW6. They also refer to a Web page [43] devoted to MCGRP benchmarks and corresponding best known upper and lower bounds that is intended to be a central resource for MCGRP researchers.

Bosco, Laganà, Musmanno, and Vocaturo [11] propose the first integer programming formulation for the MCGRP with three indices for links and two indices for nodes. They extend some inequalities originally introduced for the CARP, and discuss identification procedures for these inequalities and some relaxed constraints. They propose a branch-and-cut algorithm and present computational experiments on 12 datasets that they have constructed from classical datasets (gdb and mval) for the undirected and mixed CARP. The number of required tasks of the so-called mggdb and mgval benchmarks varies between 8 and 129, and the authors only solve instances with a number of vehicles less than 8. The optimal solution is found for 154 of 288 instances, whereas the average and maximum percentage gaps for the nonclosed instances are 2.56 and 12.82, respectively. The algorithm is also tested on the four CBMix instances that require a small number of tours. It finds the proven optimal solution for CBMix12 and CBMix23 and provides good bounds for CBMix1 and CBMix22.

In his master's thesis, Kevin Gaze [19] presents the first Branch & Price algorithm for the MCGRP. Computational experiments on all 138 mggdb instances and a selection of 20 of the 150 mgval instances produces 54 improved lower bounds and 26 improved upper bounds. Three instances were closed for the first time.

Lukas Bach et al. propose a Branch & Cut & Price for the MCGRP in [5]. This algorithm is better suited to handle larger numbers of vehicles available as they decompose the problem and are thus able to remove the vehicle dimension from the original 3-index model by Bosco et al.. They provide new optimal solutions, tighten gaps, and generally decrease run times for the mggdb class of instances. Furthermore, new upper bounds to a few CBMix instances and new best lower bounds in many cases for the CBMix and BHW instances are provided.

In [6], Bode, Irnich, Laganà, and Vocaturo present a new mathematical model for the MCGRP based on two-index variables. Their proposed solution approach is a two-phase Branch & Cut algorithm that uses an aggregate formulation to develop an effective lower bounding procedure. This procedure also provides strong valid inequalities for the two-index model. Experimental results on a subset of the mggdb and mgval instances include many new optimal solutions and improved bounds. Reasonable gaps were produced also for the larger size instances.

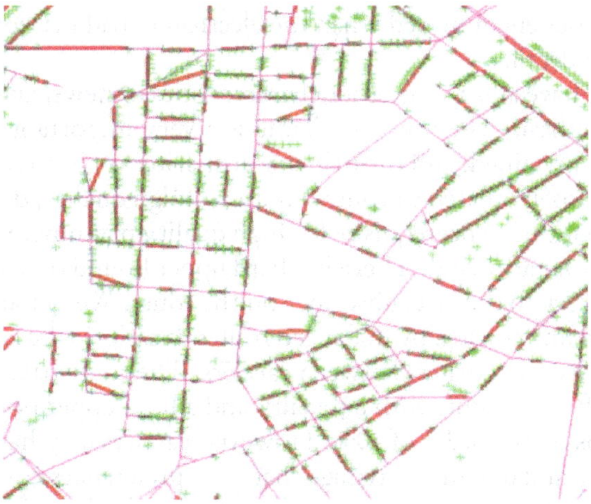

Figure 16.5. *Road topology-based aggregation of modules in a dense urban part of Oslo. Green crosses mark the original modules, and red lines mark aggregates.*

16.6 ▪ Arc routing problems in newspaper delivery

The routing problems in newspaper distribution where variants of the ARP are adequate models are of course located in the final, "last mile" carrier delivery. This may be characterized as a Street Routing Problem, with many of the complications pointed out by Bodin et al. in [7]. As demand is fundamentally node based, an arc routing problem model must be based on some form of abstraction, e.g., through demand aggregation based on links of the street network. In our experience, it is not uncommon for newspaper companies to have manually made, first level aggregates of households in their area of interest into modules that will never be split between carrier routes. Such aggregates will typically reduce the number of points in a natural planning session by an order of magnitude. Often a module corresponds to a place where the carrier stops her car or leaves her carriage to make a small subtour. The module's service time includes the walking time between the households.

Despite this first level aggregation into modules, the number of modules in an actual planning session may be in the thousands, so a second level aggregation may be useful. The aggregation heuristics must take the underlying road topology into account, and constraints on meandering, U-turns, etc. There must be limits on the duration of aggregates to allow planning flexibility. This type of aggregation will result in aggregate orders whose location is an edge or an arc. Even in dense urban areas, not all modules will become part of an aggregate. Hence, the result will be an MCGRP with side constraints and possibly multiple criteria. Despite two levels of aggregation, the number of required nodes or links in a real-life routing session will easily be in the hundreds or even thousands, so large size is a common characteristic of ARP for last mile newspaper delivery.

Figure 16.5 illustrates the result of an aggregation heuristic[14] applied to modules in a dense urban part of Oslo. Green crosses represent individual modules, and the red segments represent aggregates. We note that, although most modules have been aggregate, there is still a substantial number of individual modules. The figure also illustrates the

[14] The aggregation heuristic of the commercial solver Spider [44].

difficulties encountered when the electronic road network is not of sufficient quality and level of detail.

We have alluded to other characteristics of newspaper carrier delivery routing above, but emphasize them below. Time is a very important dimension. Delivery is severely time constrained, and the primary cost criterion is often dominated by salary that is typically closely related to route duration. High quality, detailed address and road topology information is a prerequisite for high quality planning. Good service time and travel time models are also vital. There is a hard upper bound on route duration, or, rather, the time of delivery to the last subscriber on the route. Weight and volume capacity is normally a nonbinding constraint. Balanced routes duration-wise is an important criterion to ensure flexibility and fairness between carriers. The route balance requirement may be regarded as a soft constraint, with a penalty and a hard upper bound on the maximum difference, perhaps on the order of 10%. However, carriers may have individual route duration preferences, and one may envisage that such preferences could be given as input to a solver.

The planners often want to specify the number of routes or, alternatively, minimize the number of routes for the given area of interest. Each route starts from one of a set of alternative drop-off points. Due to noise, visual impact, and other environmental concerns, drop-off points cannot be selected at will. However, the solution of an integrated optimization of route design and selection of drop-off points, i.e., a location routing problem, is certainly relevant. The choice of drop-off points may influence the first-echelon distribution, but typically not to a large degree.

The area of interest may encompass urban, suburban, and rural areas with highly varying density of deliveries that require a mixture of driven and pedestrian routes. Depending on modality, whether the carriers are subcontracted or hired, and the types of equipment involved, carrier routes should be closed or open as explained above. Choice of modality and, in case, type of car, will also determine whether a given demand aggregate may be regarded as an edge or must be modeled as two arcs. For pedestrian and bicycle routes, topography may be an important factor. In areas with steep hills, routes should not start uphill when the load is heavy.

The criteria for mode selection for routes may be complex, but accessibility and the relation between travel and service time are the most essential. Ideally, many newspaper companies would like to see functionality in a route optimization system that could automatically construct optimized carrier routes, and decide modality and drop-off points for the routes simultaneously. These decisions are closely intertwined. Even the type of link chosen to represent a demand aggregate partly depends on modality. We believe there are many interesting, open research challenges buried in these issues.

Finally, we return to the multicriteria aspect of the newspaper carrier delivery problem. We have already discussed the two most important objectives: total cost and route duration balance. In addition, there is the "visual beauty" concern, as described in Section 16.2 above. The users want compact, nonoverlapping routes, but not at any cost.

In summary, the ARP found in newspaper distribution is at the final carrier delivery. It is in our opinion a case for rich variants of the MCGRP model, after aggregation of point-based demand. There is a hard duration constraint, and traditional capacity constraints are nonbinding for most operations. The primary objective is cost that is closely related to total duration, whereas there is an important secondary objective to balance route duration. Moreover, users prefer solutions with compact, nonoverlapping routes.

We have given an account of the routing literature specifically targeted at newspaper applications in Section 16.3 above. Moreover, we have given an extensive survey of the literature on the (multi-vehicle) Mixed Capacitated General Routing Problem (MCGRP), alias the Node, Edge, and Arc Routing Problem, in Section 16.5. Let us now also point to general literature that relates to the more exotic problem characteristics described above.

The dichotomy between node and arc routing and the presence of mixed cases in real life is addressed by a paper of Oppen and Løkketangen [36]. They consider borderline cases, namely, VRPs based on a graph representing a road network, where customers are situated along the arcs. Garbage collection and newspaper delivery are among their application examples. They define "Stringed VRP" as a problem where some customers are close and form "strings" in the graph. They comment that it often seems obvious that the nodes of such a string should be visited by the same vehicle, and claim that such a string of nodes can naturally be replaced by an arc between the first and last node in the string. The paper proposes aggregation heuristics, but the aggregates are represented as nodes rather than arcs or edges.

Although important to many applications, there is surprisingly little literature on VRP with route balance aspects. We refer to work by Tsouros et al. [47], Pasia et al. [39], Borgulya [10], Jozefowiez et al. [30], and Gulczynski et al. [24]. "Visual beauty" issues are addressed in [34], [46], and [35]. The combination is a topic of [31] and [28].

16.7 ▪ Case study: A Web-based service for carrier route design

Together with inhabitants of Japan, Switzerland, and the other Nordic countries, Norwegians are among the most prolific readers of newspapers in the world. Print newspapers are read every day by over 80% of the population. However, circulation numbers have kept falling steadily since the 1990s. Norway is also the country where newspapers have the highest percentage of advertising revenues from digital versions, at some 17%. A polynomial regression analysis from 2011, for what it is worth, shows that there will be no print newspapers left in Norway by October 9, 2025.

Norwegian newspaper companies saw the writing on the wall. Ideas of consolidation, both regarding ownership and logistics, emerged. The need for a more efficient supply chain with reduced distribution costs led to a new practice where several titles utilized the same resources. Also, ideas of distributing other products through the newspaper distribution chain to create new revenues emerged. The added dynamics called for new ICT technology for carrier route management, including a system for route optimization. The idea of replacing the old carrier's book illustrated in Figure 16.1 came early, maybe even before Personal Digital Assistants (PDAs), also called palmtops, and the Internet became widespread in the mid 1990s.

In 2000, Aftenposten, the largest Norwegian subscription newspaper, took contact with SINTEF, a large contract research institute in Oslo. Their idea was to develop an industry standard system for logistics management for paper-based media products. The first critical step was to develop an electronic version of the carrier's book. The requirements on usability were extremely high, as user acceptance was absolutely crucial to the success of the whole system. Newspaper carriers of today in Norway constitute a heterogeneous group of people, with many nationalities and different levels of education and backgrounds. The PDA-based guide was to be used in a possibly very harsh and difficult working environment, with darkness, cold temperatures, and humidity.

The project started in 2001, with Aftenposten as project owner, several newspaper and distribution companies as partners, SINTEF as the research partner, and financial support by the Research Council of Norway. The main priority in the beginning was the development of the PDA-based carrier's book illustrated in Figure 16.6. A focus group of carriers was established, and this was instrumental in the research based development that proved to be successful [18]. Also, requirements for route optimization were collected, and VRP

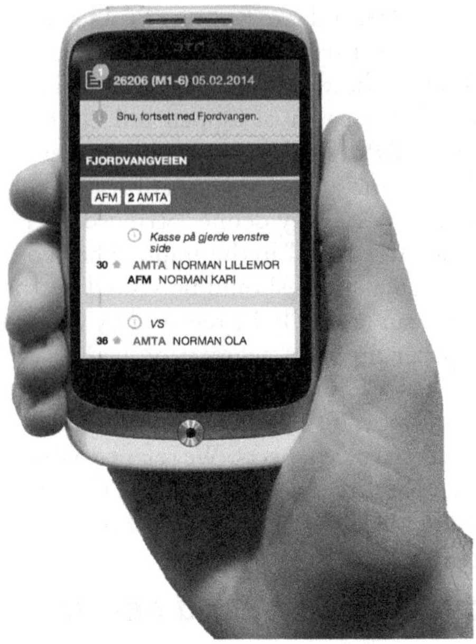

Figure 16.6. *The PDA/Smartphone-based carrier's book, courtesy of DI.*

solver prototypes based on SINTEF's industrial solver Spider were investigated on real data. A spin-off company called Distribution Innovation AS (DI) [3] was established for industrial dissemination.

In parallel, a logistics management system was developed, partly based on reengineering and partial reuse of earlier systems such as product, subscriber, and carrier route databases. Figure 16.7 illustrates the main concepts. A main design feature was a Web-based interface and a cloud computing architecture that allows ready access to the functionality from "anywhere" without software installation and hardware investment; a Web browser and a commodity PC suffice. Hence, the system became attractive also for small newspapers and distribution companies. Soon, the DI system became market leading in Norway, with customers that covered more than 50% of Norwegian households.

The electronic carrier's book of course has many advantages over the traditional one. It has Internet access, and two-way communication is possible. In principle, carrier route information could be updated while the carrier is en route, but in practice, she downloads her route a few hours before the start. The route information contains the delivery sequence, specification of products per subscriber, detailed travel and delivery directions, and aggregate information such as total number of copies for each title. Of course, the electronic carrier's book enables cost reductions. Common distribution of several products, with a more dynamic demand situation, is made easy to handle for the carrier. Likewise, more frequent route optimization (for instance, through a VRP solver) is enabled. The electronic carrier's book and the associated logistics software also facilitate increased revenues, for instance through targeted insert advertisement. Figure 16.8 shows the overall architecture of DI's Web solution.

The Web solution has been further developed over time, also through several RTD projects. A major step was the integration of a VRP solver for automatic and optimized route construction and revision. Earlier, full revision of carrier routes in Oslo could re-

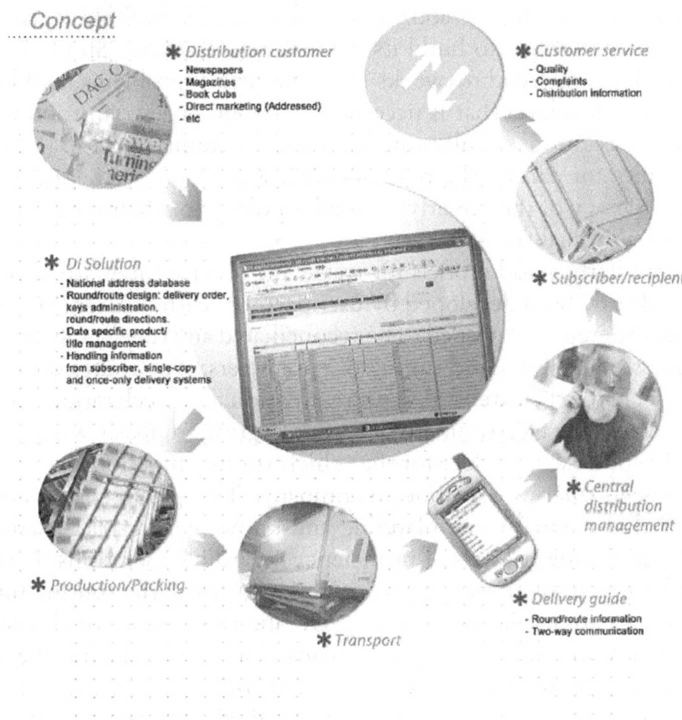

Figure 16.7. *The DI solution concept.*

Figure 16.8. *The DI Web-solution architecture.*

quire several man-years and take many months. Hence, revisions were not performed frequently enough, with less than optimal route plans and unnecessarily high costs as a result. Before this, functionality for manual revision of existing, manually planned routes, and map visualization, had been added to the Web solution. Moreover, a set of performance indicators were developed that allows the monitoring of the efficiency of routing plans to identify areas where a route revision is called for.

A challenge that became apparent with map visualization and automatic route planning was geocoding of addresses. The official household register did not have sufficient quality, so DI had to build its own address database. Similarly, the commercially available electronic road networks also had quality problems, and lacked information at the very detailed level that is necessary to design pedestrian routes automatically. The high user expectations of fully automatic route planning from Day 1 had to be lowered. Functionality for editing the road network was added, so the users could gradually increase the accuracy of the underlying road topology for their area of interest and soon reach an acceptable quality.

Similarly, good models for travel and service times are critical. Several such models had already been developed by user partners with different types of operations, but they were different and needed to be reconciliated and tuned. Observed data were the basis for regression analyses and tuning of parameters. For routes driven by car, the travel time model had to be extended, with retardation and acceleration phases for each stop.

As for the electronic carrier's book, great emphasis was put on usability, simplicity, and a good user interface for the vehicle routing functionality. The target user is a planner at a newspaper or distribution company. This group consists of people who are highly skilled at manual route planning and know their district extremely well, but typically have much lower levels of proficiency regarding operations research, computing science, and software engineering. Hence, to be accepted, the routing functionality needed to be easily accessible through a Web GUI, and the number of choices and parameters had to be at a minimum. Low quality plans would of course not be acceptable, although the manual planning functionality could be used for robustness. A user vision for response time was route revision for a major area during the time it takes to fetch a cup of coffee.

SINTEF's industrial VRP solver Spider [26, 44] was the starting point. The model had been extended, and the accompanying, uniform solution algorithm had been improved through exposure to new industrial cases. Good performance had been demonstrated also on several academic benchmarks, although running times are generally higher for a generic solver that cannot utilize all the speedup tricks of a highly specialized solver. The most important novel problem features of newspaper carrier routing that were exposed by the newspaper and distribution companies were the potentially very large number of customers, the route balance constraint, and the visual beauty criteria. In addition there were important complexities related to road network traversal such as meandering and U-turns.

The initial idea was to keep the uniform solution algorithm, but add a route imbalance penalty to the objective. A clustering objective component that at least partially addressed the visual beauty aspect was already implemented. To accommodate very large instances, a road topology–based demand aggregation framework was designed, and a concrete heuristic based on the road topology was developed. The model was extended with aggregate orders. These have a geographical extension that is a part of one or more consecutive road segments. Hence, the VRP solver was now equipped to represent arc routing problems on a mixed graph, and also MCGRPs. Very few changes had to be made on the solution algorithm. We became aware of the paper of Prins and Bouchenoua [40] and tested the solver on the 23 CBMix instances, with reasonably good average results (average error less than 1%) and six best known results.[15]

However, clear feedback from users on prototype versions for the real-life problem said that solution quality was not consistently good. As could be expected, determining generally valid weights in a scalar combination of the three major objective components

[15]Three of the instances had been improved before in [32], a paper that we were not aware of at the time.

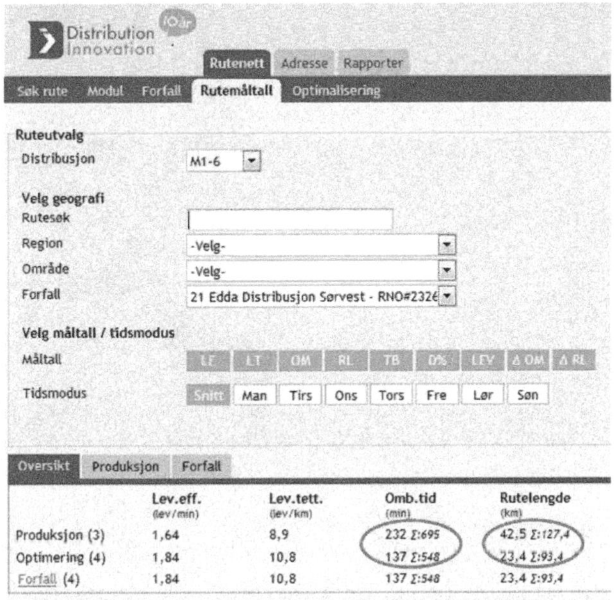

Figure 16.9. *Screenshot from the DI solution. Comparison of performance indicators.*

turned out to be difficult. A true multiobjective approach was deemed too expensive. The remedy became a new method that imposes the desired structure, i.e., balanced routes that look good in the map in an initial solution, and then performs iterative improvement without destroying that structure.

The general idea is one of cluster-first route-second, where clustering is performed by solving a capacitated clustering problem with route duration as capacity. The problem is solved heuristically by a k-means-like algorithm. There are issues regarding clustering in a non-Euclidean space that have been solved by combining real and Euclidean distances. Duration is estimated by local search with 2-opt to a local optimum. The number of clusters is either user defined, or estimated. If the estimated or given number is not adequate, the process is repeated with an adjusted number. Finally, each route is postoptimized.

The revised solution method gives high quality plans that are accepted by the users. Automatic planning of carrier routes is now taken for granted by the user companies; manual route revision is no longer performed. As the DI solution database also contains manually generated routes that have not yet been optimized, there is a good basis for comparison. Cost improvement varies widely between a few percent for routing plans that have evolved naturally over a long period of time, to more than 30% for areas where there have been drastic changes such as the addition of new products or new residential areas. Figure 16.9 shows a case where a manually generated routing plan with three routes is compared to an automatically generated routing plan for the same set of households. The user has specified one more route, but the total distance has decreased by some 27% and the total time has been reduced with approximately 22%. Figure 16.10 shows an example of an optimized routing plan that has considerable beauty in the map.

The functionality for automatic carrier route design is routinely used by more than 35 distribution companies and newspapers in Norway, Sweden, and Finland. The Norwegian users cover more than 80% of Norwegian households. Every evening, some 6,300

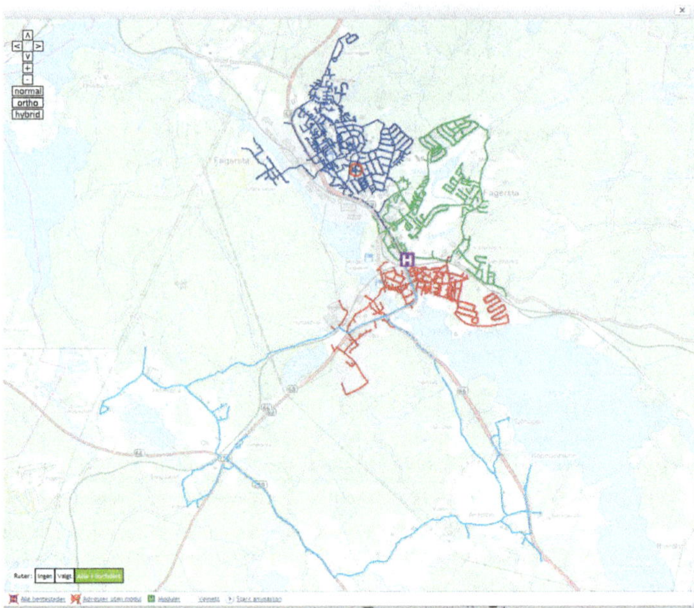

Figure 16.10. *An example of a routing plan with considerable beauty, courtesy of DI.*

carriers that perform more than a million deliveries daily download their routes for the next morning. The time to get that cup of coffee while the route was optimized may have been somewhat longer than usual, but further VRP research and hardware development will soon get us there.

The DI system is a rich source of industrial case data. With permission, SINTEF has downloaded six sets of case data and from these generated a suite of 24 MCGRP instances that have been used for experiments also in academia. A Web page devoted to MCGRP benchmarks has been established [43], and it is our hope that SINTEF in this way provides a small bridge between industry and academia.

16.8 ▪ Summary

In this chapter, our goal was to describe arc routing applications in the newspaper business. Judging from the literature that we have found, there are none. In line with this fact, we have argued that "last mile" carrier deliveries to households is fundamentally a node routing problem, as is also mail delivery and collection of household waste, contrary to statements in the introduction of many technical arc routing papers. However, abstraction and problem size reduction, e.g., by qualified aggregation heuristics that utilize the road topology, will typically produce a combined arc and node routing problem. We have emphasized the importance of the Mixed Capacitated General Routing Problem, also known as the Node, Edge, and Arc Routing Problem, as an adequate model for such applications, and provided a survey of the relevant literature. We have also provided a narrative style case description: the development of a web based system for management and optimization of carrier delivery in Norway.

The dire straits that large parts of the global newspaper business have been in the past two decades do not seem easily navigable. Circulation numbers and advertisement revenues are falling, and a good business model for digital newspapers seems hard to find. These pressures have motivated consolidation of the newspaper business, and cooperation between companies on better efficiency and utilization of the supply chain. Other products are distributed through this supply chain to increase revenues. It is relatively easy to motivate the use of good software tools for better coordination. As we have illustrated, the routing problems in question are complex, and the newspaper supply chain is in our opinion a rich source of industrially relevant and scientifically exciting research topics that has not been much exploited yet. In particular, there are aspects of last mile delivery to (or pickup from) households that render pure node routing and pure arc routing formulations inadequate. The Mixed Capacitated General Routing Problem provides flexibility to model routing problems where there is demand both on arcs and nodes. Although it has picked up lately, research on the MCGRP is scarce, and there is a need for further development of both exact methods and heuristics for this basic generalization of the CVRP and the CARP. Studies of richer variants of the MCGRP have not been reported at all, for instance, to address some of the aspects of newspaper delivery discussed in 16.2. We know that research efforts on the bi-criteria MCGRP with route balance as the second objective has started.

One may argue that the best way of tackling a computationally hard discrete optimization problem is to remove it. Personally, we like to read even yesterday's papers on newsprint, but we may belong to the last generation with such inclinations. It is difficult for us to see, however, that a supply chain that reaches every household every early morning within a strict time frame will become totally useless in the future.

Acknowledgments

We extend our sincere thanks to the Distribution Innovation AS company which has made screenshots from their Web-based carrier route design system available to us and allowed us to publish information on their case and the development of their system. We also thank Norway's largest subscription newspaper *Aftenposten* for permission to use illustrations from their carrier's manual [1].

The preparation of this chapter has been supported by the eVita and SMARTRANS programs of the Research Council of Norway under contract numbers 205298/V30 (DOMinant II), 192905/I40 (Collab), and 217108 (Respons).

Bibliography

[1] AFTENPOSTEN, *Carrier's manual*, http://www.avisbud.aftenposten.no/GetElement.do?id=614. Accessed 05/10/2012.

[2] C. ARCHETTI, K. DOERNER, AND F. TRICOIRE, *Free newspapers delivery optimization*, http://www.sintef.no/tristan, 2010.

[3] DISTRIBUTION INNOVATION AS, *Home page*, http://www.di.no. Accessed 2012-10-08.

[4] L. BACH, G. HASLE, AND S. WØHLK, *A lower bound for the node, edge, and arc routing problem*, Computers & Operations Research, 40 (2013), pp. 943–952.

[5] L. BACH, J. LYSGAARD, AND S. WØHLK, *A branch-and-cut-and-price algorithm for the mixed capacitated general routing problem*, in Routing and Scheduling Problems: Optimization using Exact and Heuristic Methods, L. Bach, Ph.D. Thesis, Aarhus University, Aarhus, Denmark, 2014, pp. 39–75.

[6] C. BODE, S. IRNICH, D. LAGANÀ, AND F. VOCATURO, *Two-phase branch-and-cut for the mixed capacitated general routing problem*, Tech. Report LM-2014-02, Johannes Gutenberg University Mainz, Mainz, Germany, 2014.

[7] L. BODIN, V. MANIEZZO, AND A. MINGOZZI, *Street routing and scheduling problems*, in Handbook of Transportation Science, R.W. Hall, ed., Kluwer, Dordrecht, The Netherlands, 1999.

[8] D. BÖHNLEIN, C. GAHM, AND A. TUMA, *A hybrid meta-heuristic for the VRPTW with cluster-dependent tour starts*, in Proceedings of the 42nd Hawaii International Conference on System Sciences, 2009, pp. 1–10.

[9] D. BÖHNLEIN, K. SCHWEIGER, AND A. TUMA, *Multi-agent-based transport planning in the newspaper industry*, International Journal on Production Economics, 131 (2011), pp. 146–157.

[10] I. BORGULYA, *An algorithm for the capacitated vehicle routing problem with route balancing*, Central European Journal of Operations Research, 16 (2008), pp. 331–343.

[11] A. BOSCO, D. LAGANÀ, R. MUSMANNO, AND F. VOCATURO, *Modeling and solving the mixed capacitated general routing problem*, Optimization Letters, 7 (2013), pp. 1451–1469.

[12] O. BRÄYSY AND G. HASLE, *Software tools and emerging technologies for vehicle routing and intermodal transportation*, accepted for publication as Chapter 12 in Vehicle Routing: Problems, Methods, and Applications, 2nd ed., P. Toth and D. Vigo, eds., MOS–SIAM Series on Optimization, SIAM, Philadelphia, 2014.

[13] M. G. VAN BUER, D. L. WOODRUFF, AND R. T. OLSON, *Solving the medium newspaper production/distribution problem*, European Journal of Operational Research, 115 (1999), pp. 237–253.

[14] W.-C. CHIANG, R. RUSSELL, X. XU, AND D. ZEPEDA, *A simulation-metaheuristic approach to newspaper production and distribution supply chain problems*, International Journal on Production Economics, 121 (2009), pp. 752–767.

[15] C. B. CUNHA AND F. MUTARELLI, *A spreadsheet-based optimization model for the integrated problem of producing and distributing a major weekly newsmagazine*, European Journal of Operational Research, 176 (2007), pp. 925–940.

[16] G. B. DANTZIG AND J. H. RAMSER, *The truck dispatching problem*, Management Science, 6 (1960), pp. 80–91.

[17] E. ERASLAN AND T. DERYA, *Daily newspaper distribution planning with integer programming: An application in Turkey*, Transportation Planning and Technology, 33 (2010), pp. 423–433.

[18] A. FØLSTAD AND O.-W. RAHLFF, *Basic user requirements for mobile work support systems: Three easy steps*, in Advanced Conceptual Modeling Techniques, Lecture Notes in Computer Science 2784, A. Olivé, M. Yoshikawa, and E. S. K. Yu, eds., Springer, New York, 2003, pp. 217–228.

[19] K. GAZE, *Exact Optimization Methods for the Mixed Capacitated General Routing Problem*, Master's thesis, NTNU, Trondheim, Norway, 2013.

[20] B. L. GOLDEN, A. A. ASSAD, AND E. A. WASIL, *Routing vehicles in the real world: Applications in the solid waste, beverage, food, diary, and newspaper industries*, in The Vehicle Routing Problem, P. Toth and D. Vigo, eds., SIAM, Philadelphia, 2002, pp. 266–280.

[21] B. L. GOLDEN, T. L. MAGNANTI, AND H. Q. NGUYEN, *Implementing Vehicle Routing Algorithms*, Tech. Report 115, Operations Research Center, MIT, Cambridge, MA, 1975.

[22] ——, *Implementing vehicle routing algorithms*, Networks, 7 (1977), pp. 113–148.

[23] B. L. GOLDEN AND R. T. WONG, *Capacitated arc routing problems*, Networks, 11 (1981), pp. 305–315.

[24] D. GULCZYNSKI, B. GOLDEN, AND E. WASIL, *The period vehicle routing problem: New heuristics and real-world variants*, Transportation Research Part E: Logistics and Transportation Review Journal, 47 (2010), pp. 648–668.

[25] J. C. A. GUTIÉRREZ, D. SOLER, AND A. HERVÁS, *The capacitated general routing problem on mixed graphs*, Revista Investigacion Operacional, 23 (2002), pp. 15–26.

[26] G. HASLE AND O. KLOSTER, *Industrial Vehicle Routing*, in Geometric Modelling, Numerical Simulation, and Optimization: Applied Mathematics at SINTEF, G. Hasle, K.-A. Lie, , and E. Quak, eds., Springer, New York, 2007, pp. 397–435.

[27] G. HASLE, O. KLOSTER, M. SMEDSRUD, AND K. GAZE, *Experiments on the Node, Edge, and Arc Routing Problem*, Tech. Report A23265, SINTEF, Trondheim, Norway, 2012.

[28] R. HE, W. XU, J. SUN, AND B. ZU, *Balanced K-means algorithm for partitioning areas in large-scale vehicle routing problem*, in Proceedings of the 2009 Third International Symposium on Intelligent Information Technology Application—Volume 03 (IITA '09), 2009.

[29] M. JAGGER AND K. RICHARDS, *Who wants yesterday's papers?*, in Between the Buttons, Decca, 1967.

[30] N. JOZEFOWIEZ, F. SEMET, AND E.-G. TALBI, *An evolutionary algorithm for the vehicle routing problem with route balancing*, European Journal of Operational Research, 195 (2009), pp. 761–769.

[31] B.-I. KIM, S. KIM, AND S. SAHOO, *Waste collection vehicle routing problem with time windows*, Computers & Operations Research, 33 (2006), pp. 3624–3642.

[32] H. KOKUBUGATA, A. MORIYAMA, AND H. KAWASHIMA, *A practical solution using simulated annealing for general routing problems with nodes, edges, and arcs*, in Proceedings of the 2007 International Conference on Engineering Stochastic Local Search Algorithms: Designing, Implementing and Analyzing Effective Heuristics, SLS'07, T. Stützle, M. Birattari, and H.H. Hoos, eds., Springer, New York, 2007, pp. 136–149.

[33] M.-K. KWAN, *Graphic programming using odd or even points*, Chinese Math., 1 (1962), pp. 273–277.

[34] Q. LU AND M. M. DESSOUKY, *A new insertion-based construction heuristic for solving the pickup and delivery problem with hard time windows*, European Journal of Operational Research, 175 (2005), pp. 672–687.

[35] P. MATIS, *DSS for the street routing problem*, Transport, 23 (2008), pp. 230–235.

[36] J. OPPEN AND A. LØKKETANGEN, *Arc routing in a node routing environment*, Computers & Operations Research, 33 (2006), pp. 1033–1055.

[37] C. S. ORLOFF, *A fundamental problem in vehicle routing*, Networks, 4 (1974), pp. 35–64.

[38] R. PANDI AND B. MURALIDHARAN, *A capacitated general routing problem on mixed networks*, Computers & Operations Research, 22 (1995), pp. 465–478.

[39] J. M. PASIA, K. F. DOERNER, R. F. HARTL, AND M. REIMANN, *Solving a bi-objective vehicle routing problem by Pareto-ant colony optimization*, in Proceedings of the 2007 International Conference on Engineering Stochastic Local Search Algorithms: Designing, Implementing and Analyzing Effective Heuristics (SLS'07), T. Stützle, M. Birattari, and H. H. Hoos, eds., Springer, New York, 2007, pp. 187–191.

[40] C. PRINS AND S. BOUCHENOUA, *A memetic algorithm solving the VRP, the CARP and general routing problems with nodes, edges and arcs*, in Recent Advances in Memetic Algorithms, Studies in Fuzziness and Soft Computing, W. Hart, N. Krasnogor, and J. Smith, eds., Springer, New York, 2004, pp. 65–85.

[41] R. RUSSELL, *A constraint programming approach to designing a newspaper distribution system*, International Journal of Production Economics, 145 (2013), pp. 132–138.

[42] R. RUSSELL, W.-C. CHIANG, AND D. ZEPEDA, *Integrating multi-product production and distribution in newspaper logistics*, Computers & Operations Research, 35 (2008), pp. 1576–1588.

[43] SINTEF, *NEARP Web pages*. http://www.sintef.no/Projectweb/TOP/nearp. Accessed 05/10/2012.

[44] ——, *Spider Web pages*, http://www.sintef.no/Projectweb/Transportation-planning/Software/Spider. Accessed 05/10/2012.

[45] S. H. SONG, K. S. LEE, AND G. S. KIM, *A practical approach to solving a newspaper logistics problem using a digital map*, Computers and Industrial Engineering, 43 (2002), pp. 315–330.

[46] H. TANG AND E. MILLER-HOOKS, *Interactive heuristic for practical vehicle routing problem with solution shape constraints*, Transportation Research Record: Journal of the Transportation Research Board, 1964 (2006), pp. 9–18.

[47] M. TSOUROS, M PITSIAVA, K. GRAMMENIDOU, AND K. TSOUROS, *Routing-loading balance heuristic algorithms for a capacitated vehicle routing problem*, in Information and Communication Technologies, 2006 (ICTTA '06), 2006.

[48] WAN-IFRA, *World Press Trends 2010*, http://www.wan-ifra.org/worldpresstrends2010.

[49] ——, *World Press Trends: Newspapers Still Reach More Than Internet* http://www.wan-ifra.org/articles/2012/04/17/world-press-trends-2011.

Index